CAPITALISM IN THE ANTHROPOCENE

CAPITALISM in the ANTHROPOCENE

Ecological Ruin or Ecological Revolution

John Bellamy Foster

MONTHLY REVIEW PRESS
New York

Copyright © 2022 by John Bellamy Foster
All Rights Reserved

Library of Congress Cataloging-in-Publication Data
available from the publisher

ISBN 978-158367-975-3 (paperback)
ISBN 978-158367-976-0 (cloth

Typeset in Minion Pro

MONTHLY REVIEW PRESS, NEW YORK
monthlyreview.org

5 4 3 2 1

Contents

Preface | 9
Introduction | 13

I. THE PLANETARY RIFT | 39

1. Marx and the Rift in the Universal Metabolism of Nature | 41
2. The Great Capitalist Climacteric | 62
3. The Anthropocene Crisis | 82
4. Crossing the River of Fire | 91
5. The Fossil Fuels War | 111
6. Making War on the Planet | 126

II. ECOLOGY AS CRITIQUE | 139

7. Nature | 141
8. Third Nature | 148
9. Weber and the Environment | 153
10. The Theory of Unequal Ecological Exchange | 201
11. Marxism in the Anthropocene | 245
12. Marxism and the Dialectics of Ecology | 277
13. Engels's *Dialectics of Nature* in the Anthropocene | 296
14. Late Soviet Ecology | 316
15. *The Return of Nature* and *Marx's Ecology* | 338

III. THE FUTURE OF HISTORY | 361

16. Capitalism and Degrowth: An Impossibility Theorem | 363
17. The Ecology of Marxian Political Economy | 373
18. On Fire This Time | 390
19. COVID-19 and Catastrophe Capitalism | 411
20. Ecological Catastrophe or Ecological Civilization | 433
21. The Capitalinian: The First Geological Age of the Anthropocene | 457
22. Conclusion: Ecology and the Future of History | 475

Notes | 493
Name Index | 633
Subject Index | 655

for Carrie Ann

Preface

This book is about the terrible reality of our times, in which humanity's future is being threatened by the capitalist system in which we live, now in conflict with the earth itself. But it is also a work of indomitable hope derived from knowledge of the strivings of human beings at critical moments in history. In 1919, Marxist philosopher Georg Lukács declared that in revolutionary situations where the fate of the world hangs in the balance, "the individual's conscience and sense of responsibility are confronted with the postulate that he[/she] must act as if on his[/her] action or inaction depended the changing of the world's destiny."[1] We live in the midst of such critical times today, and the quality and even existence of the human future will depend on the nature of our struggles, on our ability to reinvent the world by reinventing ourselves, and on our willingness to transform the destructive social conditions that surround us and threaten the future—not only for the sake of current human generations and those still to come, but also for the sake of the majority of living species on the planet.

The question of "capitalism in the Anthropocene" is one that I first took up a little over a decade ago, in *The Ecological Rift* (2010), written together with Brett Clark and Richard York. The chapters in this book were all composed in the intervening years, and were aimed, in various ways, at the issue of capitalism's war on the earth. A central theme is the continuing significance in this respect of Karl Marx's classical theory of metabolic rift. All of the previously published writings incorporated into this book have been revised, adapted, and updated for the present volume, some of them extensively.

The book is divided into three distinct parts. Part One: "The Planetary Rift" addresses today's Earth crisis viewed through the lens of Marxian ecology. Part Two: "Ecology as Critique" is concerned with various environmental debates and with questions of history, theory, and methods. Part Three: "The Future of History" focuses on ecological (and social) struggles and possible pathways to a more sustainable world.

A number of chapters in this book are drawn directly from previously coauthored work. "Crossing the River of Fire," "Marxism and the Dialectics of Ecology," and "The Capitalinian" were originally coauthored with Brett Clark, appearing in the February 2015, November 2020, and September 2021 issues of *Monthly Review*, respectively. Hannah Holleman and I coauthored two chapters: "Weber and the Environment," published in May 2012 in the *American Journal of Sociology*, and "The Theory of Unequal Ecological Exchange," published in the March 2014 issue of the *Journal of Peasant Studies*. "COVID-19 and Catastrophe Capitalism" was coauthored with Intan Suwandi, appearing in June 2020 in *Monthly Review*. Alejandro Pedregral conduced the interview in chapter 15.

Most of the research going into this book was undertaken while I was working on a number of other closely related books and research projects connected to ecosocialism, some in collaboration with other scholars. In 2011, Fred Magdoff and I wrote *What Every Environmentalist Needs to Know About Capitalism*, showing how the global ecological crisis is a product of capitalist society and requires for its solution the transcendence of the present world order. In 2016, I published, with Paul Burkett, *Marx and the Earth*, critically examining attempts by some ecosocialists to distance their analysis from classical historical materialism, rather than building on it, in an attempt to construct an ecological critique of the system. In 2020, Brett Clark and I published *The Robbery of Nature*, which was primarily directed at extending the metabolic rift analysis to issues of the expropriation of nature and human bodies (the corporeal rift), and questions of social reproduction related to gender and the environment. That same year I completed *The Return of Nature*, the product of twenty years of research, exploring socialist contributions to the development of ecological theory, principally in Britain, but also the United States, from the time of the deaths of Darwin and Marx, to the present.

Other ecological work and projects also overlapped with the writing of this book. "The Anthropocene Crisis" is a revised version of the foreword to Ian Angus's indispensable book, *Facing the Anthropocene* (2016). "Third Nature" appeared in Vijay Prashad, ed., *Will the Flower Slip Through the Asphalt* (New Delhi: Left Word, 2017). For a number of years, Brian Napoletano, Pedro Urquijo, Brett Clark, and I have been engaged in research on the spatial aspects of metabolic rift theory centered in geography, beginning with our article "Making Space in Critical Environmental Geography for the Metabolic Rift" in the *Annals of the American Association of Geographers* in 2019. Holleman, Clark, and I have produced a series of research publications, part of a book project, linking metabolic rift theory to concepts of expropriation and the question of the global environmental proletariat, including "Imperialism in the Anthropocene," "Marx and the Indigenous," "Marx and Slavery," and "Capital and the Ecology of Disease," published in the July–August 2019, February 2020, July–August 2020, and June 2021 issues of *Monthly Review*, respectively; along with "Marx and the Commons" in the Spring 2021 issue of *Social Research*. In 2017, I wrote a foreword to Magdoff and Chris Williams's pathbreaking *Creating an Ecological Society* (2017). Lastly, Lola Loustaunau, Mauricio Bentancourt, Clark, and I, building on Marx's observations on the nineteenth-century guano and "coolie" trades, coauthored an article titled "'Chinese Contract Labour, the Corporeal Rift, and Ecological Imperialism in Peru's Nineteenth-Century Guano Boom," for the *Journal of Peasant Studies* (2021).

Robert W. McChesney and I as the closest of friends have worked together on and off and have influenced each other in numerous ways for nearly half a century. During the period that I was working on the research and writing that constitutes this book we coauthored numerous publications, including *The Endless Crisis: How Monopoly-Finance Capital Produces Stagnation and Upheaval from the USA to China* (New York: Monthly Review Press, 2012), which contributed to much of the political-economic analysis here.

In the last couple of years, I have gained immensely from a regular informal seminar on Marxism and philosophy with three extremely gifted philosophy PhD students at the University of Oregon: Kenny Knowlton, Oscar A. Ralda, and Chris Shambaugh. As my *Monthly Review* research assistant during the last year, Chris has played a major role in helping prepare this book for publication.

In addition to the above-named individuals, all of whom have affected the ideas in the book, I would like to thank others at *Monthly Review* to whom I am deeply indebted, including Susie Day, Jamil Jonna, John Mage, Rebecca Manski, Martin Paddio, John J. Simon, Camila Valle, Victor Wallis, and Michael D. Yates—all of whom I work with on a regular basis and whose influence permeates this work.

Over the last decade I have also gained from interchanges with Jordan Besek, Alex Callinicos, Desmond Crooks, Martin Empson, Joseph Fracchia, Nancy Fraser, Andreas Malm, the late István Mészáros, Kohei Saito, Lauren Regan, Helena Sheehan, Eamonn Slater, and Ryan Wishart—and from frequent conversations with Saul Bellamy Foster, Linda Berentsen, and the late William Edwin Foster.

Most of all I would like to thank Carrie Ann Naumoff, to whom this book is dedicated.

—EUGENE, OREGON
SEPTEMBER 2021

Introduction

They say that it's an ill bird that fouls its own nest.
—John Ruskin

Capitalism is a system that inherently and irredeemably fouls its own nest, which has now been extended to the planet itself.[1] The scale of the problem that this poses for humanity as a whole and for all future generations almost defies imagination. As the dominant socioeconomic system in the world today, capitalism impacts the everyday lives of most people on Earth. It is so pervasive and all-encompassing that it is reasonable to ask the questions: How could it ever have come into being and in the end pass away? How could it exist in or be dependent on anything else? Isn't it easier to conceive the end of the world than the end of capitalism?[2] If today's planetary crisis requires the transcendence of capitalist relations of production, is this not, then, a hopeless case since, as we are told every day, there is no alternative to the present order? Indeed, some thinkers have now gone so far as to argue that capitalism is inextricable from what William Blake called the "web of life," now increasingly seen as capitalism's own creation.[3]

Nevertheless, capitalism is a historically transitory system. Like everything terrestrial, it is part of the Earth System. In terms of geological time, it exists today within what scientists have labeled the Anthropocene Epoch, commencing in the early post–Second World War period, during which human or anthropogenic impacts, largely engendered by capitalism itself, became the main source of Earth System change, displacing the relatively benign Holocene Epoch of the previous 11,700 years in which civilization emerged. The advent of the Anthropocene coincided with a planetary rift, as the human economy under capitalism heedlessly crossed, or began to cross, Earth System boundaries, fouling its own nest and threatening the destruction of the planet as a safe home for humanity.

Since the Anthropocene stands for the geological epoch in which human impacts play the dominant role in Earth System change, there is no going back in that respect. Human impacts will at least potentially remain the key influences on Earth System change, emanating from terrestrial sources, as long as the industrialized world exists. The choice before us, then, is between the development of a sustainable relation to the Earth System, requiring the creation of a new ecological civilization, or the continuation of the present unsustainable system, which is headed toward destruction and the termination of humanity itself. Either the world will exit the present geological *age* of the early Anthropocene Epoch, referred to here as the *Capitalinian Age*, and enter a new *Communian Age*, based on a society of substantive equality and ecological sustainability, or it will be propelled by the existing relations of production toward an end-Anthropocene extinction event, spelling absolute catastrophe for humankind and for innumerable other species on the planet.[4] Hence, humanity is now facing the greatest challenge in its long history; one that it can overcome, but only by carrying out revolutionary change in social and ecological relations on a world scale.

Capitalism Before the Anthropocene

The word *capitalism* entered the English language in the mid-nineteenth century but was preceded, beginning in the fourteenth and fifteenth centuries, by the word *capital*, initially from the Latin, meaning "head" or "first," and, eventually, in the eighteenth century, taking on its current meaning as an economic category.[5] Adam Smith's *The Wealth of Nations* in 1776 referred to "the accumulation of capital."[6] Arthur Young, writing in his *Travels in France* in 1792, described "moneyed men" as "capitalists."[7] This usage soon took on a more critical and systematic form in the socialist literature. Thomas Hodgskin, in his *Labour Defended Against the Claims of Capital* in 1825, at the beginning of the nineteenth century, wrote: "Betwixt him who produces food and him who produces clothing, betwixt him who makes instruments and him who uses them, in steps the capitalist, who neither makes nor uses them, and appropriates to himself the produce of both.... On the one side is the labourer and on the other the capitalist.... Capitalists ... can grow rich only where there is an oppressed body of labourers."[8] As cultural theorist Raymond Williams notes, Hodgskin was for the first time describing "an economic *system*" mediated by the capitalist.[9]

It is, however, the mid-Victorian novelist William Makepeace Thackeray, the author of *Vanity Fair*, who is credited with having introduced the word *capitalism* into the English language (entering the French and German languages about the same time).[10] In his novel *The Newcomes*, appearing first in serial form in 1853–55, Thackeray focuses on a family that rises into banking and wealth, while also being caught up in the British Empire, via India and profits from the Chinese opium trade. Here we encounter "the sense of capitalism" associated with "the crowd of *bourgeois*," including "city magnates and capitalists" with no other interest other than in investing in banks and railways, while constantly displaying their "cheque-books."[11]

Nevertheless, it was Karl Marx in *Capital*, written in London in German in 1867, who provided the first fully developed view of *capital as a system*.[12] In Marx's critique of political economy, capitalists were viewed theoretically as mere "personifications" of capital, and thus of the process of accumulation rooted in the exploitation of labor. For the capitalist, Marx wrote, "Accumulate! Accumulate! That is Moses and the prophets!"[13] Although *capitalism* as a category appeared for the first time in the mid-nineteenth century at the height of the British Industrial Revolution, Marx argued that "the capitalist era dates from the sixteenth century"—now sometimes called the "long sixteenth century," since capitalism could be seen as arising in nascent form as early as the mid-fifteenth century with the beginnings of a revolution in agriculture and capitalism's rise as a world system.[14] The period from the sixteenth to the late eighteenth century is generally viewed as the stage of mercantilism or merchant capitalism, during which the Dutch emerged for a time as the hegemonic capitalist power. It was also the period of the initial colonization of much of the world by the European capitalist powers, encompassing the transoceanic slave trade in the Atlantic and the Dutch East Indies and the genocide of Indigenous populations.[15]

The late eighteenth century saw maturation of the system with the Industrial Revolution in Britain, emerging as the new hegemonic power in the world economy, and the rise of the age of industry, or freely competitive capitalism, all of which was best captured in Marx's *Capital*. In this era coal arose as the principal form of energy, initiating "fossil capital," or an industrialized order marked by the exploitation of labor coupled with the energetic reliance on fossil fuels.[16] Driving this entire historical process was a class-based system geared to ever-greater capital accumulation, concentration and centralization of capital, and amassing of wealth at the top—a system in which the motor force was competition, but which led to increasing monopoly power and relative impoverishment of the population and the earth.[17]

The late nineteenth century witnessed the beginnings of monopoly capitalism, or the system of concentrated capitalism. Monopoly capitalism is also often referred to in Marxian theory, following V. I. Lenin, as the imperialist stage of capitalism (not to be confused with imperialism as a generic phenomenon encompassing the entire history of colonialism and imperialism under capitalism). Monopoly capitalism was marked by the dividing up of the entire world, symbolized by the partition of Africa by the Great Powers at the Berlin Conference in 1884. The initial rise of monopoly capitalism was associated with the destabilization of the entire capitalist world order, leading to the two World Wars, with the Great Depression of the 1930s occurring in the interim.[18]

At the end of the Second World War—the closing act of which was the U.S. atomic bombings of Hiroshima and Nagasaki—the United States emerged as the new hegemonic power in the world capitalist economy, rivaled geopolitically only by the Soviet Union, which had arisen in 1917 in a socialist revolution against capitalism. This was a period of the consolidation of monopoly capitalism, with the rise of multinational corporations and the beginnings of globalized production. In the Cold War the rivalry between the nascent Socialist Bloc, centered in the Soviet Union, and the much larger world capitalist system, with the United States as the hegemonic power, the concept of *capitalism* took on a quite different meaning, associated with the competition between rival systems. Capitalism thus came to stand ideologically for the entirety of the Western system, and for reliance on the so-called market mechanism, as opposed to the emphasis on central planning under actually existing socialism in the East.

In Cold War propaganda, and in the work of conservative economists and ideologues such as Milton Friedman, capitalism thus became equated with "freedom" (as opposed to so-called totalitarianism).[19] This was used beginning in the 1980s to justify *neoliberalism*, leading to the removal of restrictions on capital and the worldwide demolition of labor, a process that was carried to its logical conclusion in the wake of the demise of the Soviet-type economies in 1989–91, under the rubric, made famous by Margaret Thatcher, that "there is no alternative."[20] Freedom, in these terms, meant the removal of all limits on the market, that is, capital, and to the entire bourgeois order, leading to further concentration and centralization of capital and an epoch of financialization.

In reality, "capitalism," as economist Robert Heilbroner wrote, "is the regime of capital [dedicated to 'the drive to amass wealth'], the form of rulership we find when power takes the remarkable aspect of the domination, by those who control access to the means of production, of the great majority who must gain 'employment.'"[21] It is an uncontrollable, creatively destructive system, dedicated to using money capital to purchase commodities and labor power in order to produce new commodities that can be sold for *more money* (surplus value—derived from the exploitation of labor power), in an endless sequence of capital accumulation, leading to the concentration of ownership of the means of production and financial claims to wealth in ever-fewer hands.[22] Nature in this system is viewed simply as a "free gift" to capital, while the vast majority of human beings are treated as an exploitable and expendable mass from which to generate surplus for the wealthy owners.[23] The result is a system that knows no bounds, is oblivious to genuinely human needs, and is inherently unsustainable—now confronting its absolute limits in the Anthropocene.

Capitalism in the Anthropocene

The Anthropocene Epoch, according to the Anthropocene Working Group of the International Union of Geological Sciences can be seen as having begun in 1945, at the end of the Second World War, or else in the 1950s.[24] The stratigraphic markers most commonly referred to are radionuclides from hundreds of nuclear tests (and two nuclear bombings) and the development of plastics and petrochemicals. These developments are seen as having introduced a new "synthetic age."[25] Yet, this was also the time of the Cold War, the consolidation of global monopoly capitalism, and what is often referred to as capitalism's "golden age." The immediate post–Second World War period saw what environmental historians have referred to as the Great Acceleration of economic impacts on the earth to the point that planetary boundaries were being crossed, or in danger of being crossed.[26] The Anthropocene Epoch can thus be seen as having its origins in the post–Second World War era, with monopoly capitalism at a high level of globalization as its principal driver.

Since each geological epoch is divided into geological ages, the first age of the Anthropocene Epoch can be referred to as the *Capitalinian Age*, describing its *social* origins, now dominating over the stratigraphic indicators of geological change. To designate the present as a particular age of the Anthropocene Epoch is to point to the temporary, historical character of the geological age in which we now reside, which will lead either to a new geological (and social) age, stabilizing the human relation to the earth, referred to here as the *Communian*, or to an end-Anthropocene extinction event resulting in the destruction of civilization and quite possibly humanity itself.

The fact that the relation of humanity to the earth had fundamentally changed, raising issues of *Science and Survival*, as ecologist Barry Commoner titled his book in 1966, was first apparent in the early 1950s due to the global spread of radionuclides from nuclear testing.[27] This was coupled with the introduction of tens of thousands of synthetic chemicals that were the product of organic chemistry rather than the long process of evolution, and thus were carcinogenic, mutagenic, and teratogenic (leading to birth defects).[28] The result of this new "synthetic age" was the rise from the 1950s to the 1970s of the modern environmental movement.[29] Already in the early-1960s climate change was recognized by climatologists in the Soviet Union and in the United States as a possible scenario, but it was not until 1988 that James E. Hansen, then director of NASA's Goddard Institute for Space Studies, was to report definitively to the U.S. Congress that climate change was a reality, with the founding of the United Nations Intergovernmental Panel on Climate Change that same year.[30] Related warnings of planetary crisis emerged in the late 1970s and 1980s with respect to the global die-down of species, the depletion of the ozone layer, and other global ecological threats.

Indeed, the notion of the Anthropocene Epoch in geological history, expressing how human society, via capitalism, has proceeded to foul its planetary nest, has not been the revelation of a moment. Rather it can be seen as a product of a century-long discussion on the growing human impacts on the earth environment. In his *Kingdom of Man* in 1911, E. Ray Lankester, the leading British zoologist in the generation after Charles Darwin and a close friend of Marx, insisted that humanity as a result of capitalism had become a "disturber" of the ecology of the earth to such an extent that it undermined its own environment, giving rise to "nature's revenges," including new zoonotic diseases threatening humanity. The world therefore had no choice but to use science rationally to create a more sustainable relation to the planet, via science.[31]

According to Soviet geologist E. V. Shantser, writing in *The Great Soviet Encyclopedia* in 1973, the prominent Russian geologist Aleksei Petrovich Pavlov had coined the term "Anthropogenic system (period), or Anthropocene" in 1922. It is Shanster's 1973 reference to Pavlov's coinage of the term *Anthropocene* that constitutes the first documented use of the word in the English language.[32] Pavlov used the notion of the Anthropocene (or Anthropogene) to refer to a new geological *period* in which humanity was emerging as the main driver of planetary ecological change. In this way, Pavlov and subsequent Soviet geologists provided an alternative geochronology, one that substituted the Anthropocene (Anthropogenic) Period for the entire Quaternary Period in the geological time scale. Pavlov's designation of the Anthropocene/Anthropogene Period was closely connected to Vladimir I. Vernadsky's landmark publication *The Biosphere* in 1926, constituting an early proto-Earth System analysis, focusing on the biogeochemical cycles of the planet.[33] Both Pavlov and Vernadsky strongly emphasized that anthropogenic factors had come to dominate the biosphere.[34]

Although long associated with Soviet science and dialectical forms of thought, and therefore held for a long time at arm's length, the concept of the biosphere became widely accepted in the United States and increasingly intergraded into Western science beginning in the 1970s, as a result of the growing planetary environmental crisis.[35] The period starting in the early 1970s, during which the Club of Rome's famous study *The Limits to Growth* appeared, to the end of the century, saw the rapid development of what came to be understood as Earth System science, extending beyond the concept of the biosphere.[36] A pioneering contribution in the development of Earth System analysis was atmospheric scientist James Lovelock's introduction of the Gaia hypothesis in the early 1970s, arguing that life or the biosphere governed the metabolism of the entire Earth System. This received a definitive expression in Lovelock and Lynn Margulis's article, "Atmospheric Homeostasis By and For the Biosphere," in

Tellus in 1974.[37] Although Lovelock's approach was rejected by many scientists as overly teleological, the notion of life's role in generating homeostasis in the planetary metabolism was nonetheless crucial in defining Earth System science more generally. Hence, it is common to claim that Earth System science had its "start" at this point, in the understanding of the "two-way coupling between life and its planetary environment" that the Gaia hypothesis introduced.[38]

By the first decade of the twenty-first century the new Earth System analysis focused increasingly on the crossing of what are now formally designated as "planetary boundaries," such as climate change, depletion of the ozone layer, ocean acidification, disruption of the nitrogen and phosphorus cycles, disappearing ground cover (including forests), diminishing fresh water supplies, and losses in biological diversity, encompassing what came to be known by the end of the twentieth century as "the sixth extinction"—all of which were seen as the result of cumulative anthropogenic factors impacting the planet as an overall system.[39] It was the new consciousness of anthropogenic disruptions in the biogeochemical cycles of the planet associated with developing Earth System analysis that led to Nobel Prize–winning atmospheric scientist Paul J. Crutzen's famous February 2000 declaration: "We're not in the Holocene anymore. We're in the . . . Anthropocene!"[40]

The Anthropocene is generally seen as associated at the outset with the development of "anthropogenic rifts" in the biogeochemical cycles of the Earth System.[41] It is no mere accident, therefore, that the modern ecological movement arose in the 1950s and the 1960s, in response to what were then perceived as unprecedented threats to the earth's environment and human life, resulting from nuclear weapons testing and the petrochemical industry. Today this growing threat to the global environment is understood in terms of the crossing of the planetary boundaries, separating the Holocene Epoch in the geological time scale from the Anthropocene epoch—a framework that needs to be expanded today to account for the increased incidence of zoonotic diseases associated with ecosystem destruction.[42]

The Anthropocene thus represents a kind of knife's edge for humanity since the way in which the global economy impacts the earth will always have to factor into the equation as long as industrial civilization persists. The continuation of the Capitalinian Age, driven by processes of capital accumulation, will lead *of itself* to the destruction of civilization, and of the planet as a safe place for humanity. Only with an ecological and social revolution that creates a more sustainable relation to the earth, and conditions of substantive equality in society as a whole, can society move forward in the Anthropocene, rather than reaching a tipping point in its long development and an end-Anthropocene extinction event. Either the world will go in the direction of ecological civilization transcending the Capitalinian Age (or the first geological age of the Anthropocene) and generating a new geological age, the Communian, or the human story on earth will end abruptly. Key to the development is the spread of egalitarian, collective, and communal values and ways of living—that is, a fully developed socialism or a society of substantive equality and ecological sustainability. "By 'socialism,'" Heilbroner wrote, "I mean a society unmistakably disconnected from the very idea of economic determinism."[43]

For many, willing to resign humanity to its "fate," the idea of a way out of our current dilemma, fundamentally altering society in order to avoid the socioecological chasm before us, will undoubtedly sound utopian. But *utopia*, a pun coined in the sixteenth century by Thomas More meaning both "nowhere" and "good place," and therefore often seen as representing a kind of dream state or wishful projection into the future, loses its idealistic connotation in the context of a planetary *dystopia* where catastrophe, measured against historical precedents, has now become normal and threatens to become irreversible on a planetary scale, due to the inherent apocalyptic tendencies of the current mode of production.[44] Under such circumstances, only the reconstitution of society as a whole, and thus of the human relation to the earth, holds any realistic hope for the future of humanity.

Catastrophe Capitalism

It would be a serious error to reduce the entire planetary ecological emergency signaled by the rise of the Anthropocene simply to climate change, as the other critical planetary boundaries, each of which represent an Earth System emergency, are being crossed as well, with immense dangers for humanity. For example, the biodiversity crisis stemming mainly from the destruction of the world's ecosystems is arguably as serious as climate change.[45] Nevertheless, climate change currently threatens *irreversible* change on a scale and with a speed that has no counterpart. Carbon dioxide levels in the atmosphere are now "higher than at any time in at least two million years," heating up the entire earth via the greenhouse effect.[46]

Part 1 of *Climate Change 2021: The Physical Science Basis*, published in August 2021 by the United Nations Intergovernmental Panel on Climate Change (IPCC), indicated that even in the case of SSP1-1.9, the most optimistic scenario in its Shared Socioeconomic Pathways—in which carbon emissions would peak globally by 2025, a 1.5°C increase in global average temperature over preindustrial levels would be avoided until 2040, and the goal of net-zero carbon emissions would be reached by 2050—the consequences for global humanity would nevertheless be catastrophic by the measure of all historical precedents.[47] This would include various "compound" events of extreme weather, sometimes known as "global weirding" (including record heatwaves, persistent droughts, out-of-control wildfires, megastorms, torrential rainfall, unprecedented floods, glacial melts, and sea level rise), affecting every region and ecosystem of the world. The second-best scenario (also optimistic), in which global average temperature would be stabilized at less than 2°C (at around 1.7°C)—is a sort of last hope scenario and carries with it dangers disproportionately greater.[48] The other three scenarios are almost unthinkable, although more consistent with current trends, threatening the very existence of civilization and humanity itself.

Under the most optimistic scenario, or SSP1-1.9, in which 1.5°C is not reached until 2040, it is estimated that the mid-term (2041–2060) increase in global average temperature will remain in the 1.2–2°C range, with the best estimate 1.6°C, and then, near the end of the century (2081–2100), will fall back to 1.4°C due to the implementation of negative emissions technologies.[49] Yet, even in this highly optimistic scenario, extreme heat and heavy rainfall will be far more frequent. Sea level rise will persist over centuries and perhaps millennia. Ocean acidification will increase, carrying its own dangers. The best that can be hoped for at this point, therefore, according to the IPCC, is that the ultimate threat to humanity will be held off, and by the end of the century global average temperature could be reduced below 1.5°C again. Yet, even then, some of the negative effects of global heating, posing dire threats to billions of people, will nevertheless continue to play out over the twenty-first century. In the case of SSP5-8.5, the fifth and most apocalyptic scenario, resulting from the unhindered continuation of capitalist "business as usual," global average temperature is projected to increase by the end of the century by 3.3–5.7°C, with the "best estimate" at 4.4°C, spelling absolute catastrophe for humanity and innumerable species on the planet.[50]

The original "Summary for Policymakers" of part 3 of the Sixth Assessment Report, addressing "Mitigation," which is not due to be published in its final form until March 2022, was leaked by Scientist Rebellion, a branch of Extinction Rebellion, in August 2021.[51] Although subject to being redacted by governments prior to final adoption and publication, the leaked version, accepted by scientists, argues that what is required, in order to avoid climate disaster, "is fundamental structural changes at global scale," in accord with a "just transition" that ensures that "workers, frontline communities and the vulnerable are not left behind in low-carbon pathways."[52] As the leaked chapter one of part 3 of the report on mitigation of climate change indicates: "The character of social and economic development produced by the nature of capitalist society" is viewed by many political-economic critics and radical ecological thinkers "as ultimately unsustainable."[53] Indeed, a close reading of the leaked chapter one leaves little doubt that system change on a revolutionary scale is now the only remaining path to a sustainable future for humanity. As UN secretary general António Guterres exclaimed in a statement accompanying the release of part 1 of the new IPCC report, this is "a code red for humanity."[54]

In September 2021, over 200 of the world's leading medical

journals, including the *British Medical Journal*, the *Lancet*, the *New England Journal of Medicine*, the *European Heart Journal*, the *Chinese Science Bulletin*, the *National Medical Journal of India*, the *East African Medical Journal*, the *Medical Journal of Australia*, and the *Canadian Medical Association Journal*, simultaneously published an editorial titled "Promises Are Not Enough," insisting that, given global heating, "only fundamental and equitable changes to societies will reverse our current trajectory" from the standpoint of the rapid deterioration in world health conditions. As the editorial states:

> The risks to health of increases above 1.5° C are now well established. Indeed, no temperature rise is "safe." In the past 20 years, heat-related mortality among people over 65 years of age has increased by more than 50%. Higher temperatures have brought increased dehydration and renal function loss, dermatological malignancies, tropical infections, adverse mental health outcomes, pregnancy complications, allergies, and cardiovascular and pulmonary morbidity and mortality. Harms disproportionately affect the most vulnerable, including children, older populations, ethnic minorities, poorer communities, and those with underlying health problems.
>
> Global heating is also contributing to the decline in global yield potential for major crops, which has fallen by 1.8 to 5.6% since 1981; this decline, together with the effects of extreme weather and soil depletion, is hampering efforts to reduce undernutrition. Thriving ecosystems are essential to human health, and the widespread destruction of nature, including habitats and species, is eroding water and food security and increasing the chance of pandemics.
>
> The consequences of the environmental crisis fall disproportionately on those countries and communities that have contributed least to the problem and are least able to mitigate the harms. Yet no country, no matter how wealthy, can shield itself from these impacts. Allowing the consequences to fall disproportionately on the most vulnerable will breed more conflict,

food insecurity, forced displacement, and zoonotic disease—with severe implications for all countries and communities. As with the COVID-19 pandemic, we are globally as strong as our weakest member.

Rises above 1.5° C increase the chance of reaching tipping points in natural systems that could lock the world into an acutely unstable state. This would critically impair our ability to mitigate harms and to prevent catastrophic, runaway environmental change.[55]

Underlying this planetary emergency associated with the first age of the Anthropocene, or the Capitalinian, is capitalism, or the system of capital accumulation. Capitalism is, in this sense, a faulty mechanism that has disrupted the larger Earth System of which it is a part. The logic of capital accumulation, which accepts no boundary beyond itself, now threatens in the era of catastrophe capitalism the very nature of existence on earth.

Marxism and the Universal Metabolism of Nature

The point today is to change the current world order, which is pushing humanity toward a catastrophic relation to the planet; but in order to change the world it is first necessary to understand it. Here the present revival of Marxian theory, particularly the resurrection of the classical-historical materialist ecological critique, is indispensable. Although natural science has played a central role in helping us understand the developing ecological emergency, the current planetary emergency, as the concept of the Anthropocene Epoch suggests, has *social causes*, rooted in the dominant socioeconomic mode of production: capitalism. Hence, social science is key to addressing the problem. Here the most important critical insights this century have arguably emerged from ecosocialism, building on the classical historical materialist tradition, and extending its critique of capitalism's destructive socioecological metabolism to the present epoch. At the root of this critical understanding of the ecological problem is Marx's famous theory of the metabolic rift, focusing on capitalism's alienated social metabolism.[56]

Beginning in the 1850s, based on the work of his close friend and comrade, the physician and scientist Roland Daniels, as well as on Justus von Liebig's agricultural chemistry, Marx incorporated the notion of *metabolism* into his general analysis, introducing a conception of production (or the labor and production process) as constituting the "social metabolism" of humanity and nature.[57] This conception was developed further in *Capital*, particularly in the analysis of ecological crisis, with the social metabolism standing for what we today call human-ecological relations. Here it is important to note that today's ecosystem and Earth System analyses, and all form of systems ecology, have the concept of metabolism and flows of energy as their logical bases. Marx saw the social metabolism introduced by human beings in production as part of what he called the "universal metabolism of nature."[58]

In the mid-nineteenth century a soil crisis occurred in the new industrialized agriculture. Soil nutrients, such as nitrogen, phosphorus, and potassium, contained in the food and the fiber, sent hundreds and even thousands of miles to the new urban industrial centers where the population was now concentrated, ended up as pollution in the cities rather than being returned to the land, with the result that these vital "constituent elements" of the soil were lost to the soil.[59] Marx saw this as a manifestation of a contradiction between the alienated social metabolism of capitalist production and the "universal metabolism of nature," generating a "rift in the . . . social metabolism" or *metabolic rift*, which constituted the main structure of ecological crisis under capitalism.[60] The triad of concepts of the universal metabolism of nature, the social metabolism, and the metabolic rift thus gave to Marx's understanding of the ecological nature of production a complex, historically grounded conceptual structure, encompassing both Earth change and social system change, and their coevolution within the historical process. Exploring this problem in his later works, Marx engaged in extended analyses of ecological crises, or the metabolic rift, some of which were embedded in his ecological notebooks.

Although Marx wrote of the metabolism of nature and society, this was not, as some critics have charged, a "dualistic" conception, since his emphasis was on how the social metabolism, rooted in changing relations of production, historically *mediated* the dialectical relation between humanity and earth.[61]

Fundamental to this whole framework, emanating from classical historical materialism (in which Frederick Engels, as we shall see, also played a crucial role), is the notion that economic and environmental crises are two sides of a single coin associated with capitalism's exploitation of labor, on one side, and its expropriation of people and the earth, on the other. Capitalism is not only an alienated economic regime, but also, as a precondition of this, an alienated ecological regime. Industrial capitalism requires as its basis, as Marx argued, the removal of the population from the land and thus from the organic means of production. It was the expropriation of the commons as well as whole populations on a world scale (including chattel slavery) that led to "the genesis of the capitalist farmer," on the one hand, and the "genesis of the industrial capitalist," on the other.[62] This alienation from nature constituted the basis on which the alienation of human being from human being and between classes was established. This twofold alienation from nature and other human beings constitutes the source of capitalism's continuing creative destruction of the conditions of existence of humanity itself.[63]

The restoration and elaboration of Marx's theory of the metabolic rift, discussed in this book, is largely the product of the last quarter-century, during which his ecological analysis was unearthed and developed, and applied to such questions as climate change, the destruction of ocean life, questions of biodiversity, and many other areas. A crucial step in this regard was made with the publication of Brett Clark and Richard York's "Carbon Metabolism: Global Capitalism, Climate Change, and the Biospheric Rift" in the journal *Theory and Society* in 2005, applying Marx's methodological rift methodology to the analysis of climate change.[64] Since then, the expansion of this framework of analysis has become a mainstay of Marxian ecological analysis.[65]

Why did it take until the brink of the twenty-first century before Marx's overall ecological critique associated with his theory of metabolic rift was rediscovered? One answer is simply that the depth of Marx's scientific view in this respect was not widely appreciated or even perceived until the advent of the planetary ecological crisis, coinciding with the historical emergence of the Anthropocene Epoch, presented the world with a new set of challenges. As Rosa Luxemburg observed around a century ago: "Only in proportion as our movement progresses, and demands the solution of new practical problems do we dip once more into the treasury of Marx's thought, in order to extract therefrom and to utilize new fragments of his doctrine."[66]

Marx's mature ecological critique was a very late and incomplete development in his thought, when compared to the more straightforwardly economic part of his argument.[67] Moreover, in the early years of the modern socialist movement the working-class struggle was often seen one-sidedly in narrow economistic terms associated with the direct struggle between the capitalist class and the working class over exploitation in the workplace. The issue of the natural world, extending to the built environment in the cities and to issues of social reproduction, although never entirely absent (and occupying a central place in such early works as Engels's 1845 *The Condition of the Working Class in England*), was often neglected in the struggles in the late nineteenth and early twentieth centuries (outside the natural sciences themselves), prior to the rise of the modern environmental movement.[68]

This neglect of nature/ecology was especially acute in the case of those Western Marxist critical traditions that deliberately set aside Engels's analysis of the dialectics of nature, developed in *Herr Eugen Dühring's Revolution in Science* (1878; better known as *Anti-Dühring*) and the *Dialectics of Nature* (1875–82, published 1925).[69] Here it was argued that the dialectics of nature had been introduced as a foreign element into Marxism by Engels and was alien to Marx's own thought. This led to a downplaying of Marx's treatment of the natural conditions of production, in what came to be known as the "Western Marxist" philosophical tradition, with the result that the *materialist concepetion of history* was cut off from the *materialist conception of nature*.[70]

The origin of the "Western Marxist" tradition, in this sense, is usually traced to Georg Lukács's criticism in *History and Class Consciousness* (1923) of Engels's conception of the dialectics of nature. In footnote 6 of chapter one on "What Is Orthodox Marxism" Lukács inserted a short comment in which he stated:

> It is of the first importance to realise that the method is limited here to the realms of history and society. The misunderstandings that arise from Engels's account of dialectics can in the main be put down to the fact that Engels—following Hegel's mistaken lead—extended the method to apply also to nature. However, the crucial determinants of dialectics—the interaction of subject and object, the unity of theory and practice, the historical changes in the reality underlying the categories as the root cause of changes in thought, etc.—are absent from our knowledge of nature. Unfortunately it is not possible to undertake a detailed analysis of these questions here.[71]

This footnote by Lukács, consisting of less than ten lines altogether—the last line of which says, "It is not possible to undertake a detailed analysis of these questions here"—has often been exaggerated, treated as a full-blown critique, rather than a mere aside. Yet, Lukács does not go beyond declaring that the dialectic cannot be extended to the wider objective realm of nature without being divorced from its main "critical determinants" and the human subject. Moreover, elsewhere in *History and Class Consciousness* Lukács accepted the notion of "merely objective dialectics"—even if not quite in the sense developed by Engels.[72] This suggested a *hierarchy of dialectics* extending from the "so-called *objective*" realm of nature to the dialectic of nature and society, and beyond that the dialectic of society itself—a view that was to be developed in Lukács's later social ontology.[73] Further, we now know that only a few years after the publication of *History and Class Consciousness*, in *Tailism and the Dialectic* (ca. 1925), Lukács insisted on the significance of the notion of the dialectics of nature (and of the dialectics of nature and society) rooting this primarily in Marx's treatment of social metabolism—a line of thought that he was to develop over the next nearly half-century.[74]

Nonetheless, it was Lukács's apparent absolute rejection of the dialectics of nature, which came to be seen as the initial basis, or the dividing line, distinguishing the "Western Marxist" philosophical tradition from Marxism more generally, and especially from the brand of Marxism associated with the Soviet Union.[75] Dialectical conceptions were seen as only relevant to human history and the human subject—at most to nature as subsumed within human society—excluding the dialectic of nature, even in the sense of a dialectic of nature and society.[76]

The most influential reinterpretation of classical Marxism on nature, emanating from within the Western Marxist tradition—not stopping short of the criticism of Marx himself—was Alfred Schmidt's *The Concept of Nature in Marx* (1962). Schmidt's work was a doctoral thesis written under the supervision of his mentors Max Horkheimer and Theodor Adorno, whose views it also largely represented.[77] Schmidt confined the dialectic to its highest form in the identical subject-object of history, that is, the realm of human interaction, while denying its significance in the wider realm of what Engels had referred to as universal "reciprocal action" or *universal interaction*.[78] The early Marx's notion of the reconciliation of nature and humanity, upheld by later thinkers such as Bertolt Brecht and Ernst Bloch, was also criticized.[79] In rejecting Engels's notion of the dialectics of nature, Schmidt went so far as to argue—illogically on both counts—that it presented a "'bad' contradiction" since "either it retains the temporal emergence of natural forms from each other, in which case it loses its dialectical character, or it retains the dialectic and must therefore (as in Hegel) deny the existence of a history of nature."[80]

In this way, Schmidt argued that the notions of emergence, evolution, and natural history, when treated as both dialectical and historical, created a "bad contradiction," antithetical to the human-social dialectic. Marx's own notions of the dialectic of nature and society (and of the metabolism between nature and society), as opposed to those of Engels, were designated by Schmidt as related only to the relatively "primitive" pre-bourgeois phases of human history, and were thus inapplicable to the capitalist age of the "dialectic of Enlightenment" and the mastery of nature.[81] Engels, who introduced a powerful critique of the myth of the human conquest of nature, depicting the metaphorical "revenge" of nature that such an attempt at unlimited, short-sighted environmental conquest would inevitably bring about, was dismissed by Schmidt for having "consciously left out of account 'the impact of men on nature.'"[82] For Schmidt this "impact" was seen not in terms of what Marx called the "negative, i.e. destructive side" of production, generating a metabolic rift, but rather in terms of the inherent, one-sided human conquest of nature introduced by bourgeois civilization, in which nature no longer constituted a limit on society.[83]

The result of such a faulty but influential interpretation of Marx and Engels, with respect to the dialectics of nature and society, was to cut the Western Marxist tradition off from any systematic treatment of these issues: nature in effect no longer existed for Western Marxism; the reified world of capitalism had entirely subsumed it within itself. Ironically, Schmidt's book was published in the same year as Rachel Carson's *Silent Spring* (often viewed as the beginning of the modern environmental movement), while representing an interpretation that only served to hinder significant Western Marxist contributions to the environmental movement.[84]

In rejecting the dialectics of nature, the Western Marxist philosophical tradition sought to distinguish itself from Soviet Marxism, where the dialectics of nature, sometimes viewed mechanically and positivistically, became by the mid-1930s an entrenched ideology, within a narrow doctrine of "dialectical materialism," sometimes referred to pejoratively by its detractors as "diamat."[85] Nevertheless, in the 1920s and early 1930s Soviet ecological thought had been an area of enormous vitality inspired by the dialectics-of-nature conceptions, as manifested in the work of such key figures as Marxist theorist and Soviet leader Nikolai Bukharin; pioneering geneticist Nikolai Ivanovich Vavilov; physicist Boris Hessen; zoologist Vladimir Vladimirovich Stanchinskii; and geneticist and endocrinologist B. M.

Zavadovsky—along with the discoveries of Vernadsky and Oparin. Bukharin stressed the significance of the biosphere for dialectical materialism; Vavilov discovered the global sources of the germplasm of the major agricultural crops and introduced innovations in genetics; Hessen initiated the sociology of science and discussed the significance of Engels's treatment of energetics. Stanchinskii originated the modern notion of trophic levels in ecosystems. Zavadovsky provided a dialectical critique of both mechanism and vitalism. Tragically, the majority of these figures, including Bukharin, Vavilov, Hessen, Stanchinskii, and Zavadovsky, were to be victims of Joseph Stalin's purges, which decimated Soviet ecology.[86]

Fortunately, a second foundation of Marxian dialectical science and philosophy, incorporating ecological viewpoints, was to develop in the British Isles (and Australia) in the work of leading scientists such as physicist and x-ray crystallographer J. D. Bernal; evolutionary biologist and geneticist J. B. S. Haldane; biochemist, historian, and sinologist Joseph Needham; zoologist and human ecologist Lancelot Hogben; and mathematician Hyman Levy, along with philosophers of science, cultural theorists, and classicists, Christopher Caudwell, George Thomson, Benjamin Farrington, Jack Lindsay, and social archaeologist V. Gordon Childe.[87] In France, dialectical materialist conceptions, embracing ecological viewpoints, were to be advanced in the 1940s and after by Marxist philosopher Henri Lefebvre.[88] In the United States, the Nobel Prize–winning geneticist H. J. Muller was heavily influenced by dialectical materialism.

It was this tradition within Marxism that was to play a leading role in the development of ecology, helping to inspire not only some of the leading ecologists of the post–Second World War era such as Barry Commoner and G. Evelyn Hutchinson (and more indirectly Rachel Carson), but also providing many of the foundations of the dialectical biology and evolutionary theories that characterized the radical science movement of the 1970s and after, exemplified by the work of Stephen Jay Gould, Richard Levins, Richard Lewontin in the United States, and Stephen Rose and Hilary Rose in Britain. In the United States today this tradition is best represented by the work of historical-materialist epidemiologists and medical sociologists such as Nancy Krieger, Rob Wallace, and Howard Waitzkin, as well as by Marxian ecology more generally, as embodied in metabolic rift theory, in particular.[89]

Indeed, the development of a broad tradition of work in Marxian ecology, centered on Marx's notion of the metabolic rift, and related in many ways to the foundational work of Marxian scientists and philosophers of science from the 1930s and '40s, has given rise to a new dialectical-materialist-ecological synthesis, which is of fundamental importance in the context of today's planetary crisis and the reality of capitalism in the Anthropocene. In this context, the notion of "dialectical materialism," long rejected within Marxist thought, has suddenly reemerged as a vital framework—one that, to quote Needham, constitutes "so sharp an instrument that there can be no question about its value" even if it is necessary to guard against its misuse in some cases to promote a "dogmatism [that] must at all costs be avoided."[90]

Today, not only Marx's contributions to a critical ecological worldview have been rediscovered but those of Engels as well.[91] Such previously commonplace, but ill-informed, notions as Marx's supposed Prometheanism, or his extreme productivism with respect to nature, have not only been disproven, but have now been replaced by a recognition of the indispensable character of the classical Marxian analysis of the rift in the universal metabolism of nature generated by capitalism.[92]

The return of the dialectics of nature (or the dialectics of nature and society), in the face of today's ecological and epidemiological crises— the latter symbolized by the COVID-19 pandemic—is also connected to related developments associated with the feminist historical-materialist critique of social reproduction under capitalism, and the Black radical critique of racial capitalism.[93] Political theorist Nancy Fraser has played a leading role in synthesizing Marx's notion of the metabolic rift with both social reproduction and racial capitalism. She writes: "Capitalism brutally separated human beings from natural, seasonal rhythms, conscripting them into industrial manufacturing, powered by fossil fuels and profit-driven agriculture, bulked up by chemical fertilizers," thereby "introducing what Marx called a "'metabolic rift.'"[94] The metabolic rift is seen by Fraser and other critical analysts as rooted in *expropriation*—appropriation without equivalent or without reciprocity/reproduction—what thinkers as various as Liebig, Marx, and Max Weber called "*Raubbau*" or the robbery system of capitalism.[95] This is related to the fact that the modern racialized and gendered, as well as class-exploitative, labor systems are rooted fundamentally in the expropriation of bodies, lands, and use values within various "hidden abodes" outside immediate commodity exchange.[96] In this way, the dialectical critique of Marxian ecology and the distinct critiques of race, gender, class, and nation converge, pointing to a more unified revolutionary praxis for the twenty-first century.

Ecological Ruin or Ecological Revolution

The subtitle of this book, *Ecological Ruin or Ecological Revolution*, stands for an emerging global revolutionary situation now facing the world population under the conditions of *Capitalism in the Anthropocene*. This new global material situation is marked by the coincidence in our time of the struggle for freedom and the struggle of necessity. The *struggle for freedom* represents the inner-human need to be free in terms of self-activity and human development, which means being freed of the present system of accumulation rooted in exploitation and expropriation. The *struggle of necessity* is the battle for a safe environment for all individuals on the planet and the continuation of the chain of human generations. "Things have now come to such a pass," as Marx and Engels wrote in *The German Ideology*, "that the individuals must appropriate the existing totality of productive forces, not only to achieve self-activity, but, also, merely to safeguard their very existence" and that of humanity as a whole.[97]

Fossil capitalism has generated the Capitalinian Age, the first geological age of the Anthropocene, putting human civilization and human survival on the cliff's edge. Although the Anthropocene, in the sense of a geological epoch in which anthropogenic factors constitute the leading source of Earth System change, will persist as long as industrial civilization continues (while an end-Anthropocene extinction event can only be disastrous for humanity), it is nonetheless possible to generate a new sustainable relation to the earth. To achieve this, it is necessary to build on past cultures and present possibilities, through the promotion of socialist, collective, and communal values, generating a new geological (and social) age of the earth, the Communian. But this new order will have to be created socially through collective struggle, aimed at what historical materialists have called "ecological civilization," representing a dialectic of continuity and change, through which a new, more sustainable and substantively equal world order is created out of the present one, transcending capitalist institutions.[98]

This struggle for a higher society, constituting a more sustainable, more equal world, will necessarily be a revolutionary one, arising from the bottom of society, and emanating from what is referred to in this book as an *environmental proletariat*, encompassing a critical-environmental as well as critical-economic praxis.[99] This is not a question of simply tearing down what exists, so much as the "revolutionary reconstitution of society at large."[100] The human-social metabolism via production will have to be seen once more as existing within the universal metabolism of nature, requiring a path of sustainable human development. It will be necessary to transcend the unecological tendencies embedded in all past and present class-based civilizations culminating in capitalist barbarism, characterized by: (1) the extreme separation of city and country; (2) the destructive robbery or expropriation of the earth for mere "natural resources" with no concern for the sustainability or reproduction of nature, including other species; and (3) the treatment of the mass of humanity as mere captive workers or human chattel (in the widest sense of the expropriation of the bodies of human beings) at the behest of those who are the masters of this means of production and destruction. All such negative, destructive relations will need to be overcome in the process of building the house of the world anew.

The forging of a revolutionary ecological civilization out of the growing waste and destruction of our times is to be seen as a *creative act* that necessarily draws on past and present to create a more sustainable future: a healing of the metabolic rift and the construction of a new realm of social freedom. To accomplish this will require that the populations of the earth appropriate the state as their own through the exercise of their constituent power, with the object of transforming it from a structure of class (and social) domination into a negation of itself: a political command structure, controlled from below, rooted in communal-collective forms, and based on the protagonism of the people, with the goal of ultimately nurturing "the totality of capacities in the individuals themselves."[101]

If capitalism has become, as Lukács critically observed, a kind of alientated "second nature" for those caught in its grasp, what has to be put in its place, as cultural theorist Edward Said once declared, is not a return to "first nature," but rather the pursuit of a revolutionary "third nature," representing a new material-cultural reality rooted in a society governed by the associated producers able to rationally regulate their production and thus their metabolism with the earth so as to promote their own free and full development, while maintaining the earth for future generations.[102] This is the struggle of the twenty-first century.

PART I

The Planetary Rift

CHAPTER ONE

Marx and the Rift in the Universal Metabolism of Nature

The rediscovery this century of Marx's theory of metabolic rift has given renewed force to the critique of capitalism's destructive relation to the earth. The result has been the development of a unified ecological worldview transcending the divisions between natural and social science that allows us to perceive the concrete ways in which the system of capital accumulation is generating environmental crises and catastrophes.

Yet, this recovery of Marx's ecological argument has given rise to further questions and criticisms. How is his analysis of the metabolism of nature and society related to the issue of the "dialectics of nature," traditionally considered a fault line within Marxist theory? Does the metabolic rift theory—as a number of left critics have recently charged—violate dialectical logic, falling prey to a simplistic Cartesian dualism?[1] Is it really conceivable, as some have asked, that Marx, writing in the nineteenth century, could have provided ecological insights that are of significance to us today in understanding the human relation to ecosystems and ecological complexity? Does it not stand to reason that his nineteenth-century ruminations on the metabolism of nature and society would be rather "outmoded" in our more developed technological and scientific age?[2]

In the following discussion I shall attempt briefly to answer each of these questions. In the process I shall also seek to highlight what I consider to be the crucial importance of Marx's ecological materialism in helping us to comprehend the emerging Great Rift in the earth system, and the resulting necessity of an epochal transformation in the existing nature-society metabolism.

The Dialectics of Nature

The problematic status of the dialectics of nature in Marxian theory has its classic source in Georg Lukács's famous footnote in *History and Class Consciousness* in which he stated with respect to the dialectic: "The misunderstandings that arise from Engels's account of dialectics can in the main be put down to the fact that Engels—following Hegel's mistaken lead—extended the method to apply also to nature. However, the crucial determinants of dialectics—the interaction of subject and object, the unity of theory and practice, the historical changes in the reality underlying the categories as the root cause of changes in thought, etc.—are absent from our knowledge of nature."[3]

Within what came to be known as "Western Marxism" this was generally taken to mean that the dialectic applied only to society and human history, and not to nature independent of human history.[4] Engels, in this view, was wrong in his *Dialectics of Nature* in attempting to apply dialectical logic to nature directly, as were the many Marxian scientists and theorists who had proceeded along the same lines.[5]

It would be difficult to exaggerate the importance of this stricture for Western Marxism, which saw it as one of the key elements separating Marx from Engels and Western Marxism from the Marxism of the Second and Third Internationals. It heralded a move away from the direct concern with issues of a material nature and natural science that had characterized much of Marxian thought up to that point. As Lucio Colletti observed in *Marxism and Hegel*, a vast literature "has always agreed" that differences over philosophical materialism/realism and the dialectics of nature constituted the "main distinguishing features between 'Western Marxism' and 'dialectical materialism.'" According to Russell Jacoby, "Western Marxists" almost by definition "confined Marxism to social and historical reality," distancing it from issues related to external nature and natural science.[6]

What made the stricture against the dialectics of nature so central to the Western Marxist tradition was that dialectical materialism—in the sense that it was attributed to Engels and adopted by the Second and Third Internationals—was seen as deemphasizing the role of the subjective factor (or human agency), reducing Marxism to mere conformity to objective natural laws, giving rise to a kind of mechanical materialism or even positivism. In sharp contrast to this, many of those historical materialists who continued to argue, even if in a qualified way, for a dialectics of nature, regarded its complete rejection as threatening the loss of materialism altogether, and a reversion to idealist frames of thought.[7]

Ironically, it was none other than Lukács himself, who, in a major theoretical shift, took the strongest stand against the wholesale abandonment of the dialectics of nature, arguing that this struck at the very heart of not just Engels's but also Marx's ontology. Even in *History and Class Consciousness* Lukács, following Hegel, had recognized the existence of a limited, "merely objective dialectics of nature" consisting of a "dialectics of movement witnessed by the detached observer."[8] In his famous 1967 preface to the new edition of this work, in which he distanced himself from some of his earlier positions, he declared that his original argument was faulty in its exaggerated critique of the dialectics of nature, since, as he put it, the "basic Marxist category, labor as the mediator of the metabolic interaction between society and nature, is missing. . . . It is self-evident that this means the disappearance of the ontological objectivity of labor," which cannot itself be separated from its natural conditions.[9] As he explained in his well-known *Conversations* that same year, "Since human life is based on a metabolism with nature, it goes without saying that certain truths which we acquire in the process of carrying out this metabolism have a general validity—for example the truths of mathematics, geometry, physics, and so on."[10]

For the post–*History and Class Consciousness* Lukács, then, it was Marx's conception of labor and production as the metabolic relation between human beings and external nature that was the key to the dialectical understanding of the natural world. Human beings could comprehend nature dialectically within limits because they were organically *part of it*, through their own metabolic relations. Even as sharp a critic of the dialectics of nature as Alfred Schmidt in his *Concept of Nature in Marx* acknowledged that it was only in terms of Marx's use of the "concept of 'metabolism,'" in which he "introduced a completely new understanding of man's relation to nature," that we can "speak meaningfully of a 'dialectic of nature.'"[11]

The remarkable discovery in the Soviet archives of Lukács's manuscript *Tailism and the Dialectic*, some seventy years after it was written in the mid-1920s (just a few years after the writing of *History and Class Consciousness*), makes it clear that this critical shift in Lukács's understanding, via Marx's concept of social and ecological metabolism, had already been largely reached by that time. There he explained that "the metabolic interchange with nature" was "socially mediated" through labor and production. The labor process, as a form of metabolism between humanity and nature, made it possible for human beings to perceive—in ways that were limited by the historical development of production—certain objective conditions of existence. Such a metabolic "*exchange* of matter" between nature and society, Lukács wrote, "cannot possibly be achieved—even on the most primitive level—without possessing a certain degree of objectively correct knowledge about the processes of nature (which exist prior to people and function independently of them)." It was precisely the development of this metabolic "exchange of matter" by means of production that formed, in Lukács's interpretation of Marx's dialectic, "the material basis of modern science."[12]

Lukács's emphasis on the centrality of Marx's notion of social metabolism was to be carried forward by his assistant and younger colleague István Mészáros in *Marx's Theory of Alienation*. For Mészáros the "conceptual structure" of Marx's theory of alienation involved the triadic relation of humanity-production-nature, with production constituting a form of mediation between humanity and nature. In this way human beings could be conceived as the "self-mediating" beings of nature. It should not altogether surprise us, then, that it was Mészáros who provided the first comprehensive Marxian critique of the emerging planetary ecological crisis in his 1971 Deutscher Prize Lecture—published a year before the Club of Rome's *Limits to Growth* study. In *Beyond Capital* he was to develop this further in terms of a full-scale critique of capital's alienated social metabolism, including its ecological effects, in his discussion of "the activation of capital's absolute limits" associated with the "destruction of the conditions of social metabolic reproduction."[13]

Lukács and Mészáros thus saw Marx's social-metabolism argument as a way of transcending the divisions within Marxism that had fractured the dialectic and Marx's social (and natural) ontology. It allowed for a praxis-based approach that integrated nature and society, and social history and natural history, without reducing one entirely to the other. In our present ecological age this complex understanding— complex because it dialectically encompasses the relations between part and whole, subject and object—becomes an indispensable element in any rational social transition.

Marx and the Universal Metabolism of Nature

To understand this more fully we need to look at the actual ecological dimensions of Marx's thought. Marx's use of the metabolism concept in his work was not simply (or even mainly) an attempt to solve a philosophical problem but rather an endeavor to ground his critique of political economy materialistically in an understanding of human-nature relations emanating from the natural science of his day. It was central to his analysis of both the production of use values and the labor process. It was out of this framework that Marx was to develop his major ecological critique, that of metabolic rift, or, as he put it, the "irreparable rift in the interdependent process of social metabolism, a metabolism prescribed by the natural laws of life itself."[14]

This critical outlook was an outgrowth of the historical contradictions in nineteenth-century industrial agriculture and the consequent revolution in agricultural chemistry—particularly in the understanding of the chemical properties of the soil—during this same period. Within agricultural chemistry, Justus von Liebig in Germany and James F. W. Johnston in Britain both provided powerful critiques of the loss of soil nutrients in the early to mid-nineteenth century due to capitalist agriculture, singling out for criticism British high farming. This extended to robbing, in effect, the soil of some countries by others.

In the United States figures like the early environmental planner George Waring, in his analysis of the despoliation of the earth in agriculture, and the political economist Henry Carey, who was influenced by Waring, emphasized that food and fiber, containing the elementary constituents of the soil, were being shipped long distances in a one-way movement from country to city, leading to the loss to the soil of its nutrients, which had to be replaced by natural (later synthetic) fertilizers. In his great 1840 work, *Organic Chemistry and Its Application to Agriculture and Physiology* (commonly known as his *Agricultural Chemistry*), Liebig had diagnosed the problem as due to the depletion of nitrogen, phosphorus, and potassium, with these essential soil nutrients ending up in increasingly populated cities where they contributed to urban pollution. In 1842, the British agricultural chemist J. B. Lawes developed a means for making phosphates soluble and built a factory to produce his superphosphates in the first step in the development of synthetic fertilizer. But for the most part in the nineteenth century countries were almost completely dependent on natural fertilizers to restore the soil.

It was in this period of deepening agricultural difficulties, due to the depletion of soil nutrients, that Britain led the way in the global seizure of natural fertilizers, including, as Liebig pointed out, digging up and transporting the bones of the Napoleonic battlefields and the catacombs of Europe, and, more important, the extraction by forced labor of guano (from the excrement of seabirds) on the islands off the coast of Peru, setting off a worldwide guano rush.[15] In the introduction to the 1862 edition of his *Agricultural Chemistry*, Liebig wrote a scathing critique of capitalist industrial agriculture in its British model, observing: "If we do not succeed in making the farmer better aware of the conditions under which he produces and in giving him the means necessary for the increase of his output, wars, emigration, famines and epidemics will of necessity create the conditions of a new equilibrium which will undermine the welfare of everyone and finally lead to the ruin of agriculture."[16]

Marx was deeply concerned with the ecological crisis tendencies associated with soil depletion. In 1866, the year before the first volume of *Capital* was published, he wrote to Engels that in developing the critique of ground rent in the third volume, "I had to plough through the new agricultural chemistry in Germany, in particular Liebig and Schönbein, which is more important for this matter than all the economists put together."[17] Marx, who had been studying Liebig's work since the 1850s, was impressed by the critical introduction to the 1862 edition of the latter's *Agricultural Chemistry*, integrating it with his own critique of political economy.

Since the *Grundrisse* in 1857–58, Marx had given the concept of metabolism (*Stoffwechsel*)—first developed in the 1830s by scientists engaged in the new discoveries of cellular biology and physiology and then applied to chemistry (by Liebig especially) and physics—a central place in his account of the interaction between nature and society through production. He defined the labor process as the metabolic relation between humanity and nature. For human beings this metabolism necessarily took a socially mediated form, encompassing the organic conditions common to all life, but also taking a distinctly human-historical character through production.[18]

Building on this framework, Marx emphasized in *Capital* that the disruption of the soil cycle in industrialized capitalist agriculture constituted nothing less than "a rift" in the metabolic relation between human beings and nature. "Capitalist production," he wrote,

> collects the population together in great centres, and causes the urban population to achieve an ever-greater preponderance. This has two results. On the one hand it concentrates the

historical motive force of society; on the other hand, it disturbs the metabolic interaction between man and the earth, i.e. it prevents the return to the soil of its constituent elements consumed by man in the form of food and clothing; hence it hinders the operation of the eternal natural condition for the lasting fertility of the soil. . . . But by destroying the circumstances surrounding this metabolism . . . it compels its systematic restoration as a regulative law of social production, and in a form adequate to the full development of the human race. . . . All progress in capitalist agriculture is a progress in the art, not only of robbing the worker, but of robbing the soil; all progress in increasing the fertility of the soil for a given time is progress toward ruining the more long-lasting sources of that fertility. . . . Capitalist production, therefore, only develops the technique and the degree of combination of the social process of production by simultaneously undermining the original sources of all wealth—the soil and the worker.[19]

Following Liebig, Marx highlighted the global character of this rift in the metabolism between nature and society, arguing, for example, that "for a century and a half England has indirectly exported the soil of Ireland without even allowing its cultivators the means for replacing the constituents of the exhausted soil."[20] He integrated his analysis with a call for ecological sustainability, that is, preservation of "the whole gamut of permanent conditions of life required by the chain of human generations." In his most comprehensive statement on the nature of production under socialism he declared: "Freedom, in this sphere, can consist only in this, that socialized man, the associated producers, govern the human metabolism with nature in a rational way, bringing it under their collective control . . . accomplishing it with the least expenditure of energy and in conditions most worthy and appropriate for their human nature."[21]

Over the last decade and a half ecological researchers have utilized the theoretical perspective of Marx's metabolic rift analysis to analyze the developing capitalist contradictions in a wide array of areas:

planetary boundaries, the carbon metabolism, soil depletion, fertilizer production, the ocean metabolism, the exploitation of fisheries, the clearing of forests, forest-fire management, hydrological cycles, mountaintop removal, the management of livestock, agro-fuels, global land grabs, and the contradiction between town and country.[22]

However, a number of critics on the left have recently raised theoretical objections to this view. One such criticism suggests that the metabolic rift perspective falls prey to a "Cartesian binary," in which nature and society are conceived dualistically as separate entities.[23] Hence, it is seen as violating the fundamental principles of dialectical analysis. A related criticism charges that the very concept of a rift in the metabolism between nature and society is "non-reflexive" in that it denies "the dialectical reciprocity of the biophysical environment."[24] Still others have suggested that the reality of the metabolic rift itself generates an "epistemic rift" or a dualistic view of the world, which ends up infecting Marx's own value theory, causing him to downplay ecological relations in his analysis.[25]

Here it is important to emphasize that Marx's metabolic rift theory, as it is usually expounded, is a theory of ecological crisis—of the disruption of what Marx saw as the everlasting dependence of human society on the conditions of organic existence. This represented, in his view, an insurmountable contradiction associated with capitalist commodity production, the full implications of which, however, could only be understood within the larger theory of nature-society metabolism.

To account for the wider natural realm within which human society had emerged, and within which it necessarily existed, Marx employed the concept of the "universal metabolism of nature." Production mediated between human existence and this "universal metabolism." At the same time, human society and production remained *internal to* and *dependent on* this larger earthly metabolism, which preceded the appearance of human life itself. Marx explained this as constituting "the universal condition for the metabolic interaction between nature and man, and as such a natural condition of human life." Humanity, through its production, "withdraws" or extracts its natural-material use values from this "universal metabolism of nature," at the same time "breathing [new] life" into these natural conditions "as elements of a new [social] formation," thereby generating a kind of second nature. However, in a capitalist commodity economy this realm of second nature takes on an alienated form, dominated by exchange value rather than use value, leading to a rift in this universal metabolism.[26]

This, I believe, provides the basic outline for a materialist-dialectical understanding of the nature-society relation—one that is in remarkably close accord not only with the most developed science (including the emerging thermodynamics) of Marx's day, but also with today's more advanced ecological understanding.[27] There is nothing dualistic or non-reflexive in such a view. In Marx's materialist dialectic, it is true, neither society (the subject/consciousness) nor nature (the object) is subsumed entirely within the other, thus avoiding the pitfalls of both absolute idealism and mechanistic science.[28] Human beings transform nature through their production, but they do not do so just as they please; rather they do so under conditions inherited from the past (of both natural and social history), remaining dependent on the underlying dynamics of life and material existence.

The main reason no doubt that a handful of left critics, struggling with this conceptual framework, have characterized the metabolic-rift theory as a form of Cartesian dualism is due to a failure to perceive that within a materialist-dialectical perspective it is impossible to analyze the world in a meaningful way except through the use of abstraction that temporarily isolates, for purposes of analysis, one "moment" (or mediation) within a totality.[29] This means employing conceptions that at first sight—when separated out from the overall dynamics—may appear one-sided, mechanical, dualistic, or reductionist. In referring, as Marx does, to "the metabolic interaction between nature and man" it should never be supposed that "man" (humanity) actually exists completely independently of or outside "nature"—or even that nature today exists completely independent of (or unaffected by) humanity. The object of such an exercise in abstraction is merely to comprehend the larger concrete totality through the scrutiny of those specific mediations that can be rationally said to constitute it within a developing historical context.[30] Our very knowledge of nature, in Marx's view, is a product of our human-social metabolism, that is, our productive relation to the natural world.

Far from representing a dualistic or non-reflexive approach to the world, Marx's analysis of "the metabolism of nature and society" was eminently dialectical, aimed at comprehending the larger concrete totality. I agree with David Harvey's observation in his 2011 Deutscher Prize Lecture that the "universality" associated with Marx's conception of "the metabolic relation to nature" constituted a kind of outer set of conditions or boundaries in his conception of reality within which all the "different 'moments'" of his critique of political economy were potentially linked to one another. It is true also, as Harvey says, that Marx seems to have set aside in his critique of capital these larger boundary questions, leaving for later the issues of the world economy and the universal metabolism of nature.[31] Indeed, Marx's wider ecological view remained in certain respects necessarily undifferentiated and abstract—unable to reach the level of concrete totality. This is because there was a seemingly endless amount of scientific literature to pore through before it would be possible to discuss the distinct, historic mediations associated with the coevolutionary nature-society dialectic.

Still, Marx did not shirk in the face of the sheer enormity of this task and we find him at the end of his life carefully taking notes on how shifts in isotherms (the temperature zones of the earth) associated with climate change in earlier geological eras led to the great extinctions in Earth's history. It is this shift in the isotherms that James Hansen, the leading U.S. climatologist, sees as the main threat facing flora and fauna today as a result of global warming, with the isotherms moving toward the poles faster than the species.[32] Another instance of this deep concern with natural science is Marx's interest in John Tyndall's Royal Institution lectures regarding the experiments he was carrying out on the interrelation of solar radiation and various gases in determining the earth's climate. It was quite possible that Marx, who attended some of these lectures, was actually present when

Tyndall provided the first empirical account of the greenhouse effect governing the climate.[33] Such attentiveness to natural conditions on Marx's part makes it clear that he took seriously both the issue of the universal metabolism of nature and the more specific socio-metabolic interaction of society and nature within production. The future of humanity and life in general depended, as he clearly recognized, on the sustainability of these relationships in terms of "the chain of human generations."[34]

The Rift in Earth's Metabolism

All of this leaves us with the third objection to Marx's metabolic rift theory in which it is seen as outdated, and no longer of any direct use in analyzing our current world ecology, given today's more developed conditions and analysis. Thus, the criticism has been made that the metabolic rift is "outmoded as a way to describe ruptures in natural pathways and processes" unless developed further to address ecosystems and dynamic natural cycles and to take into account the labor process.[35]

Such a dialectical synthesis, however, was a strength of Marx's metabolic rift theory from the start, which was explicitly based on an understanding of the labor process as the metabolic exchange between human beings and nature, and thus pointed to the importance of human society in relation to biogeochemical cycles, and to exchanges of matter and energy in general.[36] The concept of ecosystem had its origin in this dialectical-systems approach, in which Marx's friend E. Ray Lankester, the foremost Darwinian biologist in England in the generation after Darwin and an admirer of Marx's *Capital*, was to play a leading role. Lankester first introduced the word "œcology" (later ecology) into English in 1873, in the translation that he supervised of Ernst Haeckel's *History of Creation*. Lankester later developed a complex ecological analysis, beginning in the 1880s, under his own concept of "bionomics," a term viewed as synonymous with ecology. It was Lankester's student, Arthur Tansley, who, influenced by Lankester's bionomic studies (and by the early systems theory of the British Marxist mathematician Hyman Levy), introduced the concept of ecosystem as a materialist explanation of ecological relations in 1935.[37]

In the twentieth century the concept of metabolism was to become the basis of systems ecology, particularly in the landmark work of Eugene and Howard Odum. As Frank Golley explains in *A History of the Ecosystem Concept in Ecology*, Howard Odum "pioneered a method of studying [eco-]system dynamics by measuring . . . the difference of input and output, under steady state conditions," to determine "the metabolism of the whole system." Based on the foundational work of the Odums, metabolism is now used to refer to all biological levels, starting with the single cell and ending with the ecosystem (and beyond that the Earth System). In his later attempts to incorporate human society into this broad ecological systems theory, Howard Odum was to draw heavily on Marx's work, particularly in developing a theory of what he called ecologically "unequal exchange" rooted in "imperial capitalism."[38]

Indeed, if we were to return today to Marx's original issue of the human-social metabolism and the problem of the soil nutrient cycle, looking at it from the viewpoint of ecological science, the argument would go like this. Living organisms, in their normal interactions with one another and the inorganic world, are constantly gaining nutrients and energy from consuming other organisms or, for green plants, through photosynthesis and nutrient uptake from the soil—which are then passed along to other organisms in a complex "food web" in which nutrients are eventually cycled back to near where they originated. In the process the energy extracted is used up in the functioning of the organism although ultimately a portion is left over in the form of difficult to decompose soil organic matter. Plants are constantly exchanging products with the soil through their roots—taking up nutrients and giving off energy-rich compounds that produce an active microbiological zone near the roots. Animals that eat plants or other animals generally use only a small fraction of the nutrients they eat and deposit the rest as feces and urine nearby. When they die, soil organisms use their nutrients and the energy contained in their bodies. The interactions of living organisms with matter (mineral or alive or previously alive) are such that the ecosystem is only lightly affected and nutrients cycle back to near where they were originally obtained. Also on a geological time scale, weathering of nutrients locked inside minerals renders them available for future organisms to use. Thus, natural ecosystems do not normally "run down" due to nutrient depletion or loss of other aspects of healthy environments such as productive soils.

As human societies develop, especially with the growth and spread of capitalism, the interactions between nature and humans are much greater and more intense than before, affecting first the local, then the regional, and finally the global environment. Since food and animal feeds are now routinely shipped long distances, this depletes the soil, just as Liebig and Marx contended in the nineteenth century, necessitating routine applications of commercial fertilizers on crop farms. At the same time this physical separation of where crops are grown and where humans or farm animals consume them creates massive disposal issues for the accumulation of nutrients in city sewage and in the manure that piles up around concentrations of factory farming operations. And the issue of breaks in the cycling of nutrients is only one of the many metabolic rifts that are now occurring. It is the change in the nature of the metabolism between a particular animal—humans—and the rest of the Earth System (including other species) that is at the heart of the ecological problems we face.[39]

Despite the fact that our understanding of these ecological processes has developed enormously since Marx and Engels's day, it is clear that in pinpointing the metabolic rift brought on by capitalist society they captured the essence of the contemporary ecological problem. As Engels put it in a summary of Marx's argument in *Capital*, industrialized-capitalist agriculture is characterized by "*the robbing of the soil*: the acme of the capitalist mode of production is the undermining of the *sources of all wealth*: the soil and labourer."[40] For Marx and Engels this reflected the contradiction between town and country, and the need to prevent the worst distortions of the human metabolism with nature associated with urban development. As Engels wrote in *The Housing Question*:

The abolition of the antithesis between town and country is no more and no less utopian than the abolition of the antithesis between capitalists and wage-workers. From day to day it is becoming more and more a practical demand of both industrial and agricultural production. No one has demanded this more energetically than Liebig in his writings on the chemistry of agriculture, in which his first demand has always been that man shall give back to the land what he receives from it, and in which he proves that only the existence of the towns, and in particular the big towns, prevents this. When one observes how here in London alone a greater quantity of manure than is produced in the whole kingdom of Saxony is poured away every day into the sea with an expenditure of enormous sums, and what colossal structures are necessary in order to prevent this manure from poisoning the whole of London, then the utopia of abolishing the distinction between town and country is given a remarkably practical basis.[41]

Although problems of the nutrient cycle and waste treatment, as well as the relation between country and city, have changed since the nineteenth century, the fundamental problem of the rift in natural cycles generated by the human-social metabolism remains.

Marx and Engels's approach to materialism and dialectics can therefore be seen as intersecting in complex ways with the development of the modern ecological critique. The reason that this story is so unknown can be traced to the tendency of Western Marxism to write off all of those (even leading scientists) who delved into the dialectics of nature—except perhaps as reminders of various follies and capitulations (notably the Lysenko affair in the Soviet Union).[42] Here I am referring to such important critical figures, in the British context, as Levy, Christopher Caudwell, J. D. Bernal, J. B. S. Haldane, Joseph Needham, Lancelot Hogben, and Benjamin Farrington—along with other, non-Marxian, materialists and socialists, such as Lankester and Tansley.[43] Later on we see a developing ecological critique drawing in part on Marx emerging in the work of such thinkers as Howard Odum, Barry Commoner, Richard Levins, Richard Lewontin, and

Stephen Jay Gould.[44] Although Frankfurt School thinkers made remarkable observations on the "domination of nature" by the "dialectic of the Enlightenment," as well as on the negative environmental effects of modern industrial technology, it was not there, but rather within the more adamantly materialist and scientific traditions, that the main socialist contributions to ecological thought emerged.[45]

Today we are making enormous advances in our critical understanding of the ecological rift. Marx's metabolic approach to the nature-society connection has been widely adopted within environmental thought, though seldom incorporating the full dialectical critique of the capital relation that his own work represented. A cross-disciplinary research tradition on "industrial metabolism," addressing material flows associated with urban areas, has developed in the last couple of decades. As Marina Fischer-Kowalski, founder of the Institute of Social Ecology in Vienna and the foremost representative of material-flows analysis today, noted in the late 1990s, metabolism has become "a rising conceptual star" within socioecological thought. "Within the nineteenth-century foundations of social theory," she added, "it was Marx and Engels who applied the term 'metabolism' to society."[46]

The global ecological crisis is now increasingly understood within social science in terms of the industrialization of the human-metabolic relation to nature at the expense of the world's ecosystems, undermining the very bases on which society exists. Marx's concept of "social metabolism" (also sometimes referred to as "socio-ecological metabolism") has been used by critical ecological economists to chart the whole history of human-nature intersections, together with the conditions of ecological instability in the present. This has led to analyses of modes of production as successive "socio-metabolic regimes," as well as to demands for a "socio-metabolic transition."[47] Meanwhile, a more direct linking of Marx's metabolic rift theory to the critique of capitalist society has allowed researchers in environmental sociology to carry out penetrating, historical-empirical inquiries into a whole range of ecological problems—extending to issues of unequal ecological exchange or ecological imperialism.[48]

Much of this work of course has its roots in the recognition that the world is crossing crucial "planetary boundaries" defined by the departure from the conditions of the Holocene Epoch that nurtured the growth of human civilization—a critical approach pioneered by Johan Röckström of the Stockholm Resilience Institute and leading climate scientists such as James Hansen. Here the main concern is what could be called the Great Rift in the human relation to nature brought on by the crossing of the Earth System boundaries associated with climate change, ocean acidification, ozone depletion, loss of biological diversity (and species extinction), the disruption of the nitrogen and phosphorus cycles, loss of land cover, loss of fresh water sources, aerosol loading, and chemical pollution.[49]

On Earth Day 2003, NASA released its first quantitative satellite measurements and maps of the "earth's metabolism," focusing on the extent to which the plant life on Earth was fixing carbon through photosynthesis. This data is also being used for monitoring the growth of deserts, the effects of droughts, the vulnerability of forests, and other climate-change developments.[50] The issue of the earth's metabolism is of course directly related to the human interaction with the environment. Humanity now consumes a substantial share of the global terrestrial net primary production through photosynthesis and that share is growing at unsustainable levels. Meanwhile, the disruption of the "carbon metabolism" through human production is radically affecting the earth's metabolism in ways that, if not altered, will have catastrophic effects on life on the planet, including the human species itself.[51] As Hansen describes the potential consequences of the Great Rift in the carbon metabolism in particular:

> The picture that emerges for Earth sometime in the distant future, if we should dig up and burn every fossil fuel is thus consistent with . . . an ice-free Antarctica and a desolate planet without human inhabitants. Although temperatures in the Himalayas may have become seductive, it is doubtful that the many would allow the wealthy few to appropriate this territory to themselves or that humans would survive the extermination

of most other species on the planet.... It is not an exaggeration to suggest, based on the best available scientific evidence, that burning all fossil fuels could result in the planet being not only ice-free but human-free.[52]

Marx and Socio-Ecological Revolution

It is precisely when we confront the sheer enormity of the Great Rift in the earth's metabolism that Marx's approach to the metabolism of nature and society becomes most indispensable. Marx's analysis stressed the rupture by capitalist production of the "eternal natural conditions," constituting the "robbery" of the earth itself.[53] But his analysis was unique in that it pointed beyond the forces of accumulation and technology (that is, the treadmill of production) to the qualitative, use-value structure of the commodity economy: the question of human needs and their fulfillment. The natural-material use value of human labor itself, in Marx's theory, resided in its *real productivity* in terms of the genuine fulfillment of human needs. In capitalism, he argued, this creative potential was so distorted that labor power was seen as being "useful" (from a capitalist exchange-value perspective) only insofar as it generated surplus value for the capitalist.[54]

To be sure, Marx did not himself follow out the full ramifications of this distortion of use value (and of labor's own usefulness). Although he raised the question of the qualitative, use-value structure of the commodity economy he left it largely unexamined in his critique of political economy.[55] It was generally assumed in the context of mid-nineteenth-century capitalism that those use values that were produced—outside of the relatively insignificant realm of luxury production—conformed to genuine human needs. Under monopoly capitalism, beginning in the last quarter of the nineteenth century, and with the emergence more recently of the phase of globalized monopoly-finance capital, this all changed. The system increasingly demanded, simply to keep going under conditions of chronic overaccumulation, the production of *negative* use values and the *non-fulfillment* of human needs.[56] This entails the absolute alienation of the labor process, that is, of the metabolic relation between human beings and nature, turning it predominantly into a form of waste.

The first to recognize this in a big way was William Morris, who emphasized the growth of monopolistic capital and the waste associated with the massive production of useless goods and the "useless toil" this entailed.[57] Morris, who had studied Marx's *Capital* carefully—and especially the analysis of the labor process and the general law of accumulation—emphasized more than any other thinker the direct connection between socially wasted production and socially wasted labor, drawing out the consequences of this in terms of human life and creativity and the environment itself. In his 1894 lecture "Makeshift," Morris stated:

> I noticed the other day that Mr. Balfour was saying that Socialism was impossible because under it we should produce so much less than we do now. Now I say that we might produce half or a quarter of what we do now, and yet be much wealthier, and consequently much happier, than we are now: and that by turning whatever labour we exercised, into the production of useful things, things that we all want, and by... refusing to labour in producing useless things, things which none of us, not even fools want....
>
> My friends, a very great many people are employed in producing mere nuisances, like barbed wire, 100-ton guns, sky signs and advertising boards for the disfigurement of the green fields along the railways and so forth. But apart from these nuisances, how many more are employed in making market wares for rich people which are of no use whatever except to enable the said rich to 'spend their money' as 'tis called; and again how many more in producing wretched makeshifts for the working classes because they can afford nothing better?[58]

Others, including Thorstein Veblen at the beginning of the twentieth century, and Paul Baran and Paul Sweezy in the 1960s, were to develop further the economic critique of waste and the distortion of use values in the capitalist economy, pointing to "the interpenetration effect," whereby the sales effort penetrated into production itself, destroying whatever claims to rationality existed in the latter.[59] Yet, Morris remained unsurpassed in his emphasis on the effects of the capitalist-commodity-exchange process on the qualitative nature of the labor process itself, converting what was already an exploited labor force into one that was also engaged in useless, uncreative, empty toil—no longer serving to satisfy social needs, but rather squandering both resources and lives.

It is here that Marxian theory, and in particular the critique of monopoly capital, suggests a way out of capitalism's endless creative destructiveness. It is through the politicization of the use value structure of the economy, and the relation of this to the labor process and to the whole qualitative structure of the economy, that Marx's dialectical approach to the metabolism between nature and society takes on potent form. U.S. expenditures in such areas as the military, marketing, public and private security, highways, and personal luxury goods add up to trillions of dollars a year, while much of humanity lacks basic necessities and a decent life, and the biosphere is being systematically degraded.[60] This inevitably raises issues of communal needs and environmental costs, and above all the requirement of planning—if we are to create a society of substantive equality, ecological sustainability, and freedom in general.

No transformation of the overall use-value structure of production is conceivable of course without the self-mobilization of humanity within a co-revolutionary process, uniting our multiple struggles. The combined ecological and economic contradictions of capital in our time, plus the entire imperialist legacy, tell us that the battle for such a transition will first emerge in the Global South—of which there are already signs today.[61] Yet, the underlying conditions are such that the revolutionary reconstitution of society must be truly universal in its scope and its aspirations, encompassing the entire globe and all of its peoples, if humanity is to succeed in pulling the world back from the brink of catastrophe brought on by capitalism's unrelenting creative destructiveness. In the end it is a question of the human metabolism with nature, which is also a question of human production, and of human freedom itself.

CHAPTER TWO

The Great Capitalist Climacteric

Humanity today is confronted with what might be called the Great Capitalist Climacteric. In the standard definition, a climacteric (from the Greek *klimaktēr* or rung on a ladder) is a period of critical transition or a turning point in the life of an individual or a whole society. From a social standpoint, it raises issues of historical transformation in the face of changing conditions.[1] In the 1980s environmental geographers Ian Burton and Robert Kates referred to "the Great Climacteric" to address what they saw as the developing global ecological problem of the limits to growth, stretching from 1798, the year of publication of Thomas Malthus's *An Essay on the Principle of Population*, to 2048, two hundred and fifty years later. "Applied to population, resources, and environment throughout the world," the notion of a Great Climacteric, they wrote, "captures the idea of a period that is critical and where serious change for the worse may occur. It is a time of unusual danger."[2]

The term the Great Capitalist Climacteric is used here to refer to the necessary epochal social transition associated with the current planetary emergency. It refers both to the objective necessity of a shift to a sustainable society and to the threat to the existence of *Homo sapiens* (as well as numerous other species) if the logic of capital accumulation is allowed to continue dictating to society as a whole. The current world of business as usual is marked by rapid climate change, but also by the crossing or impending crossing of numerous other planetary boundaries that define "a safe operating space for humanity."[3] It was the recognition of this and of the unprecedented speed of Earth System change due to social-historical factors that led scientists in recent years to introduce the notion of the Anthropocene Epoch, marking the emergence of humanity as a geological force on a planetary scale.[4] As leading U.S. climatologist James Hansen explains, "The rapidity with which the human-caused positive [climate] forcing is being introduced has no known analog in Earth's history. It is thus exceedingly difficult to foresee the consequences if the human-made climate forcing continues to accelerate."[5]

With the present rate of carbon emissions, the world will break the global carbon budget for a below 2°C increase in global average temperature in less than a generation, and a 1.5°C increase well before that.[6] Once we reach 2°C, it is feared, we will be entering a world of climate feedbacks and irreversibility where humanity may no longer be able to return to the conditions that defined the Holocene Epoch in which civilization developed. The 2°C "guardrail" officially adopted by world governments in Copenhagen in 2009 is meant to safeguard humanity from plunging into what prominent UK climatologist Kevin Anderson of the Tyndall Center for Climate Change has called "extremely dangerous" climate change. Yet, stopping carbon emissions prior to the 2°C boundary, Anderson tells us, will at this point require "revolutionary change to the political economic hegemony," going against the accumulation of capital or economic growth characteristics that define the capitalist system. More concretely, staying within the carbon budget determined by remaining below 2°C means that global carbon emissions must be cut by around 3 percent a year at present (and by more than 7 percent annually to stay below a 1.5°C increase). This means that in the rich countries the reductions in carbon emissions per annum must reach double digits.[7]

Yet, despite the widespread awareness of the planetary emergency represented by global warming, carbon emissions have continued to rise throughout the world. The failure of capitalism to implement the necessary cuts in carbon dioxide can be explained by the threat that this poses to its very existence as a *system of capital accumulation*. As a result, civilization is faced by a threat of self-extermination that over the long run is as great as that posed by a full nuclear exchange—and in a process that is more inexorable. The present reality of global capitalism makes it appear utopian to call for a revolutionary strategy of "System Change Not Climate Change." But the objective of stopping climate change leaves the world with no other option, since avoiding climate-change disaster will be even more difficult—and may prove impossible—if the global population does not act quickly and decisively.

Some observers have been quick to conclude that 2°C will inevitably be crossed given prevailing social reality and the failure of current climate negotiations, and that we should therefore simply accept this and shift the target, choosing to stop climate change before it reaches a 3°C or a 4°C increase. This is a view that the World Bank has subtly encouraged.[8] It is necessary, however, to account for the likely nonlinear effects of such global warming on the entire Earth System. At 2°C, the level of uncertainty, and the threat of uncontrollable Earth warming due to "slow feedbacks" and the crossing of successive thresholds (tipping points), are magnified enormously.[9] Human actions to cut greenhouse gas emissions might then come too late, not simply in the sense of an increase in catastrophic events such as extreme weather or the effects of sea level rise, but also in the even more ominous sense of humanity's loss of the power to stabilize the climate (and civilization). We do not know when and where such a global tipping point will be reached, but today's climate science tells us that a 2°C increase is more dangerous than was thought when that boundary was originally proposed. What was once believed to be "dangerous climate" change arising at 2°C is now considered to be "highly dangerous."[10] If uncontrollable global warming—driven by the reduction in the albedo effect (the reflectivity of Earth), the release of methane from the permafrost, and other slow feedbacks—were to take over, human beings would have little choice but simply to try to adapt in whatever ways they could, watching while their own future, and even more that of future generations, evaporated before their eyes.[11]

Indeed, even the 2°C guardrail approach, Hansen argues, is too conservative. If major sea level rise engulfing islands and threatening coastal cities throughout the world and displacing hundreds of millions of people is to be avoided, society needs to aim at reaching 350 parts per million (ppm) of atmospheric carbon (down from the present 420 ppm) by 2100, with the object of achieving the below 1.5°C target.[12]

As bad as all of this is, it is essential to keep in mind that climate change is only one part of the Great Capitalist Climacteric confronting the world in the twenty-first century—although related to all the others. The world economy has already crossed or is on the brink of crossing a whole set of planetary boundaries, each one of which represents a planetary emergency in its own right, including ocean acidification, loss of biological diversity, the disruption of the nitrogen and phosphorus cycles, disappearance of fresh water, land cover change (particularly deforestation), and growing pollution from synthetic chemicals (leading to biomagnification and bioaccumulation of toxins in living organisms).[13] The common denominator behind all of these rifts in the biogeochemical cycles of the planet is the system of capital accumulation on a global scale. This points to the need for truly massive, accelerated social change exceeding in scale not only the great social revolutions of the past, but also the great transformations of production marked by the original Agricultural Revolution and the Industrial Revolution: namely, a twenty-first-century Ecological Revolution.

Natural science can take us only so far on these issues. Since the source of the Great Capitalist Climacteric lies in the historical constitution of human society, necessitating a social revolution, we must turn to social science as a guide. Yet, the dominant social science has as its underlying premise—structuring its entire frame of analysis—the notion that the critique of capitalism is off limits. This is so much the case that even the name "capitalism," as John Kenneth Galbraith pointed out in *The Economics of Innocent Fraud*, was increasingly replaced in the 1980s by the "meaningless designation" of the "market system."[14] When capitalism is referred to at all today in the mainstream it is as a mere synonym for the watered-down notion of a competitive market society, viewed as the end (*telos*) of human history—both in the sense that all of history is seen as the unfolding of a natural tendency toward market capitalism, and that capitalism itself is "the end of history."[15]

The result of such ahistorical thinking is that conventional thought, with only minor exceptions, has virtually no serious social scientific analysis on which to rely in confronting today's Great Capitalist Climacteric. Those who swallow whole the notion that there is no future beyond capitalism are prone to conclude—in defiance of the facts—that the climate crisis can be mitigated within the present system. It is this *social denialism* of liberal-left approaches to the climate crisis, and of the dominant social science, that led Naomi Klein to declare in *This Changes Everything* that "the right is right" in viewing climate change as a threat to capitalism. The greatest obstacle before us, she insists, is not the *outright denialism* of the science by the far right, but rather the *social denialism* of the dominant liberal discourse, which, while giving lip service to the science, refuses to face reality and recognize that capitalism must go.[16]

If conventional social science is crippled at every point by corrupt adherence to a prevailing class reality, the postmodern turn over the last few decades has generated a left discourse that is just as ill-equipped to address the Great Capitalist Climacteric. Largely abandoning historical analysis (grand narratives) and the negation of the negation—that is, the idea of a revolutionary forward movement—the left has given way to extreme skepticism and the deconstruction of everything in existence, constituting a profound "dialectic of defeat."[17]

Although some hope is to be found in the Green theory or "ecologism" that has emerged in the context of the environmental movement, such views are typically devoid of any secure moorings within social (or natural) science, relying on neo-Malthusian assumptions coupled with an abstract ethical orientation that focuses on the need for a new, ecocentric worldview aimed at protecting the earth and other species.[18] The main weakness of this new ecological conscience is the absence of anything remotely resembling "the confrontation of reason with reality," in the form of a serious ecological and social critique of capitalism as a system.[19] Abstract notions like growth, industrialism, or consumption take the place of investigations into the laws of motion of capitalism as an economic and social order, and how these laws of motion have led to a collision course with the Earth System.

It is therefore the socialist tradition, building on the powerful foundations of historical materialism—and returning once more to its radical foundations to reinvent and re-revolutionize itself—to which we must necessarily turn in order to find the main critical tools with which to address the Great Capitalist Climacteric and the problem of the transition to a just and sustainable society. A period of self-criticism within Marxian theory, commencing in the 1960s and developing over decades, eventually gave rise to a revolution in its understanding of social-ecological conditions. Yet, like most intellectual revolutions, the new insights arose only by standing "on the shoulders of giants"—that is, based on the rediscovery and reconstruction of prior understandings, in the face of changing conditions.

The advance of Marxian ecology was the product of a massive archaeological dig in the scientific foundations of Marx's thought, allowing for the development of a much richer understanding of the relation of the *materialist conception of history* to the *materialist conception of nature*—and generating a deeper, wider social-ecological critique of capitalist society.

By the end of the last century this return to Marx's ecology had resulted in three crucial scientific breakthroughs: (1) the rediscovery of what could be called Marx's "ecological value-form analysis"; (2) the recovery and reconstruction of his theory of metabolic rift; and (3) the retrieval of the two types of ecological crisis theory embedded in his analysis. These critical breakthroughs were to generate new strategic insights into revolutionary praxis in the Anthropocene.

The Three Critical Breakthroughs of Ecological Marxism

What has often been called the Western Marxist tradition that arose in the 1920s and '30s was distinguished primarily by its rejection of the dialectics of nature and Soviet-style dialectical materialism.[20] The interpretation of Marx's approach to the relation of nature and society in the Western Marxist tradition found its most systematic early expression in Alfred Schmidt's 1962 *The Concept of Nature in Marx*, originally written as a doctoral thesis under the supervision of Frankfurt School philosophers Max Horkheimer and Theodor Adorno. Schmidt recognized the central importance of Marx's notion of social metabolism in the development of a revolutionary, new conception of nature. Yet, this was to be set aside in Schmidt's wider criticism, which attributed to Marx the same narrow instrumentalist-productivist vision purportedly characteristic of the "dialectic of Enlightenment" as a whole.[21]

In the 1970s and '80s Schmidt's overall negative assessment of Marx on nature was adopted by what has now come to be known as "first-stage ecosocialism," associated with figures such as Ted Benton and Andre Gorz.[22] Benton argued that Marx had gone overboard in his criticism of Malthus's population theory to the point of denying natural limits altogether.[23] The mature Marx (as distinguished from the Marx of the *Economic and Philosophical Manuscripts*) was thus seen as devoid of positive ecological values and as promoting a crude "Promethean" productivism. A common practice of first-stage ecosocialism was to graft both neo-Malthusian concepts and the primarily ethical standpoint of Green theory onto more traditional Marxian theory, creating a hybrid ecosocialism or what was referred to as "the greening of Marxism."[24] As Raymond Williams critically observed, the result was a tendency to "run together two kinds of thinking" associated with Green theory and Marxism, rather than going back to the roots of historical materialism to uncover its own ecological premises.[25]

It was in this context that a "second-stage ecosocialism," challenging the first, arose in the 1990s in the work of various Marxian political economists. Socialist theorists proceeded to dig into the very foundations of classical historical materialism and its value-theoretical framework. The first critical breakthrough, dramatically altering our understanding of Marx on ecology, was provided by Marxian economist Paul Burkett, who in his 1999 *Marx and Nature* recovered the ecological value-form analysis underpinning Marx's entire critique of political economy.[26] It was the early Soviet economist, I. I. Rubin, who had first emphasized the double nature of Marx's value theory as consisting of: (1) a theory of the value-form, or what Marxian economist Paul Sweezy in the United States was to call "the qualitative value problem"; and (2) a theory of the quantitative determination of value and price. The value-form analysis, focusing on the social form that value assumes and the larger qualitative aspects of capitalist valorization connecting it to class and production, was to be Marx's singular achievement—altering as well the understanding of the quantitative aspects of value.[27] In Burkett's work, Marx's value-form theory was elaborated to explain systematically for the first time the ecological value-form analysis embedded in classical historical materialism.[28]

From this standpoint, Marx's entire critique was seen as rooted in the contradictory relations between what he called "production in general," characterizing human production in all of its forms, and the historically specific capitalist labor and production process.[29] In production in general the human labor process transforms the products of nature, or natural-material use values, which constitute real material wealth. However, in capitalism, conceived as a specific mode of production, this characteristic of production in general takes a more alienated form, as the majority of workers are estranged from the means of production, and particularly the land, and are thus proletarianized—able to survive only by selling their labor power.

All value, the classical political economists argued, came from labor. But *classical-liberal* political economists saw this as a universal, transhistorical reality, whereas *Marx*, in sharp contrast, conceived it as a historically specific one, confined to capitalism. Nature was excluded, as Marx stressed, from the direct creation of value/exchange value under capitalism.[30] This is still reflected in our national income or GDP statistics, which account for economic growth entirely in terms of the *value added* of human services, measured in the form of wages or property income.[31] The capitalist calculation of value or economic growth thus has as one of its underlying premises, to quote Marx, the notion of the "free gift of Nature to capital."[32] Nature's powers are presumed by the system to be a direct gift to capital itself, for which no exchange must be made.[33] This means, in truth, that nature, or real wealth, is robbed. As the socialist ecological economist, K. William Kapp, wrote in the 1960s, "Capitalism must be regarded as an economy of unpaid costs."[34] (It should be noted here that the existence of rents for land and resources does not alter the essential fact that nature is excluded from the value calculation. Instead, rents ensure that part of the surplus produced by society is redistributed to those who are able to monopolize the "rights" to natural resources.)

The second critical breakthrough in Marxian ecology was the recovery of what has come to be known as Marx's theory of metabolic rift. Marx's adoption of the concept of metabolism to address the systemic relations of nature and society was evident beginning with his writings in the *Grundrisse* in the late 1850s and in all of his major political-economic writings thereafter—up through his 1879–80 *Notes on Adolph Wagner*. In 1850 Marx encountered what amounted to an early ecological system perspective, in the extension of the concept of metabolism (*Stoffwechsel*) to the interconnected relations of plants and animals, through *Mikrokosmos*, written by his close friend and political associate, the socialist physician-scientist Roland Daniels.[35]

Marx was later influenced, as we have seen, by the German chemist Justus von Liebig's critique of British industrial agriculture, particularly the introduction to the 1862 edition of Liebig's great work on agricultural chemistry. Liebig's virulent critique of capitalist agriculture was concerned with the nineteenth-century soil crisis. He noted that the essential soil nutrients, such as nitrogen, potassium, and phosphorus, were shipped in the form of food and fiber to the new densely populated urban-industrial centers, where they contributed to the pollution of the cities, and were lost to the soil. Hence, Liebig and Marx both referred to industrial capitalist agriculture as a robbery system, leaching the soil of its nutrients. Britain in this period was forced to make up for its robbing the soil of its nutrients by imperialistically importing bones from the Napoleonic battlefields and the catacombs of Europe, and guano from Peru, in order to obtain the natural fertilizer to replenish English fields. The global metabolic rift, according to Marx, meant that capitalism disrupted "the eternal natural condition" of life itself. It therefore produced "an irreparable rift in the interdependent process of social metabolism, a metabolism prescribed by the natural laws of life itself."[36] This rift could also be seen in the unequal ecological exchange between countries, whereby capital in the center systematically robbed the periphery of its soil and resources.[37]

Marx's overall analysis in this respect is best understood in terms of a triad of concepts discussed in his *Economic Manuscripts of 1861–1862* and *Capital*: "the universal metabolism of nature," the "social metabolism," and the metabolic rift.[38] Human beings, he argued, exist within the "universal metabolism of nature," from which they extract nature's use values, and transform these in production, that is, the "social metabolism," in order to meet their needs for subsistence and development. Yet capitalism, as a historically specific form of production, systematically alienates workers from the means of production (the land, nature, tools) thereby proletarianizing labor, and making possible capitalist exploitation and accumulation. In the process, both the soil and the worker, the "original sources of all wealth," were undermined, generating a metabolic rift. The result, Marx argued, was the necessity of the "restoration" of this metabolism, which, however, could only take place in a higher society, that is, socialism.[39]

It was with such considerations in mind that Marx introduced the most radical conception of ecological sustainability ever developed. As he wrote in *Capital*:

> From the standpoint of a higher socio-economic formation, the private property of particular individuals in the earth will appear

just as absurd as the private property of one man in other men. Even an entire society, a nation, or all simultaneously existing societies taken together, are not the owners of the earth. They are simply its possessors, its beneficiaries, and have to bequeath it in an improved state to succeeding generations, as *boni patres familias* [good heads of the household].[40]

In Marx, ecological sustainability together with substantive equality defined the entire basis of socialism/communism. "Freedom, in this sphere," he wrote, "can consist only in this, that socialized man, the associated producers, govern the human metabolism with nature in a rational way . . . accomplishing it with the least expenditure of energy and in conditions most worthy and appropriate for their human nature."[41]

The third critical breakthrough of second-stage ecosocialism was the retrieval of Marx's dual conception of ecological crisis in capitalist society. In the first form of ecological crisis, depicted in *Capital*, the focus was on natural resource scarcity. Here the problem is how increasing scarcities of resources and environmental amenities in general lead to enhanced ecological costs, thereby squeezing profit margins. This can be seen in Marx's treatment of the British cotton crisis during the U.S. Civil War, the role of resources in elevating the cost of constant capital in his theory of the tendency of the rate of profit to fall, and in his discussions of the need of capital to conserve constant capital. Increasing resource costs with the degradation of the environment can create huge problems for capitalist accumulation. Here it is evident how imperialism, in keeping the price of internationally sourced raw material prices low, helps promote capital accumulation in the center of the system.

Yet, there is also to be found in Marx a theory of ecological crisis proper, or a crisis of sustainable human development, going beyond the value calculus of the system itself—as exemplified by the theory of metabolic rift. Simply because capitalism is a robbery system, in Liebig and Marx's sense, it externalizes most of the costs of environmental (and social) degradation on nature and society without this directly affecting its bottom line. Thus such phenomena as desertification and deforestation—both of which were discussed by Marx—have implications for sustainable human development but do not enter directly into the value calculation of the commodity system. A metabolic rift that disrupts biogeochemical cycles may be fully compatible with continued accumulation. In its relative insulation from the environmental degradation that it systematically creates everywhere around it, capitalism is unique among modes of production.

As Burkett writes, "For Marx . . . capital accumulation can maintain itself through environmental crises. In fact, this is one thing that makes capitalism different from previous societies. It has the ability to continue with its competitive, profit-driven pattern of accumulation despite the damage this does to natural conditions."[42] Today we see economic growth continue while the disruptions of the biogeochemical cycles of the entire planet upon which all living beings depend for their existence do not enter into the accounting. These disruptions and rifts in fact open new profit-making opportunities for capital such as the agrichemical (fertilizers and pesticides) industry or today's carbon markets.

Most of the concrete research inspired by Marxian theories of ecological crisis in recent years has focused on the theory of metabolic rift, since it is *the crisis of sustainable human development* that defines the current planetary emergency. Moreover, the metabolic rift perspective has provided an understanding of systemic environmental changes not reducible simply to issues of scale and carrying capacity or to the economics of the system—thereby probing new dimensions of the problem. Marx's metabolic rift analysis intersects with the treadmill of production analysis (which grew out of his theory of accumulation), and at the same time relates to developments in natural science, thus tying into the most developed ecological perspectives.[43] It points to the deep contradictions associated with capital's division of nature (alongside the division of labor).

For example, the metabolic rift allows us to understand more fully the implications—of which Marx was already critical in the nineteenth century—of the attempts of the system to accelerate the growth rates of animals in factory-style production, by removing them from their ecosystems, changing their food intake, breeding, and so on. Animals are decomposed, their various body parts manipulated, converted into mere processes of production to be commodified to the *n*th degree.[44]

The metabolic rift analysis was also seen by Marx and Engels in terms of open-system thermodynamics, in the context of which, as Engels observed in 1882, humanity was "squandering" the fossil fuels associated with "past solar energy" while failing to make good use of present solar energy.[45]

Marxism and the Great Capitalist Climacteric

It is on the basis of this set of critical theoretical breakthroughs—constituting a scientific revolution in Marxian theory reaching back into the very foundations of historical materialism—that it is possible to draw five broad conclusions about the ecological and social revolution that is now necessary in the face of today's Great Capitalist Climacteric.

First, the problem threatening the global environment is the *accumulation of capital under the present phase of monopoly-finance capital*, and not just economic growth in the abstract. That is, issues of the qualitative nature of development as well as quantitative development are involved. This raises the question of the ecological value-form associated with capitalism in its monopoly-finance phase, geared to the promotion of economic and ecological waste as a stimulus to accumulation. Today the rich economies are well developed and capable of satisfying the material needs of their populations, and of emphasizing qualitative human development. Capitalism, however, requires continual value expansion and commodity consumption, with increasing throughputs of energy and materials.[46] This is promoted today by means of a massive sales effort, amounting to well over a trillion dollars a year in the United States, and through a vast outpouring of economic waste in the form of synthetic goods that are toxic to the environment.[47] As the Marxian economist Paul

Baran wrote in the 1960s, "People steeped in the culture of monopoly capitalism do not want what they need and do not need what they want."[48] On top of this vast waste system (including military waste), which drives accumulation, is a financialized superstructure that has enabled the system to transfer wealth and income more rapidly to the 0.01 percent at the top of society.[49] In the new financial architecture that has emerged the credit-debt system dominates over the entire global economy. It is this irrational system of artificially stimulated growth, economic waste, financialized wealth, and extreme inequality that needs to be overturned if we are to create a society of ecological sustainability and substantive equality.

If economic growth in the wealthy countries continues as at present—even by the standards of our current period of relative economic stagnation—there is very little or no chance of avoiding breaking the world climate budget with disastrous global consequences. It is the growth in the scale of the economy, and the destructive tendencies of our ecologically inefficient, technologically destructive society, geared to roundabout production—whereby plastic spoons are made in China and shipped to the United States where they have a lifetime use of a few minutes before reentering the waste stream, generating all sorts of toxic chemicals in the process—that are threatening the biogeochemical processes of the entire planet. Capital's social metabolic processes attempt to re-create the planet in its own image, treating all planetary boundaries as mere barriers to surmount, thus generating a global metabolic rift on a rapidly warming planet. All of this points to the need to place limits on economic growth, and specifically on the expansion of today's disaster capitalism.

Second, capitalism is suffering at present from an *epochal crisis—both economic and environmental*. This is manifested in overaccumulation, stagnation, and financialization on the one hand, and ecological rifts and disruptions, both within each and every ecosystem and on the level of the planet as a whole, on the other.[50] These two long-term structural crises of the system are not reducible to each other, except in the sense that they are both induced by the logic of capital accumulation. What we have called ecological crisis proper is largely invisible to the value accounting of the capitalist system, and is systematically given a lower priority in relation to economic imperatives. Society is constantly told that the solution to economic stagnation is economic growth by any means: usually involving the promotion of neoliberal disaster capitalism. Yet such an economic solution—which is beyond the power of the system to effect in a long-term, stable way, but only on a temporary, ad hoc basis—would be fatal to the planetary environment, which requires less, not more, expansion of the economic treadmill. The epochal crisis of economy and ecology within the capitalist system is thus likely to continue, with both fault lines widening, as long as the logic of capital prevails. This conflict between economic and ecological objectives is not a contradiction of analysis, but of the capitalist system itself.

Third, if accumulation or economic growth is to be halted in the rich countries, even temporarily, out of sheer ecological necessity, this would require *a vast new system of redistribution*. As Lewis Mumford indicated in 1944 in *The Condition of Man*, a stationary state or steady-state economy is only possible under conditions of "basic communism," a term that Mumford (after Marx) used to refer to a society in which distribution is organized "according to need, not according to ability or productive contribution."[51] There must be a vast redirection of society's social surplus to genuine human requirements and ecological sustainability as opposed to the giant treadmill of production generated by the profit system. It is by creating a society directed to use value rather than exchange value that we can find the resources to develop a world that is sustainable because it is just, and just because it is sustainable. Society will need to be reordered, as Epicurus said, and Marx concurred, according to the principle of *enough*—that is, through a rich development of human needs, applicable to everyone.[52]

Fourth, Marx provided *a model of socialism as one of sustainable human development*.[53] In order to meet the challenge of the Great Capitalist Climacteric it will be necessary to shift power to the associated producers, who, acting in accord with science and communal values, will need to regulate the complex, interdependent metabolism between nature and society according to their own developed human needs and in conformity with the requirements of the earth metabolism. In today's context, this will require what Marx called the "restoration" of the essential human-natural metabolism, healing the metabolic rift.[54] In discussing the principle of "metabolic restoration," Del Weston wrote in her book *The Political Economy of Global Warming*: "The need is for human societies to live within metabolic cycles—that is, production, consumption and waste—thereby forming part of a self-sustaining cycle in which the only new inputs are energy from the sun.... Nature, in the new economics, will be recognised as the ultimate source of wealth."[55] Moreover, given the present planetary emergency we have to move fast to create this new economics and new ecological relation to the earth, diverting resources massively to creating the new energy infrastructure that can exist within the solar budget, while at the same time promoting Mumford's "basic communism," or a society based on the principle of to each according to need.

Fifth, *the hoped-for revolutionary change can only occur through human agency*. Although it is widely recognized that the world needs an ecological and social revolution, the question remains: From whence and by what agency will such a revolution arise? Ecological Marxists suggest that we may already be seeing signs of the rise of what could be called a nascent "environmental proletariat"—a broad mass of working-class humanity who recognize, as a result of the crisis of their own existence, the indissoluble bond between economic and ecological conditions.[56] Degraded material conditions associated with intermingled economic and ecological crises are now being encountered on a daily basis by the great majority of the world's population and affecting all aspects of their lives. At the ground level, economic and ecological crises are becoming increasingly indistinguishable. Food crises, land grabs, electricity shutdowns, water privatization, heightened pollution, deteriorating cities, declining public health, increasing violence against oppressed populations—are all converging with growing inequality, economic stagnation, and rising unemployment and underemployment. In

South Africa, for example, the class struggle is now as much an environmental as an economic struggle—already exhibiting signs of an emerging environmental working class.[57] The logical result is a coming together of material revolts against the system—what David Harvey has usefully referred to as a "co-revolutionary" struggle.[58] This is best exemplified by the global environmental/climate justice movement and through the radical direct action movement that Naomi Klein calls "Blockadia."[59]

Traditional working-class politics are thus co-evolving and combining with environmental struggles, and with the movements of people of color, of women, and all those fighting basic, reproductive battles throughout society. Such an ecological and social struggle will be revolutionary to the extent that it draws its force from those layers of society where people's lives are most precarious: Third World workers, working-class women, oppressed people of color in the imperial core, indigenous populations, peasants/landless agricultural workers, and those fighting for fundamentally new relations of sexuality, gender, family, and community—as well as highly exploited and dispossessed workers everywhere.

A revolutionary struggle in these circumstances will need to occur in two phases: an *ecodemocratic phase* in the immediate present, seeking to build a broad alliance—one in which the vast majority of humanity outside of the ruling interests will be compelled by their inhuman conditions to demand a world of sustainable human development. Over time this should create the conditions for a second, more decisive, *ecosocialist phase* of the revolutionary struggle, directed at the creation of a society of substantive equality, ecological sustainability, and collective democracy. All of this points to the translation of classical Marx's ecological critique into contemporary revolutionary praxis.[60]

In the *ecodemocratic* phase, the goal would be to carry out those radical reforms that would arrest the current destructive logic of capital, by fighting for changes that are radical, even revolutionary, in that they go against the logic of capital, but are nonetheless conceivable as concrete, meaningful forms of struggle in the present context. These would include measures like: (1) an emergency plan of reduction in carbon emissions in the rich economies by 8 to 10 percent a year; (2) implementing a moratorium on economic growth coupled with radical redistribution of income and wealth, conservation of resources, rationing, and reductions in economic waste; (3) diverting military spending, now universally called "defense spending" to *the defense of the planet* as a place of human habitation; (4) the creation of an alternative energy infrastructure designed to stay within the solar budget; (5) closing down coal-fired plants and blocking unconventional fossil fuels such as tar sands oil; (6) a carbon fee and dividend system of the kind proposed by Hansen that would redistribute 100 percent of the revenue to the population on a per capita basis; (7) global initiatives to aid emerging economies to move toward sustainable development; (8) implementation of principles of environmental justice throughout the society and linking this to adaptation to climate change (which cannot be stopped completely) to ensure that people of color, the poor, women, indigenous populations, and Third World populations do not bear the brunt of catastrophe; and (9) adoption of climate negotiations and policies on the model proposed in the Peoples' Agreement on Climate Change in Cochabamba, Bolivia, in 2010. Such radical change proposals can be multiplied, and would need to affect all aspects of society and individual human development. The rule in the ecodemocratic phase of development would be to address the epochal crisis (ecological and economic) in which the world is now caught, and to do so in ways that go against the logic of business as usual, which is indisputably leading the world toward cumulative catastrophe.

The logic of the ecodemocratic phase of the struggle, if it were carried out fully, would create the conditions for an *ecosocialist* phase in which the mobilization of the population on their own behalf, and the cultural and economic changes that this brings about, would give the impetus to the creation of a society of *from each according to ability, to each according to need*.[61] The system of social metabolic reproduction would be reconstituted on a more communal basis taking into account not only present and future generations, but the

Earth System itself and the diversity of life within it. The necessary social and ecological planning would start from local needs and local communities and would be integrated with larger political-executive entities responsible for coordination and implementation in relation to these needs.

Such a society would be democratic in the classical sense of the word—rule of society by the people, the associated producers.[62] It was this that Marx had in mind when he stressed (as quoted above) that "socialized man, the associated producers, [would] govern the human metabolism with nature in a rational way . . . accomplishing it with the least expenditure of energy and in conditions most worthy and appropriate for their human nature." For Marx in the nineteenth century this was a struggle for human freedom; today, in the twenty-first century, it is a struggle for human freedom *and human survival*.

In 1980, the British Marxist historian E. P. Thompson wrote a cautionary essay for *New Left Review* titled "Notes on Exterminism, the Last Stage of Civilization." Although directed particularly at the growth of nuclear arsenals and the dangers of global holocaust from a nuclear exchange in the final phase of the Cold War, Thompson's thesis was also concerned with the larger realm of ecological destruction wrought by the system. Rudolf Bahro later commented on Thompson's ideas in his *Avoiding Social and Ecological Disaster*, explaining: "To express the exterminism-thesis in Marxist terms, one could say that the relationship between productive and destructive forces is turned upside down. Marx had seen the trail of blood running through it, and that 'civilisation leaves deserts behind it.'"[63] Today this ecologically ruinous trend has been extended to the entire planet with capitalism's proverbial "creative destruction" being transformed into a destructive creativity endangering humanity and life in general.[64]

"The dream that man can make himself godlike by centering his energies solely on the conquest of the external world," Mumford wrote in *The Condition of Man*, "has now become the emptiest of dreams: empty and sinister."[65] The result is a kind of *economics of exterminism*. Today making war on the planet is a means to the end of capital accumulation, in which the limits of the earth itself have become invisible to the narrow value calculations of the system. Turning this economics of exterminism around, and creating a more just and sustainable world at peace with the planet is our task in the Great Capitalist Climacteric. If we cannot accomplish this humanity will surely die with capitalism. The prophecy of all defenders of the current order over the last century will then be fulfilled: capitalism will mark the *end of human* history by bringing to an end human civilization—and even human existence.

The Great Capitalist Climacteric presents us with a fatal choice: *System Change Not Climate Change!*

CHAPTER THREE

The Anthropocene Crisis

For it is because we are kept in the dark about the nature of human society—as opposed to nature in general—that we are now faced (so the scientists concerned assure me), by the complete destructibility of this planet that has barely been made fit to live in.
—BERTOLT BRECHT

The Anthropocene, viewed as a new geological epoch displacing the Holocene Epoch of the last 11,700 years, is associated with an "anthropogenic rift" in the history of the planet.[1] Formally introduced into the contemporary scientific and environmental discussion by climatologist Paul Crutzen in 2000, it stands for the notion that human beings have become the primary emergent geological force affecting the future of the Earth System. Although often traced to the Industrial Revolution in the late eighteenth century, the Anthropocene is probably best seen as arising in the late 1940s and early 1950s. Recent scientific evidence suggests that the period from around 1950 on exhibits a major spike, marking a Great Acceleration in human impacts on the environment, with the most dramatic stratigraphic trace of the anthropogenic rift to be found in fallout radionuclides from nuclear weapons testing.[2]

Viewed in this way, the Anthropocene Epoch—and what can be referred to as the present geological age of the Capitalinian—can be seen as corresponding roughly to the rise of the modern environmental movement, which had its beginnings in the protests led by scientists against above-ground nuclear testing after the Second World War, and was to emerge as a wider movement following the publication of Rachel Carson's *Silent Spring* in 1962.[3] Carson's book was soon followed in the 1960s by the very first warnings, by Soviet and U.S. scientists, of accelerated and irreversible global warming.[4] It is this dialectical interrelation between the acceleration into the Anthropocene and the acceleration of a radical environmentalist imperative in response that constitutes the central theme of Ian Angus's *Facing the Anthropocene*. It is his ability to give us insights into the Anthropocene as a new emergent level of society-nature interaction brought on by historical change—and how the new ecological imperatives it generates have become the central question confronting us in the twenty-first century—that makes his book so indispensable.

Today it seems likely that the Anthropocene will come to be linked within science to the post–Second World War era in particular. Nonetheless, as in the case of all major turning points in history, there were signs of minor spikes at earlier stages along the way, going back to the Industrial Revolution. This reflects what the Marxian philosopher István Mészáros calls the "dialectic of *continuity and discontinuity*," characterizing all novel emergent developments in history.[5] Although the Anthropocene concept arose fully only with the modern scientific conception of the Earth System, and is now increasingly seen as having its physical basis in the Great Acceleration after the Second World War, it was prefigured by earlier notions, arising from thinkers focusing on the dramatic changes in the human-environmental interface brought on by the rise of capitalism, including the Industrial Revolution, the colonization of the world, and the era of fossil fuels.

"Nature, the nature that preceded human history," Karl Marx and Frederick Engels remarked as early as 1845, "no longer exists anywhere (except perhaps on a few Australian coral islands of recent origin)."[6] Similar views were presented by George Perkins Marsh, in *Man and Nature* in 1864, two years before Ernst Haeckel coined the word *ecology*, and three years before Marx published the first volume of *Capital*, with its warning of the metabolic rift in the human relation to the earth.[7]

It was not until the last quarter of the nineteenth and the early twentieth century, however, that the key concept of the biosphere, out of which our modern notion of the Earth system was to develop, arose, with the publication, most notably, of *The Biosphere* by the Soviet geochemist Vladimir I. Vernadsky in 1926. "Remarkably," Lynn Margulis and Dorian Sagan write in *What Is Life?*, "Vernadsky dismantled the rigid boundary between living organisms and a non-living environment, depicting life globally before a single satellite had returned photographs of Earth from orbit."[8]

The appearance of Vernadsky's book corresponded to the first introduction of the term Anthropocene (together with Anthropogene) by his colleague, the Soviet geologist Aleksei Pavlov, who used it to refer to a new geological period in which humanity was the main driver of planetary geological change. As Vernadsky observed in 1945, "Proceeding from the notion of the geological role of man, the geologist A. P. Pavlov (1854–1929) in the last years of his life used to speak of the *anthropogenic era*, in which we now live.... He rightfully emphasized that man, under our very eyes, is becoming a mighty and ever-growing geological force.... In the 20th Century, man, for the first time in the history of the Earth, knew and embraced the whole biosphere, completed the geographic map of the planet Earth, and colonized its whole surface."[9]

Simultaneously with Vernadsky's work on the biosphere, the Soviet biochemist Alexander I. Oparin and the British socialist biologist J. B. S. Haldane independently developed in the 1920s the theory of the origin of life, known as the "primordial soup theory." As summed up by Harvard biologists Richard Levins and Richard Lewontin, "Life originally arose from inanimate matter [what Haldane famously described as a 'hot dilute soup'], but that origination made its continued occurrence impossible, because living organisms consume the complex organic molecules needed to re-create life *de novo*.

Moreover, the reducing atmosphere [lacking free oxygen] that existed before the beginning of life has been converted, by living organisms themselves, to one that is rich in reactive oxygen." In this way, the Oparin-Haldane theory explained for the first time how life could have originated out of inorganic matter, and why the process could not be repeated. Equally significant, life, arising in this way billions of years ago, could be seen as the creator of the biosphere within a complex process of coevolution.[10]

It was Rachel Carson, in her landmark 1963 speech "Our Polluted Environment," famously introducing the concept of ecosystem to the U.S. public, who most eloquently conveyed this integrated ecological perspective, and the need to take it into account in all of our actions. "Since the beginning of biological time," she wrote,

> there has been the closest possible interdependence between the physical environment and the life it sustains. The conditions on the young earth produced life; life then at once modified the conditions of the earth, so that this single extraordinary act of spontaneous generation could not be repeated. In one form or another, action and interaction between life and its surroundings have been going on ever since.
>
> This historic fact has, I think, more than academic significance. Once we accept it we see why we cannot with impunity make repeated assaults upon the environment as we now do. The serious student of earth history knows that neither life nor the physical world that supports it exists in little isolated compartments. On the contrary, he recognizes the extraordinary unity between organisms and the environment. For this reason he knows that harmful substances released into the environment return in time to create problems for mankind.
>
> The branch of science that deals with these interrelations is Ecology. . . . We cannot think of the living organism alone; nor can we think of the physical environment as a separate entity. The two exist together, each acting on the other to form an ecological complex or ecosystem.[11]

Nevertheless, despite the integrated ecological vision presented by figures like Carson, Vernadsky's concepts of the biosphere and biogeochemical cycles were for a long time downplayed in the West due to the reductionist mode that prevailed in Western science and the Soviet background of these concepts. Soviet scientific works were well known to scientists in the West and were frequently translated in the Cold War years by scientific presses and even by the U.S. government—though unaccountably Vernadsky's *The Biosphere* was not translated into English until 1998. This was a necessity since in some fields, such as climatology, Soviet scientists were well ahead of their U.S. counterparts. Yet this wider scientific interchange, crossing the Cold War divide, was seldom conveyed to the public at large, where knowledge of Soviet achievements in these areas was practically nonexistent. Ideologically, therefore, the concept of the biosphere seems to have long fallen under a kind of interdict.

Still, the biosphere took center stage in 1970, with a special issue of *Scientific American* on the topic.[12] At around the same time the socialist biologist Barry Commoner warned in *The Closing Circle* of the vast changes in the human relation to the planet, beginning with the atomic age and the rise of modern developments in synthetic chemistry. Commoner pointed back to the early warning of capitalism's environmental disruption of the cycles of life represented by Marx's discussion of the rift in the metabolism of the soil.[13]

In 1972, Evgeni K. Fedorov, one of the world's top climatologists and a member of the Presidium of the Supreme Soviet of the USSR, as well as the leading Soviet supporter of Commoner's analysis (writing the "Concluding Remarks" to the Russian edition), declared that the world would need to wean itself from fossil fuels: "A rise in temperature of the earth is inevitable if we do not confine ourselves to the use, as energy sources, of direct solar radiation and the hydraulic energy of wave and wind energy, but [choose instead to] obtain energy from fossil [fuels] or nuclear reactions."[14] For Fedorov, Marx's theory of "metabolism between people and nature" constituted the methodological basis for an ecological approach to the question of the Earth System.[15] It was in the 1960s and 1970s that climatologists in the USSR and the United States first found "evidence," in the words of Clive Hamilton and Jacques Grinevald, of a "worldwide metabolism."[16]

The rise of Earth System analysis in the succeeding decades was also strongly impacted by the remarkable view from outside, emanating from the early space missions. As Howard Odum, one of the leading figures in the formation of systems ecology, wrote in *Environment, Power and Society*:

> We can begin a systems view of the earth through the macroscope of the astronaut high above the earth. From an orbiting satellite, the earth's living zone appears to be very simple. The thin water and air-bathed shell covering the earth—the biosphere—is bounded on the inside by dense solids and on the outside by the near vacuum of outer space.... From the heavens it is easy to talk of gaseous balances, energy budgets per million years, and the magnificent simplicity of the overall metabolism of the earth's thin outer shell. With the exception of energy flow, the geobiosphere for the most part is a closed system of the type whose materials are cycled and reused.[17]

"The mechanism of overgrowth," threatening this "overall metabolism," Odum went on to state, "is capitalism."[18] Today's concept of the Anthropocene thus reflects, on the one hand, a growing recognition of the rapidly accelerating role of anthropogenic drivers in disrupting the biogeochemical processes and planetary boundaries of the Earth System and, on the other, a dire warning that the world, under "business as usual," is being catapulted into a new ecological phase—one less conducive to maintaining biological diversity and a stable human civilization.

It is the bringing together of these two aspects of the Anthropocene—variously viewed as the geological and the historical, the natural and social, the climate and capitalism—in one single, integrated view, that constitutes the main achievement of *Facing the Anthropocene*. Angus demonstrates that "fossil capitalism," if not stopped, is a runaway train, leading to global environmental apartheid and what the great British Marxist historian E. P. Thompson referred to as the threatened historical stage of "exterminism," in which the conditions of existence of hundreds of millions, perhaps billions, of people will be upended, and the very basis of life as we know it endangered. Moreover, this has its source in what Odum called "imperial capitalism," imperiling the lives of the most vulnerable populations on the planet in a system of forced global inequality.[19]

Such are the dangers that only a new, radical approach to social science (and thus to society itself)—one that takes seriously Carson's warning that if we undermine the living processes of Earth this will "return in time" to haunt us—can provide us with the answers that we need in the Anthropocene Epoch. Where such urgent change is concerned "tomorrow is too late."[20]

Yet the dominant social science, which serves the dominant social order and its ruling strata, has thus far served to obscure these issues, putting its weight behind ameliorative measures together with mechanistic solutions such as carbon markets and geoengineering. It is as if the answer to the Anthropocene crisis were a narrowly economic and technological one consistent with the further expansion of the hegemony of capital over Earth and its inhabitants—this despite the fact that the present system of capital accumulation is at the root of the crisis. The result is to propel the world into still greater danger. What is needed, then, is to recognize that it is the logic of our current mode of production—capitalism—that stands in the way of creating a world of sustainable human development transcending the spiraling disaster that otherwise awaits humanity. To save ourselves we must create a different socioeconomic logic pointing to different human-environmental ends: an ecosocialist revolution in which the great mass of humanity takes part.

But are there not risks to such radical change? Would not great struggles and sacrifices attend any attempt to overthrow the prevailing system of production and energy use in response to global warming? Is there any surety that we would be able to create a society of sustainable human development, as ecosocialists envision? Would it not be better to err on the side of denialism than on the side of "catastrophism"? Should we not hesitate to take action at this level until we know more?

Here it is useful to quote from the great German playwright and poet Bertolt Brecht's didactic poem "The Buddha's Parable of the Burning House":

> The Buddha still sat under the bread-fruit tree and to the others,
>   To those who had not asked [for guarantees], addressed this parable:
> "Lately I saw a house. It was burning. The flame
> Licked at its roof. I went up close and observed
> That there were people still inside. I entered the doorway
>   and called
> Out to them that the roof was ablaze, so exhorting them
> To leave at once. But those people
> Seemed in no hurry. One of them,
> While the heat was already scorching his eyebrows,
> Asked me what it was like outside, whether it wasn't raining,
> Whether the wind wasn't blowing, perhaps, whether there
>   was
> Another house for them, and more of this kind. Without
>   answering
> I went out again. These people here, I thought,
> Must burn to death before they stop asking questions.
>   And truly, friends,
> Whoever does not yet feel such heat in the floor that
>   he'll gladly
> Exchange it for any other, rather than stay, to that man
> I have nothing to say." So Gautama the Buddha.[21]

It is capitalism and the alienated global environment it has produced that constitutes our "burning house" today. Mainstream environmentalists, faced with this monstrous dilemma, have generally chosen to do little more than *contemplate* it, watching and making minor adjustments to their interior surroundings while flames lick the roof and the entire structure threatens to collapse around them. The point, rather, is to *change* it, to rebuild the house of civilization under different architectural principles, creating a more sustainable metabolism of humanity and the earth. The name of the movement to achieve this, rising out of the socialist and radical environmental movements, is *ecosocialism*, and *Facing the Anthropocene* is its most up-to-date and eloquent manifesto.

CHAPTER FOUR

Crossing the River of Fire

The front cover of Naomi Klein's *This Changes Everything* is designed to look like a protest sign. It consists of the title alone in big block letters, with the emphasis on *Changes*. Both the author's name and the subtitle are absent. It is only when we look at the spine of the book, turn it over, or open it to the title page that we see it is written by North America's leading left climate intellectual-activist and that the subtitle is *Capitalism vs. the Climate*.[1] All of which is clearly meant to convey in no uncertain terms that climate change literally *changes everything* for today's society. It threatens to turn the mythical human conquest of nature on its head, endangering present-day civilization and throwing doubt on the long-term survival of *Homo sapiens*.

The source of this closing circle is not the planet, which operates according to natural laws, but rather the economic and social system in which we live, which treats natural limits as mere barriers to surmount. It is now doing so on a planetary scale, destroying in the process the earth as a place of human habitation. Hence, the change that Klein is most concerned with, and to which her book points, is not climate change itself, but the radical social transformation that must be carried out in order to combat it. We as a species will either radically change the material conditions of our existence or they will be changed far more drastically for us. Klein argues in effect for System Change Not Climate Change—the name adopted by the current ecosocialist movement in the United States.[2]

In this way Klein, who in *No Logo* ushered in a new generational critique of commodity culture, and who in *The Shock Doctrine* established herself as perhaps the most prominent North American critic of neoliberal disaster capitalism, signals that she has now, in William Morris's famous metaphor, crossed "the river of fire" to become a critic of capital as a system.[3] The reason is climate change, including the fact that we have waited too long to address it, and the reality that nothing short of an ecological revolution will now do the job.

In the age of climate change, Klein argues, a system based on ever-expanding capital accumulation and exponential economic growth is no longer compatible with human well-being and progress—or even with human survival over the long run. We need therefore to reconstruct society along lines that go against the endless amassing of wealth as the primary goal. Society must be rebuilt on the basis of other principles, including the "regeneration" of life itself and what she calls "ferocious love."[4] This reversal in the existing social relations of production must begin immediately with a war on the fossil-fuel industry and the economic growth imperative—when such growth means more carbon emissions, more inequality, and more alienation of our humanity.

Klein's crossing of the river of fire has led to a host of liberal attacks on *This Changes Everything*, often couched as criticisms emanating from the left. These establishment criticisms of her work, we will demonstrate, are disingenuous, having little to do with serious confrontation with her analysis. Rather, their primary purpose is to rein in her ideas, bringing them into conformity with received opinion. If that should prove impossible, the next step is to exclude her ideas from the conversation. However, her message represents the growing consciousness of the need for epochal change, and as such is not easily suppressed.

The Global Climacteric

The core argument of *This Changes Everything* is a historical one. If climate change had been addressed seriously in the 1960s, when scientists first raised the issue in a major way, or even in the late 1980s and early '90s, when James Hansen gave his famous testimony in Congress on global warming, the Intergovernmental Panel on Climate Change was first established, and the Kyoto Protocol introduced, the problem could conceivably have been addressed without a complete shakeup of the system. At that historical moment, Klein suggests, it would still have been possible to cut emissions by at most 2 percent a year.[5]

Today such incremental solutions are no longer conceivable even in theory. The numbers are clear. With a continuation of the present rate of carbon emissions, it is estimated that we will break the carbon budget for remaining below a 2°C increase in global average temperature in less than a quarter-century (with the time remaining before the 1.5°C carbon budget is exhausted much shorter than that).[6] Once the 2°C increase is reached (if not before), scientists fear that there is a high probability that feedback mechanisms will come into play with reverberations so great that we will no longer be able to control where the thermometer stops in the end. The reality is that if the world is still to stay below a 2°C increase—and below the more dangerous 4°C, the point at which it is believed that disruption to life on the planet will be so great that civilization may no longer be possible— real revolutionary ecological change, unleashing the full power of an organized and rebellious humanity, is required.

What is necessary first and foremost is the cessation of fossil-fuel combustion, bringing to a rapid end the energy regime that has dominated since the Industrial Revolution. Simple arithmetic tells us that there is no way to get down to the necessary zero emissions level, that is, the complete cessation of fossil-fuel combustion, in the next few decades without implementing some kind of planned moratorium on economic growth, requiring shrinking capital formation and reduced consumption in the richest countries of the world system. We have no choice but to slam on the brakes and come to a dead stop with respect to carbon emissions before we go over the climate cliff. Never before in human history has civilization faced so daunting a challenge.

Klein draws on the argument of Kevin Anderson, of the Tyndall Centre for Climate Change in Britain, who indicates that rich countries will need to cut carbon emissions by as much as 10 percent a year. "Our ongoing and collective carbon profligacy," Anderson writes, "has squandered any opportunity for 'evolutionary change' afforded by our earlier (and larger) 2°C budget. Today, after two decades of bluff and lies, the remaining 2°C budget demands revolutionary change to the political and economic hegemony."[7]

Instead of addressing climate change when it first became critical in the 1990s, the world turned to the intensification of neoliberal globalization, notably through the creation of the World Trade Organization. It was the very success of the neoliberal campaign to remove most constraints on the operations of capitalism, and the negative effect that this had on all attempts to address the climate problem, Klein contends, that has made "revolutionary levels of transformation" of the system the only real hope in avoiding "climate chaos."[8] "As a result," she explains,

> we now find ourselves in a very difficult and slightly ironic position. Because of those decades of hardcore emitting exactly when we were supposed to be cutting back, the things that we must do to avoid catastrophic warming are no longer just in conflict with the particular strain of deregulated capitalism that triumphed in the 1980s. They are now in conflict with the fundamental imperative at the heart of our economic model: grow or die. . . .
>
> Our economy is at war with many forms of life on earth, including human life. What the climate needs to avoid collapse is a contraction in humanity's use of resources; what our economic model demands to avoid collapse is unfettered expansion. Only one of these sets of rules can be changed, and it's not the laws of nature. . . .
>
> Because of our lost decades, it is time to turn this around now.

Is it possible? Absolutely. Is it possible without challenging the fundamental logic of deregulated capitalism? Not a chance.[9]

Of course, "the fundamental logic of deregulated capitalism" is simply a roundabout way of pointing to *the fundamental logic of capitalism itself*, its underlying drive toward capital accumulation, which is hardly constrained at all in its accumulation function even in the case of a strong regulatory environment. Instead, the state in a capitalist society generally seeks to free up opportunities for capital accumulation on behalf of the system as a whole, rationalizing market relations so as to achieve greater overall, long-run expansion. As Paul Sweezy noted nearly three-quarters of a century ago in *The Theory of Capitalist Development*: "Speaking historically, control over capitalist accumulation has never for a moment been regarded as a concern of the state; economic legislation has rather had the aim of blunting class antagonisms, so that accumulation, the normal aim of capitalist behavior, could go forward smoothly and uninterruptedly."[10]

To be sure, Klein herself occasionally seems to lose sight of this basic fact, defining capitalism at one point as "consumption for consumption's sake," thus failing to perceive the Galbraith dependence effect, whereby the conditions under which we consume are structurally determined by the conditions under which we produce.[11] Nevertheless, the recognition that capital accumulation or the drive for economic growth is the defining property, not a mere attribute, of the system underlies her entire argument. Recognition of this systemic property led the great conservative economist Joseph Schumpeter to declare: "Stationary capitalism would be a *contradictio in adjecto*."[12]

It follows that no mere technological wizardry—of the kind ideologically promoted, for example, by the Breakthrough Institute—will prevent us from breaking the carbon budget within several decades, as long as the driving force of the reigning socioeconomic system is its own self-expansion. Mere improvements in carbon efficiency are too small as long as the scale of production is increasing, which has the effect of expanding the absolute level of carbon dioxide emitted.

The inevitable conclusion is that we must rapidly reorganize society on other principles than that of stoking the engine of capital with fossil fuels.

None of this, Klein assures us, is cause for despair. Rather, confronting this harsh reality head-on allows us to define the strategic context in which the struggle to prevent climate change must be fought. It is not primarily a technological problem unless one is trying to square the circle: seeking to reconcile expanding capital accumulation with the preservation of the climate. In fact, all sorts of practical solutions to climate change exist at present and are consistent with the enhancement of individual well-being and growth of human community. We can begin immediately to implement the necessary changes, such as democratic planning at all levels of society; introduction of sustainable energy technology; heightened public transportation; reductions in economic and ecological waste; a slowdown in the treadmill of production; redistribution of wealth and power; and above all an emphasis on sustainable human development.[13]

There are ample historical precedents. We could have a crash program, as in wartime, where populations sacrificed for the common good. In England during the Second World War, Klein observes, driving automobiles virtually ceased. In the United States, the automobile industry was converted in the space of half a year from producing cars to manufacturing trucks, tanks, and planes for the war machine. The necessary rationing—since the price system recognizes nothing but money—can be carried out in an egalitarian manner. Indeed, the purpose of rationing is always to share the sacrifices that have to be made when resources are constrained, and thus it can create a sense of real community, of all being in this together, in responding to a genuine emergency. Although Klein does not refer to it, one of the most inspiring historical examples of this was the slogan "Everyone Eats the Same" introduced in the initial phases of the Cuban Revolution and followed to an extraordinary extent throughout the society. Further, wartime mobilization and rationing are not the only historical examples on which we can draw. The New Deal in the United States, she indicates, focused on public investment and direct promotion of the public good, aimed at the enhancement of use values rather than exchange values.[14]

Mainstream critics of *This Changes Everything* often willfully confuse its emphasis on degrowth with the austerity policies associated with neoliberalism. However, Klein's perspective, as we have seen, could not be more different, since it is about the rational use of resources under conditions of absolute necessity and the promotion of equality and community. Nevertheless, she could strengthen her case in this respect by drawing on monopoly-capital theory and its critique of the prodigious waste in our economy, whereby only a miniscule proportion of production and human labor is now devoted to actual human needs as opposed to market-generated wants. As the author of *No Logo*, Klein is well aware of the marketing madness that characterizes the contemporary commodity economy, causing the United States alone to spend more than a trillion dollars a year on the sales effort.[15]

What is required in a rich country such as the United States at present, as detailed in *This Changes Everything*, is not an abandonment of all the comforts of civilization but a reversion to the standard of living of the 1970s—two decades into what Galbraith dubbed "the affluent society." A return to a lower per capita output (in GDP terms) could be made feasible with redistribution of income and wealth, social planning, decreases in working time, and universal satisfaction of genuine human needs—a sustainable environment; clean air and water; ample food, clothing, and shelter; high-quality health care, education, public transportation, and community-cultural life—such that most people would experience a substantial improvement in their daily lives.[16] What Klein envisions here would truly be an ecological-cultural revolution. All that is really required, since the necessary technological means already exist, is people power: the democratic mass mobilization of the population.

Such people power, Klein is convinced, is already emerging in the context of the present planetary emergency. It can be seen in the massive but diffuse social-environmental movement, stretching across the globe, representing the struggles of tens of millions of activists worldwide, to which she gives, or rather takes from the movement itself, the name "Blockadia." Numberless individuals are putting themselves on the line, confronting power, and frequently facing arrest, in their opposition to the fossil-fuel industry and capitalism itself. Indigenous peoples are organizing worldwide and taking a leading role in the environmental revolt, as in the Idle No More movement in Canada. Anti-systemic, ecologically motivated struggles are on the rise on every continent.

The primary burden for mitigating climate change necessarily resides with the rich countries, which are historically responsible for the great bulk of the carbon added to the atmosphere since the Industrial Revolution and still emit the most carbon per capita today. The disproportionate responsibility of these nations for climate change is even greater once the final consumption of goods is factored into the accounting. Poor countries are heavily dependent on producing export goods for multinational corporations to be sold to consumers at the center of the world capitalist economy. Hence, the carbon emissions associated with such exports are rightly assigned to the rich nations importing these goods rather than the poor ones exporting them. Moreover, the rich countries have ample resources available to address the problem and carry out the necessary process of social regeneration without seriously compromising the basic welfare of their populations. In these societies, the problem is no longer one of increasing per capita wealth, but rather one of the rational, sustainable, and just organization of society. Klein evokes the spirit of Seattle in 1999 and Occupy Wall Street in 2011 to argue that sparks igniting radical ecological change exist even in North America, where growing numbers of people are prepared to join a global peoples' alliance. Essential to the overall struggle, she insists, is the explicit recognition of ecological or climate debt owed by the Global North to the Global South.[17]

The left is not spared critical scrutiny in Klein's work. She acknowledges the existence of a powerful ecological critique within Marxism, and quotes Marx on "capitalism's 'irreparable rift' with 'the natural laws of life itself.'" Nevertheless, she points to the high carbon emissions of Soviet-type societies, and the heavy dependence of the economies of Bolivia and Venezuela on natural resource extraction, notwithstanding the many social justice initiatives they have introduced. She questions the support given by Greece's SYRIZA Party to offshore oil exploration in the Aegean. Many of those on the left, and particularly the so-called liberal left, with their Keynesian predilections, continue to see an expansion of the treadmill of production, even in the rich countries, as the sole means of social advance.[18] Klein's criticisms here are important, but could have benefited, with respect to the periphery, from a consideration of the structure of the imperialist world economy, which is designed specifically to close off options to the poorer countries and force them to meet the needs of the richer ones. This creates a trap that even a Movement Toward Socialism with deep ecological and indigenous values like that of present-day Bolivia cannot seek to overcome without deep contradictions.[19]

"The unfinished business of liberation," Klein counsels, requires "a process of rebuilding and reinventing the very idea of the collective, the communal, the commons, the civil, and the civic after so many decades of attack and neglect."[20] To accomplish this, it is necessary to build the greatest mass movement of humanity for revolutionary change that the world has ever seen: a challenge that is captured in the title to her conclusion: "The Leap Years: Just Enough Time for Impossible." If this seems utopian, her answer would be that the world is heading toward something worse than mere dystopia: unending, cumulative climate catastrophe, threatening civilization and countless species, including our own.[21]

Liberal Critics as Gatekeepers

Confronted with Klein's powerful argument in *This Changes Everything*, liberal pundits have rushed to rein in her arguments so that her ideas are less in conflict with the system. Even where the issue is planetary ecological catastrophe, imperiling hundreds of millions of people, future generations, civilization, and the human species itself, the inviolable rule remains the same: the permanency of capitalism is not to be questioned.

As Noam Chomsky explains, liberal opinion plays a vital gatekeeping role for the system, defining itself as the rational left of center, and constituting the outer boundaries of received opinion. Since most of the populace in the United States and the world as a whole is objectively at odds with the regime of capital, it is crucial to the central propaganda function of the media to declare as "off limits" any position that questions the foundations of the system. The media effectively says: "Thus far and no further." To venture further left beyond the narrow confines of what is permitted within liberal discourse is deemed equivalent to taking "off from the planet."[22]

In the case of an influential radical journalist, activist, and bestselling author like Klein, liberal critics seek first and foremost to refashion her message in ways compatible with the system. They offer her the opportunity to remain within the liberal fraternity—if she will only agree to conform to its rules. The aim is not simply to contain Klein herself but also the movement as a whole that she represents. Thus, we find expressions of sympathy for what is presented as her general outlook. Accompanying all such praise, however, is a subtle recasting of her argument in order to blunt its criticism of the system. For example, it is perfectly permissible on liberal grounds to criticize neoliberal disaster capitalism as an extreme policy regime. This should at no time, however, extend to a blanket critique of capitalism. Liberal discussions of *This Changes Everything*, insofar as they are positive at all, are careful to interpret it as adhering to the former position.

Yet, the very same seemingly soft-spoken liberal pundits are not above simultaneously brandishing a big stick at the slightest sign of transgression of the "Thus Far and No Further" principle. If it should turn out that Klein is really serious in arguing that "this changes everything" and actually sees our reality as one of "capitalism vs. the climate," then, we are told, she has Taken Off From the Planet, and has lost her right to be heard within the mass media or to be considered part of the conversation at all. The aim here is to issue a stern warning—to remind everyone of the rules by which the game is played, and the serious sanctions to be imposed on those not conforming.

The penalty for too great a deviation in this respect is excommunication from the mainstream, to be enforced by the corporate media. Noam Chomsky may be the most influential intellectual figure alive in the world today, but he is generally considered beyond the pale and thus *persona non grata* where the U.S. media is concerned.

None of this of course is new. Invited to speak at University College, Oxford, in 1883, with his great friend John Ruskin in the chair, William Morris, Victorian England's celebrated artist, master artisan, and epic poet, author of *The Earthly Paradise*, shocked his audience by publicly declaring himself "one of the people called Socialists." The guardians of the official order (the Podsnaps of Dickens's *Our Mutual Friend*) immediately rose up to denounce him—overriding Ruskin's protests—declaring that had they known of Morris's intentions, he would not have been given loan of the hall. They gave notice then and there that he was no longer welcome at Oxford or in establishment circles. As historian E. P. Thompson put it, "Morris had crossed the 'river of fire.' And the campaign to silence him had begun."[23]

Klein, however, presents a special problem for today's gatekeepers. Her opposition to the logic of capital in *This Changes Everything* is not couched primarily in the traditional terms of the left, concerned mainly with issues of exploitation. Rather, she makes it clear that what has finally induced her to cross the river of fire is an impending threat to the survival of civilization and humanity itself. She calls for a broad revolt of humanity against capitalism and for the creation of a more sustainable society in response to the epochal challenge of our time. This is an altogether different kind of animal—one that liberals cannot dismiss out of hand without seeming to go against the scientific consensus and concern for humanity as a whole.

Further complicating matters, Klein upsets the existing order of things in her book by declaring "the right is right." By this she means that the political right's position on climate change is largely motivated by what it correctly sees as an either/or question of capitalism vs. the climate. Hence, conservatives seek to deny climate change—even rejecting the science—in their determination to defend capitalism. In contrast, liberal ideologues—caught in the selfsame trap of capitalism vs. the climate—tend to waffle, accepting most of the science, while turning around and contradicting themselves by downplaying the logical implications for society. They pretend that there are easy, virtually painless, non-disruptive ways out of this trap via still undeveloped technology, market magic, and mild government regulation—presumably allowing climate change to be mitigated without seriously affecting the capitalist economy. Rather than accepting the either/or of capitalism against the climate, liberals convert the problem into one of neoliberalism vs. the climate, insisting that greater regulation, including such measures as carbon trading and carbon offsets, constitutes the solution, with no need to address the fundamental logic of the economic and social system.

Ultimately, it is this liberal form of denialism that is more dangerous since it denies the social dimension of the problem and blocks the necessary social solutions. Hence, it is the liberal view that is the main target of Klein's book. In a wider sense, though, conservatives and liberals can be seen as mutually taking part in a dance in which they join hands to block any solution that requires going against the system. The conservative Tweedle Dums dance to the tune that the cost of addressing climate change is too high and threatens the capitalist system. Hence, the science that points to the problem must be denied. The liberal Tweedle Dees dance to the tune that the science is correct, but that the whole problem can readily be solved with a few virtually costless tweaks here and there, put into place by a new regulatory regime. Hence, the system itself is never an issue.

It is her constant exposure of this establishment farce that makes Klein's criticism so dangerous. She demands that the gates be flung open and the room for democratic political and social maneuver be expanded enormously. What is needed, for starters, is a pro-democracy movement not simply in the periphery of the capitalist world but at the center of the system itself, where the global plutocracy has its main headquarters.

The task from a ruling-class governing perspective, then, is to find a way to contain or neutralize Klein's views and those of the entire radical climate movement. The ideas she represents are to be included in the corporate media conversation only under extreme sufferance, and then only insofar as they can be corralled and rebranded to fit within a generally liberal, reformist perspective: one that does not threaten the class-based system of capital accumulation.

Rob Nixon can be credited with laying out the general liberal strategy in this respect in a review of Klein's book in the *New York Times*. He declares outright that Klein has written "the most momentous and contentious environmental book since *Silent Spring*." He strongly applauds her for her criticisms of climate change deniers, and for revealing how industry has corrupted the political process, delaying climate action. All of this, however, is preliminary to his attempt to rein in her argument. There is a serious flaw in her book, we are told, evident in her subtitle, *Capitalism vs. the Climate*. "What's with the subtitle?" he scornfully asks. Then, stepping in as Klein's friend and protector, Nixon tells *New York Times* readers that the subtitle is simply a mistake, to be ignored. We should not be thrown off, he proclaims, by a subtitle that "sounds like a P.R. person's idea of a marquee cage fight." Rather, "Klein's adversary is neoliberalism—the extreme capitalism that has birthed our era of extreme extraction." In this subtle recasting of her argument, Klein reemerges as a mere critic of capitalist excess, rejecting specific attributes taken on by the system in its neoliberal phase that can be easily discarded, and that do not touch the system's fundamental properties. Her goal, we are told, is the same as in *The Shock Doctrine*: turning back the neoliberal "counterrevolution," returning us to a more humane Golden Age liberal order. Her subtitle can therefore be dismissed in its entirety, as it "belies the sophistication" of her work: code for her supposed conformity to the Thus Far and No Further principle. Employing ridicule as a gatekeeping device—with the implication that this is the sorry fate that awaits anyone who transgresses Thus Far and No Further—Nixon states that "Klein is smart and pragmatic enough to shun the never-never land of capitalism's global overthrow."[24]

Dave Pruett in *The Huffington Post* quickly falls into step, showing how well he comprehends the general strategy already outlined by Nixon in the *New York Times*. At the same time, he indicates his readiness to pull in the reins a bit more. Thus, we find again that Klein's book is a "masterpiece," to be put on the same shelf as Rachel Carson's *Silent Spring*. And once again we learn that her subtitle, *Capitalism vs. the Climate* is a "misnomer." Resorting to a classic Cold War ploy, Pruett further insinuates that the subtitle gives "critics room to accuse Klein of advocating for some discredited Soviet-style state-regulated economy." Of course, such critics, he turns around and says, would surely be wrong. Klein's argument in *This Changes Everything* is really nothing more than a criticism of "*unbridled* capitalism—that is, neoliberalism." Moreover, the "true culprit" of her argument is even more specific than this: "*extractivism*," or the extreme exploitation of non-renewable natural resources. Still, Pruett, through his classic Cold War ploy, has with consummate skill planted in advance a lingering doubt and a warning in the mind of the reader, along with an implicit threat directed at Klein herself. If it should turn out that Klein is serious about her subtitle, and she is actually talking about "capitalism vs. the climate," then she is discredited in advance by the fate of the Soviet Union, with which she is then to be associated.[25]

Approaching *This Changes Everything* much more bluntly, Elizabeth Kolbert, writing for the *New York Review of Books*, quickly lets us know that she has not come to praise Klein but to bury her. Klein's references to conservation, "managed degrowth," and the need to shrink humanity's ecological footprint, Kolbert says, are all non-marketable ideas, to be condemned on straightforwardly capitalist-consumerist principles. Such strategies and actions will not sell to today's consumers, even if the future of coming generations is in jeopardy. Nothing will get people to give up "HDTV or trips to the mall or the family car." Unless it is demonstrated how acting on climate change will result in a "minimal disruption to 'the American way of life,'" she asserts, nothing said with respect to climate change action matters at all. Klein has simply provided a convenient "fable" of little real value. *This Changes Everything* is indicted for having violated accepted commercial axioms in its core thesis, which Kolbert converts into an argument for extreme austerity. Klein is to be faulted for her grandiose schemes that do not fit into U.S. consumer society, and for not "looking at all closely at what this [reduction in the commodity economy] would entail." Klein has failed to specify exactly how many watts of electricity per capita will be consumed under her plan. It is much easier, Kolbert seems to say, for U.S. consumers to imagine the end of a climate permitting human survival than to envision the end of two-million-square-foot shopping malls.[26]

David Ulin in the *Los Angeles Times* unveils still another weapon in the liberal arsenal, denouncing Klein for her optimism and her faith in humanity. "There is, in places," he emphasizes, "a disconnect between her [Klein's] idealism and her realism, what she thinks ought to happen and what she recognizes likely will." Social analysis, in Ulin's view, seems to be reduced to forecasting the most likely outcomes. Klein apparently failed to consult with Las Vegas oddsmakers before making her case for saving humanity. Klein's penchant for idealism, he declares, "is most glaring in her suggestions for large-scale policy mitigation, which can seem simplistic, relying on notions of fairness . . . that corporate culture does not share." Regrettably, Ulin does not tell us exactly where the kind of climate justice programs put in place by Exxon and Walmart's "corporate culture" will actually lead us in the end. However, he does give us a specious clue in his final paragraph, describing what he apparently considers to be the most realistic scenario. The planet, we are informed, "has ample power to rock, burn, and shake us off completely." The earth will go on without us.[27]

Other liberal gatekeepers pull out all the stops, attacking not just every radical notion in Klein's book but the book as a whole, and even Klein herself. Writing for the influential liberal news and opinion website, the *Daily Beast*, Michael Signer characterizes Klein's book as "a curiously clueless manifesto." It will not spark a movement against carbon, in part because Klein "rejects capitalism, market mechanisms, and even, seemingly, profit motives and corporate governance." She offers "a compelling story," but one that "creates the paradoxical effect of making this perspicacious and successful author seem like an idiot." Signer depicts her as if she has Taken Off From the Planet simply by refusing to stay within the narrow spectrum of opinion defined by the *Wall Street Journal* on the one side and the *New York Times* on the other. "For anyone who believes in capitalism and political leadership," we are informed, "her book won't change anything at all."[28]

Mark Jaccard, an orthodox economist writing for the *Literary Review of Canada*, declares that *This Changes Everything* ignores how market-based mechanisms are a powerful means for reducing carbon emissions. However, his main evidence for this contention is Arnold Schwarzenegger's signing of a climate bill in California in 2006, which is supposed to drastically reduce the state's carbon emissions to 1990 levels. Unfortunately for Jaccard's claim, a little over a week before he criticized Klein on the basis of the California experiment, the *Los Angeles Times* broke the story that California's emissions reduction initiative was in some respects a "shell game," as California was reducing emissions on paper while emissions were growing in surrounding states from which California was also increasingly purchasing power.[29] Add to this the facts that California's initiative is more state-based than capital-based, and that the real problem is not one of getting down to 1990 level emissions, but getting down to pre-1760 level emissions, that is, carbon emissions eventually have to fall to zero (or at least net zero)—and not just in California but worldwide.

Jaccard goes on to accuse Klein of wearing "'blame capitalism' blinders" that keep her from seeing the actual difficulties that make dealing with climate so imposing. This includes her failure to perceive the "Faustian dilemma" associated with fossil fuels, given that they have yielded so many benefits for humanity and can offer many more to the poor of the world. "This dilemma," which he is so proud to have discovered, "is not the fault of capitalism." Indeed, capitalist economics, we are told, is already well equipped to solve the climate problem and only misguided state policies stand in the way. Drawing upon an argument presented by Paul Krugman in his *New York Times* column, Jaccard suggests that "greenhouse gas reductions have proven to be not nearly as costly as science deniers on the right and anti-growth activists on the left would have us believe." Krugman, a Tweedle Dee, rejects the carefree Tweedle Dum melody whereby climate change, as a threat to the system, is simply wished away along with the science. He counters this simple, carefree tune with what he regards as a more complex, harmonious song in which the problem is whisked away in spite of the science by means of a few virtually costless market regulations. So convinced is Jaccard of capitalism's basic harmonious relation to the climate that he simply ignores Klein's impressive account of the vast system-scale changes required to stop climate change.[30]

Will Boisvert, commenting on behalf of the self-described "post-environmentalist" Breakthrough Institute, condemns Klein and the entire environmental movement in an article pointedly titled, "The Left vs. the Climate: Why Progressives Should Reject Naomi Klein's Pastoral Fantasy—and Embrace Our High-Energy Planet." Apparently, it is not industry that is destroying a livable climate through its carbon dioxide emissions, but rather environmentalists, by refusing to adopt the Breakthrough Institute's technological crusade for surmounting nature's limits on a planetary scale. As Breakthrough senior fellow Bruno Latour writes in an article for the Institute, it is necessary "to love your monsters," meaning the kind of Frankenstein creations envisioned in Mary Shelley's novel. Humanity should be prepared to put its full trust, the Breakthrough Institute tells us, in such wondrous technological answers as nuclear power, "clean coal," geoengineering, and fracking. For its skepticism regarding such technologies, the whole left (and much of the scientific community) is branded as a bunch of Luddites. As Boisvert exclaims in terms designed to delight the entire corporate sector:

> To make a useful contribution to changing everything, the Left could begin by changing itself. It could start by redoing its risk assessments and rethinking its phobic hostility to nuclear power. It could abandon the infatuation with populist insurrection and advance a serious politics of systematic state action. It could stop glamorizing austerity under the guise of spiritual authenticity and put development prominently on its environmental agenda.

It could accept that industry and technology do indeed distance us from nature—and in doing so can protect nature from human extractions. And it could realize that, as obnoxious as capitalism can be, scapegoating it won't spare us the hard thinking and hard trade-offs that a sustainable future requires.[31]

Boisvert here echoes Erle Ellis, who, in an earlier essay for the Breakthrough Institute, contended that climate change is not a catastrophic threat because "human systems are prepared to adapt to and prosper in the hotter, less biodiverse planet that we are busily creating." On this basis, Boisvert chastises Klein and all who think like her for refusing to celebrate capitalism's creative destruction of everything in existence.[32]

Klein of course is not caught completely unaware by such attacks. For those imbued in the values of the current system, she writes in her book, "changing the earth's climate in ways that will be chaotic and disastrous is easier to accept than the prospect of changing the fundamental, growth-based, profit-seeking logic of capitalism."[33] Indeed, all of the mainstream challenges to *This Changes Everything* discussed above have one thing in common: they insist that capitalism is the "end of history," and that the buildup of carbon in the atmosphere since the Industrial Revolution and the threat that this represents to life as we know it change nothing about today's Panglossian best of all possible worlds.

The Ultimate Line of Defense

Naturally, it is not simply liberals, but also socialists, in some cases, who have attacked *This Changes Everything*. Socialist critics, though far more sympathetic with her analysis, are inclined to fault her book for not being explicit enough about the nature of system change, the full scale of the transformations required, and the need for socialism.[34] Klein says little about the vital question of the working class, without which the revolutionary changes she envisions are impossible. It is therefore necessary to ask: To what extent is the ultimate goal to build a new movement toward socialism, a society to be controlled by the associated producers? Such questions still remain unanswered by the left climate movement and by Klein herself.

In our view, though, it is difficult to fault Klein for her silences in this respect. Her aim at present is clearly confined to the urgent and strategic—if more limited—one of making the broad case for System Change Not Climate Change. Millions of people, she believes, are crossing or are on the brink of crossing the river of fire. Capitalism, they charge, is now obsolete, since it is no longer compatible either with our survival as a species or our welfare as individual human beings. Hence, we need to build society anew in our time with all the human creativity and collective imagination at our disposal. It is this burgeoning global movement that is now demanding anti-capitalist and post-capitalist solutions. Klein sees herself merely as the people's megaphone in this respect. The goal, she explains, is a complex social one of fusing all of the many anti-systemic movements of the left. The struggle to save a habitable earth is *humanity's ultimate line of defense*—but one that at the same time requires that we take the offensive, finding ways to move forward collectively, extending the boundaries of liberated space. David Harvey usefully describes this fusion of movements as a co-revolutionary strategy.[35]

Is the vision presented in *This Changes Everything* compatible with a classical socialist position? Given the deep ecological commitments displayed by Marx, Engels, and Morris, there is little room for doubt—which is not to deny that socialists need to engage in self-criticism, given past failures to implement ecological values and the new challenges that characterize our epoch. Yet, the whole question strikes us in a way as a bit odd, since historical materialism does not represent a rigid, set position, but is rather the ongoing struggle for a world of substantive equality and sustainable human development. As Morris wrote in *A Dream of John Ball*:

> But while I pondered all these things, and how men fight and lose the battle, and the thing that they fought for comes about in spite of their defeat, and when it comes turns out not to be what

they meant, and other men have to fight for what they meant under another name—while I pondered all this, John Ball began to speak again in the same soft and clear voice with which he had left off.

In this "soft and clear voice," Ball, a leader in the fourteenth-century English Peasants' Revolt, proceeded, in Morris's retelling, to declare that the one true end was "Fellowship on Earth"—an end that was also the *movement of the people* and could never be stopped.[36]

Klein offers us anew this same vision of human community borne of an epoch of revolutionary change. "There is little doubt," she declares in her own clear voice,

> that another crisis will see us in the streets and squares once again, taking us all by surprise. The real question is what progressive forces will make of that moment, the power and confidence with which it will be seized. Because these moments when the impossible seems suddenly possible are excruciatingly rare and precious. That means more must be made of them. The next time one arises, it must be harnessed not only to denounce the world as it is, and build fleeting pockets of liberated space. It must be the catalyst to actually build the world that will keep us all safe. The stakes are simply too high, and time too short, to settle for anything less.[37]

The ultimate goal is not simply "to build the world that will keep us all safe" but to build a world of genuine equality and human community—the only conceivable basis for sustainable human development. Equality, Simón Bolívar exclaimed, is "the law of laws."[38]

CHAPTER FIVE

The Fossil Fuels War

As recently as 2010, governments, corporations, and energy analysts were fixated on the problem of "the end of cheap oil" or "peak oil," pointing to growing shortages of conventional crude oil due to the depletion of known reserves. Thus, the International Energy Agency's 2010 report devoted a whole section to peak oil.[1] Some climate scientists saw the peaking of conventional crude oil as a silver-lining opportunity to stabilize the climate—provided that countries did not turn to dirtier forms of energy such as coal and "unconventional fossil fuels."[2]

Only a few years later all of this was to change radically with the advent of what some are calling a new energy revolution based on the production of unconventional fossil fuels.[3] The emergence in North America—but increasingly elsewhere as well—of what is now termed the "Unconventionals Era" has meant that suddenly the world is awash in new and prospective fossil-fuel supplies.[4] As journalist and climate activist Bill McKibben warns:

> Right now the fossil-fuel industry is mostly winning. In the past few years, they've proved "peak-oil" theorists wrong—as the price rose for hydrocarbons, companies found a lot of new

sources, though mostly by scraping the bottom of the barrel, spending even more money to get even-cruddier energy. They've learned to frack (in essence, explode a pipe bomb a few thousand feet beneath the surface, fracturing the surrounding rock). They've figured out how to take the sludgy tar sands and heat them with natural gas till the oil flows. They've managed to drill miles beneath the ocean's surface.[5]

The new phase of environmental struggle that the Unconventionals Era has engendered is symbolized above all by the nearly decade-long struggle over the Keystone XL Pipeline, a 1,200-mile pipeline, extending from the Alberta tar sands to refineries on the U.S. Gulf Coast, designed to deliver up to 830,000 barrels of tar sands oil (diluted bitumen or dilbit) a day from Alberta to Steele City, Nebraska, where it would have connected with other pipelines in the Keystone Pipeline system, delivering oil to refineries in Texas.[6] The Keystone XL was intended to provide a shorter pipeline route within the overall Keystone system, while using larger diameter pipe, thereby greatly expanding overall pipeline capacity.[7] Ecologically, the pipeline threatened not only ecosystems along its route but also, by allowing for expanded tar sands production, endangered the biosphere itself. Tar sands extraction is estimated to be three to four times as carbon intensive as the extraction of conventional crude.[8] Failure to halt the burning of tar sands oil would mean "game over" with respect to climate change, in the words of James Hansen, director of NASA's Goddard Institute for Space Studies, and the most renowned U.S. climatologist.[9]

The Alberta tar sands, which underlie an area roughly the size of Florida, were already generating 1.8 million barrels of oil a day in 2013 and the push has been to expand this further. The Achilles' heel of tar sands production, however, is transportation. For some time now there has been a "bitumen bubble" as tar sands oil is more readily produced than transported. The inability to get the tar sands oil to ports means that it remains dependent on the U.S. market and is unable to command world prices. Tar sands oil (known on the oil markets as Western Canadian Select) traded at times in 2012 at $35 a barrel less than the price it would have received had transcontinental oil transport been readily available. This represented a loss of about a third of its value when compared to West Texas Intermediate.[10] Hence, the tar sands industry has been desperate to secure adequate transcontinental transport to support its current as well as expanded oil production. The big push has been for pipelines. Yet, there are serious environmental concerns that diluted bitumen may be more dangerous to transport in pipelines than conventional crude oil, because of increased likelihood of pipeline corrosion, and the resulting leakages. The Keystone XL Pipeline would go right over the Ogallala aquifer, the largest drinking-water aquifer in the United States, which supplies eight states.[11]

The United States witnessed militant climate demonstrations in February 2013, with upward of 40,000 people protesting in front of the White House and more than a thousand arrested in opposition to the Keystone XL Pipeline.[12] In Canada, meanwhile, the indigenous-led Idle No More has utilized a variety of strategies and tactics in fighting tar sands production, such as a hunger strike by Attawapiskat Chief Theresa Spence; rail blockades; flashmobs in malls; a giant circle dance in a large intersection in Winnipeg; and the legal defense of First Nations sovereignty rights with respect to land, water, and resources. Idle No More protests have targeted oil transport by both rail and pipeline, with the latter including opposition to Keystone XL and to the planned Enbridge Northern Gateway Pipelines Project—designed to extend around 730 miles from the Alberta tar sands to a marine terminal in Kitimat, British Columbia.[13]

Other unconventionals are also altering the terrain of the struggle. The last decade or so has witnessed dramatic new technological developments with respect to hydraulic fracturing coupled with horizontal drilling, or "fracking." Sand, water, and chemicals are injected at high pressures in order to blast open shale rock, releasing the trapped gas inside. After the well has reached a certain depth the drilling occurs horizontally.[14] Fracking has led to the rapid exploitation of vast, hitherto inaccessible, reserves of shale gas and tight oil in states across the country, from Pennsylvania and Ohio to North Dakota and

California, unexpectedly catapulting the United States once again into the position of a major fossil-fuel power. It has already led to substantial increases in natural gas production, replacing dirtier and more carbon-emitting coal in generating electricity. Together the economic slowdown and the shift from coal to natural gas due to fracking have resulted in a 12 percent drop in U.S. (direct) carbon dioxide emissions between 2005 and 2012, reaching their lowest level since 1994.[15]

Nevertheless, the negative environmental and health effects of fracking falling on communities throughout the United States are enormous, if still not fully assessed. Toxic pollution from fracking is contaminating water supplies and affecting wastewater treatment not designed to cope with such hazards. Methane leakages from fracking, in the case of shale gas, are threatening to accelerate climate change. If such leakages cannot be contained, fracked natural gas production could prove more dangerous to the climate than coal.[16] Fracking has also engendered earthquakes in the extractive areas.[17] In response to such developments, a whole new environmental resistance to fracking has arisen in communities throughout North America, Australia, and elsewhere.

A train pulling seventy-two tank cars laden with oil from fracking in North Dakota derailed and exploded in Lac-Mégantic, Quebec, on July 6, 2013, killing fifty people. Such accidents are themselves a product of the boom in unconventionals, coupled with "pipeline on rails" methods of shipping the oil (as well as the decrease of labor used in rail transport). In 2009, corporations shipped a mere 500 tank cars of oil by rail in Canada; while by 2013 this was projected to be as much as 140,000 tank cars.[18] North Dakota tight oil is also shipped by rail to Albany, New York, where it is loaded onto barges for shipment to East Coast refineries.

On April 20, 2010, an explosion in BP's Deepwater Horizon oil platform killed eleven workers and generated a huge underwater oil gusher, which dumped a total of 170 million gallons of crude oil into the Gulf of Mexico.[19] The Deepwater Horizon disaster has come to stand for the new environmentally perilous era of ultra-deepwater oil wells—offshore oil drilled at depths of more than a mile as a result of the development of more sophisticated technologies. (Deepwater oil drilling more generally involves drilling at depths of more than a thousand feet.)

Deepwater oil drilling is most advanced in the Gulf of Mexico, but is spreading in other places, such as Canada's Atlantic Coast, Brazil's offshore zone, the Gulf of Guinea, and the South China Sea. Still more ominous from an environmental standpoint is the drive by oil companies and the five Arctic powers (the United States, Canada, Russia, Norway, and Denmark) to drill deepwater wells in the Arctic—made increasingly accessible due to global warming. Meanwhile, pressure is mounting to open up the outer continental shelf off the U.S. Atlantic and Pacific coasts to offshore oil drilling.[20]

In the face of the rush by capital to extract unconventional fossil fuels in ever-greater amounts, climate activists are seeking new means of resistance. The "Do the Math" strategy of 350.org is focused on the necessary divestment in fossil fuels, to be replaced by clean energy sources. Some financial analysts have been sounding the alarm with respect to the carbon budget imposed by the red line of a 2°C increase in global average temperature—referred to as a planetary tipping point or "point of no return" with respect to climate change. Climate scientists fear that once this point is reached processes will be set in motion that will make climate change irreversible and out of human control.[21] It will no longer be possible to stop the progression to an ice-free world. Staying within the global carbon budget means that further carbon emissions are limited to considerably less than 500 billion metric tons (of actual carbon), according to Oxford climatologist Myles Allen. This means that most of the world's current proven fossil-fuel reserves cannot be exploited without initiating extremely dangerous—even irreversible—levels of climate change. And this limitation in turn threatens trillions of dollars of potential financial losses in what are now counted as fossil-fuel assets—a phenomenon known as the "carbon bubble."[22]

While capital throughout the opening decades of this century has been triumphantly celebrating its increased ability to tap fossil fuels for decades to come, climate change has continued to accelerate—symbolized by the melting of Arctic sea ice to its lowest level ever, recorded in summer 2012, with the total ice area receding to less than half the average level of the 1970s. The vanishing Arctic ice, which is melting far faster than scientists had generally predicted, suggests that the sensitivity of the Earth System to small increases in global average temperatures is greater than was previously thought. The ice loss is of particular concern since it represents a positive feedback loop to climate change, accelerating the rate of global warming as the reflectivity of the earth declines—due to the replacement of white ice with dark seawater. The melting of Arctic sea ice, and the resulting "arctic amplification" (temperature increases in the Arctic exceeding that of the earth as a whole) is generating extreme weather events in the Northern Hemisphere and worldwide through the "jamming" and redirection of the jet stream. As Walt Meier, a research scientist at the U.S. National Snow and Ice Data Center, put it, "The Arctic is the earth's air conditioner. We're losing that."[23]

The growing incidence of extreme weather events—a phenomenon sometimes referred to as "global weirding"—is symbolized by Superstorm Sandy, which in October 2012 wreaked havoc from the Caribbean to New York and New Jersey. Australia's "angry summer" of 2012–13 saw 123 separate extreme weather records broken in a mere ninety days.[24] Meanwhile a scientific report in November 2012 revealed that Greenland and west Antarctica had lost more than 4 trillion metric tons of ice over the last two decades, contributing to sea level rise.[25]

Under these circumstances the increased exploitation of unconventional fossil fuels, made possible by higher oil prices and technological developments, has catastrophic implications for the climate. No less remarkable technological developments have arisen at the same time in relation to renewable energies, such as wind and solar, opening up the possibility of a more ecological path of development. Since 2009 solar (photovoltaic) module "prices have fallen off a cliff."[26] Although still accounting for a tiny percentage of electric-generating capacity in the United States, wind and solar have grown to about 13 percent of total German electricity production by 2012, with total renewables (including hydroelectric and biomass) accounting for about 20 percent.[27] As the energy return on energy investment (EROEI) of fossil fuels has declined due to the depletion of cheap crude-oil supplies, wind and solar have become more competitive—with EROEIs above that of tar sands oil, and in the case of wind even above conventional oil. Wind and solar, however, represent intermittent, location-specific sources of power that cannot easily cover baseload-power needs.[28] Worse still, a massive conversion of the world's energy infrastructure to renewables could take decades to accomplish when time is short.

The Carbon War

The result of all these historically converging forces, dangers, and opportunities is an emerging fossil-fuels war between those who want to burn more fossil fuels and those who want to burn less. Jeremy Leggett, a leader in the carbon-divestment movement, concluded his 2001 book *The Carbon War* with the observation that the giant fossil-fuel corporations "may well enjoy minor victories along the way. But they have already lost the pivotal battle in the carbon war. The solar revolution is coming. It is now inevitable. The only question left unanswered is, will it come in time?"[29]

The main battle lines of the carbon war are clear. On the one side, there are the dominant capitalist interests that have sought to address the decline of conventional crude-oil reserves through the incessant expansion of fossil-fuel resources. This has led to actual wars in the oil-rich Middle East and surrounding regions in an effort to gain control over the world's chief remaining "cheap oil" supplies. In 2003, the United States invaded Iraq, leading to what can only be called a continuous military intervention in the oil-rich regions of the Middle East, Central Asia, and Africa by the United States and "global NATO."[30] These military incursions have been primarily related to the geopolitics of oil, and only secondarily to terrorism, weapons of mass destruction, and so-called humanitarian intervention as the main rationales.

Nevertheless, the main response of the capitalist system to the peaking of conventional crude oil has not been geopolitical expansion but rather development of the unconventionals. Not stopping with deepwater drilling, fracking, and the exploitation of tar sands oil, the fossil-fuel industry, backed by the state, is now looking toward development of oil shale and methane hydrates—offering, if these can be brought online, what seem to be truly unlimited supplies of carbon, coupled with the prospect of unthinkable, catastrophic disruptions to the Earth System.[31]

Today's business-as-usual interests refuse to accept any limits to continued expansion of fossil-fuel production. Establishment energy policymakers—as witnessed by the Council on Foreign Relations senior energy analyst Michael Levi—see shale gas from fracking as a "bridge fuel" that will allow a reduction in carbon emissions until carbon capture and sequestration technologies can be developed sufficiently to be feasible, opening the way to supposedly unlimited exploitation of coal and other fossil fuels with zero carbon emissions. The fact that "clean coal" is a fairy tale never seems to enter the analysis.[32] Most establishment energy proponents also favor biofuels as an added option, and support large hydroelectric facilities and nuclear energy, discounting the enormous ecological problems represented by all three—particularly nuclear power. Wind, solar, and biomass, in contrast, are viewed by industry as minor supplements to fossil fuels. Empirical research by environmental sociologist Richard York, published in *Nature Climate Change* in 2012, has verified that the introduction of low-carbon energy has been used mainly to supplement rather than actually displace fossil fuels within the global economy.[33]

ExxonMobil's CEO Rex Tillerson aptly summed up the overall outlook of today's fossil-fuel industry when he declared on March 7, 2013, that renewables such as "wind, solar, biofuels" would be supplying only 1 percent of total energy in 2040. He described the struggle against the Keystone XL Pipeline by "environmental groups . . . concerned about the burning of fossil fuels" as simply "obtuse," since they "misjudged Canada's resolve" (and no doubt that of the U.S.

government) to exploit the tar sands—whatever the social and environmental cost. "My philosophy," Tillerson said, "is to make money."[34]

In the United States this addiction to fossil fuels was built into the Obama administration's "all of the above" energy strategy. Washington under Obama and Trump not only promoted the extraction/production of unconventional fossil fuels in the United States and Canada, it is also actively encouraged other countries, such as China, Poland, the Ukraine, Jordan, Colombia, Chile, and Mexico, to develop unconventionals as rapidly as possible. Meanwhile, Washington used its influence in Iraq to get it to boost its crude oil production.[35] Although the Trump administration went further in this respect, both Democratic and Republican administrations have sought to expand fossil fuel production, both conventional and unconventional.[36]

Recent U.S. administrations have strongly underscored their support for coal, and have sought to give a boost to nuclear power. They have also promoted the production of fracked natural gas globally as a "transition fuel." Washington thus remains little more than a water carrier for the oil corporations and capital in general where climate policy is concerned, reflecting what Curtis White has called capitalism's "barbaric heart."[37]

On the other side is the burgeoning climate movement, propelled into massive direct action by the new threats from the unconventionals. Hansen's dire warning that it is "game over" if the Alberta tar sands oil is exploited fully—with the tar sands generating potentially enough carbon dioxide emissions to break the world's carbon budget while symbolizing the pressing need to draw a line in the sand in relation to unconventional fossil fuels—has had an electrifying effect on the movement on the ground. Mass resistance to the Keystone XL Pipeline finally led to a victory in 2021, after Joe Biden on his first day in office revoked the necessary cross-border permit for the pipeline (thereby reversing the actions of the Donald Trump administration) and TC Energy threw in the towel six months later.[38]

Yet, the enormous battle over the Dakota Access Pipeline led by the Standing Rock Sioux in North Dakota, who are resisting the transport of fossil fuels from fracking through their historic tribal lands continues, representing one of the fiercest environment struggles ever seen in North America.[39] Idle No More is fighting oil pipelines in Canada extending south, west, and east. This on-the-ground mobilization is combined with the growing fossil-fuels divestment movement. The main thrust of the climate movement has therefore shifted from demand-side initiatives aimed at reducing consumer-market demand for carbon fuels to supply-side strategies aimed at corporations and designed to keep the fossil fuels in the ground.

The shift a little over a decade ago to a supply-side struggle targeting corporations represents a maturing of the movement and a growing radicalization. Still, the more elite technocratic and pro-capitalist elements, which appear to be in the driver's seat within the climate movement in the United States, remain wedded to the continuation of today's capitalist commodity society. The prevailing strategic outlook of the U.S. climate movement is largely predicated on the technologically optimistic assumption that there are currently available concrete alternatives to fossil fuels, particularly wind and solar, which, when combined with other renewable sources such as biomass, biofuels, and limited-scale hydroelectric power, will allow society to substitute renewable energies for fossil fuels in the near term *without altering society's social relations*. The solar revolution, it is often declared, is here.[40]

This outlook has allowed the movement to narrow its opposition to the fossil-fuel industry alone, confining its demands to keeping fossil fuels in the ground, blocking the transport of fossil fuels, and divesting in fossil-fuels corporations. As McKibben has stated, "Movements need enemies," and the strategy has been to focus not on capitalism but on the fossil-fuel industry as a "rogue industry.... Public Enemy Number One."[41] This has been highly successful in sparking the growth of the movement. Yet, there are serious questions with regard to where all of this is headed. Will the current struggle metamorphose into the necessary full-scale revolt against capitalist environmental destruction? Or will it be confined to very limited short-term gains of the kind compatible with the system? Will the movement radicalize, leading to the full mobilization of its popular base? Or will the more elite-technocratic and pro-capitalist elements within the movement leadership in the United States ultimately determine its direction, betraying the grassroots resistance? These are questions for which there are no answers at present. In the current historical moment the struggle against the fossil-fuel industry is paramount—the basis of today's ecological popular front. Yet, a realistic outlook indicates that nothing short of a full-scale ecological and social revolution will suffice to create a sustainable society out of the planetary rift generated by the present-day capitalist order. The break with the relentless logic of the system cannot be long delayed.

The Revolution against the System

A realistic historical assessment tells us that there is no purely technological path to a sustainable society. Although a rapid shift to renewables is a crucial component of any conceivable path to a carbon-free, ecological world, the technical obstacles to such a transition are much greater than is usually assumed. The biggest barrier is the up-front cost of building an entirely new energy infrastructure geared to renewables rather than relying on the existing fossil-fuel infrastructure. Construction of a new energy infrastructure requires vast amounts of energy consumption, and would lead—if current consumption and economic growth were not to be reduced—to further demands on existing fossil-fuel resources. This would mean, as ecological economist Eric Zencey has explained, "an aggressive expansion of the economy's footprint in paradoxical service to the goal of achieving sustainability." Assuming the average EROEI of fossil fuels keeps falling, the difficulty only becomes worse. Ecological economists and peak-oil theorists have dubbed this the "energy trap." In Zencey's words, "The problem is rooted in the sunken energy costs of the petroleum infrastructure (which makes the continued use of petroleum energetically cheap)" even when the EROEI of such fossil fuels in the case of unconventionals is lower than wind and solar.[42] It follows that building an alternative energy infrastructure—without breaking the carbon budget—would require a tectonic shift in the direction of energy conservation and energy efficiency.

Kevin Anderson, a leading British climate scientist and the deputy director of the Tyndall Institute for Climate Research, explained in a 2012 interview with *Transition Culture* that while it is imperative that we drastically cut fossil fuel use,

> we cannot deliver [this] reduction by switching to a low carbon energy supply, we simply cannot get the supply in place quickly enough. Therefore, in the short to medium term the only major change that we can make is by consuming less. Now, that would be fine, we could become more efficient in what we consume by probably a 2–3% per annum reduction. But bear in mind, if our economy was growing at 2% per annum, and we were trying to get a 3% per annum reduction in our emissions, that's a 5% improvement in the efficiency of what we're doing each year, year on year.
>
> Our analysis [at the Tyndall Institute] for 2°C suggests we need a 10% absolute reduction per annum [in carbon dioxide emissions in the rich countries], and there is no analysis out there that suggests that this is in any way compatible with economic growth. If you consider the *Stern Report* [*on Climate Change*], Stern was quite clear that there was no evidence that any more than a 1% per annum reduction in emissions had ever been associated with anything other than "economic recession or upheaval," I think was the exact quote.[43]

In Anderson's view, the only hope is to shift rapidly from a capitalist-growth economy to a steady-state economy—or, at the very least, to place a moratorium on economic growth for several decades while society's surplus resources are devoted to the transformation of the energy infrastructure. This would require, he says, "the community approach, the bottom-up approach," with the population mobilizing on its own behalf and that of future generations to create a new "emergent" reality. Such a social and ecological transformation would necessitate a move toward social conservation, even short-term rationing. Ecological planning of production and consumption, and energy use, would be essential.[44] In the words of the Royal Society of London, one of the world's oldest scientific bodies, it is now necessary to "develop socio-economic systems and institutions that are not dependent on continued material consumption growth."[45]

If we go beyond the climate change issue and examine the entire global ecological crisis the logic behind such reasoning is inescapable. In 2009 leading Earth System scientists led by Johan Rockström of the Stockholm Resilience Center introduced what is known as the "planetary boundaries" approach to determining the "safe operating space" for human beings on the planet, using as their baseline the biophysical conditions associated with the Holocene geological epoch in earth history, during the last 11,700 years, a period that saw the rise of civilization. The global ecological crisis can thus be defined as a sharp and potentially irreversible departure from Holocene conditions.[46]

This analysis of a "safe operating space" for humanity established a system of natural metrics in the form of nine planetary boundaries. In the case of three of these—climate change, biodiversity loss, and the nitrogen cycle, part of a boundary together with the phosphorus cycle—the planetary boundaries have already been crossed. Whereas in the case of a number of other planetary boundaries—the phosphorus cycle, ocean acidification, global freshwater use, and change in land use—alarming trends suggest that these boundaries will soon be crossed as well. Climate change is therefore only one part of a much larger ecological crisis facing humanity, traceable to the exponential growth of an increasingly destructive economic order within a finite planetary system.

These considerations all point to the limitations of what appears to be the governing outlook of the climate movement, promoted by its elite technocratic elements. The current ecological popular front has its basis in its singular opposition to fossil fuels and the fossil-fuel industry, and is largely premised on the notion the solar revolution will provide the solution to the climate problem, allowing for the continuation of the current socioeconomic order with relatively few adjustments. However, stopping climate change and the destruction of the environment in general requires not just a new, more sustainable technology, greater efficiency, and the opening of channels for green investment and green jobs, it requires an ecological revolution that will alter our entire system of production and consumption, and create new systems geared to substantive equality, and ecological sustainability—a "revolutionary reconstitution of society at large."[47] It means comprehending, as Marx presciently did in the nineteenth century, the metabolic relation between society and nature based in production—and the dangers associated with capitalism's growing metabolic rift. For Marx, the very destruction of "that metabolism" in the human relation to nature "compels its systematic restoration as a regulative law of social production, in a form adequate to the full development of the human race."[48]

The materialist conception of history has often been interpreted in ways—contrary to Marx—that systematically excluded ecological conditions from the analysis. Yet an argument can be made that the working class during its most class-conscious and revolutionary periods has been just as concerned with overall living conditions—including urban and rural community and the interaction with the natural environment—as with working conditions (in the narrow sense). A clear indication of this, reflecting the times in which it was written, is provided by Engels's 1844 *Condition of the Working Class in England*, in which environmental conditions were presented as of even greater importance to the overall material conditions of the working class than factory conditions—although the root cause resided in the class basis of production.[49] In today's world, the undermining of the lifeworld of the great majority of the population is occurring in relation to both economy and environment. We can therefore expect the most radical movements to emerge precisely where economic and ecological crises converge on the lives of the underlying population. Given the nature of capitalism and imperialism and the exigencies of the global environmental crisis, a new, revolutionary environmental proletariat is likely to arise most powerfully and most decisively in the Global South. Yet, such developments, it is now clear, will not be confined to any one part of the planet.[50]

The "bottom line" in an accounting ledger is one of capitalism's most enduring metaphors. We are now facing an ecological bottom line—a planetary carbon budget together with planetary boundaries in general—that represents a more fundamental accounting. Without a thoroughgoing transformation of production and consumption, and also social consciousness and cultural forms, the world economy will continue to emit carbon dioxide on a business-as-usual basis, pushing us all the way to the redline of 2°C and beyond—to a world in which climate change is increasingly beyond our control. In Hansen's words: "It is not an exaggeration to suggest, based on [the] best available scientific evidence, that burning all fossil fuels could result in the planet being not only ice-free but human-free."[51]

Under these conditions what is needed is a decades-long ecological revolution, in which an emergent humanity will once again, as it has innumerable times before, reinvent itself, transforming its existing relations of production and the entire realm of social existence, in order to generate a restored metabolism with nature and a whole new world of substantive equality as the key to sustainable human development. This is the peculiar "challenge and burden of our historical time."[52]

CHAPTER SIX

Making War on the Planet

A short fuse is burning. At the present rate of global emissions, the world is projected to reach the trillionth metric ton of cumulative carbon emissions, breaking the global carbon budget, in less than two decades.[1] This would usher in a period of dangerous climate change that could well prove irreversible, affecting the climate for centuries if not millennia. Even if the entire world economy were to cease emitting carbon dioxide at the present moment, the extra carbon already accumulated in the atmosphere virtually guarantees that climate change will continue with damaging effects to the human species and life in general. However, reaching the 2°C increase or the global average temperature guardrail, would lead to a qualitatively different condition. At that point, climate feedbacks would increasingly come into play threatening to catapult global average temperatures to 3°C or 4°C above preindustrial levels within this century, in the lifetime of many individuals alive today. The situation is only made more serious by the emission of other greenhouse gases, including methane and nitrous oxide.

The enormous dangers that rapid climate change present to humanity as a whole, and the inability of the existing capitalist political-economic structure to address them, symbolized by the presence of Donald Trump in the White House, have engendered a desperate search for technofixes in the form of schemes for *geoengineering*, defined as massive, deliberate human interventions to manipulate the entire climate or the planet as a whole.

Not only is geoengineering now being enthusiastically pushed by today's billionaire class, as represented by figures like Bill Gates and Richard Branson; by environmental organizations such as the Environmental Defense Fund and the Natural Resources Defense Council; by think tanks like the Breakthrough Institute and Climate Code Red; and by fossil-fuel corporations like ExxonMobil and Shell—it is also being actively pursued by the governments of the United States, the United Kingdom, China, and Russia. The UN Intergovernmental Panel on Climate Change (IPCC) has incorporated negative emissions strategies based on geoengineering in the form of Bio-Energy with Carbon Capture and Storage (BECCS) into nearly all of its climate models. Even some figures on the political left, where "accelerationist" ideas have recently taken hold in some quarters, have grabbed uncritically onto geoengineering as a *deus ex machina*—a way of defending an ecomodernist economic and technological strategy—as witnessed by a number of contributions to *Jacobin* magazine's Summer 2017 *Earth, Wind, and Fire* issue.[2]

If the Earth System is to avoid 450 ppm of carbon concentration in the atmosphere and is to return to the Holocene average of 350 ppm, some negative emissions by technological means, and hence geoengineering on at least a limited scale, will be required, according to leading climatologist James Hansen.[3] His strategy, however, like most others, remains based on the current system, that is, it excludes the possibility of a full-scale ecological revolution, involving the self-mobilization of the population around production and consumption.

What remains certain is that any attempt to implement geoengineering (even in the form of technological schemes for carbon removal) as the dominant strategy for addressing global warming, subordinated to the ends of capital accumulation, would prove fatal to humanity. The costs of such action, the burden it would put on future generations, and the dangers to living species, including our own, are so great that the only rational course is a *long ecological revolution* aimed at the most rapid possible reduction in carbon dioxide and other greenhouse gas emissions, coupled with an emphasis on agroecology and restoration of global ecosystems, including forests, to absorb carbon dioxide.[4] This would need to be accompanied by a far-reaching reconstitution of society at large, aimed at the reinstitution on a higher level of collective and egalitarian practices that were undermined by the rise of capitalism.

Geoengineering the Planet under the Regime of Fossil Capital

Geoengineering as an idea dates back to the period of the first discoveries of rapid anthropogenic climate change. Beginning in the early 1960s, the Soviet Union's—and at that time the world's—leading climatologist, Mikhail Budyko, was the first to issue a number of warnings on the inevitability of *accelerated global climate change* in industrial systems based on the burning of fossil fuels.[5] Although anthropogenic climate change had long been recognized, what was new was the discovery of major climate feedbacks such as the melting of Arctic ice and the disruption of the albedo effect as reflective white ice was replaced with blue seawater, increasing the amount of solar radiation absorbed by the planet and ratcheting up global average temperature. In 1974, Budyko offered, as a possible solution to climate change, the use of high-flying planes to release sulfur particles (forming sulfate aerosols) into the stratosphere. This was meant to mimic the role played by volcanic action in propelling sulfur into the atmosphere, thus creating a partial barrier, limiting incoming solar radiation. The rationale he offered was that capitalist economies, in particular, would not be able to curtail capital-accumulation-based growth, energy use, and emissions, despite the danger to the climate.[6] Consequently, technological alternatives to stabilize the climate would have to be explored. But it was not until 1977 when the Italian physicist Cesare Marchetti proposed a scheme for capturing carbon dioxide emissions from electrical power plants and using pipes to sequester them in the ocean depths that the word *geoengineering* appeared.[7]

Budyko's pioneering proposal to use sulfur particles to block a part of the sun's rays, now known as "stratospheric aerosol injection," and Marchetti's early notion of capturing and sequestering carbon in the ocean, stand for the two main general approaches to geoengineering—respectively, solar radiation management (SRM) and carbon dioxide removal (CDR). SRM is designed to limit the solar radiation reaching the earth. CDR seeks to capture and remove carbon to decrease the amount entering the atmosphere.

Besides stratospheric aerosol injection, first proposed by Budyko, another approach to SRM that has gained influential adherents in recent years is marine cloud brightening. This would involve cooling the earth by modifying low-lying, stratocumulus clouds covering around a third of the ocean, making them more reflective. In the standard scenario, a special fleet of 1,500 unmanned, satellite-controlled ships would roam the ocean spraying submicron drops of seawater in the air, which would evaporate leaving salty residues. These bright salt particles would reflect incoming solar radiation. They would also act as cloud condensation nuclei, increasing the surface area of the clouds, with the result that more solar radiation would be reflected.

Both stratospheric aerosol injection and marine cloud brightening are widely criticized as posing enormous hazards on top of climate change itself, simply addressing the symptoms not the cause of climate change. Stratospheric aerosol injection—to be delivered to the stratosphere by means of hoses, cannons, balloons, or planes—would alter the global hydrological cycle with enormous unpredictable effects, likely leading to massive droughts in major regions of the planet. It is feared that it could shut down the Indian monsoon system, disrupting agriculture for as many as 2 billion people.[8] There are also worries that it might affect photosynthesis and crop production over much of the globe.[9] The injection of sulfur particles into the atmosphere could contribute to depletion of the ozone layer.[10] Much of the extra sulfur would end up dropping to the earth, leading to acid rain.[11] Most worrisome of all, stratospheric aerosol injection would have to be repeated year after year. At termination the rise in temperature associated with additional carbon buildup would come almost at once with world temperature conceivably rising by 2 to 3°C in a decade—a phenomenon referred to as the "termination problem."[12]

As with stratospheric aerosol injection, marine cloud brightening would drastically affect the hydrological cycle in unpredictable ways. For example, it could generate a severe drought in the Amazon, drying up the world's most vital terrestrial ecosystem with incalculable and catastrophic effects for Earth System stability.[13] Many of the dangers of cloud brightening are similar to those of stratospheric aerosol depletion. Like other forms of SRM, it would do nothing to stop ocean acidification caused by rising carbon dioxide levels.

The first form of CDR to attract significant attention from economic interests and investors was the idea of fertilizing the ocean with iron, thereby boosting the growth of phytoplankton so as to promote greater ocean uptake of carbon. There have been a dozen experiments in this area and the difficulties attending this scheme have proven to be legion. The effects on the ecological cycles of phytoplankton, zooplankton, and a host of other marine species all the way up to whales at the top of the food chain are indeterminate. Although some parts of the ocean would become greener due to the additional iron, other parts would become bluer, more devoid of life, because they would be deprived of the nutrients—nitrate, phosphorus, and silica—needed for growth.[14] Evidence suggests that the vast portion of the carbon taken in by the ocean would stay on the surface or the intermediate levels of the ocean, with only a tiny part entering the ocean depths, where it would be naturally sequestered.[15]

Among the various CDR schemas, it is BECCS, because of its promise of negative emissions, that today is attracting the most support. This is because it seems to allow nations to overshoot climate targets on the basis that the carbon can be removed from the atmosphere decades later. Although BECCS exists at present largely as an untested computer model, it is now incorporated into almost all climate models utilized by the IPCC.[16] As modeled, BECCS would burn cultivated crops in order to generate electricity, with the capture and underground storage of the resulting carbon dioxide. In theory, since plant crops can be seen as carbon neutral—taking carbon dioxide from the atmosphere and then eventually releasing it again—BECCS, by burning biomass and then capturing and sequestering the resulting carbon emissions, would be a means of generating electricity while at the same time resulting in a net reduction of atmospheric carbon.

BECCS, however, comes into question the moment one moves from the abstract to the concrete. The IPCC's median-level models are projected to remove 630 gigatons of carbon dioxide from the atmosphere, around two-thirds of the total emitted between the Industrial Revolution and 2011.[17] This would occur on vast crop plantations to be run by agribusiness. To remove a trillion tons of carbon dioxide from the atmosphere as envisioned in the more ambitious scenarios would take up a land twice the size of India (or equal to Australia), about half as much land as currently farmed globally, requiring a supply of freshwater equal to current total global agricultural usage.[18] The costs of implementing BECCS on the imagined scales have been estimated by climatologist James Hansen—who critically notes that negative emissions have "spread like a cancer" in the IPCC climate models—to be on the order of hundreds of trillions of dollars, with "minimal estimated costs" ranging as high as $570 trillion this century.[19] The effects of BECCS—used as a primary mechanism and designed to avoid confrontation with the present system of production—would therefore be a massive displacement of small farmers and global food production.

Moreover, the notion that the forms of large-scale, commercial agricultural production presumed in BECCS models would be carbon neutral and would thus result in negative emissions with sequestration has been shown to be exaggerated or false when the larger effects on global land use are taken into account. BECCS crop cultivation is expected to take place on vast monoculture plantations, displacing other forms of land use. Yet, biologically diverse ecosystems have substantially higher rates of carbon sequestration in soil and biomass than does monocrop agriculture.[20] An alternative to BECCS in promoting carbon sequestration would be to promote massive, planetary ecological restoration, including reforestation, together with the promotion of agroecology modeled on traditional forms of agriculture organized around nutrient recycling and improved soil management methods.[21] This would avoid the metabolic rift associated with agribusiness monocultures, which are less efficient both in terms of food production per hectare and carbon sequestration.

Another commonly advocated technofix, carbon capture and sequestration (CCS), is not strictly a form of geoengineering since it is directed at capturing and sequestering carbon emissions of particular electrical facilities, such as coal-fired power plants. However, the promotion of a CCS infrastructure on a planetary scale as a means of addressing climate change—thereby skirting the necessity of an ecological revolution in production and consumption—is best seen as a form of planetary geoengineering due to its immense projected economic and ecological scale. Although CCS would theoretically allow the burning of fossil fuels from electrical power plants with no carbon emissions into the atmosphere, the scale and the costs of CCS operations are prohibitive. As Clive Hamilton writes in *Earthmasters: The Dawn of the Age of Climate Engineering*, CCS for a single "standard-sized 1,000 megawatt coal-fired plant . . . would need 30 kilometers of air-sucking machinery and six chemical plants, with a footprint of 6 square kilometers."[22] Energy expert Vaclav Smil has calculated that, "in order to sequester just a fifth of current [2010] CO_2 emissions we would have to create an entirely new worldwide absorption-gathering-compression-transportation-storage industry whose annual throughput would have to be about 70 percent larger than the annual volume now handled by the global crude oil industry, whose immense infrastructure of wells, pipelines, compressor stations and storage took generations to build."[23] Capturing and sequestering current U.S. carbon dioxide emissions would require 130 billion tons of water per year, equal to about half the annual flow of the Columbia River. This new gigantic infrastructure would be placed on top of the current fossil fuel infrastructure—all in order to allow for the continued burning of fossil fuels.[24]

A Planetary Precautionary Principle for the Anthropocene

If today's planetary ecological emergency is a product of centuries of war on the planet as a mechanism of capital accumulation, fossil-capital geoengineering schemes can be seen as gargantuan projects for keeping the system going by carrying this war to its ultimate level. Geoengineering under the present regime of accumulation has the sole objective of keeping the status quo intact—neither disturbing the dominant relations of capitalist production nor even seeking so much as to overturn the fossil-fuel industry with which capital is deeply intertwined. Profits, production, and overcoming energy poverty in the poorer parts of the world thus become justifications for keeping the present fossil-capital system going, maintaining at all cost the existing capitalist environmental regime. The Promethean mentality behind this is well captured by a question that Rex Tillerson, then CEO of ExxonMobil Corporation asked—without a trace of irony—at an annual shareholders meeting in 2013: "What good is it to save the planet if humanity suffers?"[25]

The whole history of ecological crisis leading up to the present planetary emergency, punctuated by numerous disasters—from the near total destruction of the ozone layer to nutrient loading and the spread of dead zones in the ocean, to climate change itself—serves to highlight the march of folly associated with any attempt to engineer the entire planet. The complexity of the Earth System guarantees that enormous unforeseen consequences will emerge. As Frederick Engels warned in the nineteenth century, "Let us not . . . flatter ourselves overmuch on account of our human victories over nature. For each such victory nature takes its revenge on us. Each victory, it is true, in the first place brings about the results we expected, but in the second and third places it has quite different, unforeseen effects which only too often cancel the first."[26]

In the face of uncertainty, coupled with an extremely high likelihood of inflicting incalculable harm on the Earth System, it is essential to invoke what is known as the Precautionary Principle whenever the question of planetary geoengineering is raised. As ecological economist Paul Burkett has explained, the strong version of the Precautionary Principle necessarily encompasses the following:

1. The *Precautionary Principle Proper*, which says that if an action *may* cause serious harm, there is a case for counteracting measures to ensure that the action does not take place.
2. The *Principle of Reverse Onus*, under which it is the responsibility of those supporting an action to show that it is not seriously harmful, thereby shifting the burden of proof off those potentially harmed by the action (e.g., the general population and other species occupying the environment). In short, it is safety, rather than potential harm, that needs to be demonstrated.
3. The *Principle of Alternative Assessment*, stipulating that no potentially harmful action will be undertaken if there are alternative actions available that safely achieve the same goals as the action proposed.
4. All societal deliberations bearing on the application of features 1 through 3 must be open, informed, and democratic, and must include all affected parties.[27]

It is clear that geoengineering promoted in a context of a capitalist regime of maximum accumulation would be ruled out completely by a strong Precautionary Principle based on each of the criteria listed above. There is a near certainty of extreme damage to the human species as a whole arising from all of the major geoengineering proposals. If the onus were placed on status quo proponents of capitalist geoengineering to demonstrate that great harm to the planet as a place of human habitation would not be inflicted, such proposals would fail the test. Since the alternative of not burning fossil fuels and promoting alternative forms of energy is entirely feasible, and planetary geoengineering carries with it immense added dangers for the Earth System as a whole, such a technofix as a primary means of checking global warming would be excluded by that criterion, too. Finally, geoengineering under the present economic and social system invariably involves some entity from the power structure—a single multibillionaire, a corporation, a government, or an international organization—implementing such action ostensibly on behalf of humanity as a whole, while leaving most affected parties worldwide out of the decision-making process, with hundreds of millions, perhaps billions, of people paying the environmental costs, often with their lives. In short, geoengineering, particularly if subordinated to the capital accumulation process, violates the most sacred version of the Precautionary Principle, dating back to antiquity: *First Do No Harm*.

Eco-Revolution as the Only Alternative

As an extension of the current war on the planet, a regime of climate geoengineering designed to keep the present mode of production going is sharply opposed to the view enunciated by Barry Commoner in 1992 in *Making Peace with the Planet*: "If the environment is polluted and the economy is sick, the virus that causes both will be found in the system of production."[28] There can be no doubt today that it is the present mode of production, particularly the system of fossil capital, that needs to change on a global scale. In order to stop climate change, the world economy must quickly shift to zero net carbon dioxide emissions. This is well within reach with a concerted effort by human society as a whole utilizing already existing sustainable technological means—particularly when coupled with necessary changes in social organization to reduce the colossal waste of resources and lives that is built into the current alienated system of production. Such changes could not simply be implemented from the top by elites, but rather would require the self-mobilization of the population, inspired by the revolutionary actions of youth aimed at egalitarian, ecological, collective, and socialized solutions—recognizing that it is the world they will inherit that is most at stake.

Today's necessary ecological revolution would include for starters: (1) an emergency moratorium on economic growth in the rich countries coupled with downward redistribution of income and wealth; (2)

radical reductions in greenhouse gas emissions; (3) rapid phase-out of the entire fossil fuel energy structure; (4) substitution of an alternative energy infrastructure based on sustainable alternatives such as solar and wind power and rooted in local control; (5) massive cuts in military spending with the freed-up economic surplus to be used for ecological conversion; (6) promotion of circular economies and zero-waste systems to decrease the throughput of energy and resources; (7) building effective public transportation, together with measures to decrease dependence on the private automobile; (8) restoration of global ecosystems in line with local, including Indigenous, communities; (9) transformation of destructive, energy-and chemical-intensive agribusiness-monocultural production into agroecology, based on sustainable small farms and peasant cultivation with their greater productivity of food per acre; (10) institution of strong controls on the emission of toxic chemicals; (11) prohibition of the privatization of freshwater resources; (12) imposition of strong, human community–based management of the ocean commons geared to sustainability; (13) institution of dramatic new measures to protect endangered species; (14) strict limits on excessive and destructive consumer marketing by corporations; (15) reorganization of production to break down current commodity chains geared to rapacious accumulation and the philosophy of *après moi le déluge*; and (16) the development of more rational, equitable, less wasteful, and more collective forms of production.[29]

Priority in such an eco-revolution would need to be given to the fastest imaginable elimination of fossil fuel emissions, but this would in turn require fundamental changes in the human relationship to the earth and in the relationship of human beings to one another. A new emphasis would have to be placed on sustainable human development and the creation of an organic system of social metabolic reproduction. Centuries of exploitation and expropriation, including divisions on the basis of class, gender, race, and ethnicity, would have to be transcended. The historical logic posed by current conditions thus points to the necessity of a long ecological revolution, putting into place a new system of sustainable human development aimed at addressing the totality of needs of human beings as both natural and social beings: what is now called *ecosocialism*.

PART II

Ecology as Critique

CHAPTER SEVEN

Nature

"Nature," wrote Raymond Williams in *Keywords*, "is perhaps the most complex word in the [English] language."[1] It is derived from the Latin *natura*, as exemplified by Lucretius's great didactic poem *De rerum natura* (*On the Nature of Things*) from the first century BCE. The word *nature* has three primary, interrelated meanings: (1) the intrinsic properties or essence of things or processes; (2) an inherent force that directs or determines the world; and (3) the material world or universe, the object of our sense perceptions—both in its entirety and variously understood as including or excluding God, spirit, mind, human beings, society, history, culture, etc.

In his *Critique of Stammler*, Max Weber suggested that the intrinsic difficulty of "nature" as a concept could be attributed to the fact that it was most often used to refer to "a complex of certain kinds of *objects*" from which "another complex of objects" having "different properties" were excluded; however, the objects on each side of the bifurcation could vary widely, and might only become apparent in a given usage.[2] Thus, we commonly contrast humanity or society to nature while, at the same time, recognizing that human beings are themselves part of nature. From this problem arise such distinctions as "external nature" or "the environment." At other times, we may exclude only the mind/spirit from nature.

Science and art are two of the preeminent fields of inquiry into nature, with each operating according to its own distinct principles. As Alfred North Whitehead noted in *The Concept of Nature*, natural science depicts nature as the entire field of things, which are objects of human sensory perception mediated by concepts of our understanding (such as space and time).[3] Consequently, one of the two leading scientific periodicals carries the title *Nature* (the other is *Science*). Within the Romantic tradition in art, a direct influence on modern environmentalism, nature is often perceived in accordance with notions of "natural beauty" (Percy Bysshe Shelley's skylark in his poem and William Wordsworth's Lake District). However, the validity of this concept has frequently been challenged within the field of aesthetics.[4]

As a concept, nature gives rise to serious difficulties for philosophy, encompassing both ontology (the nature of being) and epistemology (the nature of thought). Since Immanuel Kant, it has been emphasized that human beings cannot perceive "things in themselves" (*noumena*) and thus remain dependent on *a priori* knowledge, which is logically independent of experience. Within academic philosophy today, it is therefore customary either to take an outright idealist stance and thus give ontological priority to the mind/ideas or to subsume ontology within epistemology in such a way that the nature (including the limits) of knowledge takes precedence over the nature of being. In contrast, natural scientists generally adopt a materialist/realist standpoint by emphasizing our ability to comprehend the physical world directly, even if mediated by the mind. Concerned with growing ecological crises, most ecological activists today take a similar stance, implicitly stressing a kind of "critical realism," as in the work of Roy Bhaskar, that rejects both mechanical materialism (e.g. positivism) and idealism.[5]

Reflecting a similar division of views, many contemporary social scientists (particularly postmodernists) emphasize that our understanding of nature is socially or discursively constructed and that there is no nature independent of human thought and actions. For example, according to Keith Tester, "A fish is only a fish if it is socially classified as one, and that classification is only concerned with fish to the extent that scaly things living in the sea help society to define itself. ... Animals are indeed a blank paper which can be inscribed with any message, and symbolic meaning, that society wishes."[6] In contrast, while recognizing the role of thought in mediating the human relation to nature, most ecological thinkers and activists gravitate toward a critical materialism/realism, in which nature (apart from humanity) is seen as existing prior to the social world, is open to comprehension, and is something to defend.[7]

With the advent of nuclear weapons in the 1940s, the world came to the sudden realization that the relation between human beings and the environment had forever changed. The human impact on nature was no longer restricted to local or regional effects; conceivably, it extended to the destruction of the entire planet in the sense of constituting a safe home for humanity. Subsequently, modern synthetic chemicals (with their capacity, like radionuclides, to bioaccumlate in organisms and biomagnify across the food chain) and anthropogenic climate change brought the human degradation of nature to the forefront of society's concerns. Book titles such as *Silent Spring, The Closing Circle, The Domination of Nature, The Death of Nature, The Vulnerable Planet, The End of Nature, The Sixth Extinction*, and *This Changes Everything* reflect a growing state of alarm about ecological sustainability and the conditions required for human survival.[8]

Compared to earlier centuries, the question of nature in the twentieth and twenty-first century has been radically transformed. No longer is nature seen as a *direct* external threat to humanity through forces like famines and disease. Instead, emerging or threatened global natural catastrophes are viewed as the *indirect* products mediated by human action itself. We now live in what scientists have provisionally designated the Anthropocene, a new geological epoch in which humanity has become the dominant geological force, disrupting the biogeochemical cycles of the entire planet. This new reality has compelled a growing recognition of the limits of nature, of planetary boundaries, and of the limits of economic growth within a finite environment.

The meteoric rise of "ecology," along with derivates like "ecosystem," "ecosphere," "eco-development," "ecosocialism," and "ecofeminism," stems from these rapidly changing interactions between capitalism and its natural environment. The concepts of ecology, ecosystem, and the Earth System have become central both to science and to popular struggle. At times, they even displace the concept of nature.

Attempts to address the enormity of the ecological problem have, however, been complicated by a resurrection of essentialist conceptions of "human nature." By subsuming the social under the "natural," such views often downplay or altogether deny the importance of a social-historical dimension in the human interaction with nature. This outlook has recently gained ground through the Social Darwinist pronouncements of socio-biologists and evolutionary psychologists. E. O. Wilson's 1978 *On Human Nature*, for instance, professes to be "simply the extension of population biology and evolutionary theory to social organization."[9] An inevitable struggle thus arises between ecological radicals who demand that society be historically transformed to create a sustainable relation to nature and more establishment-oriented thinkers who insist that possessive individualism, the Hobbesian war of all against all, and a tendency to overpopulate are all inscribed in human DNA.[10] Accompanying this revival of biological determinism has been the presumption that capitalism is a product of human nature and of the natural world as a whole. Such views deny the historical origins of alienation. In contrast, most radicals view the alienation of nature and the alienation of society as interconnected and interdependent phenomena requiring a new coevolutionary social metabolism if the world ecology as we know it is to be sustained.

Contemporary conflicts over the relationship between nature and society can be traced to the rise of capitalism and modern science in the sixteenth and seventeenth centuries. The seventeenth-century scientific revolution witnessed the emergence—most notably in Francis Bacon, but also in René Descartes—of calls for the "conquest," "mastery," or "domination" of nature. In *The Masculine Birth of Time*, Bacon metaphorically declared: "I am come in very truth leading to you Nature with all her children to bind her to your service and make her your slave."[11] In *The New Atlantis* this ambition was tied to a program for the institutionalization of science as the basis of knowledge and power.[12] Descartes also linked it to a mechanistic worldview in which animals were reduced to machines. Following Bacon, the conquest of nature became a universal trope to signify a vague mechanical progress achieved through the development of science. Nevertheless, as Bacon made clear in his famous statement in *Novum Organum*, "Nature is only overcome by obeying her." In this view, "nature" could only be subjected by following "her" laws.[13]

The domination of nature espoused by Bacon was subjected to critique during the nineteenth century through the dialectical perspectives associated with Hegel and Marx. In his *Philosophy of Nature*, Hegel insisted that even though Bacon's strategy of pitting nature against itself could yield a limited mastery, total mastery of the natural world would forever remain beyond humanity's reach: "Need and ingenuity have enabled man to discover endlessly varied ways of mastering and making use of nature," he wrote. Nevertheless, "Nature itself, as it is in its universality, cannot be mastered in this manner ... nor bent to the purposes of man."[14] For Hegel, the drive to master nature generated wider contradictions that were beyond human control. In the *Grundrisse*, Marx treated Bacon's strategy as a "ruse" introduced by bourgeois society.[15] In his *Theses on Feuerbach*, Marx rejected essentialist views of human nature outright. Human nature, he argued, is "the ensemble of the social relations."[16] Similarly, in *The Poverty of Philosophy*, he declared that history was "nothing but a continuous transformation of human nature."[17]

In his later economic writings, Marx developed an analysis of the human relation to nature as a form of "social metabolism." The social metabolism was part of the "universal metabolism of nature," which found itself increasingly in contradiction with industrial capitalist development. With the development of industrial-capitalist agriculture, the soil was being robbed of essential nutrients (e.g., nitrogen, phosphorous, and potassium), which were being shipped hundreds and sometimes thousands of miles to the new urban centers. "Instead of a conscious rational treatment of the land as permanent communal property," Marx charged, "we have the exploitation and squandering of the powers of the earth."[18] In response, he introduced the notion of an "irreparable rift in the interdependent processes of social metabolism" imposed by the very nature of accumulation under capitalism. This break with the "eternal natural condition" underlying human-social existence, he argued, demanded its "restoration" through the rational regulation of the metabolism between humanity and nature.[19] In *Capital*, he advanced what is perhaps the most radical conception of ecological sustainability yet propounded: "From the standpoint of a higher socio-economic formation, the private property of particular individuals in the earth will appear just as absurd as the private property of one man in other men. Even an entire society, a nation, or all simultaneously existing societies taken together, are not the owners of the earth. They are simply its possessors, its beneficiaries, and have to bequeath it in an improved state to succeeding generations, as *boni patres familias*."[20]

Today, radical ecologists tend to fall into two broad camps. The first consists of those who, from a deep-ecology, radical-green, or "ecologism" perspective, simply counter Baconian anthropocentrism with ecocentric philosophies.[21] Such views retain the society-nature dualism but approach it from the side of external nature, the web of life, or some kind of spiritualized nature. This general perspective has played an important role within the ecological movement. Ecofeminist thinkers, for instance, have highlighted the link between the mastery of nature and the subordination of women (often by taking the critique of Bacon as their starting point). Nevertheless, the one-sidedness of radical-green or deep-ecology perspectives often encourages misanthropic views (especially when human population growth is seen as the principal problem) and anti-science stances, where the critical role of science in understanding ecology is misunderstood.

The second broad camp consists of those who have adopted more dialectical perspectives.[22] Here the problem of nature and society is conceived as one of social metabolism, which stands for the way in which human production in a given society mediates the relation between humanity and the universal metabolism of nature. Here the goal is to transcend the "rift in the interdependent process of social metabolism" generated by capitalism, in order to create a more sustainable form of human development—inseparable from the struggle for human equality.[23] This outlook builds critically on ecological science with its emphasis on the ontological interconnectedness of all living and nonliving things. Conflict arises between a social system geared to endless accumulation and growth and the everlasting, nature-imposed, conditions of ecological sustainability and substantive equality. It is along these lines that critical scientists, eco-socialists, socialist ecofeminists, anarchist social ecologists, and many Indigenous activists have coalesced to take a stand in defense of the earth. As Frederick Engels wrote in the *Dialectics of Nature*: "Let us not . . . flatter ourselves overmuch on account of our human victories over nature. For each such victory nature takes its revenge on us. . . . Thus at every step we are reminded that we by no means rule over nature like a conqueror over a foreign people, like someone standing outside nature, but that we, with flesh, blood and brain, belong to nature, and exist in its midst, and that all of our mastery of it consists in the fact that we have the advantage over all other creatures of being able to learn its laws and apply them correctly."[24]

CHAPTER EIGHT

Third Nature

Naomi Klein's wonderful essay on the numerous ecological implications that appear almost unconsciously in Edward Said's texts, forming part of their structural background—a perfect example of what he himself famously called a "contrapuntal reading"—demonstrates that ecological themes were always just below the surface in his work, conditioning his own sense of resistance.[1]

This is hardly surprising given Said's Palestinian heritage and his identification with the struggles there and throughout the Global South. Klein goes on to use this reading of Said on ecological imperialism to comment on the entire phenomenon of a world engulfed in Earth System crisis, moving beyond the drought-ridden Palestine to Pacific Islands being submerged by sea level rise due to climate change—from which she takes her sardonic title "Let Them Drown."

It takes nothing away from Klein's remarkable argument in this respect, indeed it only serves to reinforce it, if we go on and recognize that in his last decade, particularly in his *Culture and Imperialism*, Said was drawn directly into the ecological discussion. This should not surprise us. He was altogether too sensitive a cultural critic of imperialism to fail to discern the degree to which ecology formed the background for many of the colonial and decolonial allusions to be found in writers from Austen to Yeats. In examining the literature of anti-imperial resistance in particular, and putting this into historical context, Said became acutely conscious of ecological themes. More important, he broke through the usual discussions and offered his own unique insights in this area. Relying on Alfred Crosby's *Ecological Imperialism*, Said explained:

> Wherever they went Europeans immediately began to change the local habitat; their conscious aim was to transform territories into images of what they had left behind. The process was never-ending, as a huge number of plants, animals and crops as well as building methods gradually turned the colony into a new place, complete with new diseases, environmental imbalances, and traumatic dislocations for the overpowered natives. A changed ecology also introduced a changed political system.[2]

This ecological remaking of colonial territories in the image of the colonizer's own territory was tied to the unequal development that the imperial powers imposed on most of the world. Referring to the work of Marxian geographer Neil Smith, Said saw imperialism as culminating in a process that "universally commodifies all space under the aegis of the metropolitan center."[3] This was then justified within the geopolitical ideology of imperialism in the work of thinkers like Halford Mackinder, who saw it all as a result of national conditions of fertility, differentiated ecological zones, climates, and races.[4]

Most important in Said's discussion was his treatment of what Hegel, Marx, and Lukács called "second nature," resulting from the transformation introduced by human production, counterposed to "first nature." From a position of resistance to imperialism, Said explained, such a second nature was clearly an *imperialist second nature*. There could be no return to first nature. What was required therefore was the creation of a *third nature* that would both restore (in part) what had existed before and would transform the human relation to nature into something new:

To the anti-imperialist imagination, our space at home in the peripheries has been usurped and put to use by outsiders for their purpose. It is therefore necessary to seek out, to map, to invent, or to discover a *third* nature, not pristine and pre-historical ("romantic Ireland's dead and gone," says Yeats) but deriving from the deprivations of the present. The impulse is cartographic [a kind of remapping of the land], and among its most striking examples are Yeats's early poems collected in *The Rose*, Neruda's various poems charting the Chilean landscape, Césaire on the Antilles, Faiz on Pakistan, and Darwish on Palestine—

> *Restore to me the color of face,*
> *And the warmth of body,*
> *The light of heart and eye,*
> *The salt of bread and earth . . . the Motherland.*[5]

The restoration of the land and the ecology was a constant theme of revolutionary anti-colonialism. "One of the first tasks of the culture of resistance," Said observed, "was to reclaim, rename, and reinhabit the land. And with that came a whole set of further assertions, recoveries, and identifications, all of them quite literally grounded on this poetically projected base." In this way, he pointed to a poetic aesthetic of ecological resistance in the periphery. That it was predicated on the need for a *third nature* made the "emergence of an opposition" in the periphery at the same time the articulation of a new revolutionary ecology, an alternate relation to the earth. It meant transforming "the imperialized place" of the present into a renewed and more developed social commons.[6]

Said's powerful explanation of how "space at home" had been usurped seemed to recast the images from Marx's treatment of primary—"so-called primitive"—accumulation, and what Marx had called "usurpation of the common lands" accompanying the "expropriation of the agricultural population from the land." Said also seems to have recast Marx's treatment of the alienation of nature, in referring to how imperialism had "alienated the land" and thus "alienated people from their authentic traditions."[7]

All of this gave a more radical meaning to ecological aspirations, in which the recovery of the human connection to the earth, and therefore to labor, and to human community—as well as to past traditions—played an indispensable role in the urge to resist and create a new cultural reality. The sense of expropriation, of theft, of robbery, of alienation of the earth, and estrangement from the past, Said recognized, existed among people struggling everywhere; but this alienation was especially prevalent among those seeking to cast off the imperialist yoke. Lukács, in his discussion of "lost transcendence" in *The Theory of the Novel*, as Said pointed out, had argued that "every novelistic hero . . . attempts to restore the lost world of his or her imagination." This was a reflection of the deep alienation of nineteenth-century society, in which the rifts could not be healed.[8]

Culture and Imperialism, despite its scholarly form, was meant as a work for contemporary struggle. Although much of it focused on eighteenth- and nineteenth-century English literature, it leaped forward into a more global, more resistance-based perspective of the twentieth century. Here Said identified the ecological perils now emerging on a planetary level, and the importance of ecology in mustering the needed global revolt from below. He referred to "the immense range of global forces, including what has been called 'the death of nature,'" producing the contemporary period of crisis and change. In this context, he contended:

> The two general areas of agreement nearly everywhere are that personal freedoms should be safeguarded, and that the earth's environment should be defended against further decline. Democracy and ecology, each providing a local context and plenty of concrete combat zones, are set against a cosmic backdrop. Whether in the struggle of nationalities or in the problems of deforestation and global warming, the interactions between individual identity (embodied in minor activities like smoking,

or using of aerosol cans) and the general framework are tremendously direct, and the time-honored conventions of art, history and philosophy do not seem well suited for them.... More reliable now are the reports from the front line where struggles are being fought.... The major task, then, is to match the new economic and socio-political dislocations and configurations of our time with the startling realities of human interdependence on a world scale.[9]

There can be no doubt that what Said was calling for here was the creation of a *third nature* on a global scale, a new cultural-material-reality, reflecting a sustainable relation between human beings and the earth, and a world of substantive equality. This of course was closely related to the society of associated producers as conceived by Marx.[10]

Said knew that the human and cultural resources for this change were to emerge first in the periphery, in a process of *de-imperialism*, if humanity were to have a meaningful future at all. He had a sense of being a permanent exile, but he drew from this the personal resources for an alternative vision of human liberation. "Just as human beings make their own history," he wrote in the closing paragraph of *Culture and Imperialism*, "they also make their cultures and ethnic identities. ... Survival in fact is about the connections between things."[11] It was necessary finally to heal the rifts—social, ecological, and cultural—in our disconnected world.

CHAPTER NINE

Weber and the Environment

In the last two decades classical sociology, notably classical Marxian theory, has been mined for environmental insights in the attempt to surmount the "human exemptionalism" of post–Second World War sociology. Max Weber, however, has remained an enigma in this respect. This article addresses Weber's approach to the environment, including its significance for his interpretive-causal framework and his understanding of capitalism. For Weber, sociological meanings were often anchored in biophysical realities, including climate change, resource consumption, and energy scarcity, while environmental influences were refracted in complex ways within cultural reproduction. His work thus constitutes a crucial key to constructing a meaningful post-exemptionalist sociology.

Environmental Sociology and the Enigma of Weber

Environmental sociologists have long seen ecological issues as consigned to the wilderness within sociological thought. In the first two decades following its organization as a field in the late 1970s, environmental sociology was largely defined by a persistent critique of sociology as a whole for its "aversion to the natural environment."[1] In the most influential expression of this by Catton and Dunlap, the dominant post–Second World War sociological tradition was seen as having embraced a human-exemptionalist paradigm, in which human beings in technologically advanced societies were considered exempt from natural-environmental influences.[2] An unfortunate consequence of the dominance of this human-exemptionalist paradigm, they argued, was the relative impermeability of mainstream sociology to serious environmental concerns. This led to a call for a new environmental paradigm (now sometimes referred to as the "post-exemptionalist paradigm") denying such human-exemptionalist notions.[3]

These environmental criticisms of late twentieth-century mainstream sociology were often carried over, though much more ambivalently, to the classics themselves. Environmental sociologists saw sociology as a discipline having been organized around the "'social facts' injunction," identified with Durkheim in particular, which had systematically cordoned off the realm of the social from that of the biophysical in an attempt to distinguish sociology from biology and psychology. Weber similarly had criticized social evolutionism for its elevation of biological metaphors to the level of sociological concepts, warning against drawing crude social analogues with natural evolution.[4] Likewise Marx's sharp critique of Malthus was frequently seen as a rejection of biophysical influences.[5] Classical sociologists were thus often viewed as having systematically excluded biophysical issues from their core concerns. Environmental sociologists were therefore estranged not only from twentieth-century sociology but also to a considerable extent from the founding traditions of the discipline.

Today there are signs that environmental sociology's long period in the wilderness may be coming to a close. Not only have environmental issues been gaining considerable currency in various fields within sociology, such as world-systems theory, critical theory, cultural sociology, and so on, but they have increasingly been acknowledged within sociology as a whole over the last decade, with prominent articles in the leading general sociological journals.[6] More significantly for the theoretical development of the field, perhaps, environmental sociologists have been engaged since the late 1990s in reconceptualizing the foundations of sociology to take into account green issues, attempting to construct, in this way, a post-exemptionalist sociology. Marx and Durkheim in particular, and, in a much more limited and indirect way, Weber, have been reexamined for evidence of the environmental aspects of their thought.

The most decisive break in this respect arose in relation to Marx. Beginning in the 1990s a systematic reconsideration of Marx's environmental contribution has been under way, centered in sociology.[7] At the same time, the controversy over whether Durkheim's sociological approach created a theoretical blind spot with respect to the environment has led to important discussions of the ecological nexus of his world—focusing on how his modified evolutionism contributed to the development of human ecology.[8]

Yet Weber's work, in contrast, has remained an enigma within environmental sociology. Patrick West first systematically and positively assessed Weber's work from an environmental-sociological standpoint in a dissertation written in the mid-1970s.[9] A book chapter by West based on his thesis appeared a decade later.[10] But West's writings in this area (and particularly his dissertation), composed before the organization of environmental sociology as a field, are almost completely unknown.[11]

Currently, the most prominent work related to Weber and the environment is that of Raymond Murphy, who has provided a neo-Weberian approach to environmental sociology.[12] However, this neo-Weberian perspective was itself predicated on the critical view that "the relation between social action and the processes of nature" was something that "Weber himself did not examine in any detail."[13]

Weber has thus attained an enigmatic, even paradoxical, status within environmental sociology. On the one hand, it has been argued by leading environmental sociologists that "Weber's relation to environmental sociology is the least controversial or problematic of the legacies of the 'big three.'"[14] On the other hand, these same thinkers went on to contend in the very same piece that Weberian contributions (including those of Weber himself) to the development of environmental sociology have been "relatively invisible."

Indeed, claims that "Weber had little to say about the natural environment per se" are commonplace within the literature.[15] "Max Weber," Frederick Buttel observed, "is almost never thought of as an ecological theorist."[16] "Of the classical trinity," Goldblatt contended, "Weber's work conducts the most limited engagement with the natural world."[17] Ted Benton went so far as to declare that the very "oppositions between action and behaviour, meaning and cause, interpretation and explanation" that characterize Weber's interpretive sociology have imposed "an impenetrable barrier to any project for a comprehensively naturalistic (i.e., biologically rooted) approach to the human sciences."[18] Benton and Redclift declared that Weber's sociological theory was characterized throughout by "space-time indifference," making it immune to environmental influences.[19]

Others have argued that, for the more interpretive strand of sociological thought, emanating from Weber in particular, "the reality of a situation" lies "in the definition attached to it by the participating actors," with the implication that "the physical properties of the situation" might be "ignored."[20] Based on this, Riley Dunlap stated that while "the Durkheimian antireductionist legacy suggested that the physical environment *should* be ignored ... the Weberian legacy suggested that it *could* be ignored."[21]

Yet, for all of this, Weber's broad contribution to environmental thought is not to be denied. In West's argument, "Weber did not self-consciously develop an explicit ecology theory or perspective. But a comprehensive analysis of the role of ecological factors is implicit in [his] historical and comparative studies ... [which] provide rich contributions to a sociological human ecology."[22] Robert J. Antonio recently declared: "Although Weber was no ecologist, he grasped the tension between capitalist growth and the environment."[23] In his new biography of Weber, Joachim Radkau referred at one point to Weber's "social ecology."[24]

Given the comparative neglect of Weber's environmental contributions within environmental sociology, it is ironic that probably the best-known statement by a sociologist referring to environmental factors is to be found in his famous declaration in *The Protestant*

Ethic and the Spirit of Capitalism: "This [modern economic] order is now bound to the technical and economic conditions of machine production which today determine the lives of all the individuals who are born into this mechanism . . . with irresistible force. Perhaps it will so determine them until the last ton of fossil fuel is burnt."[25] One could of course view this as a mere rhetorical flourish, unrelated to any substantive concern with the environment. Yet both the existence of "natural limits" on production and the "heedless consumption of natural resources," particularly "coal and ore," were important themes in his overall *Weltanschauung*.[26] Another concern of Weber's was the robbing of the soil.[27] Indeed, his critical view of naturalism/positivism did not prevent him from stating that "it is entirely proper" for sociology as a discipline "to take into account the physical and chemical balance sheets" of energy and natural resources.[28] Within ecological economics Weber's contributions to the sociology of energy are well recognized, though this has only rarely penetrated into sociology itself.[29] Perhaps the most startling indication of Weber's environmental perspicacity, from today's perspective, was his emphasis in *Economy and Society* and elsewhere on adaptation to "climatic changes" as of crucial importance in the history of human development.[30]

Nevertheless, understanding the role that environmental factors played in Weber's thought constitutes a considerable conceptual challenge for sociological theory (and environmental sociology in particular). Although Weber, "unlike his contemporary, Durkheim," Albrow remarked, "had no reluctance to admit the causal significance of non-social factors for social processes," the way in which this fits into his interpretive sociology and theory of rationalization still remained to be explained.[31] As Martin Albrow stated with respect to Weber's concept of rationality: "Population trends, resource limitations, health factors . . . all provide either the boundaries or the material for rational action but are outside the prescriptive rules of rationality."[32] From this perspective, then, the key to applying a Weberian approach to biophysical conditions appears to be understanding how "rationality and irrationality are locked in a dialectical embrace."[33] Before Weber's environmental insights can be addressed directly it is thus necessary to explain the theoretical status of environmental factors within his interpretive sociology and causal analytics.

What emerges from such an investigation, we will contend, is a much wider conception of the systematic character and richness of Weber's *verstehende Soziologie*—including the complex causal analysis associated with Weber's overall approach to comparative-historical change.[34] Weber's sociology can be seen as striving constantly for a balance between causes and interpretation, biophysical dynamics and meanings, nature's constructions and society's constructions, the material and the cultural. Perhaps nothing so clearly illustrates the complex, interactive character of Weber's thinking than the attention he devotes to environmental influences and how they are refracted within cultural forms.

Indeed, environmental discussions play a large, though far from determining, role in Weber's comparative-historical analysis of why societies came to differ from one another and, specifically, in his analysis of the origins and development of capitalism. A key element in the rise of industrial capitalism, he makes clear, was the discovery of the process of coking coal, without which industrialism in the modern sense would have been virtually impossible. Weber's analysis of the environmental conditions of capitalism, in fact, places heavy emphasis on the energy-intensive and fossil-fuel-intensive nature of the system, which could eventually place limitations, he suggested, on its further development. Weber was thus perhaps the first thinker to underscore the way in which a particular energy regime both enabled and constrained the development of capitalism. Indeed, Weber depicted capitalism at various points in his work as a major driver of environmental change, with notable repercussions for the future of society.

Weber's Interpretive Sociology and the Environment

Although environmental conditions often seem to stand outside Weber's sociology operating as external parameters, a more accurate way of characterizing his approach in this respect, as West observed, would be in terms of significant environmental-social interactions. What needs to be explained, however, is how this was integrated with the interpretive structure and the causal analytics of Weber's theoretical methodology.

In "Some Categories of Interpretive Sociology," Weber usefully observed: "The relevance for interpretive sociology of processes devoid of subjective 'meaning' [such as environmental factors]... lies exclusively in their role as 'conditions' and 'consequences' toward which meaningful action is oriented, just as climatic or botanical conditions are relevant for economic theory."[35] However, once such environmental conditions and consequences have entered into human history and are no longer "devoid of subjective meaning," they are no longer mere external causes and consequences but become a part of cultural life. Structures of meaning and causal connections create a complex intellectual framework in which significant natural-environmental events are "anchored" in cultural-historical processes.

Thus in one of his principal methodological works, *Roscher and Knies*, Weber sought to account for the interrelationship of physical-environmental factors and the complexes of causality and meaning that characterize social life, using the example of the Black Death/bubonic plague.[36] "Was the meaning of the Black Death for social history," he rhetorically asked, "'contained' in the bacteria and the other causes of infection?" The answer was obviously no. What made the Black Death socially meaningful was that it contributed to "historically *significant* consequences anchored in our 'cultural values.'"[37]

Drawing on the views of the influential German psychologist and philosopher Wilhelm Wundt, Weber insisted that the cultural reproduction of environmental events included "new properties," not reducible to environmental conditions, in which they were anchored.[38] As Weber put it, "The *meaning* we ascribe to phenomena [environmental or otherwise]—that is, the relations which we establish between these phenomena and 'values'—is a logically incongruous and heterogeneous factor which cannot be 'deduced' from the 'constitutive elements' of the event in question." The chief significance of the Black Death for the cultural domain lay not in the "discovery of laws, e.g., bacteriological laws," but rather the way "we ascribe historical 'meaning'" to it as an event.[39]

In order to make this methodological point clear, Weber referred a number of times, in both *Roscher and Knies* and *Economy and Society* to "the incursion of the Dollart [Dollard]" in the medieval and early modern Netherlands (near the Dutch-German border)—with storm floods leading to breaks in the sea defenses, massive loss of life and land, and the migrations that resulted from this. The cultural results, he insisted, were not "'contained' in the geological and meteorological causes which produced this phenomenon." Rather such geological and meteorological events (like the bacteriological event represented by the Black Death) end up "anchoring" cultural history, insofar as they enter into human action and meaning.[40] What Weber called "the discursive nature of our knowledge" of the social, cultural, and historical sciences is thus not infrequently attached to environmental events, which become part of the content of the cultural realm, incorporated into "the causal explanation of cultural-historical 'facts.'"[41]

Another way of looking at this is in terms of Weber's use of the concept of "refraction," in which interests (material or otherwise) are seen as being refracted within cognitive culture.[42] The importance of "refraction" (or a "refractive effect") as a concept in understanding Weber's methodology was emphasized by Warner and Smelser, while West applied it directly to Weber's analysis of environmental-cultural linkages.[43] As Smelser and Warner indicated, "Refraction suggests the contingent or switching function of ideas that was of great importance to Weber."[44] In this way, they associated it with Weber's famous metaphor of the "switchman," whereby ideas "switch" the route taken—altering the original direction derived from other more elemental forces—and thus end up becoming forces in themselves.[45]

Here we use the concept of refraction in a somewhat different but related way in order to indicate the manner in which environmental causes are refracted through a cultural lens. What were originally material-environmental influences assume an altered form (refracted or bent as in light) and take on new content within the realm of cultural meanings and social interaction.

For example, in *Ancient Judaism,* Weber argued that due to harsh environmental conditions, Bedouins and semi-nomads were caught in a "selective struggle for existence," which favored certain cultural forms. Their lives revolved around camel breeding and control of oases and trade routes. This outcome was thus refracted in a complex way within cultural (including religious and political) institutions.[46]

This whole understanding of the cognitive refraction of environmental causes/interests within the cultural prism of a historical society thus constitutes the primary conceptual basis for Weber's approach to the environmental-cultural nexus. "The forces of nature become an intellectual problem," Weber wrote, "as soon as they are no longer part of the immediate environment"—that is, as soon as they are viewed at a distance, through a process of objectification.[47] In this way significant aspects of the environment become intellectualized and part of the cultural domain and are given specific historical meaning but only through the objectification of nature itself.

It was perfectly rational from an environmental perspective, Weber suggested, to seek to ascertain "which specific concrete elements in the particular cultural phenomena are determined by climate or similar geographical factors."[48] Weber had no doubt that environmental factors had a causal impact on human culture and vice versa. As Stephen Kalberg states, "Weber viewed *geography* [environment] as not only capable of setting distinct parameters to social actions—ones that, moreover, could remain effective over long periods of time—but also as itself constituting a causal force."[49] Yet, equally important was how such environmental factors, if they came to bear on a culture, were then refracted in complex ways within the culture itself.

The complex, interpenetrating causality here, with environment and culture seen as mutually determining, was crucial to Weber's overall perspective, as was his emphasis on the confrontation of reason (interpretation) and reality (empirical causes). Sociology, he stressed, was not to be conceived as an "empirical science of concrete *reality,*" but rather as the "confrontation of empirical reality with the ideal-type."[50] It was therefore both interpretive and causal-analytic. Perhaps nowhere else is this complex framework of Weber's thought more evident than in his understanding of the environmental-cultural interface.

The epistemological sophistication of Weber's treatment of the environment can be seen in his extraordinarily nuanced analysis of the concept of "nature" in his *Critique of Stammler*. "In ordinary discourse," Weber writes:

> The word "nature" is used in several ways. (1) Sometimes it refers to "inanimate" nature. (2) Or sometimes it refers both to "inanimate" nature and to all "organic" phenomena that are not distinctively human. (3) Or sometimes it refers to both these objects and, in addition, to those organic characteristics of a "vegetative" or "animal" sort which men and animals share. . . . In each of these three senses . . . nature is invariably conceived as a complex of certain kinds *of objects,* a complex that is distinguished from another complex of *objects* which have different properties.[51]

Based on such close scrutiny of the concept, Weber insisted on the fundamental "ambiguity of the concept of 'nature.'"[52] However, the theoretical import of this ambiguity led not, in his view, to the rejection of the concept itself, if properly handled, but rather to the rejection of "naturalistic" or positivistic attempts to cordon off "nature" from society. What he objected to especially was the attempt to construct "an absolutely strict and mutually exclusive conceptual distinction between the object's 'nature' and 'social life.'"[53]

Weber dealt with the complexity and ambiguity of nature in *The Religion of India,* where he wrote: "Before the cosmos of nature we think: it must still—be it to the analyzing thinker, be it to the observer contemplating the total picture and its beauty—have some sort of 'last word' to say as to its 'significance.' . . . Whether there is such a 'last word' as to the meaning of nature is a metaphysical indeterminable."[54] In other words, the overriding significance of nature was not to be doubted. But the cognitive domain mediated its cultural impact. Here Weber expressed his epistemological sophistication, in neo-Kantian terms. Nature in its pure state, or the realm of the noumena (the Kantian "thing in itself"), was unknown and unknowable; nevertheless, human sense perception allowed us to explore empirical phenomena as mediated by the categories of the understanding and human reason.[55]

Steeped as he was in neo-Kantian epistemology, Weber saw the "conflation of laws of nature and 'categories'" (of understanding) as philosophically naive.[56] Nevertheless, this did not exclude realism of a more crucial kind. Nature was both something external to society (first nature), and in that sense not entirely knowable—that is, in its pure form as "the thing in itself," independent of human cognitive powers. At the same time, it was part of society/culture (second nature), where it was interwoven with cultural meanings. Here nature becomes truly part of the human world. As Weber put it, "the outside world which is relevant for economic theory may in the particular case be 'nature' (in the sense of ordinary language [that is, first nature]) or it may be 'social environment' [second nature]."[57] Second nature was a hybrid "man-made product" interpenetrating with society.[58]

Weber's insistence that what are often taken to be the impermeable barriers between the biophysical/natural and cultural/sociological realms governing human action are actually quite porous is made explicit in his *Critique of Stammler* with the example of Robinson Crusoe, as depicted in Defoe's novel. Weber objected to Stammler's contention that the actions of Robinson Crusoe on his island, since they were carried out by an isolated individual in relation to his environment, were merely "natural" and "technical" and thus could be relegated to the realm of natural science rather than social science. Rather, for Weber the constellation of causes governing what Crusoe did on his island was both environmental and social, while the meanings attached to the environment were social and thus belonged to the domain of sociology. Thus Weber pointed out that if Robinson Crusoe, concerned with the "reforestation" of his island, were to choose to make certain "marks" on trees, this is a social meaning (the legacy of the society from which he came) that reflects the complex interpenetration of environmental and social causes. It thus lies within the social realm, as well as being related to factors outside of it—that is, ecological conditions.[59]

Weber's interpretive approach, combined with what Kalberg has called his "radical multicausality," formed the basis of his interpretive-causal approach to environmental issues, that is, the contingent anchoring of the cultural in the biophysical and vice versa, so often revealed in his comparative-historical study of society.[60] This complex cultural refraction of environmental causes within social meaning/interpretation and multicausality is evident to varying degrees in all of his major comparative-historical works: *The Agrarian Sociology of Ancient Civilizations, Ancient Judaism, The Religion of China, The Religion of India, The Protestant Ethic and the Spirit of Capitalism,* and *The General Economic History.* It is also present at certain points in *Economy and Society.*

In line with what we take to be the general thrust of Weber's environmental-sociological contributions, it is possible to designate two broad, comparative ideal-typical social epochs, corresponding to different phases of history/modernization: (1) traditional-organic and (2) rational-inorganic. Thus the analysis of Weber's chief environmental insights in what follows will be divided into two parts, reflecting these two phases of cultural-material development. For Weber, the traditional-organic phase can be seen as encompassing a wide variety of pre-industrial-capitalist societies; while the rational-inorganic phase is associated with the rise of industrial capitalism. As we shall see in the following discussion, it is the reliance on "inorganic" sources of energy (fossil fuels), along with energy-intensive and high-resource consumption, that, for Weber, distinguishes the environmental context of industrial capitalism. In this conception, capitalism emerges as the major driver not only of the rational-inorganic phases of development but also of growing natural-resource constraints.

The Traditional-Organic Era in Human History: The Environment and Non-Industrial Society

The ideal-typical distinction between the traditional-organic period in world history, governed by natural cycles, and the rational-inorganic world, in which the "organic cycle of simple peasant existence" no longer dominates human awareness, was a thread running throughout Weber's work.[61] He saw the "rational systematization" (and disenchantment) of the "total life pattern" as antithetical to "the lot of peasants," which was "so strongly tied to nature, so dependent on organic processes and natural events."[62]

These observations on the dissolution of traditional-organic life were closely linked to the notion that rational industrial capitalism depended on "substituting inorganic raw materials and means of production for organic raw materials and labor forces."[63] Such liberation from natural limits was, however, only possible under specific historical conditions that would not persist.

Given that Weber saw the role of environmental factors taking on quite different meaning for society in the traditional-organic and rational-inorganic (or nonindustrial and industrial) eras, his historical inquiries were divided into these two periods—conceived as ideal-typical generalizations intended to guide our inquiries into empirical history. In terms of his major substantive historical works this means that such studies as *Ancient Judaism, The Religion of China, The Religion of India,* and *The Agrarian Sociology of Ancient Civilizations,* as well as most of the first three parts of the *General Economic History,* relate primarily to society at a time when traditional-organic relations were, in his view, predominant; while the later parts of *The General Economic History* and *The Protestant Ethic and the Spirit of Capitalism* belong to the rational-inorganic era.

In relation to the traditional-organic era, Weber thus explores a wide variety of environmental-cultural relations, including the effect of climate on religion in Palestine; the role of hydraulic bureaucracies in Mesopotamia, Egypt, and China; the effects of rain-fed agriculture in Europe; and the deforestation associated with early industrialization (and the smelting of iron with charcoal) primarily in Britain. In relation to the rational-inorganic era, he discusses the "fateful unity" of coal and iron; the robbery of the land by capitalist agriculture; the destruction of the organic cycle of life; the sociology of energy; and the rationalization and "disenchantment of the world"—all of which were to be exemplified by the United States.

Ancient Judaism and Climatic Conditions

Weber's comparative approach to environmental-cultural interactions is most explicit in his *Ancient Judaism,* which offers what Radkau has called a "social ecology of the Jewish religion."[64] This work begins with a consideration of general historical and climatic conditions. For Weber, Palestine and the surrounding regions offered a laboratory with respect to environment-cultural relations. Ancient Palestine in the period from the settlement of Israel to the Division of the Monarchy (from approximately the thirteenth to the tenth centuries BCE)—lay precariously between the two great civilizations of Mesopotamia and Egypt, both of which intruded on its history. The nearness to Egypt raised the question as to why Egyptian culture had not penetrated more deeply into Judaic beliefs. Weber explained this as mainly due to "profound differences in natural environmental conditions" underlying the social orders. "The Egyptian corvée state, developing out of the necessity of water regulation and the construction works of the kings," was seen by the "inhabitants of Palestine as a profoundly alien way of life." Thus the separation of the two realms was "based on natural and social differences." Just as ancient Egyptian culture was the refractive effect of the environmental conditions of the Nile, cultural life in ancient Palestine was the refractive effect of rain-fed agriculture and stock breeding.[65]

Palestine itself afforded "important climatically determined contrasts in economic opportunities."[66] These varied from fairly settled or semi-settled peasant agriculture and stock breeding of goats, sheep, and cattle on the mountain slopes and plains to the nomadic existence characteristic of Bedouin tribes in the marginal and desert lands to the east and south. Given that irrigation-based agriculture was limited, peasant farmers, and even more so herders who engaged in stock breeding in the mountainous areas, were dependent on rainfall, which varied dramatically, seasonally and annually. The entire region was prone to numerous natural depredations, including violent storms, which eroded the sandy soil, and droughts. During droughts, herders purchased grain from Egypt or were forced to migrate. Life was therefore, in Weber's words, "meteorologically precarious."[67] He so much identified the social, cultural, and economic developments of the tribes of Israel with the land from which they sprang that when describing the eventual coalition of peasants and herdsmen against urban patricians he wrote: "With slight inaccuracy one might say: it was the struggle of the mountain against the plain."[68]

Climatic variance also created sharp cultural differences between the Hebrew tribes and the surrounding Bedouins, located mainly to the south and east, where the "sterile desert . . . has been and is a place of horror and demons."[69] "Naturally given contrasts in economic conditions," he observed, "have always found expression in differences of the social and economic structure." The desert Bedouins, as distinct "from the settled Arab," were nomadic "tent-communities," without any kind of real state organization, engaged in camel breeding and occupying oases and caravan routes.[70]

Environmental factors within Palestine and the surrounding regions were refracted in Judaic religious doctrines, which showed strong evidence of the natural conditions in which they arose. To illustrate some of the religious implications, Weber contrasted Yahweh with the god Baal (standing in fact for numerous local deities). Like the "Babylonian god, Bel, Lord of the Fertile Soil," the Palestinian Baal was a fertility god attached to the earth, "lord of the land, of all of its fruits."[71] Juxtaposed to this, Yahweh, worshipped by the Jews, was primarily a god from afar—a "rain god," a god of thunderstorms, and a "war god." Yahweh showed his "sovereign might and greatness in the events of nature." Indeed, "he was originally a god of the great catastrophes of nature." The biblical stories of military victory, such as the parting of the Red Sea and the devastation of the Egyptian armies, were viewed by Weber as likely emanating from natural catastrophes (ebb-tide, volcano, etc.) which were then refracted in particular religious beliefs: in Yahweh as a god of wrath. "This historically [and climatically] determined peculiarity of God [Yahweh]" was "fraught with consequences [extending] into times when the early Christian doctrine of natural law emerged."[72]

Although there are places in *Ancient Judaism*, as Radkau has noted, that appear to point to a kind of "ecological determinism," the predominant notion is that "natural conditions do not determine forms of human life but contain several different opportunities: instead of ecological determinism, then, a possibilism that corresponds to our present state of knowledge."[73]

Hydraulic Bureaucracy

The best-known, but also most controversial, of Weber's treatments of environment-culture interactions is his discussion of hydraulic civilizations in Asia. Weber drew on a set of prevailing theses on "Oriental Despotism," the Asian mode of production, and hydraulic society. Although such notions were central to much of nineteenth and early twentieth-century European thought, particularly Marx, Weber, and Karl Wittfogel, they are largely rejected today.[74]

The idea that the dependence of Asian agriculture on the construction and regulation of navigable canals and irrigation systems led to extensive public works and systems of centralized, state-bureaucratic power was first suggested by Adam Smith and John Stuart Mill and adopted by Marx in June 1853, in an article for the *New York Tribune:*

> Climate and territorial conditions, especially the vast tracts of desert, extending from the Sahara through Arabia, Persia, India and Tartary to the most elevated Asiatic highlands, constituted artificial irrigation by canals and waterworks [as] the basis of Oriental agriculture. As in Egypt and India, inundations are used for fertilizing the soil of Mesopotamia, Persia, etc. . . . Hence an economical function devolved upon all Asiatic Governments, the function of providing public works. This artificial fertilization of the soil, dependent on a Central Government, and immediately decaying with the neglect of irrigation and drainage, explains the otherwise strange fact that we now find whole

territories barren and desert that were once brilliantly cultivated, as Palmyra, Petra, the ruins in Yemen, and large provinces of Egypt, Persia and Hindustan.[75]

Marx was later to expand this interpretation in the *Grundrisse, Capital,* and his *Ethnological Notebooks* into a larger theory of the "Asiatic mode of production"—a term, however, that he used only one time in 1859.[76] In *Capital* Marx briefly discussed the role of irrigation in "the domination of the priests as the directors of agriculture" and the way in which this was related to the development of astronomy and the management of agricultural systems.[77] In *Anti-Dühring* Engels returned to the original hydraulic civilization notion, which, except for this brief mention in *Capital,* had been deemphasized by Marx for twenty years.[78] Lawrence Krader has conceptually divided Marx and Engels's treatment of the Asian mode into 24 separate elements, with the hydraulic civilization element as only one of these.[79] It is clear that the central purpose of the concept of the Asian mode of production in Marx's theory was to provide a comparative-historical explanation for why capitalism had not developed in Asia as in Europe. In doing so he ended up focusing primarily on the issue of the village community as opposed to hydraulics.[80]

But for some later social theorists the hydraulic civilization argument was to loom particularly large: most notably in the writings of Weber and Wittfogel. It was Wittfogel, going beyond both Marx and Weber, as Krader has noted, who "made the hydraulic interpretation of the Oriental society into the central one," leading to a "hypostatization of water control" in what amounted to an environmentally determinist argument.[81] Today scholars have abandoned this view as based on faulty Eurocentric preconceptions.[82]

Weber's own approach to the analysis of Asian societies, though not beyond reproach from the present-day standpoint, was complex, multicausal, and based on varied sources. Nevertheless, central to much of his analysis was the development of what he called "'hydraulic' bureaucracy" (sometimes referred to as "irrigation bureaucracy"), which he incorporated as a central component in his overall comparative cultural interpretation.[83] Here Weber focused on the need in Mesopotamia, Egypt, China, and Ceylon, and to a lesser extent India—viewed as great river civilizations existing within arid or semi-arid climates—for extensive engineering works related to irrigation, canals, dams, and dikes. This led in turn to state bureaucracies and royal power.[84] For Weber most ancient civilizations, particularly in the East, were "riparian in character."[85]

The most obvious, and at first sight perplexing, feature in Weber's claims about hydraulic bureaucracy in Asia is the seemingly strong causal determinacy of many of his statements. Thus he claimed that, in Mesopotamia, Egypt, and much of China, irrigation was an absolute "necessity" imposed by an arid or semiarid environment, a question of winning land back from the desert. The "Mesopotamian and Egyptian subject," Weber noted, "hardly knew rain."[86] The lack of rainfall led directly to a bureaucratic state with irrigation as its "prerequisite."[87] In *Economy and Society* he wrote: "The *necessity* of river regulation and an irrigation policy in the Near East and Egypt, and to a lesser degree also in China, *caused* the development of royal bureaucracies."[88] Elsewhere in the same work he asserted: "In Mesopotamia irrigation was *the sole source* of the absolute power of the monarch."[89] In *The Religion of China*, he was equally emphatic: "Political subjection to princely power was *determined by* river control [in China] in the manner of Egypt and the Middle East."[90] In *The Religion of India* he insisted that Ceylon's "kingship [was] *based upon* a magnificent irrigation system."[91] Such statements raised the issue of what Radkau has called the "paradoxical ecological determinism" that occasionally seemed to appear in Weber's writing.[92]

However, despite such strong, deterministic-sounding statements, which showed the causal importance he placed on these environmental factors, it would be a serious mistake if one were to interpret him as a rigid thinker in this respect. Weber should not be confused, as some have done, with Wittfogel.[93] For Weber, as we have seen, environmental causes never gave rise to a simple determinism in which an environmental event is adequate to produce a particular cultural result. Rather, such material causes were refracted in complex ways within a given culture. Hence, the somewhat exaggerated statements on the role of environmental factors in the development of state formation in Asia arose not from determinism as such but, rather, from the comparative-historical perspective underlying his studies in the sociology of religion. Contrasting ideal types were being drawn between two different forms of civilizational rationality, attributable in part to varying environmental influences, distinguishing Asia, where rainfall was sparse and irrigation necessary, and Europe (and Palestine), where rain-fed agriculture was common. Thus Weber compared the "relatively individualist activity of clearing virgin forest" in the rain-fed agriculture of Europe to the state-dominated building of irrigation canals in Mesopotamia, Egypt, central and southern China, and Ceylon.[94] Despite deterministic-sounding statements with respect to hydraulic civilizations, there is no doubt, particularly if entire texts are examined, that Weber's understanding of the complex chain of cultural meanings through which such conditions were refracted was a multicausal one.

European Rain-Fed Agriculture, Forest Clearances, and Landholdings

In Weber's comparative-historical conception of European development such key geographical factors as the "position of the Mediterranean as an inland sea, and the abundant interconnections through the rivers, favored . . . the development of international commerce" and mercantilist development in Europe, as opposed to the "decisively inland commerce" of China and India. Nevertheless, industrial development in Europe was to occur not on the seacoasts but in the interior regions, once cleared of forests. "Capitalism in the West," he wrote, "was born in the industrial cities of the interior, not in the cities which were centers of sea trade."[95] In Weber's overall comparative-historical perspective on East versus West, "the rationalization of the irrigation economy in the ancient Orient" was anchored in the state-patrimonial bureaucracy. "By contrast, acquisition of new land through the clearing of forests in Northern Europe favored the manorial system and therefore feudalism."[96]

As in John Locke's theory of property, the clearing and cultivation of the earth converted it into landholdings.[97] Thus Weber defined "land" as opposed to the earth or soil as a social artifact created "by virtue of clearing or irrigation."[98] For the "Oriental economy—China, Asia Minor, Egypt—irrigation husbandry became dominant, while in the West where settlements resulted from the clearing of land, forestry sets the type."[99] Forest clearings to increase cultivable land therefore constitute an integral part of Weber's theory of agricultural and community development.

In discussing the role of forest clearings in generating the "economic *milieu*" of Germanic agricultural development, Weber explained that land settlement in the Germanic region took the village form.[100] These villages were associated with a very large tract of land called the "mark," which included wood and wasteland as commons. There was a head official of the mark, usually preempted by the king or lord, and a "wood court" representing those that originally had equal land allotments associated with the various communities.[101]

The rise of the manor and seignorial property increased demand for servile labor to further land appropriation through forest clearings. The lords of the manors "regularly appropriated to themselves the common mark and often the common pasture."[102] The great Peasant War in Germany, beginning in 1525, was waged against this usurpation, with the peasants demanding free access to woodlands and pasture. These, however, "could not be granted as the land had become too scarce, and fatal deforestation would have resulted as in Sicily."[103]

As markets emerged for agricultural goods, and the commercial interests of the bourgeoisie developed, the manor system, which was "originally directed toward using dependent land and dependent labor force to support an upper-class life," gave way to the two forms of plantation and estate.[104] "With the dissolution of the manors and of the remains of the earlier agrarian communism through consolidation, separation, etc., private property in land" was established, and much of the population permanently dislocated. This transformation of the countryside "was bound up with the development of industry and trade."[105] These changes associated with nascent capitalist development "disrupted the 'natural' rhythms of pre-modern means of production and consumption in the traditional household."[106] Cooperative village agriculture (the old German mark) "bound to place, time and organic means of work" was completely dissolved, as the epoch of wood gave way to the age of iron and coal, associated with the transition to industrial capitalism.[107]

Deforestation: From the Epoch of Wood to the Age of Iron

For Weber, a revolutionary transformation in the role of forests, setting off a deep-seated ecological crisis, played a critical role in the transition to industrial capitalism. In the precapitalist period, land was cleared primarily to advance agriculture or enlarge the landholdings of the lord. Now suddenly forests and the land in general were sites of accelerated resource extraction necessary to feed industry. "Capitalism," Weber wrote, "extracts produce from the land, from the mines, foundries, and machine industries."[108]

The mercantilist period in the sixteenth, seventeenth, and early eighteenth centuries saw rapid deforestation in Europe, and particularly in Britain, where the smelting of iron with charcoal intensified demand for wood. This was the great ecological crisis that occurred at the very moment that Europe was on the verge of an industrial revolution. As Weber put it: "Until the 18th century the [iron smelting] technique was determined by the fact that smelting and all preparation of iron was done with charcoal. The deforestation of England resulted. . . . Everywhere [where industrialization was taking place] the destruction of the forests brought the industrial development to a standstill at a certain point," threatening the nascent industrial takeoff.[109] "However energetic landowners and farmers might be in afforestation," historian T. S. Ashton was later to write, "they could hardly hope to keep pace with this development: in Malthusian language, though the supply of charcoal might at best increase in arithmetical proportion, the needs of industry increased in geometrical proportion."[110]

In early industrial England, verses were sung celebrating John

Wilkinson, a pioneer in the new coked iron and steel: "That the wood of old England would fail, did appear, / And tough iron was scarce because charcoal was dear, / by puddling and stamping he cured the evil, / So the Swedes and Russians may go to the devil."[111] The last line refers to the imports of wood in the eighteenth century to supply charcoal for the iron mills in England—or else the importing of iron directly—prior to smelting of iron with coal. In France as well as England protests against the overtaxing of forests in response to the demands of ironworks arose.[112] So serious, Engels indicated, was the shortage of wood for charcoal in the eighteenth century until the means of smelting iron with coal became widespread, that the English were forced when the environmental crisis peaked to "obtain all their wrought iron from abroad."[113]

Energy analyst and historian Vaclav Smil has recently explained the severity of the charcoal-smelting crisis facing the nascent industry in the period of charcoal-based iron smelting:

> During the early eighteenth century a single English blast furnace, working from October to May, produced 300 t[ons] of pig iron. With as little as 8 kg of charcoal per kilogram of iron and 5 kg of wood per kilogram of charcoal, it needed some 12,000 t[ons] of wood. . . . In 1720 60 British furnaces produced about 17,000 t[ons] of pig iron, requiring about 680,000 t[ons] of trees. Forging added another 150,000 t[ons], for a total of some 830,000 t[ons] of charcoaling wood. . . . Already in 1548 anguished inhabitants of Sussex wondered how many towns would decay if the iron mills and furnaces were allowed to continue (people would have no wood to build houses, watermills, wheels, barrels, and hundreds of other necessities), and they asked the king to close down many of the mills. . . . Widespread European deforestation was to a large degree a matter of horseshoes, nails, axes (and mail shirts and guns).[114]

Too late to save England's forests, coked coal was introduced in the smelting process in the early eighteenth century, becoming widespread in England only late in the century. "Germany," Weber remarked, "was [only] saved from this fate [deforestation] by the circumstance that in the 17th and 18th centuries it was untouched by capitalist development."[115]

For Weber the discovery of the process for smelting iron with coal constituted what he called the "fateful union of iron and coal," without which, in his view, the Industrial Revolution was scarcely conceivable. Indeed, "the victory" of the Industrial Revolution, he emphasized, "was decided by coal and iron," in particular the "coking of coal . . . and the use of coke in blast furnace operation."[116] The dramatic introduction of a coal-smelting process for iron anchored the Industrial Revolution in particular environmental-technological conditions, in which coal was king. Today historians concur with Weber regarding the limits of charcoal-based iron smelting, the crisis this posed for nascent industry, and the dire consequences if it had persisted: "An impossible amount of woodland would have been needed if iron producers had continued to use charcoal by the year 1850."[117] "The forest [land in the British Isles]—or what was left of it—was saved only by coal, a fuel more suitable for industry than charcoal."[118] For Weber the shift from charcoal smelting to coke smelting represented a critical historical turning point, without which the emergence of industrial capitalism and the rational-inorganic phase of development would have been blocked.

The Rational-Inorganic Era in Human History: The Age of Coal, Iron, and Industrial Capitalism

Weber's best-known definition of modern capitalism is the one provided in his 1920 "Prefatory Remarks" to his *Sociology of Religion*. There he wrote:

> We will define a capitalistic economic action as one which rests on the expectation of profit by the utilization of opportunities for exchange, that is on (formally) peaceful chances of profit. . . . In modern times the Occident has developed . . . a very different

form of capitalism which has appeared nowhere else: the rational capitalistic organization of (formally) free labour. . . .

Rational industrial organization, attuned to a regular market . . . is not, however, the only peculiarity of Western capitalism. The modern rational organization of the capitalistic enterprise would not have been possible without two other important factors in its development: the separation of business from the household, which completely dominates modern economic life, and closely connected with it, rational book-keeping.[119]

Weber thus treated capitalism (along with modern bureaucracy) as representing the fullest development of formal rationality or rationalization. This was consistent with his argument in *The Protestant Ethic and the Spirit of Capitalism*. In his *General Economic History* Weber, however, went somewhat further, providing what Randall Collins has called his "full theory of capitalism as a historical dynamic."[120] Hence, environmental factors enter in at a causal level, with the rational organization of the modern industrial enterprise anchored in environmental-technological conditions.

Modern industrial capitalism, associated with machine development and rational calculation, was, according to Weber's description of it in *The General Economic History,* anchored in "the age of iron," which was just as much the age of coal, "the most valuable and most crucial of all products peculiar to the western world."[121] Since coal was viewed as an inorganic or nonrenewable form of energy, modern capitalism was, in Weber's conception, an age dependent on "substituting inorganic" for "organic" materials/energy.[122] A similar observation on the shift to "inorganic energies" from an earlier reliance on human and animal (or physiological) energy was made by Weber's contemporary, the German chemist and energetics theorist, Wilhelm Ostwald.[123] Weber's central distinction here between "traditional-organic" and "rational-inorganic" phases in the development of energy was to be elaborated upon decades later in the United States by Lewis Mumford, who differentiated between the "ecotechnic" and "paleotechnic" phases of civilization.[124] More recently, Collins has referred to this transformation from the traditional-organic to the rational-inorganic, as presented in Weber's analysis of historical development—in terms of the shift, at the time of the Industrial Revolution, from "agrarian to inanimate-energy-based technologies."[125]

Today this change, highlighted by Weber, is commonly described as the shift from biomass to fossil fuels as the primary form of energy. In the world at large, 1,000 million metric tons of biomass were consumed as fuel in 1800, as opposed to 10 million metric tons of coal. By Weber's day, in 1900, 1,400 million metric tons of biomass were consumed globally but coal consumption rose to 1,000 million metric tons, and oil had made its appearance, accounting for 20 million metric tons.[126] Industrial development has come to be identified with this shift to "inorganic" materials/energy in the form of coal and petroleum. In Weber's view, this broad transformation paralleled the development of modern, rational chemistry, represented by Justus von Liebig, and its introduction of new synthetic chemicals.[127]

Coal was seen as crucial to the rise of industrial capitalism, in the eyes of nineteenth and early twentieth-century observers, not simply because of its role in powering industry through steam engines—though its importance in that respect was indisputable—but even more so because coked coal was the basis of blast furnace technology for the smelting of iron. In 1869 coal consumption by the iron and steel industries in Britain was greater than the combined coal consumption of both general manufacturers and railroads.[128]

In Weber's conception coal was as important and indispensable as the revolutionary technologies that it made necessary. Rather than seeing coal as the basis of the steam engine, with the latter as the object, he dramatically turned this on its head: arguing that the steam engine, used first in mining, "made it possible to produce the amount of coal necessary for modern industry." Yet for Weber, even the railroad, "the most revolutionary instrumentality known to industry," was a manifestation of "the age of iron" and coal.[129]

So significant was coal for the rise of industrial capitalism in Weber's view that it entered into his comparative-historical interpretation of world civilizations. Anthracite coal, he noted, was used in ancient times in China. Yet, he argued (in ways that later would open him to charges of Eurocentrism) that its further use was hindered by the prevalence "of a superstructure [in Chinese society] of magically 'rational' science," consisting of such beliefs as geomancy or earth divination. Mining was thought to "incense the spirits" while smoke from burning coal "magically invested whole areas.... The magic stereotyping of technology and economics, anchored in this belief . . . completely precluded the advent of indigenous modern enterprises in communication and industry" in China. The barrier to the critical rise of king coal in China as opposed to Europe was therefore a product of the former's lack of demagification/disenchantment. "To overcome this stupendous barrier" to industrialization in China, Weber claimed, "Occidental high capitalism had to sit in the saddle aided by the mandarins who invested tremendous fortunes in railroad capital."[130]

The burning of fossilized coal in blast furnaces, and its use as a means to steam power, therefore constituted, for Weber, a major transformation in human society, liberating it from its traditional relation to nature and providing a crucial environmental precondition for the rise of industrial capitalism. As he wrote in his *General Economic History*:

> In the first place, coal and iron released technology and productive possibilities from the limitations of the qualities inherent in organic materials; from this time forward industry was no longer dependent upon animal power or plant growth. Through a process of exhaustive exploitation, fossil fuel, and by its aid iron ore, were brought up to the light of day, and by means of both men achieved the possibility of extending production to a degree which would have previously been beyond the bounds of the conceivable. Thus, iron became the most important factor in the development of capitalism; what would have happened to this system or to Europe in the absence of this development [made practical by the introduction of coked coal in iron smelting] we do not know.[131]

For Weber, "The mechanization of the production process through the steam engine liberated production from the organic limitations of human labor." The *relative* energetic significance of human energy" for production was thereby diminished.[132] This was accompanied, in an increasingly industrialized agriculture, by the accelerated "liberation of the peasants" from the land and the dissolution of the organic relation to the earth.[133] In the age of industrial capitalism, Weber declared, the machine is no longer "the servant of the man," but rather "the inverse relation holds."[134]

It is a characteristic feature of Weber's complex theory of human-environmental interactions that, in contrast to those who were to adopt the crude human-exemptionalist notion that humanity had conquered nature and history by means of fossil fuels, iron, and machinery, he was to reveal the deeper alienation and instability in these same processes. Not only did he repeatedly emphasize, as we shall see, that human beings were becoming "servants to machines," he also recognized the resource limitations of an industrial capitalism increasingly dependent on fossil fuels and the rapid consumption of natural resources.

At the same time, Weber managed to elude the simplistic, quasi-Malthusian notion that the development of modern industrial capitalism was to be explained primarily by the effects of population growth. "It is a widespread error," he contended,

> that the increase of population is to be included as a really crucial agent in the evolution of western capitalism. In opposition to this view, Karl Marx made the assertion that every economic epoch has its own law of population, and although this proposition is untenable in so general a form, it is justified in the present case. . . . The growth of population in Europe did indeed favor the development of capitalism, to the extent that in a small population the system would have been unable to secure the necessary labor force, but in itself it never called forth that development.[135]

Weber hammered home this point by arguing that China in the same period (the eighteenth and nineteenth centuries) saw "a population growth of at least equal extent [to that of the West]—from 60 or 70 to 400 millions, allowing for the inevitable exaggerations." This growth, however, took place in different strata, under a different system than the West, making "China the seat of a 'swarming mass of small peasants.'" Because "capitalism went backward in China and not forward" in this period, the masses did not become a modern proletariat.[136] From a theoretical standpoint, this meant that population growth was not an adequate cause of capitalist development. When it came to the environmental preconditions of capitalism, Weber thus emphasized energy and resources over population. Moreover, it was capitalism's relentless consumption of energy and resources that was to constitute its main environmental constraint.

Raubbau *and the Heedless Consumption of Natural Resources*

Weber, more than most social theorists of his day, was acutely aware of what he called the "dissolving effects of capitalism"—both materially and culturally—on the previous organic relations with respect to land and resources.[137] As a major contributor to rural sociology in Germany, he recognized the importance of the disruption of the soil nutrient cycle, first described by Liebig and analyzed in social terms in Marx's theory of metabolic rift.[138] Under conditions of modern agriculture, Weber argued, it was no longer adequate to assume that the "agricultural product" was the result of natural soil quality and the work of farmers, and nothing more. Rather, means of production such as "improved tools, modern buildings, or artificial fertilizer," were increasingly necessary, independent of the farmer. Artificial fertilizers were essential in industrialized agriculture because "even the nutrients in the soil" were no longer "produced by the farmer with the aid of the gifts of nature within the natural soil, but far away in machine and tool factories, 'potash mines,' Thomas blast furnaces, fitters' workshops, and the like," and imported to the farm.[139] "Capitalism," he argued, "shifts the imputation of the yield of agricultural land from the place of direct agricultural production to the workshops where the agricultural implements, artificial fertilizers, etc., are produced."[140] In industrial capitalism—the age of coal, iron, and synthetic fertilizer—agriculture was increasingly dominated by inorganic, inanimate forms of energy.

In the natural organic cycle, soil nutrients (chiefly nitrogen, phosphorus, and potassium) formed the basis of plant cultivation. However, as Liebig had pointed out, with the growth of industrial capitalism and the shift of populations to the cities beginning in the 1840s, soil nutrients were increasingly being shipped in the form of food and fiber to the urban centers where they eventually became sources of pollution rather than being returned to the soil. As a result, the soil was continually robbed of vital nutrients, what Liebig called earth robbery, *Raubbau*, or the robbery system, *Raubsystem*.[141]

The primary limitation on cultivation in Europe in the nineteenth century was lack of nitrogen for fertilizer, followed by phosphorus shortages. When artificial sources for these two minerals were secured, and when the agricultural yield reached a certain level, potassium became a major constraint on agricultural productivity (in accordance with Liebig's famous "law of the minimum"). Hence potassium was the last of the three great mineral fertilizers exploited. The Germany of Weber's day played a leading role in addressing this natural limitation, beginning in the 1870s, with its potassium (or potash) mines. It was no accident therefore that potash mines were to be singled out by Weber as a prime example of the external sources of fertilizer essential to industrialized agriculture.[142]

The concept of *Raubbau* recurred numerous times in Weber's work, playing a key role in his conception of the break from the organic conditions of existence. He saw such "land-robbing agriculture (*Raubbau*)" as particularly characteristic of agriculture in the United States (as opposed to Europe), since the very abundance of "virgin soils" in the former made it possible for farmers, often in distressed circumstances, to use up the soil and move on.[143]

The slave plantation system in the antebellum South in the United States, he argued, was an extreme version of a soil "culture [that] was exploitative.... The system required cheap land and the possibility of constantly bringing new land under tillage." This contributed to the crisis of the slave-based plantation system and helped generate the conditions leading to the Civil War.[144] During his trip to Booker T. Washington's Tuskegee Institute in 1904, Weber questioned the extreme exploitation of the land that characterized even the postbellum South, remarking that the farmers' training was aimed at the "conquest of the soil" as "a definite ideal."[145]

Weber was also concerned with sustainability in relation to forests. He celebrated the German forests as a lasting treasure of German culture, having a role in the development of the German character, and argued for their preservation.[146] He compared the well-managed German forests, which were "nurtured with all the care that the highly developed technic of forestry has made possible" to the "primitive forestry conditions" that prevailed in the United States, where forests were simply cleared away in anticipation of their further exploitation.[147]

Although Germany was relatively rich in raw materials, with coal reserves that could, Weber suggested, outlast those of Britain by centuries, Germany did not have the same advantage as Britain (and some parts of the United States) of coal and iron mines that were close together, facilitating industrialization. Key raw materials necessitated rational management in order to not "hasten unnecessarily the exhaustion of mines."[148]

During his trip to the United States in September–November 1904, Weber provided a general historical view of natural resource constraints within modern capitalism and their relation to cultural development. He was invited along with other German social and natural scientists—the social scientists included Werner Sombart and Ernst Troeltsch, and the leading natural scientist was Ostwald—to present a paper at the Universal Exposition of the Congress of Arts and Science in St. Louis, commemorating the Louisiana Purchase.[149] Weber's talk, presented in German to a small audience on September 21, 1904, dealt with the question of rural society and the overall social structure of capitalism in the United States and Germany.[150]

What was most remarkable about Weber's St. Louis presentation was his adoption of a line of argument that paralleled Frederick Jackson Turner's frontier thesis (first introduced in 1893). Turner was famous for contending that with the closing of the frontier, U.S. society would come to resemble the more densely populated, class societies of Europe.[151] Echoing this, Weber claimed that scarcity of land and natural resources would eventually impinge on capitalism in the United States, which would no longer have the outlets of free soil and boundless raw materials. As a result, the United States, which had hitherto been constrained primarily by the effects of racism and ethnocentrism, would increasingly come to resemble the older societies of Europe, where economically related class and status issues dominated. Thus Weber introduced his own environmentally nuanced interpretation of "American exceptionalism" ahead even of Sombart.[152]

Like Turner, Weber was concerned not just with the disappearance of free land (or the frontier) but also with the depletion of supplies of coal, iron ore, and other natural resources.[153] "We must not forget," Weber wrote, "that the boiling heat of modern capitalist culture is connected with the heedless consumption of natural resources, for which there are no substitutes. It is difficult to determine how long the present supply of coal and ore will last." If the timeline governing the inevitable exhaustion of key raw materials was uncertain, the end of frontier land was on the immediate horizon. "The utilization of new farm lands will soon have reached an end in America; in Europe it no longer exists."[154]

It was this awareness of the overall problem of natural resources and energy under capitalism that would form the environmental-sociological basis of Weber's comparison of German and American rural life. After "all the free land has been exhausted," the United States, he wrote, will eventually confront "increased density of population and rising land values" and "the so-called 'law of decreasing productivity of the land.'" This would lead to higher rents and a sharpening of capitalist social relations and class divisions. Over a longer period, the inability to continuously revolutionize agriculture by "substituting inorganic raw materials [fossil fuels] and mechanical means of production for organic raw materials and labor forces," could also intensify social divisions. In short, "America will one day also experience the effects of such [social] factors—the effects of modern capitalism under conditions of completely settled old civilized countries." With "the areas of free soil... now vanishing everywhere in the world" the distinction between old world and new would give way before the "dissolving effects of capitalism."[155]

This fundamental perspective on ecological constraints was evident in many of Weber's concrete observations during his 1904 trip to the United States. In the course of his travels he wrote of the pollution, filth, environmental degradation, and wasted resources. In the state of New York, the "natural beauty" of many of the sights was subject to "shameful disfigurement." In Chicago, he noted, the pollution from the burning of "soft coal" was so severe that "one can see only three blocks ahead—everything is haze and smoke, the whole lake is covered in a huge pall of smoke from which the little steamers suddenly emerge and in which the sails of the ships putting out to sea quickly disappear." The stockyards were characterized by endless filth and an "'ocean of blood.' ... There one can follow the pig from the sty to the sausage and the can."[156]

It was, however, Weber's trip to Muskogee in Indian Territory, in present-day Oklahoma, that gave rise to his most powerful environmental indictments while in the United States. Three days after his presentation in St. Louis Weber announced in a letter to Georg Jellinek his plan to travel "*perhaps* to Oklahoma and Texas, instead of to [Theodore] Roosevelt" for a White House reception.[157] Weber's wife, Marianne, who accompanied him to the United States, but not to Indian Territory because of what she called its "primitive" state, explained his motivations (while employing racial terms not characteristic of Weber himself and that do not appear in his letters from Oklahoma): "Here it was still possible to observe the unarmed subjugation and absorption of an 'inferior' race by a 'superior,' more intelligent one, the transformation of Indian tribal property into private property, and the conquest of the virgin forest by the colonists."[158]

Weber sent two letters to his mother from Muskogee, one of the main commercial centers in Indian Territory, containing detailed sociological descriptions of the conditions, including environmental relations. "In no other location in his correspondence does Weber have as much to say about 'nature' as in his Indian Territory commentaries."[159] Much of Weber's discussion focused on the fate of Indian Territory and the Indians themselves. He was concerned with how the privatization of Indian land was being imposed on the Five Civilized Tribes (Cherokee, Chickasaw, Choctaw, Creek, and Seminole) forcibly relocated to Oklahoma in the 1830s via the Trail of Tears—and on some twenty other tribes that had at various times been removed to the area of present-day Oklahoma. Weber was equally caught up, however, in the related issues of environmental change. Comparing what he saw to romantic conceptions of wilderness in James Fenimore Cooper's Leatherstocking tales and Ludwig Ganghofer's *The Silence of the Forest* (the peak of German sylvan Romanticism), Weber proclaimed, with evident misgivings, that soon "the last remnant of 'Romanticism' will be gone."[160] In a dramatic description that encompasses both the tragedy of the Indians and the rise of the oil fields, he wrote:

> Nowhere else does the old Indian Romanticism [*Indianerpoesie*] blend with the most modern capitalistic culture as much as it does here right now. The newly built railroad from Tulsa to McAlester first runs along the Canadian river for an hour through veritable virgin forest [*Urwald*], although one must not imagine it [*sich vorstellen*] as the "Silence in the Forest" with huge tree trunks.... The large rivers, like the Canadian River, have the most Leatherstocking Romanticism [*Poesie*]. They are in an utterly wild state.... But the virgin forest's hour has struck even here.... [In occasional clearings] the bases of the trees had been smeared with tar and ignited. They are dying off, stretching their pale smoky fingers into the air in a confused tangle.... And suddenly it begins to smell like petroleum: one sees the tall Eiffel Tower-like structure of the drilling holes, right in the middle of the forest, and comes to a "town."[161]

The first oil wells in the vicinity of Muskogee had appeared only the year before but already dominated the environment, creating a booming, camp-like atmosphere. Weber wrote of the constant "stench of the petroleum and the fumes" and the "primitive state" of the streets, "usually doused with petroleum twice each summer to prevent dust and smelling accordingly." He mentioned in both letters that the more romantic aspects of this world were fast passing away and constituted a true loss: "This is a more 'civilized' place than Chicago. It would be quite wrong to believe that one can behave as one wishes. . . . Too bad; in a year this place will look like Oklahoma (City), that is, like any other American city. With almost lightning speed everything that stands in the way of capitalistic culture is being crushed."[162]

Weber's letters from Indian Territory reveal his enormous ability to integrate causal analysis at an empirical level, accounting for environmental changes, with his larger interpretive vision of capitalist cultural development. In fact, Weber's account here of the "lightning speed" in which all that "stands in the way of capitalistic culture" is simply "crushed" (with reference in particular to the environment and Native Americans) reads like a precursor to the "treadmill of production" perspective of contemporary environmental sociology.[163] Nowhere else perhaps does Weber point so forcefully to capitalism as a driving force for environmental change.

The Sociology of Energy

Weber's emphasis on the energy-intensive, fossil-fuel-dependent, and high-resource-consumption character of capitalism led to intensive studies in the economics and sociology of energy. Although his work in this area is celebrated by ecological economists, it is little known to sociologists. Yet without an understanding of Weber's approach to the sociology of energy it is impossible to comprehend fully his theoretical rendition of the way in which capitalism, as a specific cultural formation, is anchored in environmental conditions.

During his journey to the United States in 1904 Weber became well acquainted with Ostwald. Besides being a leading chemist, Ostwald was especially well known for his advocacy of energetics as the key to a universal theory of culture. In St. Louis Ostwald presented a paper—quite likely with Weber in attendance—on the methodology of science in which he advanced the Comtian view of a hierarchy of sciences, with the three great divisions of mathematics (theory of order; theory of numbers, or arithmetic; theory of space, or geometry), energetics (mechanics, physics, and chemistry), and biology (physiology, psychology, and sociology). "Mathematics, energetics, and biology," he wrote, "therefore embrace the totality of the sciences," with sociology as the final, most epiphenomenal of the sciences.[164] Such views were anathema to Weber and led to his critique of Ostwald's energetics and conception of science five years later, in 1909, the same year that Ostwald received the Nobel Prize in chemistry.

In 1909 Ostwald published *Energetic Foundations of a Science of Culture,* which sought to establish the energetic bases of all culture. In this context, he addressed issues of energy scarcity/abundance; the application of energy concepts to all aspects of life, including psychology, language, and so on; and the issue of the Comtian hierarchy of the sciences as seen from the standpoint of energetics. A key part of his analysis was his chapter "Raw Energy." Here he attacked prevailing views of energy scarcity, claiming that given "the enormous capital from the energy of the sun" humanity was at present making "use of only a disappearingly small portion—much like the rich child who inherited a fortune but is not capable of using more than it spends for nutrition, clothing and shelter." The various untapped sources of energy, even taking into account entropy, were "so extraordinarily great," he observed, "that we do not need to worry about the exhaustion of fossil fuels. In the few centuries that separate us from this event" the different forms of solar energy could easily fill the gap—before "the legacy [of fossil fuel] is completely exhausted."[165]

Ostwald emphasized that human beings were then using the sun's available energy, mainly by two means: first, "planting [*Bestockung*] of a part of the land with fields, meadows, and forests, and through the use of plants raised there for chemical storage. A second, and presently much smaller part, rests on the use of the water quantities raised by the sun's rays that pour down from the mountains for the driving of mechanical motors."[166] At the time Ostwald was writing, the latter capture of energy mainly took the form of water mills, while hydroelectric power was only just coming into use. Ostwald insisted that the main means for expanding energy availability was through the construction of hydroelectric power facilities using the recent developments in electrical transmission to transfer this energy to more distant locales and constructing large dams or "gigantic reservoirs" to ensure that this energy was stored and available on a non-seasonal basis. Dependence on fossil fuels and energy scarcity could be a thing of the past. "In terms of the utilization of energy," he wrote, "humanity remains thoroughly stuck in childhood. The used part of the annual intake is still in comparison to the entire supply so extremely small that the danger of it not being sufficient later does not at all exist."[167] Not only was it possible to have a "more complete capture of the energy stream," but also "through the improvement of the efficiency of the process of the transformation of the already captured raw energies" it was possible to achieve more with less. Indeed, it was "not improbable that in the future humanity might even find its pleasure in leading a comfortable life with a lesser consumption of energy and will consider the gluttonous consumption of raw energy in contemporary life to be a blamable barbarism."[168]

Ostwald's views are significant in today's context both because of their emphasis on solar energy as the energy of the future and because of the ironic counterpart to that: his contention that energy was superabundant to human society and if harnessed properly could lead to endless economic expansion. His work arose out of a long tradition, going back to Herbert Spencer, which, in the words of Rosa, Machlis, and Keating, claimed that "the ability to harness more and more energy to production lay at the foundation of the evolution of societies." For Ostwald, according to the above analysts, "the greater the coefficient of useful energy obtained (in the transformation) the greater a society's progress."[169]

Weber reacted strongly to Ostwald's argument, declaring in a letter in May 1909 that he "'dreaded' Ostwald's 'energetic sociology.'"[170]

This led to a full-blown critique in the form of an extensive review of Ostwald's book, which Weber titled "'Energetic' Theories of Culture." Although Weber's critique of Ostwald is known as one of his most important methodological papers and is frequently referred to in Weber studies, sociologists have treated it almost entirely in terms of a critique of Comtian-style positivism and hierarchy of the sciences, while ignoring for the most part Weber's strong engagement with energetics.[171] This contrasts sharply with the response of ecological economists, who draw extensively on Weber's critique of Ostwald.[172]

Weber's critique of energetics was remarkable as the work of an economic sociologist who challenged the views of a Nobel Prize–winning chemist on his own ground: thermodynamics. Adopting a perspective that we would now call ecological economics, Weber displayed a startling grasp of issues related to natural science and energy specifically. In general, he objected that the "positivist" project of Ostwald was "influenced . . . by the (supposedly) 'exact' sociological method derived from the work of Comte and Quetelet," nurtured at Ernest Solvay's institute. This led Ostwald to the crude (and indeed absurd) reduction of all revolutions in culture to energetics.

Weber's objection to Ostwald's views, however, went beyond questions of methodology and extended to Ostwald's treatment of energy itself. For Weber, Ostwald's account of the potentially unlimited supply of energy emanating from the sun, which human beings had not yet tapped fully, was questionable if taken to the extreme of denying resource scarcity. Economics, after all, was "bound up with the application of *scarce* material means," including limited natural resources. Weber thus strongly questioned Ostwald's claim that a "'squandering of our inheritance' [with respect to energy and natural resources] seems totally unthinkable."[173] Not only was Weber extremely skeptical about the end of dependence on fossil fuel, but he argued—anticipating in this respect the founder of modern ecological economics, Georgescu-Roegen—that the entropy law could be seen as applying to essential raw materials as well as energy as such, so that the squandering, for example, of iron ore and copper, could prove crucial in limiting production and enforcing conditions of scarcity.[174] Thus Ostwald's views of energetic abundance were naive in that "the indispensable chemical and form-energy of every substance used for production, transmission, and utilization of the most important energies that are used is *equally* irretrievably dissipated. This, after all, is the case with all free energy according to the law of entropy."[175]

Ostwald's expectations of the elimination of energy scarcity were further compromised, according to Weber, by his failure to take into account the "energy ladder," representing different qualities and compositions of energy, which were bound in various ways to space and time. In opposition to this, Weber argued that even if there were such a thing as a *"perpetuum mobile"*—and if free energy were theoretically available at a given rate and no cost, constraints on energy use (scarcity) would disappear only if the energy were available in (1) the appropriate form, (2) everywhere, (3) at every time and in each time differential, (4) in unlimited quantity, and (5) in the appropriate direction for the desired effect. In other words, even if Ostwald's notion of the expansion of technological "apparatuses" to capture energy from the sun appeared theoretically to make energy superabundant, real constraints of space and time in relation to production would still inevitably apply.[176]

In terms of the energetics of production, Weber pointed out that Ostwald was mistaken in assuming that the absolute importance of human energy in production was decreasing and that human energy was less thermodynamically efficient than other forms of energy used in production, such as the electric dynamo. As Weber put it, "It is . . . completely wrong to say that 'advanced' culture . . . is identical with an absolute [as opposed to relative] *diminution* of the use of *human* energy."[177] With respect to the efficiency of human energy, Weber stated that if it were possible to compute all of the energy components (kinetic, chemical, and other forms of energy) that went into machine weaving of textiles (including dissipated energy) and compare that to human weaving, it would be found that the latter was more thermodynamically efficient (though more expensive in terms of *economic* unit costs).

Indeed, basing himself on thermodynamic/physiological studies conducted since the late nineteenth century, Weber explained that "the 'primitive' tool that man is given by nature, the human muscle" has greater thermodynamic efficiency "for the utilization of the energy set free through the biochemical oxidation process than the best generator can ever attain."[178] As Martinez-Alier points out, "Steam engines, at the time Weber was writing, had efficiencies as low as five percent, whilst the human body can convert food energy into work with an efficiency on the order of twenty percent, as has been known since the 1860s."[179]

Such elementary truths, Weber argued, completely destroyed Ostwald's attempts to generate an energy theory of value. "Even a dilettante like Ostwald could ultimately see that the relationship between need and cost simply cannot be defined 'energetically' and this is so even when one makes allowances for his totally worthless discussion of the economic concept of value."[180]

Weber's sensitivity to environmental issues reflected his critique of one-sided notions of progress under capitalism. His intense dislike of Ostwald's views was engendered not simply by Ostwald as a personification of positivism, but even more as a personification of a crude productivism. For Weber, Ostwald's ultimate failing was his inability to recognize that there were other possible forms of social action and meaning beyond productivist ones. As Weber stated elsewhere in a discussion of technology: "How else could a chemist of Ostwald's importance hold exclusively technological ideals of life and view all cultural development as a process of saving energy if his whole science were really not exclusively dependent on the requirements and the progress of modern technology in our factories, and through this ... to the utmost extent on capitalist-economic conditions?"[181]

Ostwald's energetics, Weber contended, were rooted in the economic drives of capitalism. For this reason, he pointedly raised the issue of Sombart's critique of the Reuleauxian concept of technology in which Sombart claimed technology had moved from a situation in which the instrument was the servant of the human being to one in which the human being is the servant of the machine.[182] Likewise,

Weber went on in his critique of Ostwald to attack what he called a "fanaticism for 'productivity'"—directed specifically at Solvay.[183] In Ostwald's case, Weber saw this, as we have seen, as tied to "capitalist economic conditions."[184] The crime of thinkers like Ostwald, Solvay, and Comte was to promote in the crudest positivistic, productivist manner possible the heedless consumption of resources and energy associated with rationalized industrial capitalism and its "disenchantment of the world."

The Disenchantment of the World

If rationalization was the defining theme in Weber's view of modernity, his notion of "the disenchantment of the world" (*Entzauberung der Welt*— literally "demagification of the world") constituted an important, if somewhat controversial, element in his critique of a rationalized modernity.[185] "The absence of the gods, the 'disappearance of the sacred,'" was presented by Weber, according to Lukács, "as the real physiognomy of our times, which is necessary to accept as a historical inevitability but which invokes in us an infinite melancholia and a profound nostalgia for the good times when there was 'science of the true, good and beautiful,' when there were 'sacred things.'"[186]

In the work of later critical theorists, such as Horkheimer and Adorno, the concept of "disenchantment" not only became the means of questioning "the dialectic of enlightenment" but also stood for the contradictions inherent in the human "conquest of nature."[187] It is not surprising therefore, as Murphy noted, that Weber's concept of the disenchantment of the world is often seen as having deep ecological implications.[188]

Most analysts of disenchantment in Weber's thought have approached it from a purely cultural angle, seeing it as a kind of mirror image of the growth of calculation, formal rationalization, and the disappearance of magic—all factors that he emphasizes in defining the concept. Yet some commentators have recognized that it is also connected to his references, at the end of *The Protestant Ethic*, to an iron cage (steel casing) and to the eventual burning up of fossil fuels (as the inorganic substance of modern mechanization).[189] Still others have noted that there is a direct relation between Weber's concept of the disenchantment of the world and his allusions to the loss of connection to "organic life."[190]

Weber first employed his notion of "the disenchantment of the world" in 1913 in "Some Categories of Interpretive Sociology," using it thereafter in numerous works.[191] He made a point of inserting it into the final edition of *The Protestant Ethic and the Spirit of Capitalism*, published fifteen years after the original.[192] There he gave the disenchantment process a millennia-long timeline. Thus he referred to the "great historic process in the development of religions, the elimination of magic from the world which had begun with the old Hebrew prophets, and in conjunction with Hellenistic scientific thought, had repudiated all magical means to salvation as superstition and sin."[193]

Within German Romantic literature and philosophy it was Friedrich Schiller in the eighteenth century who most powerfully conveyed this sense of disenchantment in his poem "The Gods of Greece": "Insensible of her maker's glory / Like the dead stroke of the pendulum / She slavishly obeys the law of gravity / A Nature shorn of the divine [*Die entgötterte Natur*]."[194] Where the gods previously held sway, there was now only the insensible law of gravity.

For Weber, as he indicated in *The Protestant Ethic* with regard to the concept of "rationalization," the notion of the disenchantment of the world was to be viewed as "an historical concept" that covered "a world of different things" and thus carried contradictions within itself.[195] Weber employed this "historical concept" in two main, overlapping senses: (a) a narrower technical meaning of *demagification*, related principally to his comparative-historical sociology of religion; and (b) a broader, more philosophical concept associated with the German Romantic tradition, embodying the loss of connection with nature as a realm of meaning, that is, as a process of *disenchantment*. It is the latter, more philosophical sense of the term that has sometimes been seen as constituting the central element—the negative aspect of rationalization—in Weber's critique of modernity.[196]

The logic of the broad tendency to rationalization/disenchantment in Western capitalism was tied, in Weber's tragic conception, to the exhaustion of fossil fuels, which would constrain cultural development. Although environmental forces do not seem to affect directly Weber's discussion in *The Protestant Ethic and the Spirit of Capitalism*, the illusion of a culture entirely free of such constraints, as Guenther Roth cogently observed, is broken at the end of the work when Weber explains: "We must [now] worry about what will happen when 'the last ton of fossil fuel has been used up.' . . . Suddenly we may find ourselves in a position in which Weber's '*apprehensions*' with respect to the environment and 'material constraints' are activated and the whole 'evanescent superstructure' becomes 'powerless before the geographic, demographic, and economic substructures of long duration.'" [197]

In *The Protestant Ethic* Weber stressed, in line with Sombart, that the technical-rational expansion of the "productivity of labor" had "relieved" the production process "from its dependence on the natural organic limitations of the human individual."[198] Here was embodied a powerful ecological critique, which was to be developed further in *Economy and Society*, of Taylorism or "scientific management," which he depicted as a process in which "the individual is shorn of the natural rhythm as determined by his organism."[199] For Sombart, writing of the ecologically destructive aspects of formally rational capitalism: "Modern culture has alienated us from nature, has put a layer of asphalt between ourselves and nature so that nature can at best be merely an object of aesthetic enjoyment."[200] The "rationalization and intellectualization" of modern society, Weber wrote in a similar fashion in *Economy and Society*, "parallel the loss of the immediate relationship to the palpable and vital realities of nature, because the work is done largely within the house and is removed from the organically determined quest for food."[201]

Recognition of Weber's sensitivity to environmental changes and their refracted effect on culture can therefore help attune us more fully to the significance of his master theme of rationalization/disenchantment. For Weber, the demise of the traditional-organic world of preindustrial capitalist society and its replacement with a rational-inorganic one of modern capitalism was an overarching frame characterizing his thought. The coupling of the traditional with the organic, and the formally rational with the mechanical and inorganic, appeared repeatedly in his works. Embedded within this was a deep critique of any unilinear notion of progress.

Weber and the Classical Foundations of Post-exemptionalist Sociology

The foregoing analysis brings us back to two critical questions raised at the beginning of this article: (1) How do we account for the fact that leading environmental sociologists have characterized Weber's environmental contributions as "relatively invisible"?[202] And (2) what do Weber's insights into the environment and society teach us with regard to the needed transformation of environmental sociology and sociology as a whole, in our post-exemptionalist age, symbolized by global warming, when we realize all too fully the dangers of the human degradation of the environment?

The Question of Invisibility of the Environment in Weber

How are we to account for the invisibility of Weber's consideration of the natural environment in the eyes of so many sociologists? One possible explanation lies in the fact that the two masterworks that receive the most attention from Weber scholars and sociologists in general— *The Protestant Ethic and the Spirit of Capitalism* and *Economy and Society*—appear at first sight to be completely detached from environmental questions.

In the case of *The Protestant Ethic* we have already seen that this detachment (which at first sight seems to conform to Dunlap's contention that the environment "could" be ignored in the Weberian tradition) was based on very special assumptions.[203] The reference to the using up of fossil fuels at the end of his treatise underscores Weber's key assumption, presented in his *General Economic History*, that in industrial society coal, viewed as an inorganic material, had temporarily detached society from organic materials (a view that merged with his notion that rationalization had detached society from the cultural basis of organic life). Given Weber's passionate denunciation in 1904 (while he was working on *The Protestant Ethic*) of the reckless wasting of natural resources in the United States, and his Turner-like thesis on the same occasion, which suggested that environmental scarcity would come back to haunt the country, placing it in a more European context once the conquest of nature could no longer substitute for the conflict between classes, it is clear that the environment always constituted a background condition in his analysis.[204]

However, this same argument might be seen as less applicable to *Economy and Society*, which was not principally a historical work but rather a grand theoretical-taxonomic treatise aimed at providing a master framework, or set of ideal-typical patterns and domains, for analyzing society. Although Weber made, as we have seen, important environmental statements in this treatise, mentioning such factors as climate change, at no point does the environment enter into the basic structure and organization of *Economy and Society* itself. As a result, some have argued that this is clear evidence of the relative unimportance of environmental influences in Weber's thought.[205]

This conclusion is directly refuted, however, by the wider context in which *Economy and Society* was written and of which it was a part. *Economy and Society* was part of a larger, multivolume work, *Grundriss für Sozialökonomik* (*Outline of Social Economics*), of which Weber was the editor. It thus entailed a certain division of labor. As Roth pointed out, there were three distinct parts in the *Outline of Social Economics* that dealt with "'Economy and . . . ': (1) 'Economy and Nature'; (2) 'Economy and Technology'; and (3) 'Economy and Society.'"[206] It was Weber's self-assigned task to write the third part. "In the section on 'economy and nature' Alfred Hettner contributed 'The Geographic Conditions of the Human Economy,' a systematic treatise on the surface of the earth, the coasts, the mountains and seas, the quality of the land, crops and animals, and the climate, concluding with an historical overview of 'The Geographic Course of Economic

Culture.'"[207] Since this was part of a larger, multivolume work supervised by Weber as editor (in which he wrote the concluding part) Roth came to the conclusion that there was "no basic difference" in the theoretical importance accorded to environmental-geographical factors between the geographically oriented world-systems theorist Fernand Braudel and Weber—a fact corroborated by Weber's many discussions of environmental influences in his more historical writings.[208] Here, again, the "relative invisibility" of Weber's environmental discussions remarked upon by some environmental sociologists is explained by looking at the overall context of Weber's work. Both *The Protestant Ethic and the Spirit of Capitalism* and *Economy and Society* can be seen as belonging to a general Weberian perspective in which environmental causes and environmental-cultural relations were significant. The natural environment, for Weber, constrains and channels social development, while also enabling it in a complex process of cultural refraction.

Weber and Post-Exemptionalist Sociology

If mainstream sociology, even in our time, has had difficulty incorporating environmental issues into its canon and still often exhibits a human exemptionalism where environmental conditions are concerned—seeing them as unimportant or outside the proper domain of sociology—Weber offers us one lesson after another on how a post-exemptionalist sociology might be developed. His treatment of environmental causes that were significant for human society stretched from climate change, natural disasters, natural resource exhaustion, and soil robbery (*Raubbau*) to deforestation, fossil fuel exhaustion, pollution, and the passing away of the relatively pristine nature-society relations in Indian Territory. His analysis of the sociology of energy, in which he challenged the ideas of the world's leading scientific exponent of energetics, was among the most advanced in his time. His recognition of ecological crisis, in the context of the rapid deforestation in Europe resulting from charcoal smelting, constituted an important contribution to the history of environmental development in relation to the Industrial Revolution. The complex interface between such analysis of environmental causes and Weber's interpretive sociology helps us to understand more fully his central, comparative-historical theme of the transformation from traditional-organic to rational-inorganic society. His critique of one-dimensional capitalist notions of progress, so evident in his environmental analyses, lays bare the crude assumptions of postwar human-exemptionalist analysis. Above all, an understanding of how environment and culture interpenetrated through a complex process of cultural refraction, which gave added cultural meaning to environmental events, is crucial to understanding the wider dimensions and scope of Weber's thought.

The theoretical approach Weber introduced opens the way to a more powerful sociological vision that is anchored in biophysical realities and better suited to the examination of environmental questions. Indeed, there are times when his environmental observations seem startlingly prescient. "It is impossible to infer from the . . . natural environment alone," Weber cautioned in *Economy and Society*, how peoples, even at a given level of technological development, will adjust. In the face of "such factors as climatic changes, inroads of sand [desertification] or deforestation . . . human groups have adapted themselves in widely different ways," depending on numerous causal factors and "structures of interests."[209] Today with climate change (not to mention desertification and deforestation) constituting a dominant global reality, Weber's sophisticated outlook, which addressed human economic and cultural adaptation to climatic changes, is especially relevant. We once again are faced, due to global warming, with conditions that he described (in relation to the ancient Middle East) as "meteorologically precarious."[210] His profound insights into the anchoring of culture in environmental conditions (and on the effects of culture on the environment) can be used to explore these issues more fully.

The natural environment, for Weber, is refracted through a cultural lens, which gives it social meaning, but these meanings are complex and are as likely (or more so) to constitute an iron cage as a definite way forward. Weber's non-teleological perspective is definitely at odds with the outlook of today's ecological modernization theorists, who see solutions to ecological problems in terms of a further stage in the modernization process (sometimes called "reflexive modernity"), and sometime seek to present their views of ecological reform as a further development on Weber.[211] What might be called Weber's "refracted materiality" represents a critical perspective that denies to modernization any simple, harmonious reflexivity. While ecological modernization theorists suggest that capitalism can finally exempt itself, if not from environmental influences, at least from their main constraints on development, through a more reflexive modernity, Weber's outlook is clearly immune to all such exemptionalist notions.

For Weber the mismatch of today's cultural norms and environmental realities—as evidenced by the 2004 tsunami in the Indian Ocean, which perversely killed many thousands of people due to a lack of adequate societal recognition and preparation for such an eventuality—would have come as no surprise. As he implied in relation to the Dollard incursion, the inability of humanity to protect itself in the face of environmental disasters has long been part of our cultural history and constitutes a domain of meaning, even if a negative one. Likewise Weber's example of the Black Death as a pandemic carrying significant social meanings demonstrates his concern with environmental crises capable of challenging whole societies. It is easy to see the relation between his references to the Black Death and the COVID-19 pandemic today.

The most important discovery with regard to Weber's environmental analysis uncovered here is the extent to which it entered into his critique of modern, rational-inorganic capitalism—its origins, development, and (perhaps) decline. Weber's work was notable for its understanding of historical capitalism as energy intensive and resource dependent and its foreshadowing of the contradictions that this posed for the system. On a number of occasions he questioned the permanence of machine capitalism on this basis. His concern with energy and resource scarcity led him to refer in his critique of Ostwald to the "fanaticism for 'productivity'" and productivism brought on by "capitalist economic conditions."[212] His understanding of the destruction of the soil—*Raubbau*—overlapped with Marx's theory of the metabolic rift. During his tour of Indian Territory, Weber noted with respect to capitalism's effect on the environment and the lives of Native Americans that "with almost lightning speed everything that stands in the way of capitalistic culture is being crushed."[213] This evoked a view similar to the contemporary treadmill of production theory in environmental sociology—but one that was even more forceful in emphasizing the role of capitalism as a driver of environmental change. For Weber it was essential to recognize "the dissolving effects of [rational-inorganic] capitalism" with respect both to the preexisting natural environment and traditional-organic societies.[214]

Nor could such change be seen, as in the case of Wundt, as simple progress: the displacement of "the peoples of nature" by the peoples of history, in the inevitable "progression" of the latter.[215] Rather, as Weber emphasized in his critique of Wundt, one should reject any such "metaphysical . . . belief in 'progress.'"[216]

In Weber we thus find some of the strongest classical foundations for the construction of a post-exemptionalist sociology, one in which culture is seen as anchored in material existence and environmental causes generate important, refracted effects on the world of social meaning. It is possible on the basis of his work, and that of other classical theorists (notably Marx), to "bring nature back in"—constructing a sociology fully equipped to address the human-environmental challenges of the twenty-first century.

CHAPTER TEN

The Theory of Unequal Ecological Exchange

A world-system analysis of the ecological rift generated by capitalism requires as one of its elements a developed theory of the unequal ecological exchange between center and periphery. After reviewing the literature on unequal exchange (both economic and ecological) from Ricardo and Marx to the present, a new approach is provided, based on a critical appropriation of systems ecologist Howard Odum's "emergy" (spelled with an *m*) analysis. Odum's contribution offers key elements of a wider dialectical synthesis, made possible in part by his intensive studies of Marx's political-economic critique of capitalism and by Marx's own theory of metabolic rift.

The search for a meaningful theory of ecological imperialism has become in many ways the holy grail of the ecological critique of the capitalist world system. From somewhat different but related standpoints, world-systems sociologists, development theorists, systems ecologists, ecological economists, environmental sociologists, and environmental historians have all been searching for a consistent approach to this core problem, which is tied up with such crucial issues as the metabolic rift, unequal ecological exchange, ecological debt, ecological footprints, the resource curse, embodied carbon, and global environmental justice.

Over the last decade, two bodies of work have emerged in sociology addressing ecological imperialism: (1) metabolic-rift analysis and (2) studies of unequal ecological exchange, sometimes called "ecologically unequal exchange."[1] In the first of these, as Schneider and McMichael state, "Marx's concept of the 'metabolic rift' . . . in the context of an international peasant mobilisation embracing the science of ecology . . . has become the focal point of attempts to restore forms of agriculture that are environmentally and socially sustainable," transcending relations that are widely viewed to be the product of ecological imperialism.[2] Works in this tradition include such important contributions as: John Bellamy Foster, "Marx's Theory of Metabolic Rift," and *Marx's Ecology*; Jason W. Moore, "Transcending the Metabolic Rift"; Burkett, *Marxism and Ecological Economics*; Rebecca Clausen, "Healing the Rift"; Hannah Wittman, "Reworking the Metabolic Rift"; Foster et al., *The Ecological Rift*; Mindi Schneider and Philip McMichael, "Deepening, and Repairing, the Metabolic Rift"; Ryan Gunderson, "The Metabolic Rift of Livestock Agribusiness"; and Ricardo Dobrovolski, "Marx's Ecology and the Understanding of Land Cover Change."[3]

Alongside this research on the metabolic rift, a second, related literature emerged consisting of a number of pioneering empirical-historical studies directed at the unequal ecological exchange relations between core and periphery of the capitalist world economy, in an attempt to gauge the ecological disadvantages systematically imposed on the periphery.[4]

The obvious question that arises from these two literatures, taken together, is to what extent is unequal ecological exchange a source of the global metabolic rift, with the "free environment" of the periphery being sacrificed on the altar of the gods of profit and accumulation in the center? The present contribution attempts to help develop an answer to this question by providing the basis for a more comprehensive theory of unequal ecological exchange-since much of the problem at present, we argue, lies in the under-theorization of this key concept.[5]

The possibility of a more comprehensive theoretical and empirical approach to the unequal–exchange issue, we believe, is offered through a critical engagement with the work of systems ecologist Howard T. Odum. In a series of studies over a two-decade period (1983–2002),[6] Odum developed an illuminating theory of what he called "imperial capitalism," in which embodied energy (*emergy*) exchanged was shown typically to be "several hundred percent higher" per dollar for the peripheral, primary resource-exporting countries than for their core counterparts.[7]

Our approach to Odum's analysis, we stress at the outset, is not an uncritical one, but rather a "critical appropriation," somewhat akin to Marx's response to the physiocrats.[8] Odum's sophisticated scientific approach to systems ecology can be seen as falling prey at times to a kind of physico-reductionism, when viewed from the wider standpoints of natural evolution and historical society. But like much of systems theory it can also be viewed as a holistic attempt to break out of the crude reductionism that has plagued so much of modern science.[9] "Systems science," at its best, is "a technical application of holistic, materialist philosophy."[10]

Odum's theory of unequal ecological exchange, we will argue, is largely free from the reductionism that plagues his overarching systems theory, since the former rests mainly on the extent to which nations or regions draw on their "free environment" in their economic activity, and the global inequalities with which this is associated. It thus is closely related to Marx's analysis of capitalism's metabolic rift. Although there is no single measure for unequal ecological exchange in all of its manifold historical and qualitative dimensions—how do we begin to quantify the loss of even a single species?—Odum's *emergy* (spelled with an m) approach sought to create a common metric for energy used in production and to explore relations of unequal ecological exchange in this regard. Odum thus provided a mode of analysis and an empirical indicator with which to gauge the vast ecological gains realized by the center capitalist states, and the corresponding losses inflicted on the periphery.

While at times Odum seemed to argue for an "energy theory of value," he explained on numerous occasions that emergy analysis was not meant as an energy theory of economic/monetary value, but rather an attempt to gauge *real* wealth in terms of energy.[11] The goal here was to find a material-ecological basis for the critique of the capitalist economy and orthodox, neoclassical economics. The success of this endeavor was necessarily limited, since the world of nature and of production in general is so complex and variegated as to raise fundamental problems of incommensurability facing anyone attempting to bring it within a single measure, such as energy accounting. Nevertheless, Odum's analysis highlighted the ecological-economic contradictions of the capitalist system and pulled the legs out from under neoclassical economics' scientific claims.

What makes an approach to unequal ecological exchange drawing critically on Odum's analysis especially intriguing, in our view, is that it represents a crucial interface between ecological science and Marxian theory. Odum, along with David Scienceman, was engaged in a continual dialogue with Marxian theory for around two decades, from 1983 on, resulting not only in in-depth studies on the work of Marx himself, but also of various Marxist theorists of unequal (economic) exchange, such as James Becker and Samir Amin.[12] Attention to Marx's work grew rather than diminished as Odum's analysis advanced, with particular reference to the logic of unequal-exchange relations. The Marx-Odum connection, moreover, was made possible by Marx's own deep concern with the problem of the metabolic rift between human society and the natural environment, and hence the way in which his work was embedded in an ecological critique.

In this chapter, we thus seek to point to the general theoretical foundations of what we are referring to tentatively as a Marx-Odum dialectic in the treatment of unequal ecological exchange—tied to a more general critique of capitalism's metabolic rift.[13] In order to do this it is necessary to trace the theoretical development of both unequal economic exchange and unequal ecological exchange from the nineteenth century to the present, and show how Odum's analysis, though emerging out of physical science, represents a serious attempt to interface with Marxian and world-system analysis. In this way we hope to contribute to the eventual development of a broader world-system/ earth-system analysis.

The Theory of Unequal Economic Exchange

Although our chief concern here is with unequal ecological exchange rather than unequal economic exchange, the logical and historical relationship between the two is so intertwined as to necessitate a brief history of the latter. The issue of unequal economic exchange, particularly in international transactions, was a problem intrinsic to classical political economy.[14] Most of the major contributors to classical political economy—Smith, Ricardo, J. S. Mill, and Marx—wrote extensively on colonialism and the question of the pillage during the mercantilist era of what is now known as the Third World.[15] For liberal political economists such as Smith, Ricardo and Mill criticism of colonial practices was part of a general theoretical defense of free trade.

The theory of unequal exchange itself arose, ironically, out of the Ricardian theory of international trade.[16] Ricardo's famous theory of comparative advantages in international exchange relations was originally illustrated using a two-production sector, two-country model: wine and cloth in Portugal and England. Portugal, in Ricardo's example, produced both wine and cloth more efficiently, that is, with less total labor time than England, and thus had an absolute advantage over England in the production of both commodities. Nevertheless, Portugal had a comparative advantage in making wine over cloth, since it was most efficient in producing the former, while England had a comparative advantage in producing cloth over wine. Under these circumstances, both countries would be best off, he demonstrated, if they each specialized in trading that product in which they were relatively most efficient—in Portugal's case wine, in England's cloth. The result would be to provide the maximum benefit in terms of the total use values produced (cloth and wine) for both countries.[17] This theory still remains the basis of mainstream international trade theory, repeated in every introductory economics textbook.

In presenting his theory of comparative advantages and international trade, Ricardo inverted his usual economic standpoint, developing an argument that was based not on value generated in production and the formation of prices of production, but rather on supply and demand. Rooting his argument in the then-realistic assumption of the international immobility of capital and labor, Ricardo saw trade in the international realm as dictating to production rather than the other way around.[18] This inversion of the argument of classical economics diverted attention from the fact, well recognized by Ricardo, Mill, and Marx, that the reality behind the Ricardian comparative advantage theory was one of unequal exchange (associated with differing productivities and labor intensities in different countries). Hence, Ricardo himself acknowledged as part of his theory that trade would result in one country receiving less labor for more, while the other country would be gaining more labor for less, reflecting the greater intensity and productivity of labor in one country as opposed to the other.[19] "All that this theory [the Ricardian theory of comparative advantages in international trade] allows us to state," Samir Amin sums up, "is that, at a given moment, the distribution of levels of productivity being what it is, it is to the interest of the two countries to effect an exchange, even though it is unequal."[20]

For Marx, although he did not write his planned volume on the world economy and crises, and did not develop his ideas fully on the subject, the reality of unequal economic exchange was obviously of great importance.[21] "Even according to Ricardo's theory," Marx noted, "three days' labour of one country can be exchanged against one of another country."[22] In this case, the richer country exploits the poorer one, even where the latter gains by the exchange, as John Stuart Mill explains in his *Some Unsettled Questions*. In international trade, Marx observed in *Capital*, "the privileged country receives more labour in exchange for less," thereby obtaining "surplus profit," while, inversely, the poorer country "gives more objectified labour in kind than it receives."[23] Likewise: "Two nations may exchange according to the law of profit in such a way that both gain, but one is always defrauded. . . . One of the nations may continually appropriate for itself a part of the surplus labour of the other, giving back nothing for it in the exchange."[24] Related to this was the fact that "the profit rate is generally higher there [in the periphery] on account of the lower degree of development, and so too is the exploitation of labour, through the use of slaves and coolies, etc."[25] Cheaper imports could thus raise the rate of profit in the metropolitan countries by reducing the costs of subsistence or of constant capital. It was therefore possible to see "how one nation can grow rich at the expense of another" even under conditions of free trade, and the more so where monopolies and colonial relations apply.[26]

If Marx laid the foundations of unequal exchange analysis, the articulation of a definite theory of unequal exchange is usually thought as having emerged with the work of the Austrian Marxist Otto Bauer, who argued:

> The capital of a more highly developed region has a higher organic composition, which means that in this more advanced area a larger quantity of constant capital corresponds to the same size of wage fund (variable capital) than in the backward area. Now Marx has taught that owing to the tendency to equalization of the rate of profit, it is not the labor of each of the two areas respectively that produces the surplus value taken by each area's capitalists: the totality of the surplus value produced by the workers of both areas will be shared between the capitalists of those two areas not in proportion to the amount of labor contributed in each but in proportion to the amount of capital invested in each. Since in the more highly developed area there is more capital to the same amount of labor, this area appropriates a larger share of the surplus value than would correspond to the amount of labor it has contributed. . . . Thus, the capitalists of the more highly developed areas not only exploit their own workers but also appropriate some of the surplus value produced in the less highly developed areas. If we consider the prices of commodities, each area receives in exchange as much as it has given. But if we look at the values involved we see that the things exchanged are not equivalent.[27]

Bauer's argument necessarily departed from Ricardian theory of foreign trade premised on the international immobility of capital by pointing to the competitive equalization of profit rates between regions or countries, which could only occur on the basis of capital mobility. This approach to unequal exchange associated with the effects of differing organic compositions of capital has been called "unequal exchange in the broad sense."[28]

The notion that unequal exchange derived from differences in organic composition was one that received considerable support among Marxist economists, with Henryk Grossman notably following the main lines of Bauer's argument. "International trade," Grossman argued,

> is not based on an exchange of equivalents because, as on the national market, there is a tendency for the rate of profit to be equalized. The commodities of the advanced capitalist country with the higher organic composition will therefore be sold at prices of production higher than value; those of the backward country at prices of production lower than value.[29]

Such non-equivalent exchange theory played a significant role in Soviet debates on the economic relations between developed and underdeveloped regions.[30]

In the 1970s, a different but related theory of unequal exchange appeared in the work of Arghiri Emmanuel and Samir Amin that viewed unequal exchange in its most appropriate designation not as arising primarily through differences in organic composition between countries, but rather from differences in wage levels and rates of surplus value—in those cases where the differences in wages were greater than the differences in productivities. For Emmanuel, this "narrower" conception of wage-based unequal exchange was seen as rooted in the international mobility of capital and the international equalization of profits—together with the international immobility of labor.[31] Transfer of free or "hidden" value from low-wage to high-wage countries was viewed as occurring by means of the price mechanism.[32] Emmanuel's analysis stipulated that free trade conditions applied and strictly excluded monopoly as a consideration. "As for the actions of the monopolies, of which the Marxist authors talk so much, this question is as remote from our subject as any other form of direct plunder of the underdeveloped countries by the rich and strong ones."[33]

Emmanuel's rejection of monopoly and plunder as factors, given that his goal was to demonstrate the existence of unequal exchange even under free-trade conditions, made his theory less historically relevant, and led to a shift toward a more realistic, if less logically tight, theory of unequal exchange. This could already be seen in the broad tradition of dependency and world-system theory associated with Baran, Sweezy, Frank, and Wallerstein.[34] It was in this wider, historical sense of unequal exchange that leading Marxist thinkers such as Paul Baran, Paul Sweezy, and Ernesto "Che" Guevara referred to unequal exchange in the early 1960s. Writing in 1964, Baran and Sweezy explained:

> Unequal relations between the developed and underdeveloped countries result in the establishment of terms of trade which greatly favor the former at the expense of the latter. In this way wealth is transferred from the poor countries to the rich.[35]

Yet, their analysis did not stop with mere trade but emphasized the manifold ways in which monopolistic multinational corporations created a net flow of surplus from the underdeveloped countries to the developed countries. Che wrote of "prices forced on the backward countries by the law of value and the international relations of unequal exchange that result from the law of value." The "so-called deterioration of the terms of trade" was "nothing but the result of the unequal exchange between countries producing raw materials and industrial countries, which dominate markets and impose the illusory justice of equal exchange of values." Che argued that "monopoly capital" now dominated the world, imposing its wider forms of exploitation and unequal exchange.[36]

For Wallerstein, "unequal exchange" develops from a quasi-monopoly system involving "politically strong" core states and their economically strong corporations and is not easily distinguished from "plunder."[37] It was in this wider, historical sense—extending Emmanuel's earlier free-trade-based analysis to account for a reality in which monopoly played a central role—that unequal exchange became a generally accepted part of world-system theory.[38]

The synthesis of these various traditions was left to Amin, who stressed that Emmanuel's work derived its importance from its focus on global wage inequality and the problem of "international value."[39] Viewing actual historical conditions in terms of a world of increasingly "generalized monopolies" (monopolistic multinational corporations), Amin emphasized the tendency toward equality in organic composition of capital (that is, productivities) worldwide, since the same technology was increasingly being employed everywhere. This nonetheless was accompanied by wage inequality, unequal rates of surplus value, and higher profits in the periphery than the center. These conditions pointed to a theory of unequal exchange as a global transfer of value or "imperial rent."[40] Trade inequities were accompanied by numerous other forms of surplus extraction from the periphery—all, however, rooted finally in the wage differentials between the Global North and Global South.

Much of today's imperial rent remains disguised by exchange rates (as indicated by the difference between market-value exchange rates and purchasing-power parities).[41] Nevertheless, unequal exchange can be shown to be broadly measurable in order of magnitude.[42] Unequal economic exchange/imperial rent rests ultimately on the fact that the differences in wages between center and periphery are greater than the productivities, allowing extensive capture by the center economies of value created in the periphery.[43] This embodies the fundamental characteristic of all unequal economic exchange: the exchange of more labor for less.

The Theory of Unequal Ecological Exchange

Just as unequal economic exchange theory postulated the exchange of more labor for less, unequal ecological exchange theory had as its basis the exchange of more ecological use value (or nature's product) for less. Unequal ecological exchange was first raised as a major issue in the work of Liebig and Marx. From the 1840s to the 1860s, the great German chemist Justus von Liebig introduced a critique of industrial agriculture as practiced most fully in England, referring to this as a condition of *"Raubbau"* or the *"Raubsystem,"* a system of robbery or overexploitation of the land and agriculture at the behest of the new industrial capitalism emerging in the towns.[44] In Liebig's view, the elementary soil nutrients, nitrogen, phosphorus and potassium, were being removed from the soil and sent to the cities in the form of food and fiber, where they ended up contributing to pollution rather than being recirculated to the soil. The result was the systematic robbing of the soil of its nutrients. English agriculture, then, tried to compensate for this by importing bones from the catacombs and battlefields of Europe and guano from Peru. "Great Britain," Liebig wrote,

> deprives all countries of the conditions of their fertility. It has raked up the battlefields of Leipzig, Waterloo, and the Crimea; it has consumed the bones of many generations accumulated in the catacombs of Sicily; and now annually destroys the food for future generations of three millions and a half of people. Like a vampire it hangs on the breast of Europe, and even the world, sucking its lifeblood.[45]

Marx developed Liebig's approach into a more systematic ecological critique of capitalism by designating the robbery of the earth as "an irreparable rift in the interdependent process of social metabolism," or *metabolic rift*.[46] Such conditions were, for Marx, the material counterpart of the capitalist organization of labor and production. It constituted the alienation of the "metabolic interaction" between humanity and the earth, that is, of the "universal condition" of human existence.[47]

The metabolic rift under capitalism was connected to unequal ecological exchange. England, as the leading capitalist country at the center of a world-system, Marx stated, was "the metropolis of landlordism and capitalism all over the world," drawing on the resources of the globe, with nations in the periphery often reduced to mere raw material providers. "One part of the globe" is converted "into a chiefly agricultural [and raw material] field of production for supplying the other part, which remains a preeminently industrial field." Thus a whole nation, such as Ireland, could be turned into "mere pasture land which provides the English market with meat and wool at the cheapest possible prices." Indeed, Ireland was reduced by imperialist means to "merely an agricultural district of England which happens to be divided by a wide stretch of water from the country for which it provides corn, wool, cattle and industrial and military recruits." The resulting "misuse" of "certain portions of the globe" in the periphery of the system is thus determined by the accumulation imperatives of the center.[48] Marx illustrated the absolute robbery involved in the appropriation of the natural wealth of the one country by another by stating, "England has indirectly exported the soil of Ireland, without even allowing the cultivators the means for replacing the constituents of the exhausted soil."[49] Like Liebig, Marx pointed to the fact that England was forced to import guano in massive quantities from Peru (in a world-system of exploitation that also involved importing Chinese labor to dig the guano) in order to make up for the loss of nutrients in English fields.[50]

Marx saw production as a flow of both material use values and exchange values or, simply, values. He used the term "metabolism" (*Stoffwechsel*) to refer to the material exchange (the exchange of matter-energy) that always accompanied monetary exchange of value.[51] Such material exchange was associated with the production of use values, representing the material conditions of production in general, as opposed to exchange value (value). A social "use value" is quite literally for Marx a "piece of natural material adapted to human needs by means of a change in its form."[52] It was this twofold aspect of his analysis—as material-physical and value-related—that allowed Marx to perceive the contradictions between use value and exchange value and between the accumulation process and natural-material conditions.[53] In Marxian theory, this has been understood as constituting the dual value problems: the "qualitative value problem" and the "quantitative value problem."[54] Unequal economic exchange is mainly concerned with a quantitative value problem related to exchange-value relations (and a break in this at the international level), while unequal ecological exchange is chiefly concerned with use-value relations and real wealth (including the contradictions between use value and exchange value).[55]

Marx emphasized that human production still employed "many means of production which are provided directly by nature and do not represent any combination of natural substances with human labour."[56] Such direct products of nature, the result of nature's work, were treated under capitalism, he pointed out (following the classical political economists who had preceded him), as "free gifts" that did not enter into the value process of the system.

Consistent with this, Marx drew a distinction between real "wealth," to which both nature and labor contributed, and value, where only labor was taken into account.[57] It was the inherently one-sided nature of the value calculus of capitalist production that led to the robbing of nature. that is, the failure to provide for the full "restoration" of what had been taken from the earth. "Capitalist production," he wrote, "only develops ... by simultaneously undermining the original sources of all wealth—the soil and the worker."[58]

Recent scholarship has shown the extent to which Marx integrated his political economic analysis with the new conception of thermodynamics appearing in his time, as reflected in his argument on metabolism.[59] It is hardly surprising, therefore, that most early forms of ecological economics were heavily indebted to Marx.[60]

Marx's treatment of unequal exchange, imperialism, and the global metabolic rift meant that the notion of unequal ecological exchange arose periodically in Marxian political economy, although its role within the Marxian critique was minor prior to the 1970s.

For Eduardo Galeano, production was so organized in colonial and neocolonial Latin America as to constitute "a sieve for the draining-off of natural wealth" to the benefit of the colonizers.[61] Emmanuel contended that the advanced capitalist countries were using up the ecological commons, ridding "themselves of their wastes by dumping them into the sea or the air," which was possible because they were the only ones doing it."[62] Amin commented explicitly on "a whole series of 'unequal exchanges'" related to ecological factors existing side by side with the unequal exchange of labor.[63] Such "other forms of unequal exchange" were for Amin crucial to understanding the role that the extraction of natural resources from the periphery played in the overall analysis of imperialism:

> The capitalist system makes use of the precapitalist forms of appropriation that are current in the countries of the periphery in order *not* to pay for the upkeep of the land. Systematic destruction of soils is a major factor of long-term impoverishment for the dependent countries.[64]

If Marxian political economy naturally led to theories of unequal ecological exchange, what is generally seen as the "Malthusian" tradition (related specifically to carrying capacity) generated an approach that was in many ways overlapping.[65] In 1965, Georg Borgstrom, a food scientist at Michigan State University, published his book *The Hungry Planet*, which devoted a chapter to what he called "ghost acreage." What allowed some countries to overshoot their available land or ecological base was the import of food from elsewhere—other countries or the sea. Such "ghost acres" permitted wealthy countries, like the Netherlands, to develop population density and industrialized production while having an inadequate agricultural base (and allowed them to draw on tropical products). In the poorer countries, meanwhile, committing this "ghost acreage" to production for export to the rich countries decreased the food acreage available for local subsistence. The idea of "ghost acres" in terms of land was, Borgstrom argued, aimed at providing a "commensurate gauge" with which to record ecological usage. Thus he was concerned with "devising methods whereby the use of commercial fertilizers and the energy inputs" used in agriculture could be "computed in corresponding terms and added to the ghost acreages."[66] This approach can be regarded as a forerunner of ecological footprint analysis.

THE THEORY OF UNEQUAL ECOLOGICAL EXCHANGE

Within sociology, the issue of unequal ecological exchange is often seen as being brought to the forefront by Stephen Bunker's 1985 study, *Underdeveloping the Amazon: Extraction, Unequal Exchange, and the Failure of the Modern State*. Bunker incorporated "mode of extraction" as the counterpart of the "mode of production." The unequal exchange of energy and materials occurred to the detriment of extractive economies or "extreme peripheries." He wrote:

> There are multiple inequalities in international exchange. One, certainly, results from the differential wages of labor. Another, however, is in the transfer of the natural value in the raw resources from periphery to center. . . . The outward flows of energy and the absence of consumption-production linkages combine with the instability of external demand and with the depletion of site-specific natural resources to prevent the storage of energy in useful physical and social forms in the periphery, and leave it increasingly vulnerable to domination by energy-intensifying social formations at the core. Finally, if the resources do not renew themselves naturally, the inequality of the exchange is intensified by the loss of resources and by the disruption of associated natural energy flows in the periphery itself.[67]

Christopher Chase-Dunn saw Bunker's analysis as key in developing a theory of unequal ecological exchange, observing:

> Use values are lost to the [underdeveloped, extractive] region both through exports of the resources and through the disruption of the ecosystems from which they are extracted. Unequal exchange of labor is accompanied by the unequal exchange of matter and energy.[68]

Still, despite the insights of such varied analysts, the problem of developing a coherent theoretical and empirical approach to the issue of unequal ecological exchange has remained. Some world-system theorists usefully argued that the dominant nineteenth-century world-system was a dissipative structure imposing entropy on its periphery.[69] But clear conceptual frameworks illustrating and operationalizing this were lacking.

A major breakthrough came with the development of ecological footprint analysis in the 1990s.[70] The ecological footprint was devised as the inverse of the old carrying capacity notion of ecology. Instead of asking, as in the analysis of carrying capacity, how much population or environmental load a particular unit of land would support, the ecological footprint inverted the question, asking how much land was required to support a particular environmental load, or a given population with a given per capita consumption. Land, measured in hectares, thus became a common metric for the extent of environmental services that went into providing a given consumption level on an indefinite basis.

Ecological footprint analysis has facilitated inquiries into the ecological impacts of nations by capturing the larger "footprint" extending beyond national borders.[71] This made it possible to determine the extent to which a given region or country overshot its own land/resources, relying on an environmental deficit or overdraft or, alternatively, "environmental load displacement" with respect to the rest of the globe.[72] By providing a basis, however limited, for measuring ecological consumption from the individual level all the way up to the world system, ecological footprint analysis, as Amin has insisted, made possible a more trenchant use-value-based critique of capitalist accumulation.[73]

The ecological footprint has inspired considerable empirical research, mainly within sociology, directly aimed at assessing unequal ecological exchange.[74] Ecological footprint analysis demonstrates that larger footprints are primarily a function of economic development, and do not match the ecological carrying capacities of particular nations. The more developed countries have larger ecological footprints but less domestic environmental degradation within their borders, while less developed countries have smaller footprints and more environmental degradation within their borders.[75] The obvious explanation for these disproportionalities in environmental impact is that the center capitalist countries rely heavily on importing resources from countries of the periphery, and engage in various forms of outsourcing of production and environmental load displacement.[76]

Yet, while useful in demonstrating the uneven ecological impact of nations, in terms of the environmental loads they require to support their consumption and the uneven appropriation of global environmental space, the ecological footprint does not in itself measure actual material exchanges, the use-value transfer, or the spatial origins of goods consumed.[77] World-system analysts and environmental sociologists concerned with unequal ecological exchange have therefore sought to connect the dots, employing price-based data to show that those less-developed countries with a relatively higher level of exports to developed countries have, at the same time, smaller ecological footprints and suffer disproportionate environmental degradation. In such analysis it is presumed, based on historical experience, that exports from the periphery are heavily weighted to natural-resource exports.[78] However, the price-based data used in broad cross-national studies do not generally allow for disaggregation of the types (physical character) of goods traded, while exports from the Global South are increasingly manufacturing-based, calling into question this assumption. The whole argument relies on broad inferences from price relations without a direct consideration of transfers of real wealth.

Indeed, the theory of unequal ecological exchange that emerged from such studies has been devised in a somewhat Procrustean manner to fit the empirical data available. Although recent studies in environmental sociology and ecological economics strongly point to the existence of ecologically unequal exchange, there are serious problems at the level of both empirical analysis and, more important, underlying theory. Existing approaches have relied on data in which the ecological (indeed physical) content of the goods is unknown and quantitative measures are in terms of prices rather than goods. As a result, very little is actually revealed in most current empirical studies of ecologically unequal exchange about the ecological nature of the exchange itself in terms of matter, energy, resources, etc. The theory, which is harnessed to such empirical data, is vague and roundabout, drawing large generalizations about environmental load displacement, while failing to engage directly with what would logically constitute the core element in any theory of unequal ecological exchange: the exchange of more ecological wealth for less. For example, Andrew Jorgenson tells us in a roundabout way what we already know:

> Developed countries with higher levels of resource consumption externalize their consumption-based environmental costs to less-developed countries, which increase levels of environmental degradation within the latter.... The majority of extracted materials as well as agricultural products and produced goods [of underdeveloped countries] are exported to and consumed in more developed countries.[79]

Despite the pioneering nature of such analyses, we learn little or nothing here about the processes involved or the real extent of the unequal exchange.

In short, the standard analyses of "ecologically unequal exchange" are dependent largely on ecological-footprint analysis, arising from the traditional notions of carrying capacity. This is then coupled with an examination of trade relations in price terms (mostly with respect to the directionality of trade). All of this represents an attempt to establish broad correlations—as opposed to a historical-theoretical examination of the structures and processes of unequal ecological exchange within the world-system. Despite the fact that the concepts of unequal economic exchange and unequal ecological exchange both arose from classical Marxian theory, there is no direct recourse to classical Marxian analysis beyond the inconsistent allusions of Bunker, who rejects the labor theory of value (along with neoclassical theory) in favor of an undefined theory of "natural value."[80] As a critique, the standard ecologically unequal exchange perspective therefore remains theoretically undeveloped, failing to make full use of the crucial use-value and exchange-value distinction of the classical Marxian value-based perspective.

The main obstacle confronting empirical analysis within this theoretical domain is of course the problem of incommensurability: the lack of a common metric beyond price.[81] The problem in conceiving processes of unequal exchange, as Hornborg put it, is that

> most trade statistics are in monetary units, rather than invested labor time, energy, or hectares.... If invested energy or hectares were counted instead of dollars, the significance of imports from the south would be recognized as much greater than that suggested by monetary measures.[82]

In essence, the problem becomes the lack of a common metric (or a number of related common metrics) with which to begin to analyze unequal ecological exchange. It is here that Odum's analysis takes on significance.

Odum and Real Wealth Analysis

Howard Odum and his older brother Eugene Odum are generally considered the foremost systems ecologists of the late twentieth century, having largely created the field.[83] They coauthored *The Fundamentals of Ecology*, the foundational text in systems ecology, which "created a generation of ecosystem ecologists—as distinct from plant ecologists and animal ecologists—who were prepared mentally and technically to contribute to the environmental decades."[84] What was previously a narrow, technical field was brought into the mainstream of biological analysis.[85] Central to their work was the use of the concept of metabolism to refer to all biological levels from the cell to the ecosystem.[86]

In the final decades of his life, from 1983 to 2002, Howard Odum developed a method for measuring the total work of ecosystems embodied in commodities resulting from economic and ecosystem processes. This provided a way of calculating the extent of natural wealth (in energy terms) exchanged between countries, or the loss of a country's natural endowment through commodity trade. He called this embodied ecosystem work—measured in terms of the energy required to produce or sustain a commodity, natural resource, or entire national economic system *emergy*. Emergy thus was meant, in Odum's conception, to provide a common energetic metric for measuring real wealth/use values.

For Odum, unequal ecological exchange arose as a result of "imperial capitalism."[87] Trade relations, it was shown, resulted in some countries exchanging more emergy (embodied energy) for less. Given large inequalities in ecological exchange, it was impossible for peripheral countries to foster long-term development that was ecologically sustainable, relying on exports, so long as unequal ecological trade relationships persisted. The analysis was constructed with close analytical attention to Marxian value theory and Marxian theories of unequal exchange, which were used as ways of getting at the somewhat parallel considerations of emergy analysis.

Odum made it clear, repeating this again and again, that he was not attempting to construct an energy theory of economic value.[88] Rather, in a manner somewhat parallel to Marx's theory, the analysis pointed to circuits of material use value and exchange value (abstract value) that were in contradictory relation to each other (economic value moving in a circular flow, energy/emergy within a thermodynamically open system) resulting in the robbing of the earth and the failure to provide for the replacement of lost ecological wealth in a system dominated by the accumulation of labor values.[89]

The key to Odum's theory of unequal ecological exchange was the emergy concept. The emergy nomenclature—emergy, transformity, empower, emvalue, emdollar—and the conceptual innovations accompanying it were introduced by Scienceman in collaboration with Odum, beginning in 1983, following Scienceman's study of Odum's *Systems Ecology*.[90] The original motivation for devising the new terminology was to avoid confusions that had arisen in Odum's theory through his use of the concept of "embodied energy."[91] This notion allowed for numerous interpretations and appeared to conflict with the way the concept of energy was commonly used in science to refer to available energy or exergy.[92] Moreover, the concept of "embodied energy" was often confusing, since "embodied," in the sense that Odum used it, meant something more like the effects of a jelly bean entering a body than a bullet, that is, the energy was utilized and dissipated.[93] The essential idea of Odum's "embodied energy" was one of the past energy, no longer physically present in the same form or degree, that went into making an object or product—an approach roughly analogous to Marx's concept of value as arising from dated inputs of labor. All of this led to the introduction of the emergy category. As Odum explained:

> In 1983, the term EMERGY, spelled with an "M," was suggested by David Scienceman for our concept [of embodied energy] and emjoule or emcalorie as the unit.... EMERGY is defined as the energy of one kind required directly and indirectly to produce a service or product.... For example, the production of green plants can be expressed in solar emjoules, which includes the solar energy required to make all the inputs to the plant, such as rain, wind, nutrients, cultivation efforts, seeds, and so forth.[94]

In essence, "EMERGY, a measure of real wealth, is the work previously required to generate a product or service."[95] The "m" in emergy was meant to symbolize energy memory, or the fact that this was an accounting system aimed at total energy inputs over time.[96]

Emergy analysis was aimed at a method that would take the various forms of energy that went into the making of a product or service and transform them into "units of one kind."[97] Crucial to this was the concept of "transformity," defined as "the EMERGY of one type required to make a unit of energy of another type," usually measured in solar emjoules.[98] "Because EMERGY evaluation traces what was required for a product back to a common form of energy, it is a way of showing how the requirements for different products compare."[99]

Goods with higher transformity represented dated inputs of emergy (including entropy or dissipated energy) that went into their production. Higher transformity was associated with the emergence at higher levels of production of more useful products in forms more accessible to human beings. One cannot eat sunlight or crude oil, but one can eat potatoes grown with the aid of such energy sources.[100]

Thus "work increases the utility of energy while degrading and dispersing part of that energy."[101] For example, it is well known that it takes about 4 calories of coal to generate a calorie of electric power, giving electricity "a higher transformity," associated with greater usefulness, and higher quality—even though available energy has been lost in the process.[102]

Odum's emergy concept and his notion of transformity are especially indebted to Lotka's development of the maximum power principle as a law of thermodynamics.[103] However, because Lotka did not specify systems principles based on qualities of energy, Odum modified Lotka's statement of this principle by placing energy of each level on a common basis using the concept "empower," or power as a representation of higher levels of energy transformity. "Prevailing systems," he declared, "are those whose designs maximize empower by reinforcing resource intake at the *optimum* efficiency."[104] Systems must operate according to principles dictated by the "universal energy hierarchy" which "provides transformities for quantitatively relating energy on one scale to that of another."[105]

Odum and Marx: Toward a Dialectical-Ecological Synthesis

The significance of Odum's analysis is brought out most fully in a comparison with Marx. In a remarkable cross-fertilization of ideas between physical science and social science, Odum and Scienceman developed the emergy-transformity framework while conducting a decades-long investigation into Marxist political economy, and particularly the labor theory of value. The close connection between Odum's ecological critique and Marxian political economy is reflected in the overlapping critiques of mainstream (today neoclassical) economics with its subjective theory of value.

Odum and Scienceman viewed emergy analysis as a form of real (ecological) wealth accounting and employed the concept of "emergy value" or "emvalue" to distinguish this approach from the labor theory of value or "lavalue," which they saw as a related "donor" or production-based theory of value connected to one energetic input-labor, as well as from other economic forms of value.[106] "In the Odum terminology," Scienceman wrote, "use value, being the bodily form of a commodity, would refer to the value (emvalue) in solar emergy content."[107] In other words, emergy value (or emvalue) referred to use value or "real value" (real wealth). It was not to be confused with economic or monetary value.[108] Odum referred to emvalue as "a second value, the contribution of real wealth, how to use real wealth," which was distinct from "market value" or economic value.[109] "Emergy," he stressed, "measures natural value—real wealth."[110] Not only was money not a measure of real wealth, the relation was often an "inverse" one, with prices "being lowest when [ecological] contributions are greatest."[111] The whole analysis pointed to a notion of "emvalue in a value added hierarchy" that resembled Marx's analysis but was oriented instead to real wealth, seen as in contradiction with the labor-value (or human services) basis of the capitalist economy.[112]

Odum saw great significance in Marx's linking of his approach to labor value to thermodynamics, which was being developed in his time and was integrated into his theory.[113] Marx himself wrote: "Creation of value is transformation of labour-power into labour. Labour-power itself is energy transformed to a human organism by means of nourishing matter."[114] For Odum, Marx's theory was an attempt to explain wealth/value creation under capitalism encompassing both value and exchange value. The more physical or use-value side of Marx's analysis was viewed as having an energetic or ecological character.[115] Thus abstract labor in Marx's theory was depicted by Odum and Scienceman, following Heilbroner, as "weighted by some as yet inadequately explained calculus," which for Odum and Scienceman clearly represented its emvalue.[116] Marx's approach, viewed from this perspective, was a "donor" value parallel to "emergy value."[117] Marx saw economic value as "coming from human hours contributed"; Odum saw "emergy value [as] derived from the resource contributions."[118] In both cases, the focus was on production (natural and social).

Here it is crucial to interject that a certain ambiguity remained in Odum's interpretation of Marx's economic analysis in terms of physics/energetics. At times he (and Scienceman) criticized Marx's value theory for not being a kind of pure physics—as if historical economic forms could be reduced straightforwardly to energetics in the manner of Podolinsky, whom Engels had criticized.[119] More often, Odum seemed to recognize that Marx's distinction between value (exchange value) and wealth (use value), and the contradiction this represented for capitalism, constituted the real strength of Marx's theory. Indeed, this same contradiction was repeatedly raised in Odum's own analysis, which set real wealth against exchange value and made this the sharp edge of his critique of neoclassical economics. As indicated above, Odum's reduction of qualitative/scale distinctions with respect to the energy hierarchy to a single common metric (emergy), though useful in the analysis of unequal exchange, also lends itself to reductionism if it leads to ignoring other dimensions of nature/ reality. For life, there is no single metric. Thus Martinez-Alier is quite right in cautioning that the use of a concept "like emergy," aside from the inherent difficulties in

> calculation and application, would only account for one aspect of the link between extraction of resources and the environment. The important point is not the difficulty of calculation. The essential point ... is that incommensurability applies not only to money value but also to physical reductionism. Can "biopiracy" be reduced to energy calculations?[120]

It is naturally impossible to measure fully the impact in energy/ emergy terms of the extinction of a single species, such as the golden toad or the Javan tiger.

Yet, with these qualifications in mind, it is nonetheless clear that the conceptual approach offered by ecological systems theory has much to offer. One must be wary of energy reductionism, but energy flows are nonetheless crucial to developing a comprehensive approach to unequal ecological exchange. Odum's systems ecology, though open to question for the reductionism it sometimes encouraged, is too revealing in scientific terms to be disregarded.

The strength of Odum's approach is revealed in his deep engagement with Marxian political economy. In a letter written to Engels on July 6, 1863, Marx provided a diagram (which he referred to as an "Economic Table") of his reproduction schemes for capitalist production, distinguishing this from Quesnay's early "Tableau economique."[121] Odum and Scienceman translated this diagram of Marx's reproduction schemes into an energy-systems language diagram.[122] They then went on to develop a deep analysis of Marx's political economy, transposing his systemic view of the capitalist economy into emergy-systems language/diagrams/equations and modeling it under different conditions (for example, steady state, expanded reproduction), running various computer simulations. This was most fully developed in their 26-page chapter "An Energy Systems View of Karl Marx's Concepts of Production and Labor Value."[123] In this view, "Marx was basically trying to introduce a labor transformity scale [to explain the capitalist economy], based on an intermediate (labor energy) source rather than an original (solar energy) source."[124]

Although it is clear that Odum and Scienceman cannot be characterized in any sense as Marxists in their overall worldviews, their research into the Marxian system was thoroughgoing, reaching beyond Marx's *Capital* into the wider Marxian treatment of value theory, the transformation of values into prices of production, the reproductive schemes, unequal exchange, and the role of nature in capitalist production. In the process they scrutinized the work of such thinkers as Amin, Becker, Carchedi, Cleaver, Foley, Goodwin and Punzo, Heilbroner, Howard and King, Krause, Lonergan, Martinez-Alier, Morishma, Rubin, Samuelson, Seton, and Wolff.[125]

Marx, Odum and Scienceman noted, had stipulated that all production was based on nature and labor, the ultimate sources of wealth.[126] Yet, in a capitalist economy, as depicted by Marx, the "region" of labor values defined the realm of commodity production. Capitalism, in its value relations, thereby excluded nature (independent of labor) as a source of value.[127] Here Odum and Scienceman appear to have accepted the labor theory of value as operative in the "region" of capitalist economics in the manner depicted by Marx, while arguing (as Marx himself did) that the realm of real wealth was much larger, encompassing nature's work (emvalue).

For Marx, value in the system of economic accounting that characterized capitalism was the result of the addition of labor (or in Odum's system, the addition of human services) to what nature has already provided *gratis*. As we have seen, Marx, like the classical economists who preceded him, referred to the production of nature itself, independent of labor, as a "free gift" to capitalism in that it did not enter into the (economic) value-added of the system.[128] For Marx, however, this was a contradiction of the system itself, constituting a form of robbery or overexploitation *(Raubbau)* generating a metabolic rift.[129]

In classical political economy, the contradiction between use value and exchange value was commonly viewed in the form of the famous Lauderdale Paradox (named after the early classical economist the Earl of Lauderdale), whereby the expansion of *private riches* was seen as based to a considerable degree on the destruction of *public wealth*. For example, the destruction of certain crops by landowners in order to artificially inflate their market prices represented the despoliation of real public wealth (use values) for purposes of enhancing private riches (exchange values). This was viewed not as a rare instance, but as an intrinsic feature of a capitalist economy.[130] Odum followed Marx and other classical economists in incorporating the Lauderdale Paradox into his analysis, thereby pointing to a global capitalist destruction of natural wealth for private enrichment.[131]

In general, Odum seemed to argue that Marxian theory, in emphasizing labor power rather than energetic inputs in general, had failed to develop an adequate analysis of the role of real wealth in production, requiring that this be put on a more scientific basis through emergy analysis.[132] However, in various places Odum offered a more subtle interpretation of Marx, seeming to recognize that the latter was depicting what was a real contradiction of the capitalist economy between the accumulation system and nature, a contradiction that Odum also recognized in his own critique of neoclassical economics.[133] Indeed, Odum's position was in many ways more similar to Marx's than the former realized, since Marx theorized the limitations of the law of value under capitalism, given that it did not incorporate nature's role in the creation of wealth.[134]

Odum and Scienceman's sharpest criticism of Marx was directed at Marx's argument that since the price of labor was lower in rural, especially underdeveloped, regions, workers were highly exploited there.[135] "Emergy evaluation," they wrote, "indicates a different interpretation. . . . Emergy values for products from rural countries in relation to price are higher than in developed countries in relation to price because more of the support of labor comes from the landscape without payment."[136]

In this respect, however, Odum and Scienceman underestimated Marx and Marxian theory. Marx and Engels explicitly indicated that workers could be paid less than the value of labor power for long periods of time only in those cases where the reproduction of labor was supported by marginal access to land, that is, ecological resources. In Marxian terms, such labor under capitalism becomes the basis of superprofits arising from "profits by deduction," that is, deductions from the price of labor or the value of labor power.[137] Such superprofits were made possible by the fact that wages did not cover the full cost of reproduction (the value of labor power) of the workers. What made this possible was most clearly described by Engels in the second edition of *The Housing Question,* where he explained that kitchen gardening and small-scale agriculture had allowed German workers to be paid extremely low wages generating exceptionally high profits, which amounted to a "deduction from normal wages," that is, from the level of wages necessary for workers who had no access to land and the ability to grow their own food.[138] Hence, like Odum, Marx and Engels argued that exceptionally low wages in rural areas were due to nature's subsidies.

It is evident from all of this that Marx's critique and Odum's emergy analysis have a certain affinity. Both focus on the contradiction between use value and exchange value. Odum provided a concrete way of understanding the inequalities and losses in real wealth imposed by the capitalist system. As a non-historical systems-theory model, however, his treatment was dependent on a somewhat artificial impetus, giving direction and purposefulness to the analysis—Lotka's maximum power, or Odum's maximum empower.[139] Hence, the energetic parameters of the system were necessarily conceived in mechanical, universalist terms. As with many systems-theory approaches, Odum's analysis represents

> the attempt of a reductionist scientific tradition to come to terms with complexity, non-linearity and change through sophisticated mathematical and computational techniques, a groping toward a more dialectical understanding that is held back both by its philosophical biases and the institutional and economic contexts of its development.[140]

Where its holistic/systemic outlook and attempted dialectical break with reductionism offers new critical insights, such an approach can be cautiously utilized. Where its break with a mechanistic scientific tradition remains incomplete, and where reductionism is reproduced within its analysis, it needs to be subjected to critique. In Odum's overall systems ecology, a rose, a butterfly, an ecosystem and a symphony orchestra can be evaluated in terms of the maximum empower principle and hence optimum efficiency from an energetic standpoint. This may tell us something about each of these objects from the standpoint of physics, but the resulting information is limited by the narrowness of the measure adopted.[141] Learning from such a systems ecology approach is one thing; falling prey to the reductionism to which it can potentially lead is another. In any Marxian analysis, ecological materialism must take theoretical precedence over energetics, much as historical materialism, as Amin argues, takes theoretical precedence over the law of value.[142]

Odum and the Theory of Unequal Ecological Exchange

None of the limitations of Odum's overall systems ecology, however, prevent us from drawing upon his approach to unequal ecological exchange. Although the general structure of Marx's labor-value theory of capitalist production was a source of inspiration for Odum, it was the Marxian theory of unequal economic exchange that was of most concrete interest, helping him to develop his own theory of unequal emergy exchange. While Marxian theorists used Marx's labor-value concept "to show large imbalances where trade was based on market prices," Odum and Scienceman suggested that systems ecology "had shown [similar] large imbalances using emergy" analysis.[143] A key work, comparing the two approaches to unequal exchange, influencing Odum himself, was provided by Stephen Lonergan in a review of the unequal exchange literature.[144] Lonergan showed that in international trade, in the Marxian approach, more labor was traded for less, while in Odum's analysis more emergy value (or emvalue) was traded for less. In both cases, prices deviated from "values" (though in Odum's case "emvalue" or emergy was directed at real wealth or use value), creating a global transfer to the benefit of the developed countries. Thus "recent empirical work suggests that developed economies import more labour value than they export, and, similarly, they may also import more embodied energy than they export."[145]

Although discussing the work of Emmanuel and Amin, Lonergan highlighted the analysis of unequal economic exchange developed by Becker in his *Marxian Political Economy*.[146] Becker's work emphasized the first (broader) form of unequal exchange theory within Marxism, focusing on differences between organic compositions, and how this affected exchanges between predominantly urban and predominantly rural areas—an approach that was then extended to Global North/South relations. "The law of unequal exchange," Becker wrote,

> ensures that within the less developed countries most departments [of production] will experience on the average unfavorable terms in their exchanges with countries the majority of whose departments of production will experience better-than-average terms. It is not the famous—or infamous—law of comparative advantage that determines commodity flows and their relative rates of exchange. The strains of a mutual harmony of interest, sung so sweetly by economic apologists, are

now and again drowned out by the noise of exchange inequalities and inequities.[147]

It was the very rigorous argument on unequal exchange presented by Becker that seems to have had the biggest influence, within the Marxian secondary literature, on Odum's further development of his own analysis.[148]

In order to understand Odum's theory of unequal ecological exchange, it is necessary to look more closely at his method of emergy analysis. The process for calculating emergy begins with drawing an energy system diagram for the system under study. Odum suggests that experts on a process gather round a table and list all the elements contributing to the system. For example, if you wanted to calculate the emergy of corn, you would draw an energy system diagram illustrating the inputs required to grow corn under the particular conditions. Their relation to one another via energy pathways is also indicated via energy systems' diagramming notation.

Once the system diagram is completed, each input becomes a line item in an emergy evaluation table. In this table, the raw energy data for each line item, found in already existing literature, is multiplied by its previously published or currently calculated transformity (according to the method laid out in Odum's *Environmental Accounting*) to arrive at the solar emergy of each item. In this way, the items may be summed and other indices may be calculated to look at the quantities in relation to one another and to compare systems. Calculations are included in the table according to the needs of the particular study. Emergy-per-dollar calculations are used to relate economic and ecological indicators. The Center for Environmental Policy has now published emergy calculations for the natural resource base of 134 national economies.[149] Actual global maps of emergy use along various dimensions, providing comparative perspectives, are now available.[150]

In emergy analysis, the dispersion and degradation of energy are accounted for in the diagrams used to delineate energy systems. Odum and Arding write: "The definition often used in elementary physics and engineering courses that energy is the ability to do work is incorrect. Degraded energy can't do any work. The work that potential energy [exergy] can do depends on its position in the energy hierarchy."[151]

Available energy or exergy is thus the "potential energy capable of doing work and being degraded in the process."[152] Emergy sums all the previous potential energy inputs in the series of energy transformations required to produce any given output. Exergy analyses, which measure available energy, are thus not as comprehensive as emergy. However, data measuring exergy can be converted to emergy data by multiplying by the correct transformities.[153]

While not proposing emergy as a price/exchange-value determinant, Odum did relate emergy to money and thus commodity values via several indicators that are used to assess, from a real-wealth standpoint, the long-term viability, equity, and sustainability of economic processes like production, extraction, and trade. Having a working knowledge of these concepts is essential to an understanding of Odum's analysis of unequal ecological exchange. Key concepts include the following:

- *The emergy investment ratio* is the "ratio of purchased emergy to the emergy from the local free environment."[154] "The ratio for an area is set by the state of development of the economy using nonrenewable resources."[155] A competitive emergy investment ratio for a rich developed country such as the United States is around 7 to 1, while for many peripheral economies the ratio is 1:1 or less.[156]
- *The emergy/$* (emergy/money) *ratio* is the emergy used from all sources in an economy divided by the Gross Domestic Product (GDP) for that year. A high relative emergy/money ratio means that such countries, usually rural and undeveloped, are drawing heavily on "direct environmental resource inputs not paid for."[157] The exports of such countries include higher levels of emergy for the international dollars received, and have lower relative ecological purchasing power.[158] An emdollar "is the emergy contribution that goes to support one dollar of gross domestic product."[159]

- *The emergy exchange ratio* "is the ratio of EMERGY received for EMERGY delivered in a trade or sales transaction. . . . The area receiving the larger EMERGY receives the larger value and has its economy stimulated more. Raw products such as minerals, rural products from agriculture, fisheries, and forestry, all tend to have high EMERGY exchange ratios when sold at market price. This is a result of money being paid for human services and not for the extensive work of nature that went into these products."[160] To assess trade between countries or local sales, "the relative benefit is determined from the exchange ratio. . . . A local economy is hurt when the new development takes more EMERGY than it returns in buying power. Keeping the product for home use raises the standard of those living at home."[161]

Odum utilized these ratios and indicators in developing his theory of unequal ecological exchange. "Free trade," he wrote, is "an ideal based on the assumption of equitable trade. . . . But free trade made developed countries rich, with high standards of living, leaving less developed countries devastated."[162] Developed economies (and urban areas) generally have much higher emergy investment ratios than less developed countries (and rural areas). In other words, the former rely more heavily than the latter on purchased emergy (brought in from outside), and less on the work of the free environment.[163] Developed countries, where reliance on the work of the free environment is less and where emergy is largely purchased, have low emergy-money ratios. Conversely, less developed (rural) countries, in which the free environment plays a larger role in the economy, have high emergy-money ratios.[164] As a result, a developed country's currency, when converted into international dollars (foreign exchange) and used to purchase products in an underdeveloped country, has a far greater emergy-buying power per dollar than in its own domestic economy, while the inverse is true for an underdeveloped economy when purchasing the products of a developed economy—the local currency when converted into international dollars and used to purchase products in a developed economy has considerably less emergy-buying power than at home. A poor country that borrows from a rich country and has to pay back in local currency converted into international dollars loses emergy-buying power through the exchange. Thus, "in the 1980s Brazil paid back 2.6 times more real wealth [measured in emergy terms] than it received with a foreign loan." Odum sums this up by saying:

> When an environmental product is sold from a rural state to a more developed economy, there is a large net EMERGY benefit to the developed buyer for two reasons: (1) the EMERGY of environmental products is higher than that in the money paid for the processing services; and (2) the EMERGY/money ratio is much greater in the rural state supplying the product than in the purchasing economy.[165]

The emergy exchange ratio is thus heavily biased against poor rural countries. Odum found that in the 1980s and early 1990s the unequal exchange of real wealth (emergy received/emergy exported) in trade between nations was extraordinary.[166] Thus, the Netherlands, West Germany, and Japan all had emergy exchange ratios of 4 or above, receiving four times as much emergy in exchange as they exported; the United States had an emergy exchange ratio of 2.2; India had one of 1.45; and, lower down on the scale of development, Liberia and Ecuador had emergy exchange ratios of 0.151 and 0.119, respectively.

The basis of this inequality is the fact (already emphasized by Marx) that "no money is paid to the environment for its extensive work," and this sets up the basis of a global *Raubbau* in which underdeveloped countries are systematically robbed of real wealth.[167] As Odum put it in his criticism of "imperial capitalism," the entire system of "global investing bleeds net emergy benefits from less developed areas to developed areas because of the imbalance in emergy-money ratios."[168]

To make matters worse, economies that specialize in the export of primary resources are specializing in those products that have high net emergy yields (defined as the emergy yield minus the emergy used to process the product). Fossil fuels are examples of commodities with high net emergy yields.[169] In purchasing such primary products, buyer nations thus gain more in real wealth terms than seller nations. Consequently, "developed nations receive much more real wealth [in such exchanges] than they export or pay for."[170]

From this standpoint, poor countries would be better off using their own resources to benefit the local population rather than selling them off at prices that leave nothing for ecological reinvestment at home. Along with the loss to the local population, poor countries are not compensated enough under the current terms of trade to do restorative work ensuring long-term ecosystem survival in areas degraded to supply the export market.[171]

Odum and J. E. Arding provided an intensive study into unequal emergy exchange with respect to Ecuador, allowing us to see more fully how emergy analysis can contribute to a comprehensive understanding of ecological imperialism.[172] Although their report focused on shrimp mariculture and export from Ecuador to wealthy countries like the United States, it also looked at Ecuador's overall position with respect to emergy exchange. "The ratio of purchased EMERGY to free Environmental EMERGY [the emergy investment ratio] within Ecuador was only 0.09, much less than the values of 7 or more in developed countries."[173] The emergy buying power of a U.S. dollar in Ecuador was found to be 3.6 times that in the United States. This meant:

> If money is borrowed by Ecuador from the US and used to buy products in the United States and later paid back from Ecuadorian currency converted on international currency exchange, 3.6 times more buying power is paid back. This is equivalent to an interest rate of 360%. Little wonder that investments by developed countries in underdeveloped countries have caused financial depression in underdeveloped countries.[174]

All together, the emergy received/emergy exported ratio for Ecuador in the early 1990s was 0.20 as opposed to 2.2 for the United States. Thus, Ecuador sent five times as much emergy abroad as it received, reflecting net ecological losses. In terms of shrimp mariculture, the emergy of the shrimp being sent to foreign buyers was about four times what was received back in emergy buying power via international dollars.[175]

The bulk of Ecuador's resource exports, of course, were in the form of oil, which represented seven times as much emergy exported as in the case of shrimp.[176] "Oil from the Amazon is pumped over the mountains and down to a shipping terminal on the Pacific Ocean for export."[177] This means the Ecuadorian Amazon region suffers most as a result of the export of oil.

Odum and Arding demonstrated that the natural wealth of Ecuador was drained through the mechanisms of international trade and debt to benefit the importing countries:

> Energy, minerals, and information are the real wealth. It takes energy to concentrate the minerals needed by an economy. It takes energy to maintain and process information. When resources are abundant and cheap, there can be abundant wealth and a high standard of living. If resources and basic products are imported cheaply, abundant wealth is imported.... Countries that sell their energy [fuels] give away their EMERGY 6 for 1 or worse. The benefits to countries that buy their fuels depend on the EMERGY ratio of their trade transaction.[178]

Greenpeace used this analysis in the anti-shrimp-farming campaign of the early 1990s, when it sent a letter to the Ecuadorian president citing Odum and Arding's study of shrimp mariculture and unequal ecological exchange.[179]

The research on international inequalities in emergy use and emergy exchange continues to expand.[180] Looking at emergy exports over time, and the inadequate compensation received for this wealth transfer, Devincenzo King analyzed the ecological debt owed to five focal countries in the Sahelian region of Sub-Saharan Africa (Burkina Faso, Mali, Mauritania, Niger and Senegal) due to cumulative net-emergy exports that have enriched wealthier countries.[181] According to this analysis, Sub-Saharan countries paid off all international debt in emergy terms by the early 1990s (in the cases of Mauritania, Niger and Senegal, by the 1970s) and should now be allowed to use their resources to develop internally. Indeed, in emergy terms, the Sahelian countries are shown to be net creditors, rather than debtors. Further, King reports that these nations experienced an emergy inequity factor (EIF, the ratio of the official exchange rate to the emergy based equitable exchange rate) in their trade with the United States that increased dramatically between 1970 and 2000, rising by the beginning of the new millennium to an EIF that gave the United States more than a 10:1 advantage in emergy (real wealth) trade with all of these countries.[182]

To be sure, despite the attempt at comprehensiveness in accounting for all energy inputs and exchanges, emergy analysis remains a unitary indicator, one that, though particularly useful, is unable to capture all dimensions of an enormously complex and dynamic relationship of environmental exploitation, degradation, and unequal exchange inflicted on the periphery by the center. It cannot by itself, for example, account for all aspects of the long-term ecological destruction of the nineteenth-century guano trade in Peru, which robbed that country of an invaluable resource with incalculable effects, and which was the basis of the social and ecological devastation and long-term underdevelopment (enforced by military conquest) of that country up to the present time.[183] Nevertheless, the analysis of unequal ecological exchange in emergy terms can be a valuable indicator—the best we have—of the vast extent of the center-periphery environmental *Raubbau*. As an analytical tool it also helps us understand the processes involved in unequal ecological exchange and can be used in conjunction with quite different indicators, such as ecological footprint analysis, to give us a more complete picture of ecological imperialism as a major factor in the modern capitalist world-system.

The strength of Odum's analysis lies in the fact that it provides a basis for recognizing the ecological conditions and contributions of Third World peoples and subsistence populations, who often are seen as "counting for nothing" in the ruling system of economic accounting.[184] In emphasizing that "developed nations receive much more real wealth than they export or pay for," Odum was defending the struggles of Indigenous and peasant populations and rural peoples in general against the insatiable accumulation tendency of "global capitalism," which he characterized "as a large-scale analog of weed overgrowth."[185] Solutions to the global ecological dilemma, he argued, were often to be found in Indigenous and peasant societies. "Policies about population and development appropriate to low-energy restoration," Odum observed, "may be like those formerly found in low-energy cultures like the Yanomamo Indians of Venezuela."[186] He pointed to Kerala in India as an example of "social progress without economic growth."[187]

Marx, Odum, and the Discourse of Unequal Ecological Exchange: Theoretical Challenges

Our treatment of Odum's emergy theory has explored the question of a Marx-Odum dialectic in the analysis of unequal ecological exchange, building on and supplementing other analyses of this phenomenon. Both Marxian political economy and Odum's systems ecology are highly critical of neoclassical economics and the dominant doctrine of free trade. Moreover, Odum's work dramatizes a split in ecological economics between, on the one hand, a radical approach, exemplified by Odum and Marxian ecological analysis, which stress the contradiction between use value (real wealth) and exchange value (economic value), and, on the other hand, an increasingly dominant approach that seeks to find ways of internalizing the externalities, aligning ecology with price data, which is more in line with neoclassical environmental economics.

Indeed, the division that developed in ecological economics is best seen in regard to the distinct approaches adopted by Robert Costanza and Odum.[188] A former student of Odum at the University of Florida, Costanza was a cofounder of the International Society of Ecological Economics and was chief editor of the Society's economic journal, *Ecological Economics*, from its beginning in 1989 until 2002. Odum was a member of the board of *Ecological Economics* at its inception. In the early 1990s, however, there was a deep struggle regarding the question of emergy/real wealth versus market value. The differences that arose led Costanza to remove Odum and a number of other natural scientists from the board in 1992, and articles affirming the concept of emergy were virtually banned in the journal.[189]

The basis of this dispute preceded by a number of years the founding of *Ecological Economics*. Costanza used an embodied-energy, input-output approach employing price-based data to argue for an "energy theory of value," which claimed that "calculated embodied energy values ... show a very good empirical relation to market-determined dollar values."[190] Costanza's approach was sharply criticized by none other than Georgescu-Roegen, the founder of ecological economics. Quoting Engels on the impossibility of an energy theory of value, Georgescu-Roegen pointed out that Costanza relied on "an input-output table with money values instead of real [energy] data" and left the reader "at a loss about how" the various factors "have been converted into energy."[191] Georgescu-Roegen claimed that an "embodied energy" theory of value as proposed by Costanza was an "extreme falsification of actuality."[192]

Likewise, Herman Daly launched a major attack on Costanza's attempt to construct an energy theory of value and his attempt to demonstrate that prices were based in such energy values.[193] According to Daly, Costanza's results in this respect were given in the assumption, built into his approach, of an energy theory of economic value and in no way proved the former. Indeed, as Costanza himself had admitted, his results were just as consistent with a labor theory of value—a possibility he dismissed, however, by exclaiming: "Can anyone seriously suggest that labor creates sunlight?"[194] In Daly's view, Costanza's "empirical result (or analytical imposition) that market prices closely reflect embodied energy is taken as a sanctification of the market within the framework of the energetic dogma."[195] Finally, Daly expressly objected to Costanza's argument that since energy values supposedly were good predictors of market values, in those cases where markets exist, they could then be employed "to determine "market values" where markets do not exist, for example, in ecological systems."[196] The weaknesses of Costanza's approach were further highlighted from a Marxist perspective by Burkett.[197]

Criticizing Costanza's embodied energy theory of value from a physical science rather than an economic standpoint, Odum insisted that the focus of ecological economics should be on real-wealth accounting, which could not be derived from money-based categories. Nor was it legitimate to add energy of different forms and qualities without converting to emergy of one kind first.[198] All of this demanded an emergy approach directed at use values, with such real-wealth flows constituting a contradictory "countercurrent" to monetary flows. Crucial to Odum's analysis, as we have seen, was the recognition that "much of the contribution of environment to society has no corresponding circulation of money."[199]

This split in ecological economics over such issues as (1) the emergy concept vs. Costanza's embodied energy approach and (2) real wealth/use value versus market/exchange value was carried over into the analysis of unequal exchange. Alf Hornborg, a cultural anthropologist, who has played a leading and generally positive role in the discussion of unequal ecological exchange, launched an attack in *Ecological Economics* (under Costanza's editorship) on Odum's approach. Hornborg asserted that "emergy" was a "metaphysical" concept like the labor theory of value.[200] Hornborg went on to disparage the labor theory of value for its failure to demonstrate the correspondence of values and market prices, not understanding that such a correspondence was contrary to Marx's own analysis,[201] and as a "normative" theory of value (also a misconception).[202] Instead, value, Hornborg declared, was "subjective, cultural, and contextual."[203] The chief object of Hornborg's attack, however, was not Marx, but rather Odum, who was criticized for providing in his emergy analysis a "normative" view in the form of "an energy theory of [economic] value" that "echoes Marx."[204] Moreover, Odum was characterized as offering an approach similar to the early twentieth-century Technocracy movement in the United States, which had proposed an energy theory of value.[205]

In our view, these criticisms of Odum by Hornborg completely missed the mark. Central to Odum's analysis was the stipulation, as we have seen, that "market values are inverse to real-wealth contributions from the environment," since no monetary payments are given for nature's work. Indeed, this constituted the very core of his theory of unequal exchange.[206] But by failing to critique Costanza for claiming that there was a rough correlation between embodied energy and price, and attributing this view, wrongly, to Odum, Hornborg erroneously arrived at the conclusion that emergy analysis blocked an understanding of the inverse relation between energy flows and price.[207] Confusing Odum with Costanza and targeting the former rather than the latter in this respect, Hornborg characterized Odum's analysis as "nothing less than a way to legitimate, by and large, world market prices as they are."[208]

All of this ignored Odum's repeated insistence that his concern was not with market value but real wealth, picturing these as separate, contradictory circuits in ways analogous to Marx's argument. Thus, in the very work on which Hornborg concentrated his fire,[209] Odum and Arding had stated in no uncertain terms: "EMERGY value [emvalue] is not meant to be used for market value." They added: "Some confuse EMERGY concepts with the technocrat movement of the 1930s, which used energy as the basis of value and proposed to pay people with energy certificates in place of money.... Technocrats wanted to substitute energy value for money, whereas EMERGY value is not meant to be used for market value, but for larger scale [ecological] evaluation of the economy" and planning.[210] Odum's position here was related, as we have seen throughout this contribution, to the distinction that Marx had made between wealth/use value and value in his critique of the capitalist economy.[211] In Odum's case, the analysis is so removed from a theory of economic exchange value that, as Brolin notes, there is no discussion of the formation of market prices to be found anywhere in his work.[212] Odum and Scienceman's analysis thus was, as we have noted, formally consistent with the classical labor theory of value, and was concerned with drawing out the more radical implications of this for the theory of real wealth.

In opposing Odum's concept of emergy as the basis of unequal exchange analysis, Hornborg proposed to substitute the concept of exergy, or available energy, as the basis for such a theory. He insisted that exergy was superior to emergy in the analysis of unequal ecological exchange since it was clear that the more money attached to a product the less available energy was associated with it.[213] Yet there was a fundamental flaw in this argument. Since all production and all exchanges involving physical elements in all places and all times involve losses of available energy, given that this is a fundamental law of physics, this represents a universal problem. The mere inverse relation between flows of money and exergy can hardly constitute a meaningful theory of unequal ecological exchange from a social standpoint, since it follows inexorably from the entropic condition governing all production and thus applies invariably where production and exchange, involving monetary transactions, occur. The problem is somewhat analogous to that of the broader unequal economic exchange theory based on inequalities in organic composition but on a wider scale. This is so much a part of any system of production and exchange that "unequal exchange" in these terms loses its significance. To make unequal ecological exchange a meaningful concept, it has to be based in social-economic power differentials.

In our view, the theory of unequal economic exchange developed on the basis of classical economics, and later expanded by Marxian and world-system theory to account for unequal ecological exchange as well, sets the stage for the development of a wider dialectical synthesis between ecological science, Marxian political economy, and environmental social science. Specifically, we need a Marxian/world-system analysis that draws critically on Odum's systems approach to unequal ecological (emergy) exchange and the destruction of real wealth by capitalist production. It is possible, we believe, to link this up with theories based on Marx's metabolic rift analysis.[214]

Such an approach, we are convinced, would allow for a theoretical and empirical deepening of the analysis of unequal ecological exchange already approached in various ways in the work of such important thinkers as Amin, Bunker, Clark, Hornborg, Jorgenson. Odum's method of analysis gives us a powerful way of analyzing unequal ecological exchange and ecological debt that complements and supplements ecological footprint analysis, and for which there is now extensive data for 134 countries.[215]

The strength of Odum's analysis, as we have seen, was rooted in the recognition of what Sweezy called the "qualitative value problem," that is, the role of use value and the contradiction between use value and exchange value within capitalist production.[216] By criticizing the capitalist economy from the standpoint of use value (via emergy analysis), Odum pointed to the need for an external ecological assessment of production as a means for social and ecological planning—one not subordinated to market pricing.

"In ecology," Murray Bookchin observed, "the Newton of... thermodynamics, or more properly, energetics, is Howard Odum."[217] Odum was also a major critic of capitalism, neoclassical economics, and ecological imperialism. His critique benefited from a deep and extended inquiry into Marx's environmental analysis. He was clear that the capitalist system of accumulation must in our age give way to what he called "a prosperous way down" in which the economy would need to be redirected to sustainable production, environmental (and energy) justice and social equality.[218] Historical conditions, Odum argued, pointed to the need for a stationary state (or steady-state) economy more conducive to the implementation of "socialistic ideals about distribution" on a world scale.[219] It is here, therefore, that we find one of the most important points of convergence between ecological science and environmental social science. Most crucial, however, from a world-system approach to ecology, is the opportunity that this provides to clarify the historical conditions of ecological as well as economic inequities between center and periphery. It is here, as we have seen, that Odum's analysis helps us understand some of the key dimensions of the problem, as orders of magnitude. In order to move toward the kind of contraction and convergence that is needed worldwide today in areas such as climate change, it is important to recognize the centuries of unequal exchange and the enormous ecological debt owed to the periphery—both of which are highlighted by Odum's analysis.

Odum's systems-ecology critique of imperial capitalism provides the necessary means for the synthesis of the metabolic rift and unequal ecological exchange literatures. As Clark and Foster argue, unequal ecological exchange, defined as "the disproportionate and undercompensated transfer of matter and energy from the periphery to the core, and the exploitation of environmental space within the periphery for intensive production and waste disposal," is dialectically connected to Marx's concept of metabolic rift.[220] In Odum's view, Marx's theory pointed in the right direction by emphasizing the "metabolic rate of labour," and thus a larger human-nature metabolism.[221] Recent work on Marx's concept of metabolic rift has demonstrated the larger ecological implications of Marx's metabolic critique, in relation to which Odum's work (and particularly his approach to unequal ecological exchange) can be viewed as a partial complement.[222] With the resulting Marx-Odum dialectic of unequal exchange as its basis, it is possible to envision a more critical global agroecology, supporting the international peasant mobilization over land resources,[223] and converging with the incipient rise of what has been called a nascent "environmental proletariat."[224]

It is important, however, to insert a word of caution. A dialectical analysis must be the final object of any critique of the capitalist order and its ruling ecological regime. Odum's emergy approach, evolving out of systems ecology and physics, provides us with a powerful critical tool. But neither ecology nor society, as we have seen, can be reduced to a single measure, whether it be labor values or emergy.

The danger of reification is an inherent product of capitalism. If we are compelled to search for means of commensurability in the analysis of use values or real wealth, it is only to highlight the narrowness of capitalist value analysis, its overexploitation of nature, and the unequal impact on the world's population—as well as for the purpose of helping to form a new historical system in which the associated producers are able to "govern the human metabolism with nature in a rational way."[225] There is no single universal metric that holds the key to the human relation to nature. It is a complex, contingent, and coevolutionary relation that we nonetheless have the power to affect.

What we have referred to as the Marx-Odum dialectic with respect to unequal ecological exchange attains its ultimate significance in enabling us to comprehend the means of socially transcending the metabolic rift, the rift in nature and society that finds its highest expression in capitalism itself. For Marx:

> It is not the *unity* of living and active humanity with the natural, inorganic conditions of their metabolic exchange with nature, and hence their appropriation of nature, which requires explanation or is the result of a historic process, but rather the *separation* between these [natural,] inorganic conditions of human existence and this active existence, a separation which is completely posited only in the relation of wage labor and capital.[226]

The rift in "the metabolic exchange with nature," together with the dialectical movement through which the elemental unity is "restored," represents, then, for Marx nothing other than the capitalist alienation of nature together with its eventual transcendence. Humanity is itself an emegent part of nature. However, under capitalism, this relationship becomes a one-sided expropriation and alienation of the "universal metabolism of nature" in the name of capital accumulation.[227] The analysis of unequal ecological exchange has a vital role to play in the critique of capitalism's degradation of the earth.

"The justice of nature," Epicurus (341–271 BCE) wrote, "is a pledge of reciprocal usefulness, neither to harm one another nor be harmed." Today this principle must be applied to all of our social relations and (to the degree to which is rational) to all of our ecological relations as well.[228]

CHAPTER ELEVEN

Marxism in the Anthropocene

The designation of the Anthropocene as a new geological epoch by natural scientists—although not yet officially adopted within the scientific community—can be seen as a "second Copernican Revolution," fundamentally altering the way in which human beings perceive their relation to the earth.[1] In many ways the core idea behind the notion of the Anthropocene—the view that human beings have become a major geological force disrupting the Earth System—has been around for a long time. It is an idea, moreover, in which socialist thinkers have played a critical role from the start. Marx and Engels declared in the 1840s that there were no parts of the globe, except perhaps in the case of a few recently arisen coral islands, that were untouched by human beings.[2] The word "Anthropocene," itself, and the notion of a new Anthropocene (or Anthropogene) Period, a term substituted for the entire Quaternary, was introduced in the 1920s (and into English in the early 1970s, in a translation from the Russian) in the analysis of the Soviet geologist Aleksei Pavlov.[3] Working parallel with Soviet geochemist Vladimir I. Vernadsky, who wrote his great work *The Biosphere* a few years later, Pavlov insisted that humanity in the twentieth century was more and more becoming a geological force altering the entire biosphere.[4]

In the early 1970s, U.S. socialist ecologist Barry Commoner came to a related conclusion, but one tailored to his own age. In his book *The Closing Circle,* Commoner insisted that a fundamental break in the human relation to the planet through production had occurred in the Second World War period with the rise of atomic energy and the expansion of synthetic chemicals, leading in the direction of the accelerated degradation of ecological conditions.[5] In 1970, Vernadsky's concept of the biosphere, long neglected in the West, was the subject of a special issue of *Scientific American*.[6]

The Anthropocene crisis, as Clive Hamilton and Jacques Grinevald observe, "is a new anthropogenic rift in the natural history of planet Earth."[7] It represents the transformation of quantitative change in production over the course of human history into a qualitative leap, and at the same time a crisis in the human relation to the Earth System. This is dramatized by the now famous charts of natural-physical and social change depicting the Great Acceleration since 1945 (or 1950), whereby all major measurements of biological and social change are shown to follow a hockey-stick pattern, including the well-known increase in carbon dioxide emissions.[8] Hence, the geological "golden spike" depicting the Anthropocene is now increasingly identified with the Great Acceleration in the human disruption of the planet in the post-1945 period, the most definitive stratigraphic traces of which are to be found in fallout radionuclides from nuclear weapons testing.[9]

It is no accident that the Great Acceleration after the Second World War, leading to what scientists, since atmospheric scientist Paul Crutzen recoined the term, are now commonly calling the Anthropocene Epoch, was paralleled by the development of the global environmental movement in the same period.[10] The environmental struggle from the 1950s on commenced with the protests led by scientists over atmospheric nuclear testing, and then extended into such areas as pesticides and more general ecological concerns, with the publication, in particular, of Rachel Carson's *Silent Spring*.[11] In the more than half a century that has followed, the environmental movement has increasingly focused on what is seen as the *planetary emergency* as global ecological contradictions have worsened. The world today is thus in the midst of a Great Climacteric—a transition period of immense consequence—represented by the advent of the Anthropocene, coupled with the emergence of what could be called the Age of Ecological Enlightenment.[12]

How are Marxian thinkers, and the left more generally, responding to the advent of the Anthropocene crisis, that is, the reality of a new anthropogenic rift in the Earth System, and how is this related to changing historical conditions arising from human production? Indeed, what intellectual resources does Marxism have to offer with which to address these new conditions and new perils? There is no easy answer to these questions. Rather, Marxian thought in this area, while developing rapidly and moving toward a higher synthesis, is still in many ways in a state of bifurcation brought on by long-standing divisions within socialist theory, largely attributable to the Cold War, and by the rise more recently of New Left perspectives, associated with social constructionism and postmodernism. This chapter will show that although the relation between Marx's political economy and his ecology is now largely clarified as a result of the debates of the last two decades, and while his extraordinary ecological critique is now widely recognized, the debate has now shifted to the dialectics of nature and society itself. This has led to a widening gulf in ecological left analyses between those committed to ecological dialectics and those committed to a radical social-monist outlook.

Marxian Ecological Thought in the Anthropocene (1950–)

If we look over the history of Marxian analyses of ecology in the English-speaking world since the Second World War, we see a number of key developments and controversies, centering on the status of Marx's own ecology, dividing first-stage and second-stage ecosocialist analysis. Moreover, today the decades-long controversy between first-stage and second-stage ecosocialism is being superseded by a more far-reaching debate on the dialectics of ecology and the relation of this to revolutionary praxis.

The rise everywhere of ecological thinking, in what we now understand as the advent of the Anthropocene Epoch in the period after the Second World War, led to a pre-figurative Marxian environmental perspective in the 1960s and 1970s, reflected in the work of figures like K. William Kapp, Barry Commoner, Virginia Brodine, Herbert Marcuse, Paul Sweezy, Howard Parsons, Charles Anderson, and Allan Schnaiberg.[13] Here socialism and the radical environmental movement were seen as organically connected, resulting in major environmental contributions on the left.

The negative dialectic of the domination of nature, associated with Max Horkheimer and Theodor Adorno, began slowly to infiltrate into the English-speaking world in the 1970s due to the translation of Alfred Schmidt's *The Concept of Nature in Marx*, originally published in Germany in 1962.[14] Characteristic of developments in this sphere in the late 1970s, 1980s, and 1990s was the "Western Marxist" rejection of the dialectics of nature (which came to be associated with Engels rather than Marx) and hence a distancing from not only Soviet-style Marxism but also all connections between Marxism and natural science. In Schmidt's interpretation, following Horkheimer and Adorno, the Enlightenment domination of nature, to which Marx himself was said to have fallen prey, pointed to a Weberian-like iron cage from which there was no escape.[15] Presenting what he presumed to be Marx's mature perspective, Schmidt declared: "We should... ask, whether the future society will not be a mammoth machine, whether the prophecy of the *Dialektik der Aufklärung* [Horkheimer and Adorno's *Dialectic of Enlightenment*], that 'human society will be a massive racket in nature,' will not be fulfilled rather than the young Marx's dream of a humanization of nature, which would at the same time include the naturalization of man."[16] Adorno, writing in this same vein, was to opine that Marx "underwrote something as arch-bourgeois as the program of an absolute control of nature."[17]

The criticism of Marx on nature coupled with rejection of the dialectics of nature gave rise to two disparate traditions in the 1980s and 1990s. The first of these was the growth of what has been referred to as "first-stage ecosocialist thought" in the writings of figures like Andre

Gorz, Ted Benton, Robyn Eckersley, James O'Connor, Donald Worster, Joel Kovel, Daniel Bensaïd, and Daniel Tanuro.[18] This was characterized by a negative assessment of Marx on ecology, and an attempt to link with more mainstream Green-Malthusian conceptions.

The second influential tradition to emerge in the 1980s and 1990s was the "production of nature" perspective of radical geography, associated in particular with thinkers like Neil Smith and Noel Castree.[19] Here Schmidt's negative critique of the "domination of nature" was replaced by the more positive view of the "production of nature." The result was a left social constructionism and social monism, merged with political-economic perspectives, in which nature was seen as subsumed within society. Due to its hyper-social constructionism, the production of nature perspective increasingly came to overlap with a postmodernist approach more distant from classical Marxism—notably the work of Bruno Latour, with its emphasis on the "hybridity" of society and nature.[20]

The opening decade and a half of the twenty-first century saw a sharp break with first-stage ecosocialism, emanating from attempts to reconstruct Marx's ecology, in what came to be known as "second-stage ecosocialism."[21] In this new wave, Paul Burkett and John Bellamy Foster, but also figures such as Elmar Altvater, Brett Clark, Peter Dickens, Andreas Malm, and Richard York, sought to go back to the foundations of Marx and Engels's own ecological conceptions in their classical critique of political economy.[22] The most dramatic discoveries of this period were the uncovering of Marx's ecological value analysis and his theory of metabolic rift. Recently we have seen related developments in Marxist ecofeminism in the work of Ariel Salleh and Pamela Odih.[23] This new approach, based on Marxism's classical foundations, was couched largely in opposition to first-stage ecosocialists, and thus came to be known as a second-stage ecosocialism or ecological Marxism. This gave rise eventually to a third-stage ecosocialism, which increasingly took this new theoretical perspective into the realm of ecosocialist praxis through the integration of Marx's original ecological critique with the investigation of the developing ecological rift in the Earth System.[24] This contributed to the emergence of a more revolutionary ecological movement, exemplified by the ecosocialist organization System Change Not Climate Change in the United States.

Today the discoveries of second-stage ecosocialists, who created a kind of "modern synthesis"—connecting classical Marxism dialectically with the modern ecological critique emanating in large part from ecological science—are widely accepted.[25] The rediscovery of the ecological value-form character of Marx's political economy, his conception of metabolic rift, and his recognition of unequal ecological exchange (and ecological imperialism) have all shifted the ecological debate globally in more revolutionary directions. Few involved in ecosocialist discussions today doubt the importance of Marx's foundational contributions to the ecological critique of capitalism.[26]

Yet the general convergence of views within ecosocialism on Marx's ecology, particularly around Marx's theory of metabolic rift, has only served to bring to the fore the conflict with the various forms of hyper-social-constructionist monism now developing in Marxian, post-Marxian, and postmodernist circles.[27] Such analyses emphasize the growing *unity* in ecological relations as nature is subsumed within capitalist society. They are thus at odds with the viewpoint of most radical environmentalists and ecosocialists. The production of nature perspective, which has gained influence during the past three decades, primarily within radical geography, represents a kind of parallel current, largely independent of the fierce debates that have taken place within environmentalism and ecosocialism.[28] It contends that almost all other left approaches to environmental nature-society questions (including that of Marx himself) are characterized by Cartesian dualism.[29]

Related to this are the radical social constructionist theorists of hybridity (sometimes referred to as "relational" theorists), who see a world populated by networks of machines, artifacts, cyborgs, etc., or as Latour says "monsters." These thinkers have likewise insisted that Marxism is fatally flawed—with Marx himself accused of having fallen prey, despite his dialectical perspective, to the nature-society dualism. In this view, Marx failed to perceive the emergence of a hybrid world, as depicted in Latour's actor-network-theory (ANT). As Latour said in a talk for the ecological-modernist Breakthrough Institute where he is a senior fellow, the object today should be to "Love Your Monsters."[30] In this view, "imbroglios" or technological monsters, modern versions of Mary Shelley's Frankenstein, are a normal part of our relation to nature, and we should accept them and their consequences, while rejecting environmentalism in favor of "political ecology" that consciously internalizes or bundles nature.[31] Latour thus demonstrates an affinity for Nordhaus and Shellenberger's whole notion of a "post-environmentalism" that does not challenge capital accumulation and unlimited economic growth, or accept the existence of natural limits, but rather places its emphasis on machines/technology, coupled with the magic of the market mechanism, as the complete solution.[32]

Hence, the Western left's growing interface with monism/hybridism has resulted in the emergence of an epistemic rift between ecological Marxism and radical social monism. Latourian Marxists have increasingly engaged in a critique of those numerous ecological Marxists who today root their analysis in Marx's metabolism theory.[33]

Dialectical Ecology versus Radical Ecological Monism

It has become common for the postmodernist left, and even for some Marxian theorists connected to the production of nature/social constructionist/hybridist traditions, to claim that environmentalists, including ecosocialists, are purveyors of a crude catastrophism or, in Neil Smith's words, "left apocalypticism," insofar as they subscribe to the notion that nature or the Earth System is something that can be degraded.[34]

To understand the deep theoretical differences that manifest themselves here it is necessary to recognize the degree to which the philosophical tradition commonly known as "Western Marxism" estranged itself, via its rejection of the dialectics of nature, not only from nature and natural science, but also from the Marxian concept of the alienation of nature.[35] The result is an approach to dialectics within Western Marxism that is largely idealist in character, and thus closed—restricted to notions of subject-object identity and all-embracing internal relations, while excluding all natural processes.[36]

Environmental analysis influenced by the tradition of "Western Marxism" thus exhibits a tendency to forsake materialist dialectics and critical realism for a kind of anthropocentric monism. If Cartesian dualism is to be rejected, in this view, the only alternative is to adopt an outlook more closely related to Leibniz, with his emphasis on preestablished harmony—famously parodied by Voltaire in *Candide*—than to Spinoza, and more closely related to Spinoza than to Marx or even Hegel. What is missing in this turn to social monism is the understanding of complex mediations between nature and society within a dialectical concept of totality.[37] The result is to exclude the possibility of a society of sustainable human development in line with Marx's conception of socialism.[38]

The radical severance from the historical-dialectical concept of nature evident in the new postmodernist-influenced left perspectives can be quite severe. For a Hegelian-Lacanian-Marxian philosopher such as Slavoj Žižek, even the growing recognition of the ecological problem does not entitle Marxian thinkers to resurrect Engels's own dialectics of nature. Instead, Engels-derived dialectical materialism is said to be an inherently anti-ecological philosophy. Referring to the frequent contention of Marxian ecologists that materialist dialectics, because "it locates human history in the general frame of an all-encompassing 'dialectics of nature'... is much more appropriate for grasping the ecological problematic," Žižek rhetorically queries: "But is this really so? Is it not, on the contrary, that the dialectical-materialist vision with its 'objective laws of nature' justifies a ruthless technological domination over and exploitation of nature?"[39]

Here the materialist dialectic (and materialist science more broadly) becomes the enemy. Not only the Marxian dialectics of nature, but any meaningful materialist conception of nature, is denied. In accord with Schmidt, Žižek pronounces: "We should therefore reject the young Marx's celebration of the subject's productive powers or potentials, of its essential nature," and his equation of naturalism and humanism, including the roots of this in ancient Greek thought.[40] The reason Žižek gives for this rejection is that "humanity *is* anti-nature."[41] Ecology, under capitalism, has become "a New Opium of the Masses."[42] Hence, "the ideological aspect of ecology should . . . be denounced" along with the idea of the potential development of a sustainable relation to nature. Questioning the notion that "architecture should be in harmony with its natural environment," Žižek insists that "architecture is by definition anti-nature, an act of delimitation against nature." Humanity, to be sure, is a "part of nature," but "there is no nature," he suggests, apart from humanity and human knowledge.[43]

Indeed, for many social constructionists, radical postmodernists, and left idealists, the problem of nature is essentially eliminated through its subordination to society. Neil Smith introduced his argument on the production of nature in his *Uneven Development* by saying that "there can be no apology for the anthropomorphism of this perspective."[44] Likewise critical theorist and radical social constructionist Steven Vogel, in his *Against Nature*, criticizes Georg Lukács and Herbert Marcuse for their alleged dualist views with respect to society and nature and insists on the need of critical theory to adopt "something like anthropocentrism."[45]

Such views lead to an abstract anthropomorphic holism/social monism. Nature is seen as becoming progressively anthropogenic in a unifying way, without alienation and without rifts. There is no need for a *dialectics of nature and society,* or even for natural science in the usual sense, since natural processes are now to be treated as internal to the social dialectic. Anything that smacks of contradictions between capitalism and nature, we are told, can be dispatched as a form of dualism, one that can be ultimately traced within Marxism to Marx himself.[46]

All of this has generated a widening gulf between ecological Marxism and left ecological monism. The last decade and a half, as noted, has seen the reemergence of Marx's classical ecological perspective, reaffirming its role in the critique of political economy. The debate between first-stage ecosocialism and second-stage ecosocialism, insofar as this relates to Marx's own analysis, has largely been settled, in favor of the latter, building on Marx's foundational view. Socialist thinkers have taken this forward to develop a powerful critique of the rift in planetary boundaries characterizing the Anthropocene. This new critical perspective has then been connected to on-the-ground movements. Not only has the ecological nature of Marx's value theory been uncovered, but so has his concept of ecological crisis proper, the metabolic rift—along with his notions of social metabolism and the universal metabolism of nature.[47] Marx's understanding of how capitalism robbed the soil, on an international, not just a national basis, has been developed into an analysis of unequal ecological exchange.[48] The excursions of Engels into the dialectics of nature, it is now recognized, led to a critique of capitalism's unsustainable relation to nature (also to be found in Marx's analyses in the *Grundrisse* and *Capital*). Engels's development of what is now known as gene-culture coevolution, it was discovered, prefigured the main twentieth-century discoveries in human evolution.[49] More recent work has emphasized Marx and Engels's explorations of thermodynamics, and Marx's sensuous aesthetics, showing the full range of their ecological thought.[50] For Marx, a major ecological contradiction such as anthropogenic desertification, arising from historical class society and continuing under capitalism, could be seen as "an unconscious socialist tendency," demanding the revolutionary restoration of essential natural conditions.[51]

Not only Marx and Engels but, as we are now beginning to understand, a long list of socialist thinkers contributed to ecology in the period between Marx's death and the rise of the Anthropocene. This included, in Britain alone (where the Marx-Darwin connection was strongest), figures like E. Ray Lankester, William Morris, H. G. Wells, J. B. S. Haldane, J. D. Bernal, Lancelot Hogben, Hyman Levy, Joseph Needham, Arthur Tansley, and Christopher Caudwell.[52]

However, this new dialectical understanding of socialist ecology, in which dialectics is central to the understanding of the mediation of nature and society through production (in its broadest sense), has recently come into conflict with an emphasis in left social constructionist circles on the development of a social monism subsuming nature within society/capitalism. Such a "radical monism"[53] or "monist and relational"[54] outlook is seen either as characteristic of Marx himself, or as a way out of Marx's own supposed dualism. Failing to see history, in its totality, in the Marxian view, as a process of dialectical mediation and change in the metabolism of nature and society, such analyses all too often promote idealist notions of holism, monism, and harmony, arising from capitalism's interaction with nature—or else a hybridity, where humanity and society are seen as intermeshed or bundled together in ever new ways.

By these means the alienated antagonism of capitalism toward the natural world and natural processes surrounding it (and of which it is a part) is conjured away. Marx's conception of the rift in the metabolism of nature and society is itself classified as a dualistic view. It is as if material existence were no longer the issue, and the questions of the Anthropocene, the Great Acceleration, and the Great Climacteric did not comprise the fundamental challenges of the twenty-first century.[55]

This radical social monism subsumes the environment within society—in effect abandoning the dialectic of nature and society by reducing the former to the latter. The anthropocentrism characteristic of such perspectives often goes hand in hand with a form of economic reductionism, in which *ecological* crises are seen as existing only insofar as they represent *economic crises for capital*.[56] In fact, in the new, fashionable postmodernist left perspectives, all of the characteristic forms of bourgeois thought reappear—even as they purport to transcend Marx.

As István Mészáros explained in *The Social Determination of Method*—the first volume of his magisterial *Social Structures and Forms of Consciousness*—bourgeois thought historically has formalistically counterposed dualism against idealized notions of unity, universality, and harmony as a fundamental antinomy—moving perpetually from one perspective to the other, with each contributing to the reproduction of an alienating ideology.[57] "The interminable succession of philosophical dualisms and dichotomies in the writings conceived from the point of view of capital's political economy,"

Mészáros observes," . . . remains thoroughly unintelligible without the manifold practical dualisms and antinomies of the socioeconomic order which the dualistic methodologies of this tradition both express and help to sustain."[58] Nor is it a simple matter of substituting an abstract monism or holism for these dualistic conceptions, since they are embedded in the structure of the dominant order itself. Thus, in the end, the answer to dualism is not an abstract monism, constituting dualism's dialectical twin, but rather a conception of revolutionary praxis extending to the metabolism of nature and society.

Naturally, even Marxist theorists have trouble overcoming these antinomies. Hence Jean-Paul Sartre made the extraordinary claim that Marxism "*is dualist because it is monist.*"[59] The irreducibility of material being to thought, and the recognition that thought was a product of particular forms of material practice, were in Sartre's interpretation of Marxism, invitations to a new ontological monism which gave rise, in turn, to a new epistemological dualism: a dualism no longer between thought and being, but rather between being and truth.[60] All of this was, however, the product of Sartre's own search for the closure of the subject-object dialectic, and a product of his vehement rejection of the dialectics of nature. The result was a perpetual antinomy of dualism and monism, which proved inescapable in his terms. "The dialectic," he wrote, "is precisely a form of monism. . . . Nature is the monism of materiality."[61]

Sartre, who was far from ecological in his perspective, deplored what he called "the violence of matter," and declared that "any philosophy that subordinates the human to what is Other than man . . . has hatred of man as both its basis and its consequence."[62] In this sense, Sartre's existential monism was associated with the annihilation of nature's exteriority (or, as Bhaskar would say, *alterity*), and of any ground of materiality that was not human.[63] "He," the human being, Sartre wrote in his essay "Materialism and Revolution," "is completely in Nature's clutches, and at any moment Nature can crush him and annihilate him, body and soul."[64]

Unable to reconcile necessity and freedom in these terms, or to accept an open-ended, materialist dialectic, Sartre opted—in a perpetual wheel of contradiction—for dualism as a necessary moment of monism. What he sought to transcend, by embracing both ends of the opposition, although in the name of a higher existential monism, was the abstracted, metaphysical reality of both dualism and monism. Monism (like dualism), taken by itself, is undialectical—a problem that Sartre tried unsuccessfully to overcome through his own "dialectical monism."[65] Yet, the only authentic answer to this from a historical materialist perspective, as Marx himself indicated, is the cessation of any resting point and with it any final closure: the recognition of the unending materialist dialectic of nature and society. For Marx "the only immutable thing is the abstraction of movement: *mors immortalis*" (immortal death, Lucretius).[66]

Social Monism as World Ecology: Bensaïd and Moore

The crude monism being offered today on the left as an alternative to dualism has none of Sartre's dialectical sophistication or deep revolutionary commitment, and is based rather on the mechanical assertion of monism as the answer to dualism, coupled with notions of hybridity and "bundling." Here Marxism is turned into a simple inverse of Sidney Hook's Cold War polemic against dialectical materialism/naturalism. Hook claimed that Marxism had been transformed into a crude "monistic theory," by which he meant the positivistic subordination of society to nature.[67] Today, however, this has been inverted—with left theorists, influenced by postmodernism, increasingly arguing that Marx adopted a *social monist* philosophy in his rejection of the dualist Enlightenment worldview, subsuming nature within society.[68]

This is the stance taken by the French Marxist philosopher (and first-stage ecosocialist) Daniel Bensaïd in his *Marx for Our Times,* where it is claimed that Marx put forward "the principle of a radical monism" in his *Economic and Philosophic Manuscripts,* and that "the classical philosophical antinomies (between materialism and idealism, nature and history) are resolved in this radical monism."[69] For Bensaïd, Marx was not a materialist any more than an idealist; he was rather committed to a philosophical monism as his way of transcending both. Here we are told that in Marx's famous argument in the *Economic and Philosophic Manuscripts* on the merging of naturalism and humanism, he was not only rejecting Cartesian dualism, but offering radical monism in response.

With respect to Marx's concept of nature, Bensaïd recognizes, with Schmidt, that "for Marx, nature is irreducible to a social category."[70] But he gets around this by arguing that Marx's monism was one of "a general process of hybridization," resulting in the creation of "'hybrid objects' (simultaneously natural and social)." Hence, Marx is seen as a precursor to Latour. Playing on Latour's famous title, *We Have Never Been Modern*, Bensaïd says of Marx, "He, too, was never modern."[71]

Rather than seeing Marx's critique of dualism as both materialist and dialectical, and aimed at a mediated totality, thereby linked to revolutionary praxis, Bensaïd simply substitutes monism for dualism. Moreover, the monism here is one of postmodernist hybridization. Nature no longer exists except as a collection of socially generated hybrids. If Bensaïd remains a radical thinker, it is in a left postmodernist context in which all dialectical approaches to the human relation to nature are abandoned in favor of an eclectic hybridism. Engels, meanwhile, is criticized by Bensaïd for allegedly rejecting the second law of thermodynamics (though in truth Engels simply questioned the dubious corollary of the heat death of the universe).[72] For Bensaïd, all of this is emblematic of the failure of Engels's dialectics of nature, which stood opposed to Marx's alleged radical monism.

World-ecology theorist Jason W. Moore argues similarly for what he calls a "monist and relational view," in opposition to the dualism of nature and society—confusing such monism with a dialectical perspective.[73] Moore bases much of his analysis on what he calls a "singular metabolism."[74] In this way, he departs from Marx's own complex, dialectical understanding of "the universal metabolism of nature," conceived as a totality, of which the social metabolism is a dialectically (and historically) mediated part.[75] In contrast, Moore opts for a "singular metabolism," conceived on a monist basis, or a "metabolism liberated from dualisms"—one characterized by a "nature-in-humanity" that is simultaneously a "humanity-in-nature," constituting a "double internality."[76] The object here is to dissolve the real nature-society antagonism of the capitalist alienation of nature, by postulating the subsumption of all natural processes within an "abstract social nature" or—what amounts to the same thing—their bundling together under the impetus of human-historical processes.[77]

In order to escape any tinge of dualism—or the irreducibility of both nature and society—Moore relies on a strategy of what could be called *discursive bundling*. He either utilizes hyphens, combined with the preposition "in," meant to suggest internal relations, for example, "capitalism-in-nature" and "nature-in-capitalism," or he relies on various metaphors such as bundles, hybrids, and webs.[78] The historical process, we are led to believe, can be regarded as little more than a process of bundling (and unbundling) of society-nature. Thus civilizations, Moore declares, in line with Latour, "are bundles of relations between human and extra-human natures.[79] These bundles are formed, stabilized and periodically disrupted," and make up the "web of life" or the "world ecology." He queries: "If Nature and Society are the *results* of this messy bundle of relations, what do we call the bundle itself? My term for this is the *oikeios*"—an ancient Greek term that Moore employs to refer to world ecology.[80] Ontologically, then, in the manner of the neutral monism of thinkers such as Latour, the world is seen as made up of bundled particularities.[81] None other than Marx himself, Moore claims, saw the world as "'bundled' in a world-ecological sense"—as supposedly evidenced by his treatments of the intertwining of external nature and society.[82] The implication of course is that the bundling process constitutes the essence of the Marxian dialectic, conceived in social-monist and singular terms.

With this Latourian Marxist and neutral monist outlook as his basis, Moore proceeds to criticize—under the cover of a rejection of the "Cartesian binary"—all those Marxian ecological theorists who have adopted the conceptual framework of Marx's metabolism theory.[83] For Marx, the "social metabolism" (the labor process) under capitalism is a particular, alienated form of the "metabolism of nature and society," occurring *within* the "universal metabolism of nature."

In some cases, this takes the form of an actual "rift" in the process of "metabolic interaction."[84] Such a conception, Moore claims, is a "Cartesian binary," since it posits "two metabolisms, one Social and one Natural."[85] (Here he seems to think that one cannot speak, as Marx did, of a metabolic relation of humanity to the earth through production, that is, a social metabolism, while also recognizing the universal metabolism of nature within which this social metabolism necessarily exists.)

"The Marxist metabolism school," by which he means second-stage ecosocialist thinkers like Burkett and Foster, is to be doubly condemned, Moore contends, for supposing that capitalism's alienated social metabolism gives rise to various metabolic rifts—as this would suggest a still deeper epistemological dualism on their part.[86] In opposition to this, Moore substitutes his own "singular metabolism," which is nothing other than the idealized capitalist notion of the market expanded to encompass the entire web of life.[87] This view adamantly rejects the whole notion of "natural limits," or the idea that in numerous cases ecological "limits are outside of us,"[88] constituting insuperable barriers to production—as in Marx's own underscoring of the "eternal natural condition for the lasting fertility of the soil."[89]

To point to antagonistic relations between capitalism and nature (or to conceive of nature as apart from society even by means of abstraction) is for Moore to fall prey to the "Cartesian divide."[90] In such cases, he claims, the *bundled monist character of reality*, which capitalism above all has brought into being, is denied.[91] Nature or the web of life has become so inseparable from capitalism in his world-ecology view that he can write: "Capitalism internalizes—however partially—the relations of the biosphere," while at the same time contending that the forces of capital configure *"the biosphere's internalization of capitalism's process."*[92] What is systematically excluded from this world-ecological analysis is what Moore derides as "the metabolic fetish of Green materialism," with its "narrowly biophysical" conception of Earth System flows, seen as relatively autonomous from capitalist processes.[93]

In this abstract conception, in which capitalism is more real than nature, there is no longer an ontology of nature (or an ontology of being); there is only the ontology of the market. The environment, following the bourgeois view, is thus reduced to little more than a set of inputs or "cheaps" (food, labor, raw materials, and energy) to the economy.[94] The whole question of *ecological crisis* is seen simply as the basis of *economic crisis*. It is manifested almost invariably as one of "underproduction," reflected in scarcity—understood in commodity price terms as various degrees of cheapness.[95] With increasing shortages of raw materials, prices tend to rise, threatening the economy through falling profits. Nevertheless, the capitalist world ecology is eternally triumphant, internalizing more and more of its environment, thereby reaffirming its existence as the one, *singular metabolism*. "Capital and power (and more than this, of course) unfold in the web of life, a totality that is shaped by manifold civilizational projects," uniting all human and extra-human relations by means of its universalizing value relations.[96]

Moore thus warns of "the fetishization of natural limits" characteristic of the environmental movement, and tells his readers that to focus on the rift (or rifts) that capitalism creates in the biogeochemical processes of the planet "gives us only one flavor of crisis—the apocalypse."[97] In the same vein, we are told that "it would be mystifying to say that the limits of capitalism are ultimately determined by the biosphere itself, although in an abstract sense this is true."[98] Instead, it would be better to follow Latour in insisting capitalism is infinitely adaptable in its production (or co-production) of "bundles of human and extra-human nature," allowing it to surmount any putative global ecological catastrophe.[99]

Attacking the so-called dualism of ecological Marxian theorists who put capitalism's alienation of nature at the very center of their analysis, Moore contends that it is the "Cartesian binary" of these thinkers that keeps them from understanding that "value-relations, which are themselves co-produced, make that [world-ecological] coherence" that constitutes capitalism's main achievement. "It is easy to talk," he expounds, "about the 'limits to growth' as if they were imposed by (external) Nature. But the reality is thornier, more complex—and also more hopeful."[100] Ecological problems in today's world should not be viewed as constituting so much a threatened "cataclysm," in the manner of those focusing on the dangers of climate change or the sixth extinction, but rather should be perceived as simply "the 'normal'" operation of capitalism's socioeconomic cycles within the web of life. After all, "history is replete with instances of capitalism overcoming seemingly insuperable 'natural limits'"—so why not, it is suggested, at the level of the Earth System itself?[101] Engels's metaphorical reference to the "revenge" of nature, arising from ecological catastrophes brought on by human action, is rejected by Moore as itself a dualistic (rather than dialectical) view.[102]

The result of all of this, as *Molecular Red*'s author McKenzie Wark notes in a critique of Moore, is to produce "a variant of social reductionism."[103] Indeed, we are suddenly back in the world of "idealistic Monism," of which philosophers like C. E. M. Joad complained in the 1930s—though this time in the form of capitalism's supposed infinite social constructionism.[104] As Wark rightly observes, the scientific conception of an objective world of nature, that is, the Earth System itself, simply vanishes behind "the 'socially constructed' interiors of culture" that constitute Moore's capitalist "world ecology."[105] Here the issue of the human alienation of nature in a commodified society vanishes.

For Roy Bhaskar in *Dialectic: The Pulse of Freedom*, "There are times when it is essential to disconnect, separate, distinguish, and divide."[106] The proposition that "differentiation is a necessary condition of totality and diversity of unity" is one that "all good dialecticians have understood" throughout the history of philosophy. Complicating this, according to Bhaskar, is "the characteristically subjectivist totalizing idealism of Western Marxism."[107] In the name of combating Cartesian dualism (as well as Soviet dialectical materialism), Western Marxism has commonly projected an abstract, hypostatized reality in which the larger material world outside society is almost entirely absent, except as the product of the social domination (or social production) of nature.

"Monism and subject-object identity theory," Bhaskar contends, are associated with "the anthropic fallacy,"[108] whereby *being* is reduced to *human being*, and the *objective world* to *society*.[109] But in today's age of epochal ecological crisis, to fall prey to such a narrow anthropic monism could prove fatal to a majority of the world's species, not excluding humanity itself.

Where the global ecological climacteric is concerned, there can be no doubt that the driving force behind today's growing rift in the biogeochemical cycles of the earth is capitalism. In the face of the very real bifurcation of the world in the Anthropocene by capitalism's alienated social-metabolic reproduction, to focus on the truisms that, in the end, the world is all one, and that human production inevitably creates new hybrid forms of human-nature linkages (as if this in itself transcends natural processes and laws), is to downplay the real depths of the crisis in which the world is now placed. As Marx pointed out in the *Grundrisse*:

> It is not the *unity* of living and active humanity with the natural, inorganic conditions of their metabolic exchange with nature, and hence their appropriation with nature, which requires explanation or is the result of historic process, but rather the *separation* between these inorganic conditions of human existence and this active existence, a separation which is completely posited only in the relation of wage labour and capital.[110]

In the Marxian view updated for our time, capitalism has not only inverted the world, it threatens to drive a stake through its heart. The world is not moving under capitalism toward the *unity* of humanity and nature but toward a dangerous *separation*—one, though, that represents, in the alienated context of class society, an "unconscious socialist tendency," in that it gives rise to the necessity of revolutionary human intervention.

The Production of Nature: Smith and Castree

Social-constructionist monism, which systematically excludes the alienation of nature under capitalism from its analysis, has recently entered ecosocialist discussions mainly through the work of thinkers like Bensaïd and Moore. But it has its deepest development in the "production of nature" school of Marxian geography, introduced into the academy by Neil Smith and Noel Castree. As Smith puts it, "Nature itself is not much of a Marxist category."[111] The natural world can therefore be dissolved for the most part as a category, since "nature," he says, "is nothing if it is not social."[112] In today's world, Smith tells us, "the production of nature becomes capitalized 'all the way down.'" We are thus experiencing "the real subsumption of nature" within capitalism.[113] Nor is this to be seen in negative or contradictory terms, since capitalism plays a vital, progressive role in this respect. "The historical production of nature," Smith declares, represents "the unity of nature toward which capitalism drives."[114]

In Smith's view, both the Frankfurt School (meaning Horkheimer, Adorno, and Schmidt) and the ecological movement are to be condemned for their "fetishism of nature" and "nature idolatry."[115] In contrast, the development of Marxian social science through the production of nature perspective, he suggests, provides the more universal outlook lacking not only in environmentalism but also in natural science with its idolatry of the "so-called laws of nature."[116] For Smith the dualism still prevalent within the environmental movement and the ecological sciences—insofar as they neglect to adopt the production of nature perspective—leads to a "left apocalypticism" that fails to recognize the unifying relation of capitalism with respect to nature.[117]

Opposing the language of ecology, even when it is understood to be largely metaphorical, Smith insists that "the ambition to 'save nature' is utterly self-defeating insofar as it reaffirms the externality (otherness) of a nature with and within which human societies are inextricably intermeshed."[118] Even the current focus on global warming is decried as evidence of dualism. "In the end, the attempt to distinguish social [anthropogenic] vis-à-vis natural contributions to climate change," Smith writes,

> is not only a fool's debate but a fool's philosophy: it leaves sacrosanct the chasm between nature and society—nature in one

corner, society in the other—which is precisely the shibboleth of modern Western thought that the "production of nature" thesis sought to corrode.[119]

Fiercely opposed to the direction in which the environmental and scientific debate on climate change was going, Smith stressed his own historical skepticism: "One does not have to be a 'global warming denier,'" he wrote, ". . . to be a skeptic concerning the ways that a global public is being stampeded into accepting wave upon wave of technical, economic, and social change, framed as necessary for immediate planetary survival."[120] In these terms, both "saving nature" and what he dubbed a back-to-nature "saviour environmentalism" were to be decried.[121]

According to Castree, Smith's leading follower within radical geography, the admittedly "hyper-constructionist," even Promethean, thrust of Smith's own analysis can be seen in his assumption that "nature becomes internal to capitalism in such a way that the very distinction implied by using these terms is eroded and undermined"—to be replaced by a concept of "socio-nature" as in Swyngedouw.[122] Marx, Castree points out, avoided falling prey to "the monistic doctrine of universal nature."[123] However, Smith, in promoting his own Marxian production of nature perspective as a corrective to Marx, steered toward a kind of social or anthropocentric monism—or a "monism centered on the labor process."[124] He therefore arguably fell into the opposite trap of hyper-social constructionism. In Smith's monist outlook, *universal society*, in effect, replaced *universal nature*.

For this reason, Castree claims—looking at matters from the Latourian-Marxist perspective he now favors:

> Smith gives us an explanatory monism, which far from *resolving* the problems of dualism, gives capitalism all the power in the society-nature relation and therefore *erases* nature altogether in the guise of making nondualistic theoretical space for it. That is, in Smith, the capital "side" of the relation with nature seems to swallow up the latter altogether.[125]

Castree's own solution to this dilemma, going beyond Smith himself, is to link Smith's production of nature theory with Latour's ANT and its bundles of hybrids.[126]

This monistic production-of-nature outlook, in the forms presented by Smith, and as modified by Castree, became the basis for criticizing all forms of Cartesian "dualism"—directed not simply at liberal analysis, but also most ecological and most socialist thought. Thus Smith leveled the accusation that Marx advanced "a certain version of the conceptual dualism of nature."[127] While Castree, for his part, argued that ecological Marxist theorists "*reintroduced* nature's putative separateness."[128]

Castree today clearly seeks to interface with Latour's neutral monist ANT and notions of hybridity as a way out of the dualism of nature and society, while also going beyond Smith's production of nature perspective.[129] Latour's neutral monism relies on what he calls "infralanguage"—bundling things together and conceiving things in terms of shifting imbroglios.[130] Hence, Latour's overall analytical approach is analogous to Leibniz's monads with their internal relations. Each discursively bundled entity contains within it a "complete recapitulation of all possible actions" and thus constitutes a kind of windowless monad, referring to the whole.[131]

The New Left hybrid theories are fond of references to cyborgs, quasi-objects, bundles, and imbroglios: anything that suggests the blurring of boundaries between humans, animals, and machines.[132] In the Anthropocene, however, such a perspective easily takes on a reactionary frame insofar as it removes sharp contradictions, replacing them with nebulous imbroglios. The result of such "deconstructive erudition," to adopt a Smith phrase, is to undermine all genuine radical praxis, implicitly supporting the status quo.[133]

Not surprisingly, then, Latour has officially joined the hypercapitalist Breakthrough Institute and its project of the ecological modernizing of the accumulation of capital. Likewise Castree has recently praised the theoretical perspective of the Breakthrough Institute, represented by Michael Shellenberger and Ted Nordhaus's

Break Through, seeing it as overlapping with the visions of Smith and Latour.[134] Thus, Castree writes: "Certain strands of environmental and body-politics operative outside universities are now [like Smith] dispensing with 'nature' as an ontological referent."[135] For post-environmentalists such as Shellenberger and Nordhaus, as in the cases of Smith, Castree, and Moore, capitalism has simply subsumed nature. Traditional environmentalism, even in left terms, no longer makes sense from such an anthropocentric-monist perspective, since the web of life is now synonymous with capitalism.

It should be noted that Latour, though denying the alienated mediation of nature and society under capitalism, and thus proposing (like Smith) to dispense with nature's ontological status altogether (along with that of society), nonetheless remarks at one point that "if dualism will not do, monism will not do either."[136] Yet, his emphasis on bundling (adopted as well by thinkers such as Moore) has, as we have seen, long been the characteristic method of neutral monism, which seeks to replace the dualisms of the mental and physical and the social and natural with bundled particularities.

Indeed, Latour's own commitment to a strong social constructionism, if not entirely denying realism, is not to be doubted. He criticizes what he calls "the 'bad' philosophy of ecology" of the environmental movement and science. Instead, he opts for a "political ecology" in which all human and non-human relations are simply political.[137] "Political philosophy abruptly finds itself with the obligation to *internalize* the environment." The result is that there is no longer an external environment.[138] Environmentalism thus lacks any definite referent in nature. Although Latour acknowledges that ANT has been criticized for re-creating "'that night when all cows are grey' ridiculed by Hegel," he says that his analysis leads in "exactly the opposite direction," and is altogether more uplifting. "Instead of 'sinking into relativism,' it is relatively easy to float upon it."[139] His role as a senior fellow of the Breakthrough Institute, however, exposes this as a mere rationalization of the hegemonic capitalist domination of nature, which he is now clearly content to "float upon."

The Nature of Environmental History: Cronon and Worster

The monistic-idealist outlook preferred by many left ecological thinkers is also evident in radical environmental history over the past two decades. This is exemplified by the work of thinkers like William Cronon and Donald Worster. Cronon is well known for his social constructionist insistence on the cultural mixing of nature and society to the extent that the former, in any kind of pure form, largely disappears—even as a necessary abstraction—going against the viewpoint of radical ecologists. Criticizing deep ecology, he insists that we must abandon the "set of bipolar moral scales in which the human and the nonhuman, the unnatural and the natural . . . serve as our conceptual map for understanding and valuing the world."[140] Instead we must embrace the cultural context of nature. Although Cronon's position certainly represents an intellectually rational strategy for a cultural historian, it carefully avoids the question of ecological sustainability, while largely subsuming natural history under cultural history. The trick here, for such practitioners, is always to show how much of nature can be reduced to culture—and not how much of culture is dependent on the natural world. In this conception, the radical environmental movement is portrayed as the product of a defunct modernism, which adheres to the Enlightenment "dualism" of nature and society.[141] This is to be displaced by a more postmodernist understanding—one that revels in the cultural relativism of a world that is so intermeshed that nature and culture can no longer be distinguished. Culture is thereby taken to be the sole reality. The end result is an approach that excludes any ecological critique rooted in the capitalist alienation of nature.

Adopting an idealist and anti-materialist approach, Donald Worster in his history of ecological science, *Nature's Economy*, criticizes the great early twentieth-century founding figures of materialist ecology within science: British botanist Arthur Tansley, famous for introducing the ecosystem concept, and British zoologist Charles Elton, equally famous for developing a reflexive view of animal ecology. In

Worster's terms, these thinkers are to be dismissed as mechanistic, while the work of idealist-teleological thinkers such as U.S. botanist Frederick Clements and General Jan Christiaan Smuts in South Africa are to be celebrated as representing holism.[142] Clements proposed the teleological concept of ecology as a "super organism" reflected in successions of plant communities—a view that was enormously influential in ecological science but was later rejected by Tansley in favor of a materialist ecology based on ecosystems. Smuts, who coined the term "holism" (as well as "apartheid"), conceived of it as part of an idealist ecological racism, which he sought to put into practice as head of the South African state.

It was the ecological racist "holism" of Smuts and his followers that induced Tansley and other socialist scientists, such as Hogben, to counter Smuts with a materialist, coevolutionary approach to nature and the nature-society relation. Nevertheless, Worster singles out the empire-building General Smuts—known for arresting Gandhi, his murderous mass attacks on black populations in which he pioneered in the aerial bombing of one's own population, and for his role in conceiving apartheid— as representing an overall philosophy of ecological holism.[143] Yet, in the same book Worster sees no contradiction in taking Marx and Engels to task for allegedly lacking a holistic "feeling for nature" or for "environmental preservation."[144]

All of this suggests that Marx's dialectical-historical view, rather than the recourse to an abstract monism or holism, or postmodernist hybridism, constitutes the only meaningful critical response to the dualism of bourgeois society. Marx's own anti-dualism was clearly demonstrated in his critique of Proudhon, where he remarked that "one finds with him from the beginning a *dualism* between life and ideas, between soul and body, a dualism which recurs in many forms."[145] For Marx the dualisms of bourgeois society were a product of an alienated relation to production, and hence to social metabolism. This necessitated the transcendence of existing historical forms, responding to the crises and contradictions of capitalist society. It was his recognition of the metabolic rift between nature and society that led Marx to shift his attention to intensive ecological studies, particularly in the last two decades of his life, and that helped spur Engels's explorations of the dialectics of nature.

So little did Marx subscribe to an anthropocentric monism denying the objective force of nature that he extended his studies deep into the evolutionary, paleontological record, taking notes on the role of isotherms (climate zones) in species extinction—*prior to the origin of the human species*.[146] Likewise Engels delved extensively into cosmology.[147] If science was a human product, clearly not all of what science studied, in Marx and Engels's view, was the product of humanity.

Monism first arose as a major movement in the late nineteenth century, primarily as an accompaniment to Social Darwinism and mechanistic materialism, although adopted by some idealists such as Smuts. Among the leading names associated with the early monist movement were Ernst Haeckel, Eugen Dühring, Enrico Ferri, Georgi Plekhanov, and Smuts.[148] Haeckel, Dühring, Ferri, and Smuts all developed it in a racist direction, pointing toward fascism in the cases of the first three, and apartheid in the case of the last. Plekhanov promoted a "monist interpretation of history," which represented a kind of mechanical materialism, although couched as "dialectical materialism."[149] For all of these thinkers, with the partial exception of Plekhanov, monism had to do with naturalistic determinism. Abstract "monism" of this type was strongly criticized by both Engels and Lenin.[150]

Today's social monism, or what Bhaskar has called "historicized anthropomorphic monism," associated with numerous Western Marxist and postmodernist left thinkers, comes from the opposite pole, from that of thinkers like Haeckel, Dühring, Ferri, and Smuts.[151] Rather than subsuming *society in nature*, it subsumes *nature in society*. In doing so, however, it tends to suppress the real ecological contradictions of capitalism, thus dismissing them as "catastrophism" and "apocalypticism"—as in the writings of Smith, Castree, and Moore.

Indeed, Smith, in the words of Castree, argued that "the environmental sciences (and the wider ecological movement) have been co-opted by it [neoliberal environmentalism]. To the extent that they reify 'nature' and talk of things like 'mass extinction,' the sciences of environment are today a depoliticizing force."[152] From this standpoint, the close relation between the environmental movement and ecological science was nothing but a dead end.

The Return of a Dialectics of Nature/Ecology

Western Marxism's critique of Engels's dialectics of nature had its source in a famous footnote in the young Georg Lukács's *History and Class Consciousness* that questioned the validity of a dialectics going beyond the direct subject-object relation of human consciousness and human history.[153] There, Lukács seemingly severed dialectics from any conception of external nature, outside of human action. Yet, even in *History and Class Consciousness*, Lukács had insisted on the possibility of a limited, "merely objective dialectics of nature,"[154] conforming to the Doctrine of Being and the Doctrine of Essence, the first two subdivisions of Hegel's *Logic*—depicting "a dialectics of movement witnessed by the detached observer."[155] Such an "objective dialectics of nature" constituted a crucial critical perspective, even if falling short of a full subject-object dialectic as in the human sciences. A few years later, as shown by his recently discovered *Tailism* manuscript, Lukács was to insist that not only had he *not* rejected the dialectics of nature in its entirety in *History and Class Consciousness*, but that Marx's concept of the metabolism of society and nature through labor production offered the key ontological-epistemo-logical basis for such an outlook.

Lukács extended this view in his later writings, including his famous 1967 preface to *History and Class Consciousness*. As he put it there, the "basic Marxist category, labour as the mediator of the metabolic interaction between society and nature, is missing" in the original argument in *History and Class Consciousness*.[156] Moreover, not only Marx's metabolism argument, but also scientific experimentation (as Engels had suggested), provided the basis for a materialist dialectics of nature. At the same time Lukács declared in his *Conversations* that since human life is "based on a metabolism with nature, it goes without saying that certain truths which we acquire in the process of carrying out this metabolism have a general validity—for example, the truths of mathematics, geometry, physics, and so on."[157] Marx's analysis of social metabolism, according to Lukacs, incorporated the "reciprocal relationship between man and nature" as an "insuperable precondition" of social reproduction.[158] "The natural boundary" to human production, in this conception, "can only retreat, it can never fully disappear."[159]

Lukács's emphasis on the "objective dialectic" of materialism—divorced from the idealist Hegelian subject-object dialectic with its promise of complete reflexivity within a closed circle—was carried forward by István Mészáros, Lukács's assistant and younger colleague. Mészáros was to emerge as one of the great Marxian theorists of the late twentieth century through his magisterial works *Marx's Theory of Alienation* and *Beyond Capital*. Mészáros conceived the "conceptual structure" of Marx's theory of alienation in terms of the triadic relation of humanity-production-nature, with production constituting a form of mediation (metabolism) between humanity and nature.[160] In this way human beings could be conceived as the "self-mediating" beings of nature.[161]

It should not surprise us therefore that it was Mészáros who was to provide the first comprehensive Marxian critique of the emerging planetary ecological crisis in his 1971 Deutscher Prize Lecture—published a year before the Club of Rome's *Limits to Growth*—in which he argued that the waste-based accumulation characterizing U.S. monopoly capitalism could not be expanded globally without breaking the ecological budget of the entire planet.[162] In *Beyond Capital*, he was to develop this further in terms of a full-scale critique of capitalism's alienated social metabolism, including its ecological effects—in his discussion of "the activation of capital's absolute limits" associated with the "destruction of the conditions of social metabolic reproduction."[163] In sharp contrast to hyper-social constructionists on the left, who charged the environmental movement with having succumbed to a "fetishism of natural limits," Mészáros early on incorporated the objective-historical conditions of the materialist dialectic and science in order to confront the problem of the ecological rift.[164]

In Marx's analysis, social metabolism stood for the labor and production process (and the process of social reproduction in the broadest sense), whereby humanity transformed its material relations with nature in a coevolutionary manner, involving *both* labor and nature.[165] The commodity was constituted not only by internal relations via exchange value and value (or the crystallization of abstract labor), but also by what, from a social standpoint, were largely external (environmental) relations, related to use value. Marx's notion of the universal metabolism of nature made it clear that the *social* metabolism was a set of relations within this *universal* metabolism. Under capitalism this was ultimately an alienated relation, reflecting "an irreparable rift in the interdependent process of social metabolism, a metabolism prescribed by the natural laws of life itself."[166] As David Harvey has noted, the "universality" of Marx's conception of "the metabolic relation to nature" constituted a kind of outer (as well as inner) set of conditions or boundary in his conception of reality that allowed him to link all the "different moments" of his critique of political economy—and indeed his ecological critique—together.[167]

It is precisely this materialist dialectics, as we have seen, that allows us to transcend the simple dualistic and monistic views of the world and to explore its complexity and contradictions as they arise dynamically, and as they emerge from real-world antagonisms. If Kant sought, unsuccessfully, to transcend Cartesian dualism by suggesting that the world was *mind-like,* thus setting the stage for the tradition of absolute idealism and the Hegelian dialectic that followed, Marx and Engels, as materialists, saw the mind as *world-like,* and advanced a materialist dialectic in response.[168] Such an approach follows Hegel in recognizing that "the true is the whole," but acknowledges that the whole in this sense cannot be grasped immediately; instead it requires an analysis of mediations and contradictions, in which the internal and external, the mediated particular and totality, the social and the natural are grasped in their fluid motion.[169] "'Inside' and 'outside,'" as Richard Levins wrote, "are not properties of nature but of science."[170] Yet, in our investigations, which depend on such abstractions, we cannot afford to ignore one or the other.

The present ecological crisis is forcing us to reconsider once again the notion of the dialectic of nature—most convincingly presented in recent decades by thinkers like Levins, Richard Lewontin, and Stephen Jay Gould. "Dialectical materialism," in the sense of these thinkers—see, for example, Lewontin and Levins's use of the concept in *Biology Under the Influence*—does not stand for the dogmatic, mechanical views that were sometimes crudely advanced in the Soviet Union under this label. Rather it harkens back to theories of dynamics, complexity, contradiction, emergence, and transformation in the analysis of the world at large, embodied in the work of Marx and Engels (and Lenin), and exemplified in the discoveries of socialist scientists and cultural theorists in the 1930s and 1940s in Britain, in particular. Included among the scientists (and philosophers of science) were figures such as Bernal, Haldane, Needham, Hogben, Levy, Benjamin Farrington, and V. Gordon Childe; and among the cultural theorists figures such as Caudwell and George Thomson.[171] Also related to these were the more Fabian-style ecological scientists Lankester and Tansley.[172]

Further, it is important not to ignore the very real conceptual breakthroughs (not without historical contradictions) of Soviet ecological thinkers, some of whom, such as Nikolai Bukharin, Boris Hessen, and Nikolai Vavilov, died in the Stalinist purges, but also including many others who made crucial advances. A short list of these would include: Pavlov, who introduced the term Anthropocene; Vernadsky, who had an immense influence on Soviet thought through his magisterial work, *The Biosphere*; Alexander I. Oparin, who simultaneously with Haldane introduced the modern materialist theory of life's origin; Nikolaevich Sukachev, the developer of biogeocoenosis as a more sophisticated form of ecosystem analysis; Mikhail Budyko, the leading Soviet climatologist and the key discoverer of accelerated global warming through his analysis of the albedo effect; and Ivan T. Frolov, the pioneering philosopher of late-Soviet ecology.[173]

Globally, Marxian (and socialist) theory has a rich history of ecological thought to draw upon—though most of this rich tradition is scarcely known to those who consider themselves Marxists, as a result of Western Marxism's subsequent alienation from the dialectics of nature and science. This was further complicated by Cold War divisions, in which all Soviet contributions were condemned out of hand as allegedly products of a monolithic "Stalinism." Consequently, the critical ecological discoveries of Soviet science were ignored by Western thought generally, even if incorporated into the core body of science.

The recovery on a higher level, of the dialectics of nature—to be seen as connected to the dialectics of society—is a vital task for Marxian ecological theorists today, who are seeking to explore the ecological contradictions of the Anthropocene, and to pave a way to a truly revolutionary praxis. The seeds of a more comprehensive dialectical ecology—a full historical materialist critique rooted in the materialist conception of nature as well as the materialist conception of history—already exist at present. As Caudwell wrote in the mid-1930s in his *Illusion and Reality*—shortly before dying in 1937 (age twenty-nine) at his machine gun while he covered the retreat of his comrades in the British battalion of the International Brigade in the Spanish Civil War:

> But men cannot change Nature without changing themselves. The full understanding of this mutual interpenetration or reflexive movement of men and Nature, mediated by the necessary and developing relations known as society, is the *recognition* of necessity, not only in Nature but in ourselves and therefore also in society. Viewed objectively this active subject-object relation is science, viewed subjectively it is art; but as consciousness emerging in active union with practice it is simply concrete living—the whole process of working, feeling, thinking and behaving like a human individual in one world of individuals and Nature.[174]

It is not a crude mechanistic or idealistic monism, any more than dualism, that Marxian theory offers in relation to the crisis of the Anthropocene, but rather an open-ended materialist dialectical outlook aimed at totality but without closure, revealing both the limitations and the possibilities of our time. What it points to is the need to create a new earthly existence—the object of which will no longer be the conquest of nature but a world of sustainable human development.[175] "Freedom," Engels wrote, "is the insight into necessity."[176] Today the freedom of necessity is best exemplified by ecological revolution.[177]

Such an ecological revolution must be aimed at creating a new "ecological civilization," going beyond capitalist society.[178] What is required is social action that will generate a more collective, egalitarian, and sustainable—and therefore socialist—mode of global production. An ecological civilization conceived in this way will necessarily reverse the "rift in the interdependent process of social metabolism" between nature and society, and bring about the "restoration" of that essential relation—while meeting no less essential human needs.[179] From this perspective, humanity has yet to face its greatest historical challenge.

CHAPTER TWELVE

Marxism and the Dialectics of Ecology

Does critical Criticism believe that it has reached even the beginning of a knowledge of historical reality so long as it excludes from the historical movement the theoretical and practical relations of man to nature, i.e. natural science and industry?
—KARL MARX AND FREDERICK ENGELS

The recovery of the ecological-materialist foundations of Karl Marx's thought, as embodied in his theory of metabolic rift, is redefining both Marxism and ecology in our time, reintegrating the critique of capital with critical natural science.[1] This may seem astonishing to those who were reared on the view that Marx's ideas were simply a synthesis of German idealism, French utopian socialism, and British political economy. However, such perspectives on classical historical materialism, which prevailed during the previous century, are now giving way to a broader recognition that Marx's materialist conception of history is inextricably connected to the materialist conception of nature, encompassing not only the critique of political economy, but also the critical appropriation of the natural-scientific revolutions occurring in his day.

What Georg Lukács called Marx's "ontology of social being" was rooted in a conception of labor as the metabolism of society and nature.

In this view, human-material existence is simultaneously social-historical and natural-ecological. Moreover, any realistic historical understanding required a focus on the complex interconnections and interdependencies associated with human-natural conditions.[2] It was this overall integrated approach that led Marx to define socialism in terms of a process of sustainable human development—understood as the necessity of maintaining the earth for future generations, coupled with the greatest development of human freedom and potential. Socialism thus required that the associated producers rationally regulate the metabolism of nature and society. It is in this context that Marx's central concepts of the "universal metabolism of nature," "social metabolism," and the metabolic rift have come to define his critical-ecological worldview.[3]

Marx's approach in this respect is inseparably related to his ecological value-form analysis. Central to his critique of capitalist commodity production was the contradiction between use value, representing production in general, and exchange value (as value, the crystallization of abstract labor). Moreover, Marx placed great emphasis on the fact that natural resources under capitalism are treated as a "free gift of Nature to capital," and hence do not enter directly into the production of value.[4] It was on this basis that he distinguished between wealth and commodity value. Wealth consisted of use values and was produced by both nature and labor. In contrast, the value/exchange value of the capitalist commodity economy was derived from the exploitation of human labor power alone. The contradiction between wealth and value thus lies at the core of the accumulation process and is directly associated with the degradation and disruption of natural conditions. It is this ecological contradiction within the capitalist value and accumulation process that serves to explain the system's tendency toward ecological crises proper, or the metabolic rift. The system in its narrow pursuit of profit—on ever-greater scales—increasingly disrupts the fundamental ecological processes governing all life, as well as social reproduction.

The rediscovery of Marx's metabolism and ecological value-form theories, and of their role in the analysis of ecological crises, has generated sharply discordant trends. Despite their importance in the development of both Marxism and ecology, neither idea is without its critics. One manifestation of the divergence on the left in this respect has been an attempt to appropriate aspects of Marx's social-metabolism analysis in order to promote a crude social "monist" view based on such notions as the social "production of nature" and capitalism's "singular metabolism."[5] Such perspectives, though influenced by Marxism, rely on idealist, postmodernist, and hyper-social constructivist conceptions that go against any meaningful historical-materialist ecology and tend to downplay (or to dismiss as apocalyptic or catastrophist) all ecological crises—insofar as they are not reducible to the narrow law of value of the system. All of this is connected to the persistence of anthropocentrism, human exemptionalism, and capitalocentrism within parts of the left in the face of the present planetary emergency.[6]

In what follows, we provide brief discussions of some of the major breakthroughs in Marx's ecology by examining the conceptual structure of Marx's metabolism theory, its relation to his ecological value-form theory, and some of the consequences in terms of ecological crises. We then offer a critical appraisal of currently fashionable social-monist attempts to reduce Marx's ecological analysis to a "singular metabolism" expressing the internal logic of the market or the law of value.[7] We conclude with an account of the centrality of dialectics to ecology in the Marxian conception.

The Conceptual Structure of Marx's Metabolism Theory

The complexity that characterizes Marx's metabolism theory is best viewed against what István Mészáros has called the "Conceptual Framework of Marx's Theory of Alienation," which set the basis for all of Marx's thought. For Mészáros, Marx's analysis takes a triadic relationship: *humanity—labor/production—nature*. Human beings necessarily mediate their relationship to nature through labor-production. However, capitalist-class society creates a whole set of second-order mediations associated with commodity exchange, resulting in a further alienated triadic relationship: *alienated humanity—alienated labor/production—alienated nature*, which is superimposed on the first. Capitalist political economy focuses on this second alienated triangle, accepting it in its immediacy devoid of any concept of alienation; while natural science within capitalist society, according to Mészáros, focuses principally on the relation of alienated nature to alienated production aimed at the ultimate domination of nature. From this position results the estranged role of natural science in bourgeois society. As Mészáros writes, the "intensified 'alienation of nature'—e.g. *pollution*—is unthinkable without the most active participation of the Natural Sciences in this process."[8]

This same conceptual framework, though viewed ecologically, is evident in Marx's treatment of the universal metabolism of nature, the social metabolism, and the metabolic rift in *Capital* (and in his *Economic Manuscript of 1861–1863*). For Marx, the labor-and-production process was defined as the metabolism of nature and society. Hence, the conceptual framework underlying Marx's thought, in these terms, was a non-alienated triadic relationship: *humanity—social metabolism—universal metabolism of nature*. The social metabolism, in this conception, was actual productive activity, constituting an active interchange of humanity via labor with the whole of nature, that is, the universal metabolism, though concretely taking specific historical forms and involving distinct processes.

With the emergence of second-order mediations associated with commodity production (the reduction of land and labor to commodity-like status), there is superimposed on this fundamental metabolic relation, a triangle of *alienation of humanity—alienation of "the interdependent process of social metabolism" (the metabolic rift)—alienation of nature's universal metabolism*.[9] The metabolic rift is therefore at one with what the young Marx, in his "Comments on James Mill's *Elements of Political Economy*," called the "alienated mediation" of "human species-activity" under capitalism.[10]

Bourgeois natural science increasingly takes an ecologically modernizing form, as it is forced to address the rift in the social metabolism brought about by the capitalist political economy and the estrangement of science this engenders. So-called technological fixes are generally proposed and employed, such as carbon capture and sequestration, without actually addressing the systemic roots of the ecological problem. However, insofar as capitalism is only able to shift such ecological contradictions around, it eventually creates a wider rift in the universal metabolism of nature, with effects far beyond the immediate processes of production, raising the question of capitalism's absolute limits. It is this framework that constitutes the core of Marx's ecological crisis theory, with its emphasis on the anthropogenic metabolic rift engendered by the system of production. The result is ever wider and deeper ecological challenges and catastrophes, representing the ultimate market failure of the capitalist system.

This overall framework is illustrated by Marx's discussion of the nineteenth-century soil crisis, which was the context in which he introduced the concept of the metabolic rift. Humanity has necessarily been engaged in agriculture throughout the history of civilization, in the triadic form of *humanity—agriculture—soil*. The history of civilization is dotted with examples of agriculture turning in non-sustainable directions, degrading the soil. However, with the development of industrialized agriculture under capitalism, new commodity relations emerge, disrupting this eternal-natural relationship in qualitatively new ways, resulting in a more systematic and intensive metabolic rift in agriculture, whereby the return of essential nutrients (e.g., nitrogen, phosphorus, and potassium) to the soil is disrupted. This leads to "an irreparable rift in the interdependent process of social metabolism, a metabolism prescribed by the natural laws of life itself."[11]

In response to this disruption of the natural conditions governing the reproduction of the soil—a product of bourgeois society's extreme division between town and country—natural scientists in the nineteenth and twentieth centuries were brought in to develop means of addressing this rift, resulting first in the international guano and nitrate trade, and then in the development of industrial fertilizers. The guano and nitrate trade disrupted whole ecologies and generated wars of imperial domination.[12] The development of industrial fertilizers, while also contributing to the creation of chemicals used in warfare, became more and more a prop for the expansion of capitalism. This technical solution, which ignored the underlying system of alienated nature and alienated society, has resulted in a vast fertilizer run-off, degrading waterways and causing dead zones in oceans worldwide. The development of chemical fertilizer on a global industrial basis thus served to shift the rift in the social metabolism between human beings and the soil to a wider, all-encompassing rift in the universal metabolism of nature, crossing major planetary boundaries and disrupting the fundamental biogeochemical processes of the biosphere.[13]

The Capitalist Law of Value and the Destruction of Nature

Marx's social-metabolic critique can be better understood if put in the context of his ecological value-form theory. In Marx's explanation of the commodity value system under capitalism (and in classical political economy in general), wealth consists of use values, which have a natural-material basis tied to production in general. In contrast, value (based on abstract social labor) under capitalism is derived solely from the exploitation of labor power, and is devoid of any natural-material content. Nature is thus deemed by the system as a "free gift . . . to capital." This contradiction gives rise to what is known as the Lauderdale Paradox, named after James Maitland, 8th Earl of Lauderdale, an early nineteenth-century classical political economist. Lauderdale pointed out that the accumulation of private riches (exchange value) under capitalism generally depends on the destruction of public wealth (use values), so as to generate the scarcity and monopoly essential to the accumulation process.[14] Under these conditions, accelerated environmental degradation destroying the commons is an inherent consequence of capital accumulation, and even serves as a basis for further accumulation, as new industries, such as waste management, are created to cope with the effects.

Capitalism is therefore an extreme form of dissipative system, one that is rapacious in its exploitation of natural powers—including what Marx liked to call the "vital forces" of humanity itself. In its constant drive for more surplus value it maximizes the throughput of energy and resources, which are then dumped back into the environment: "*Après moi le déluge!* is the watchword of every capitalist and every capitalist nation."[15] What distinguished Marx's ecological value-form analysis in this respect was the recognition that the degradation and disruption of nature under capitalism were intensified by a system of commodity production that based its value calculations entirely on labor and what labor had produced, while treating nature, insofar as it was not incorporated into the capitalist labor and production process, as a realm of non-value.[16]

Marx initially drew his concept of the universal metabolism of nature, and its relation to social and ecological reproduction, from the work of his friend and revolutionary comrade, the socialist physician Roland Daniels. In his 1851 work *Mikrokosmos*, Daniels applied the concept of metabolism in a systems-theory fashion to explain the interconnected relations between plants and animals.[17] Marx built on Daniels's conception, as well as the work of the German chemist Justus von Liebig, to develop his own notion of social-metabolic reproduction and the metabolic rift.[18] In writing *Capital* and in the period that followed, he became more and more concerned with ecological crises. After reading the botanist Carl Fraas's studies of the destruction of the soil and desertification over the long history of class-based civilizations, Marx argued that this process had in many ways only intensified and expanded under capitalism—and had consequently become "irreparable" under the modern system of alienated-labor production. From this he concluded that ecological destruction under capitalism represented an "unconscious socialist tendency"—in the sense that it pointed to the need for a revolutionary break with the system.[19]

In Marx's analysis, therefore, the concept of metabolism becomes the basis of a theory of the ecological aspects of human historical development, pointing to a metabolic rift under capitalism, requiring the "restoration" of a non-alienated social metabolism in the face of capitalist degradation, and the development of a society of substantive equality and ecological sustainability, namely socialism. None of this took away from Marx's political-economic critique of capitalism as a system of exploitation of labor power. Rather, in Marx's conception, capitalism undermined "the original sources of all wealth—the soil and the worker."[20]

Ecology and Social Monism: The Subsumption of Nature

The power of Marx's conception of social metabolism lies in the fact that it anticipated modern ecosystem and Earth System analyses, both of which were based on the metabolism concept—and had concrete links at the formative stage in the development of these ideas within socialist ecology.[21] Marx's general materialist approach anticipated and in some ways influenced many of the great advances in ecology in the late nineteenth and early twentieth century.[22] Moreover, his ecological critique, which was tied to his general political-economic critique of capitalism, is the most developed dialectical-systems theory perspective available to us today for understanding the enormously complex role of capitalism in the degradation of both labor and nature.

Nevertheless, a number of theorists, arising out of Marxian and other left traditions, have sought to take another path, emphasizing the *unifying* role of capitalism with respect to ecology, such that capitalism is seen as constitutive of the web of life itself, downplaying the alienation of nature as a dialectical counterpart of the self-alienation of humanity. This social-monist (and essentially idealist) approach is justified as an attack on Cartesian dualism. The objective tendency (if not intent) is to derail the ecological Marxism associated with the ecosocialist movement, especially its materialist dialectic.

Much of social-monist analysis has its epistemological roots in Western Marxism's categorical rejection of the dialectics of nature—inspired by a famous footnote in Lukács's *History of Class Consciousness* (one he partly contradicted elsewhere in the book and completely disavowed later) in which he questioned Engels's conception of the dialectics of nature.[23] Beginning with Maurice Merleau-Ponty's *Adventures of the Dialectic* and developing in the works of many other authors, this rejection of the dialectics of nature, and with it both nature as an object of analysis and natural science itself, became a defining feature of Western Marxism as a distinct philosophical tradition. This reinforced an idealist, subject-object dialectic confined to humanity, the human world, and the human-historical sciences.[24]

This was associated with the popularity on the left of abstract-idealist, hyper-social constructivist, and postmodernist readings of Marxism that defined themselves in opposition to materialism, and particularly dialectical materialism. Turning to the question of the environment, radical thinkers have increasingly promoted an anthropocentric social monism, in which nature is seen as completely internalized by society. Thus leading left geographer Neil Smith refers to capitalism's "real subsumption of nature all the way down." He writes: "Nature is nothing if not social." Social scientists, he contends, should therefore reject natural science's idolatry of the "so-called laws of nature" and decry the "left apocalypticism" and "fetishism of nature" identified with the environmental movement.[25] Extending Smith's logic, world-ecology theorist Jason W. Moore declares that capitalism appropriates and subsumes nature "*all the way down, across, and through.*"[26]

For such thinkers, "first nature," nature as preceding and relatively independent of society, has been completely absorbed by "second nature," nature as transformed by society.[27] Hence, nature (including the laws of nature) no longer exists as a reality in and of itself, or as an ontological referent, but retains only a shadowy existence within socially constructed "hybrids" or "bundles" constructed by the capitalist world-ecology.[28] This view rejects notions of the conflict between capitalism and ecology, the metabolic rift, and the alienation of nature as forms of Cartesian dualism.[29] Any suggestion that capitalist commodity production necessarily disrupts basic ecological processes is characterized as an apocalyptic vision—an accusation carried over to natural scientists and radical ecologists, perceived as the principal enemies of the social-monist worldview.

A close critical look reveals the deep contradictions associated with this social-monist perspective, including a social determinism that extends to the erasure of nature. For example, Moore proposes to counter the "dualism" of nature and society that he attributes to ecological Marxism with a "monist and relational view," whereby the "bundling" of nature and society signifies their unified existence.[30] He contends that "capitalism internalizes—however partially—the relations of the biosphere," while the forces of capital construct and configure "the biosphere's internalization of capitalism's process." Or, as he puts it elsewhere: "Capitalism internalizes the contradiction of nature as a whole, while the web of life internalizes capitalism's contradictions."[31] At every point, nature becomes merely the internal relation of capitalism, effectively ceasing to exist on its own, even as a necessary abstraction.

In his efforts to avoid dualism—while also evading any open-ended materialist dialectics—Moore proposes that the world consists of "bundles of human and extra-human nature," constituting an abstract "web of life" defined primarily in social-cultural terms.[32] In this largely discursive approach, such bundles are "formed, stabilized and periodically disrupted."[33] Indeed, "all *agency*," he declares, "*is a relational property* of specific bundles of human and extra-human nature."[34] All that exists, as in the philosophy of neutral monism, consists of "bundled" forms.[35]

The big bugbear for such theorists is dualism. Left geographers Neil Smith and Erik Swyngedouw go so far as to claim that Marx was himself a dualist. "Given Marx's own treatment of nature," Smith asserts, "it may not be unreasonable to see in his vision also a certain version of the conceptual dualism of nature." "The social and the natural," Swyngedouw writes, "may have been brought together and made historical and geographical by Marx, but he did so in ways that keep both as a priori separate domains."[36] To overcome what he sees as Marx's dualism of society and nature, Swyngedouw proposes an all-encompassing hybridism in the form of a singular "socionature."

For radical geographer Noel Castree, reflecting on the views of Smith (on whom Castree bases his own analysis), "nature becomes internal to capitalism in such a way that the very distinction implied by using these terms is eroded and undermined."[37] Capitalism holds all power over nature and "seems to swallow up the latter altogether."[38] Hence there is no longer any nature as such, in the sense of the object of natural science. As Moore puts it, "green materialism" was "forged in an era when nature still did count for much"—which, he implies, is no longer the case.[39] As a result, environmentalism lacks any definite referent in nature, and environmental concerns are themselves problematic—a view emphasized above all by anti-left French sociologist Bruno Latour.[40]

The resulting absurdities can be seen in Moore's endorsement of critical geographer Bruce Braun's attack on Marxian ecological economist Elmar Altvater for adhering in his analysis to the second law of thermodynamics, basic to physics.[41] For Moore, in contravention of natural science: "The 'law of entropy'... operates within specific patterns of power and production. It is not determined by the biosphere in the abstract. From the standpoint of historical nature, entropy is reversible and cyclical—but subject to rising entropy within specific civilizational logics."[42] In this strange social-monist view, entropy is subject to society, which is supposedly capable of *reversing* or *recycling* it—thereby turning back or bending the arrow of time.

Such left thinkers go so far as to exempt humanity altogether from nature's laws, arguing that "nature and its more recent derivatives like 'environment' or 'sustainability,' are 'empty' signifiers."[43] Although "'Nature' (as a historical product) provides the foundation, social relations produce nature's and society's history."[44]

From this essentially anti-environmentalist perspective, couched in post-Marxist or postmodernist terms, radical environmentalists, including the entire Green movement, are criticized for perceiving a conflict between nature and capitalist society, and are said to be prone to an "apocalyptic imaginary," feeding "ecologies of fear"—depicted as "clouded in [the] rhetoric of the need for radical change in order to stave off immanent catastrophe."[45] Smith chides climate scientists who "attempt to distinguish social [anthropogenic] vis-à-vis natural contributions to climate change" for contributing to "not only a fool's debate but a fool's philosophy: it leaves sacrosanct the chasm between nature and society—nature in one corner, society in the other."[46]

The general skepticism of Smith and other left thinkers toward discussion and action on climate change amounts to an acquiescence to the status quo, and to the distancing from environmental concerns. Moore attributes what he calls "the metabolic fetish of Green materialism" (a term he uses for ecological Marxists) to its "biophysical" conception of the Earth System. Not only Swyngedouw but even Alain Badiou and Slavoj Žižek argue that "ecology has become the new opium for the masses"—a formulation repeated word for word and strongly endorsed by all three thinkers.[47]

In a turn away from ecological science, Moore warns against the "fetishization of natural limits."[48] Directly contradicting some of the world's leading climate scientists, members of the Anthropocene Working Group, he asserts: "The reality is not one of humanity [that is, society] 'overwhelming the great forces of nature.'" Rather, he suggests that capitalism has an apparently infinite capacity for "overcoming seemingly insuperable 'natural limits'"—hence there is no real rift in planetary boundaries associated with the Anthropocene, and, implicitly, no cause for concern.[49] At worst, the system's appropriation of nature ends up increasing natural resource costs, creating a bottom-line problem for capital, as "cheap nature" grows more elusive.[50] Capitalism itself is seen as a world-ecology that is "unfold[ing] in the web of life," innovating to overcome economic scarcity whenever and wherever it arises.[51]

Moore adopts the term "web of life" (a term that has been employed in various contexts for centuries) to suggest that he is addressing ecological concerns. However, the phrase is used primarily as a metaphor for capitalism's subsumption of nature. The world in its entirety—natural and social—is depicted as simply a collection of bundled, entwined relationships, in which capital predominates. This position in many ways resembles that of ecological modernization and "green capitalism" scholars, who propose that environmental sustainability can be achieved by internalizing nature within the capitalist economy, bringing everything under the logic of the market.[52]

Indeed, Moore has recently gone so far as to laud the ecomodernist Breakthrough Institute founders Ted Nordhaus and Michael Shellenberger—leading ideologues of capitalist markets, high technology (including nuclear and geoengineering), and accelerated economic growth—as providing a superior analysis of environmental problems. We are told that their ideas represent a "powerful critique" to which ecological Marxists, with their focus on the supposedly "dualistic" concepts of the metabolic rift, the ecological footprint, and the Anthropocene, are "vulnerable." The latter's mistake, Moore argues, echoing the Breakthrough Institute, is a "Green critique" that concentrates on "what capitalism *does* to nature" rather than—as in the work of Nordhaus and Shellenberger, and Moore himself—on "how nature *works* for capitalism." Indeed, the task before us, he declares, is that of "Putting Nature to Work."[53]

Such an analysis rejects a critique based on alienation of labor and nature and the rift in the social metabolism. It paves over the contradiction between an alienated humanity and alienated nature and normalizes received ideology. Moore substitutes for Marx's complex notion of a "rift in the interdependent process of social metabolism," what he instead calls a "singular metabolism of power."[54] "The problem," he writes, is not "metabolic *rift*, but metabolic *shift* [a concept he took from those he criticizes] . . . . Metabolism becomes a way to discern *shifts* (provisional and specific unifications) not *rifts* (cumulative separation)."[55] The result—in conformity with Smith's notion of "the unity of nature to which capitalism drives"—is an all-out denial of Marx's conception of the "alienated mediation" of the social metabolism of humanity and nature under capitalism.[56]

In the one-dimensional perspective of such social-monist thinkers, there is no reason to analyze the interpenetration, interchange, and mediation of nature-society relations. Natural cycles and processes are not seen as relatively autonomous from society, even by force of abstraction, but are subsumed within society; hence they are no longer seen as legitimate subjects of analysis. In the place of the complex dialectic of nature and society, we are left only with a "dialectical bundling," in which reality is reduced to a series of socially constructed *assemblages* of things or processes.[57] For Moore, the notion of world-ecology simply means capitalism writ large, inscribed in everything. It is itself a "web of life," which is nothing but a collection of bundles (that is, commodities). The notion of the Earth System simply disappears.

Marx, in contrast, clearly indicated that nature and society are irreducible. One cannot and should not be subsumed within the other. The choice here is not between monism and dualism. Rather, an open-system, materialist dialectic—focused on mediation and totality and taking into account the heterogeneous character of reality and integrative levels—provides the only meaningful critical-realist basis for analysis.[58] Moreover, this cannot be accomplished by mere contemplation but requires the unification of theory and practice, in the context of the working out of real material relations.

Dialectical Realism and the Reunification of Marxism

Within Marx's critique of political economy resides his deep concern with addressing the alienation of nature. As he wrote in the *Grundrisse*:

> It is not the *unity* of living and active humanity with the natural, inorganic conditions of their metabolic exchange with nature, and hence their appropriation of nature, which requires explanation or is the result of historic process, but rather the *separation* between these inorganic conditions of human existence and this active existence, a separation which is completely posited only in the relation of wage labour and capital.[59]

Marx's conceptual framework of the universal metabolism of nature, social metabolism, and metabolic rift provides the means to address this separation. It serves as the basis to develop an open-ended dialectic of nature that accounts for internal and external relations. It also illuminates how the alienation of nature and the creation of a metabolic rift in relation to the universal metabolism of nature are intertwined with the system of capital.

Social metabolism encompasses human labor and production in relation to the larger biophysical world. Labor is, according to Marx, a necessary "metabolic interaction" between humans and the earth.[60] Following Marx, Lukács explained that the foundation of labor "is the metabolism between man (society) and nature," since these relations are "the basis of man's reproduction in society, as their insuperable preconditions."[61] "However great the transforming effect ... of the labour process," he observed, "the natural boundary can only retreat, it can never disappear."[62] The interchange between humanity and nature is, for Marx, a permanent condition of life itself and of society. The "labour process is first of all a process between man and nature . . . the metabolism between [humanity] and nature"—and can never lose that fundamental character.[63]

The rise of capitalism introduced distinct second-order mediations associated with the specific form of commodity production and the ceaseless pursuit of capital accumulation. Private property and wage labor alienated not only humanity and the productive process, but nature itself. As indicated above, this took the form of an alienated mediation, generating a metabolic rift between society and nature. The ecological crisis, or the "irreparable rift in the interdependent process of the social metabolism," can therefore only be fully addressed by means of a critical or dialectical realism.[64]

By the very fact of its active engagement in labor and production, humanity is also involved in the social metabolism of human beings and nature, and in the formation of a "second nature." Nevertheless, the universal metabolism of nature, that is, nature in its wider, dynamic, and universal sense—"first nature"—remains. A dialectical-realist perspective requires a comprehensive account of both internal and external relations, rather than confining analysis to only internal dynamics. It raises the crucial question of the distinction between open and closed dialectics. As Fredric Jameson explains, with respect to the closed idealist dialectic of Western Marxism, which is also limited by its exclusion of the dialectics of nature:

The notion of the dialectic, with a definite article—of dialectics as a philosophical system, or indeed as the only philosophical system—obviously commits you to the position that the dialectic is applicable to everything and anything... Western Marxism... stakes out what may be called a Viconian position, in the spirit of the *verum factum* of the *Scienza Nuova*; we can only understand what we have made, and therefore we are only in a position to claim knowledge of history but not of Nature itself, which is the doing of God.[65]

In contrast, a materialist dialectic is inherently open, not closed. It accepts no closure: no human domain completely separate from nature—and no domain of God. It does not rest its case on the Viconian position that we, as human beings, can understand the world simply because we have *made it*—a position that necessarily excludes everything humanity has *not made*. Rather, from a materialist-realist perspective, the universe is complex, changing, and only partially affected by human action. Human society remains an emergent part, of the realm of physical existence on earth (captured in the ancient Greek word *physis*), not its entirety.

Hence, from this perspective, it is impossible even to begin to address the dynamics of the environment while following Western Marxism in rejecting the dialectics of nature. In a chapter of his *Ontology of Social Being*, titled *Marx* (published in English as a separate book), Lukács, attempting to reunify Marxian analysis, writes:

> For Marx, dialectical knowledge has a merely approximate character, and this is because reality consists of the incessant interaction of complexes, which are located both internally and externally in heterogeneous relationships, and are themselves dynamic syntheses of often heterogeneous components, so that the number of effective elements can be quite unlimited. The approximate character of knowledge is therefore not primarily something epistemological, though it of course also affects epistemology; it is rather the reflection in knowledge

of the ontological determinacy of being itself; the infinity and heterogeneity of the objectively operative factors and the major consequences of this situation, i.e. that scientific laws can only fulfill themselves in the real world as tendencies, and necessities only in the tangle of opposing forces, only in a mediation that takes place by way of endless accidents.[66]

Dialectical-critical realism serves as a basis for analyzing material relations, especially those associated with capitalism's "alienated mediation" of humanity and nature. To reject the notion of metabolic rift and substitute bundles, "double internalities," and capitalism's supposed unification of nature is to return Marxian theory to a pre-Hegelian idealism, a speculative philosophy that resembles nothing so much as Leibniz's system, with its windowless monads and static "best of all possible worlds."[67] The newly fashionable social-monist and hybridist conceptions take as their basis the fetishism of immediate appearances, which is then used to *re-reify* social theory, arriving at an uncritical actualism. This leads to the error that Alfred North Whitehead called "the fallacy of misplaced concreteness."[68]

Here it is useful to take note of Lukács's warning against "epistemologically rooted empiricist fetishization" that did not take into account "deeper contradictions and their connections with fundamental laws." He argued that a closed dialectic, akin to the kind now being advanced by today's social monists, invariably rests "on this objectifying and rigidifying fetishization, which always arises when the results of a process are considered only in their ultimate and finished form, and not also in their real and contradictory genesis. Reality is fetishized into an immediate and vacuous 'uniqueness' and 'singularity,' which can thus easily be built up into an irrational myth."[69]

The irrational myth in question here is the concept of a "singular metabolism" that, in postulating the complete subsumption of nature into society, disregards ecological processes as such, and even natural science.[70] The resulting argument, itself dualistic, that the ecological movement must choose between an abstract monism and a crude dualism—associating the dialectic with the former—is a trap that simply affirms bourgeois ideology in a new form. Neither monism nor dualism is consistent with a dialectical method, which necessarily transcends both. In the words of environmental philosopher Richard Evanoff:

> Rather than dichotomise humanity and nature (as with dualistic theories) or identify humanity and nature (as with monistic theories), a dialectical realist perspective suggests that while nature does indeed provide the material resources that sustain human life, culture is neither determined by nature nor does it need to subsume the whole of nature to sustain itself. Nature is constituted by human culture in the sense that human interactions transform and modify the natural environment in significant ways, but natural processes nonetheless can and do continue in the absence of human interaction, suggesting that a measure of autonomy for nature can and should be both preserved and respected.[71]

Referring to Marx's metabolic rift, Naomi Klein rightly observes that "Earth's capacity to absorb the filthy byproducts of global capitalism's voracious metabolism is maxing out."[72] The capitalist juggernaut is driving the accumulation of greenhouse gases in the atmosphere, creating by this and other means an anthropogenic rift in the metabolism of the Earth System, with far-reaching consequences beyond the immediate conditions of production. Global climate change has its counterpart in ocean acidification, which has dramatic effects, for example, on marine calcifiers, who must use more energy to produce biogenic calcium for shell and plate formation.[73] These species are the base of an extensive food web, so what happens to them has widespread ramifications on a biospheric scale. Additionally, ocean warming and acidification are contributing to coral bleaching and collapse. These extensive coral ecosystems play a central role in creating a nutrient-rich environment and maintaining marine biodiversity.[74] Ocean acidification is recognized as a driver of previous mass extinctions and a contributing factor in the current mass extinction.

Marx's conceptual framework of metabolic analysis serves as a powerful basis to understand this rift in the Earth System associated with capitalism's expansion. Although capitalism attempts to address such ecological rifts through technological fixes, all of this leads to a larger, cumulative structural crisis within the universal metabolism of nature—given the continuing contradictions that constitute the system.[75] Marx warned that human history could be ruined and shortened as a result of an alienated metabolism that undermined the bases of life.[76]

Within Marx's critique of capital and alienated metabolism resides the affirmative conception of *metabolic restoration*—a nonalienated social metabolism that operates within the "everlasting nature-imposed condition of human existence."[77] Metabolic restoration necessitates confrontation with "the social antagonism between private property and labor," in order to uproot the alienation associated with the system of capital.[78] Such materialist grounding helps facilitate a complex, dynamic analysis, informing how productive activities can be managed in relation to the larger biophysical world. As critical realist Roy Bhaskar wrote, "We survive as a species only insofar as second nature respects the overriding constraints imposed upon it by first nature. From this nature, although it is always historically mediated, we can never, nor will ever, escape."[79]

In the nineteenth century, Engels stressed that "freedom does not consist in the dream of independence from natural laws, but in the knowledge of these laws." In fact, "real human freedom" requires living "an existence in harmony with the laws of nature that have become known."[80] A sustainable, coevolutionary ecology requires that the associated producers rationally regulate the social metabolism of nature and society, in the service of advancing human potential. It is this that constitutes Marx's most developed, most revolutionary definition of socialism.

CHAPTER THIRTEEN

Engels's *Dialectics of Nature* in the Anthropocene

In "The Part Played by Labour in the Transition from Ape to Man" from his *Dialectics of Nature*, Frederick Engels declared: "Everything affects and is affected by every other thing."[1] Today, two hundred years after his birth, Engels can be seen as one of the foundational ecological thinkers of modern times. If Karl Marx's theory of the metabolic rift is at the heart of historical-materialist ecology today, it nonetheless remains true that Engels's contributions to our understanding of the overall ecological problem remain indispensable, rooted in his own deep inquiries into nature's universal metabolism, which reinforced and extended Marx's analysis. As Paul Blackledge has stated in a recent study of Engels's thought, "Engels's conception of a dialectics of nature opens a place through which ecological crises" can be understood as rooted in "the alienated nature of capitalist social relations."[2] It is because of the very comprehensiveness of his approach to the dialectic of nature and society that Engels's work can help clarify the momentous challenges facing humanity in the Anthropocene Epoch and the current age of planetary ecological crisis.

Racing to Ruin

Some intimation of the contemporary significance of Engels's ecological critique can be gained by commencing with Walter Benjamin's celebrated 1940 aside, often quoted by ecosocialists, from the "Paralipomena" (or side notes) to his "On the Concept of History." There, Benjamin stated: "Marx says that revolutions are the locomotive of world history. But perhaps it is quite otherwise. Perhaps revolutions are an attempt by passengers on this train—namely, the human race—to activate the emergency brake." In Michael Löwy's well-known interpretation of Benjamin's statement: "The image suggests implicitly that if humanity were to allow the train to follow its course—already mapped out by the steel structure of the rails—and if nothing halted its headlong dash, we would be heading straight for disaster, for a crash or a plunge into the abyss."[3]

Benjamin's dramatic image of a runaway locomotive and, hence, the necessity of conceiving of revolution as a pulling of the emergency brake, recalls a similar passage in Engels's *Anti-Dühring*, written in the late 1870s, a work with which Benjamin, like all socialists in his day, was familiar. Here, Engels had indicated that the capitalist class was "a class under whose leadership society is racing to ruin like a locomotive whose jammed safety-valve the driver is too weak to open." It was precisely capital's inability to control "the productive forces, which have grown beyond its power," including the destructive effects imposed on its natural and social "environs," that was "driving the whole of bourgeois society towards ruin, or revolution." Hence, "if the whole of modern society is not to perish," Engels argued, "a revolution in the mode of production and distribution must take place."[4]

Engels's earlier metaphor differed slightly from Benjamin's later one, in that the object was to open the safety valve in order to prevent a boiler explosion and crash—a fairly common cause of train wrecks in the mid- to late nineteenth century.[5] If the system can be seen as "racing to ruin," revolution here is less about simply stopping the forward momentum than exerting control over the out-of-control forces of production. Indeed, Engels's ecological and economic argument was not predicated, as would be the case today, on the notion that there was too much production in relation to the overall carrying capacity of the earth, a perspective that was scarcely present at the time he was writing. Instead, his chief ecological concern had to do with the wanton destruction wrought by capitalism on local and regional environments—even if on an increasingly global basis. The visible effects of this were evident in industrial pollution, deforestation, the degradation of the soil, and the general deterioration of the environmental conditions (including periodic epidemics) of the working class. Engels also pointed to the devastation of whole environments (and their climates), as in the ecological destruction that played such a big role in the fall of ancient civilizations, due mainly to desertification, and the environmental damage imposed by colonialism on traditional cultures and modes of production.[6] Like Marx, Engels was deeply concerned with British colonialism's "Victorian Holocausts," including the generation of famine in India through the destruction of its ecology and hydrological infrastructure, and the ruinous expropriation and extermination inflicted on Ireland's ecology and people.[7]

It is true that we can also find in these same pages, in which the question "ruin or revolution" is raised, the most productivist (and, in this sense, seemingly Promethean) passage to be found anywhere in Marx and Engels's works.[8] Thus, Engels declared in *Anti-Dühring* that the advent of socialism would make possible the "constantly accelerated development of the productive forces, and . . . a practically unlimited increase of production itself."[9] However, in the context in which Engels was writing, this presents no particular contradiction. The view that a future society, released from the irrationality of capitalist production, would allow for what, by nineteenth-century standards, would have seemed like an almost unlimited development of production was of course practically universal among radical thinkers at the time. This was a natural reflection of the still low level of material development in most of the world at the time of the Industrial Revolution, when set against the still immeasurably vast scale of the earth itself. World manufacturing production was to increase by "about 1,730 times" in the hundred and fifty years between 1820, when Engels was born, at the time of the early nineteenth-century Industrial Revolution, and 1970, when the modern ecological movement was born, at the time of the first Earth Day.[10] Moreover, in Engels's analysis (as in Marx's), production was never viewed as an end in itself, but rather as a mere means to the creation of a freer and more equal society, dedicated to a process of sustainable human development.[11]

Two centuries after his birth, the depth of Engels's understanding of the systematic nature of capitalism's destruction of the natural and social environment, together with his development of a dialectical naturalist perspective, makes it, along with Marx's work, a starting point for a revolutionary ecosocialist critique today. As Marxist anthropologist Eleanor Leacock noted, Engels, in the *Dialectics of Nature*, sought to develop the conceptual basis for understanding "the complete interdependence of human social relations and human relations to nature."[12]

The Revenge of Nature

Ecological problems are the product of the interrelation of system and scale. In Engels's analysis, it is system that is emphasized above all. In his great work, *The Condition of the Working Class in England*, written while he was still in his early twenties, he focused on the destructive environmental and epidemiological conditions of the Industrial Revolution in the large manufacturing towns, particularly Manchester. He highlighted the horrendous ecological conditions imposed on workers by the new industrial factory system, evident in pollution, toxic contamination, physical deterioration, periodic epidemics, poor nutrition, and high working-class mortality, all associated with extreme economic exploitation. *The Condition of the Working Class in England* remains unique in its powerful indictment of the "social murder" inflicted by capitalism on the underlying population at the time of the Industrial Revolution.[13] Marx, for whom Engels's book was the starting point for his own epidemiological studies in *Capital*, was on this basis to designate "periodical epidemics," along with the destruction of the soil, as evidence of capitalism's metabolic rift. In Germany, Engels's treatment of the etiology of disease in *The Condition of the Working Class in England* exercised an influence that extended well beyond socialist circles. Rudolf Virchow, the German doctor and pathologist, famous as the author of *Cellular Pathology*, referred favorably to Engels's book in his own pioneering work in social epidemiology.[14]

This understanding of the material conditions of capitalist class society as environmental, as well as economic, was evident in all of Engels's work. Moreover, in constantly seeking to merge materialist and dialectical perspectives of nature and society, Engels eventually arrived at the thesis that "nature," of which human beings were an emergent part, was the "proof of dialectics"—a statement that today is better understood if we say that *ecology is the proof of dialectics*.[15]

In Engels's developed evolutionary-ecological perspective, evident in his mature works such as *The Dialectics of Nature* and *Anti-Dühring*, what distinguished human beings from nonhuman animals was the role of labor in transforming and mastering the environment, making it possible for "man" to become the "real, conscious, lord of nature, because he now [in a future society] becomes master of his own social organisation."[16] Nevertheless, along with this tendency toward greater mastery of nature in some respects, already exhibited under capitalism, was concealed a systematic tendency toward expanding ecological crises, since all attempts at the conquest of nature in defiance of natural laws of limits could only lead, in the end, to ecological catastrophes. This could be seen first and foremost in the mid-nineteenth century in the ecological devastation unleashed by colonialism. As he exclaimed:

> What cared the Spanish planters in Cuba, who burned down forests on the slopes of mountains and obtained from the ashes sufficient fertilizer for *one* generation of very profitable coffee trees—what cared they that the heavy tropical rain afterwards washed away the unprotected upper stratum of the soil, leaving behind only bare rock! In relation to nature, as to society, the present mode of production is predominantly concerned only

with the immediate, the most tangible result; and then surprise is expressed that the often remote effects of actions to this end turn out to be quite different, are mostly quite the opposite in character.[17]

For Engels, the starting point for a rational approach to the environment was to be found in Francis Bacon's famous maxim that "nature is only overcome by obeying her"—that is, by discovering and conforming to her laws.[18] Yet, in Marx and Engels's view, the Baconian principle, to the extent that it was applied in bourgeois society, was primarily treated as a "ruse" for conquering nature so as to bring it under capital's laws of accumulation and competition.[19] Science was made into a mere appendage of profit making, viewing nature's boundaries as mere barriers to be surmounted. Instead, the rational application of science in society as a whole was only possible in a system in which the associated producers regulated the human metabolic relation to nature on an unalienated basis, in accordance with genuine human needs and potentials and the requirements of long-term reproduction. This pointed to the contradiction between, on the one hand, science's own dialectic, which more and more recognized our "oneness with nature" and the associated need for social control, and on the other hand capitalism's myopic drive to accumulation *ad infinitum*, with its innate uncontrollability and neglect of environmental consequences.[20]

It was this deep, critical-materialist perspective that led Engels to stress the senselessness of the prevailing notion of the *conquest of nature*—as if nature were a foreign territory to be subjected at will, and as if humanity did not exist in the midst of the earth's metabolism. Such an attempt to conquer the earth could only lead to what he referred to, metaphorically, as the "revenge" of nature, as various critical thresholds (or tipping points) were crossed:

> Let us not, however, flatter ourselves overmuch on account of our human victories over nature. For each such victory nature takes its revenge on us. Each victory, it is true, in the first place

brings about the results we expected, but in the second and third places it has quite different, unforeseen effects which only too often cancel out the first. The people who, in Mesopotamia, Greece, Asia Minor, and elsewhere, destroyed the forests to obtain cultivable land, never dreamed that by removing along with the forests the collecting centres and reservoirs of moisture they were laying the basis for the present forlorn state of those countries. When the Italians of the Alps used up the pine forests on the southern slopes, so carefully cherished on the northern slopes, they had no inkling that by doing so they were cutting at the roots of the dairy industry in their region; they had still less inkling that they were thereby depriving their mountain springs of water for the greater part of the year, and making it possible for them to pour still more furious torrents on the plains during the rainy seasons. . . . Thus at every step we are reminded that we by no means rule over nature like a conqueror over a foreign people, like someone standing outside nature—but that we, with flesh, blood and brain, belong to nature, and exist in its midst, and that all our mastery of it consists in the fact that we have the advantage over all other creatures of being able to learn its laws and apply them correctly.[21]

Through conscious action in accord with rational science, human beings were capable of rising to a considerable extent above "the influence of unforeseen effects and uncontrolled forces," perceiving "the more remote consequences of our interference with the traditional course of nature." Yet, even with respect to "the most developed peoples of the present day," there could be seen to be "a colossal disproportion between the proposed aims and the results arrived at," such "that unforeseen effects predominate and . . . the uncontrolled forces are more powerful than those set in motion according to plan." Class-based commodity economies achieved "the desired end only by way of exception," more often producing "the exact opposite." Hence, a rational, scientific, and sustainable approach to the human relation to nature and society under capitalism was impossible.[22]

It is significant that this same general standpoint on capitalism and ecology articulated by Engels was to be echoed a few decades later by Ray Lankester, who was Charles Darwin and Thomas Huxley's protégé, Marx's close friend (and Engels's acquaintance), and the leading British biologist in the generation after Darwin. Lankester was a Fabian-style socialist who had read and been influenced by Marx's *Capital*. In his 1911 book, *The Kingdom of Man*—which brought together his 1905 Romanes Lecture at Oxford, "Nature's Insurgent Son," his 1906 presidential address to the British Association for the Advancement of Science, and his article "Nature's Revenges" focusing on the African sleeping sickness—Lankester insisted that the growing human dominion over the earth was giving rise, in contradictory fashion, to an increased potential for planetary-scale ecological disasters. Thus, in his chapter on "Nature's Revenges," he referred to humanity as the "disturber of Nature" and thus as the creator of periodic epidemic diseases threatening humanity along with other species. "It seems to be a legitimate view," Lankester wrote, "that every disease to which animals [including the human animal] (and probably plants also), are liable, excepting as a transient and very exceptional occurrence, is due to Man's interference."[23] Moreover, this could be traced to a system dominated by "markets" and "cosmopolitan dealers in finance" who undermined any rational and scientific approach to reconcile nature and human production.[24] Lankester was later to develop this argument further, writing systematically on "The Effacement of Nature by Man."[25]

Like the later Marx and Engels, Lankester saw the "Kingdom of Man" as ushering in a permanent ecological knife-edge state for humanity, engendered by capitalism, that would, if natural conditions were trampled over by rapacious capital accumulation, lead to catastrophic human environmental decline. If it were not to destroy the very bases of its existence, humanity therefore had no choice but to control its production, superseding the narrow dictates of capital accumulation and adopting the dictates of a rational science in line with coevolutionary development.

The Dialectics of Nature and History

Engels's ecological insights are inseparable from his inquiries into the dialectics of nature from which they arose. Yet, the very first principle of what came to be known as the philosophical tradition of Western Marxism was that the dialectic could not be said to apply to external nature, that is, there was no such thing as what Engels referred to as "so-called *objective* dialectics" beyond the active realm of the human subject.[26] Dialectical relations, and even the objects of dialectical reasoning, were thus confined to the human-historical sphere, where the identical subject-object could be said to apply, since all nonreflexive (transfactual) reality outside of human consciousness and human action was excluded from the analysis.[27] But with the complete rejection of the dialectics of nature within the Western Marxist tradition, the extraordinary power of Engels's explorations in this area and the enormous influence they exerted on evolutionary and ecological thought within the natural sciences and on Marxism were lost, except to a relatively small number of left scientists and dialectical materialists. Unable to see dialectics as related to material nature, the Western Marxist philosophical tradition tended to relegate both natural science and external nature itself to the realm of mechanism and positivism. The result was to create a deep chasm between the dominant post–Second World War conception of Marxian philosophy in the West and natural science (and between Western Marxism and the materialist conception of nature) at the very moment, ironically, that the ecological movement was emerging as a major political force.[28]

Restoring the insights of classical historical materialism in this area thus requires the recovery, at some level, of Engels's conception of the dialectics of nature.[29] This requires, in turn, rejecting superficial and often poorly conceived summary dismissals of Engels's approach to the dialectics of nature, usually polemicizing against his three broad dialectical "laws" that he derived from G. W. F. Hegel and to which he gave new *materialist* significance: (1) the transformation of quantity into quality and vice versa, (2) the identity or unity of opposites, and

(3) the negation of the negation.[30] In writing on "Engels's Philosophy of Science," Peter T. Manicas, for example, has complained of the "very nearly vacuous" nature of these laws.[31] Yet, in Engels's analysis, these were not meant as narrow, fixed laws in the positivistic sense, but rather, in today's terminology, as broad, dialectically conceived "ontological principles," equivalent to such basic propositions as the principle of the uniformity of nature, the principle of the perpetuity of substance, and the principle of causality. Indeed, Engels's approach to dialectics challenged in various ways the understanding of these very same principles as they were advanced by the science of his day.[32]

Perhaps the most succinct and penetrating assessment of Engels's contributions to the dialectics of nature provided by a natural scientist can be found in a 1936 pamphlet entitled *Engels as a Scientist* by the celebrated Marxist scientist J. D. Bernal, professor of physics and x-ray crystallography at Birkbeck College, University of London. Bernal depicted Engels as a philosopher and historian of science, one who could not "be said to have been an amateur" given the range of the scientific contacts he had developed in Manchester, and who had reached a level of analysis that far exceeded that of the professional philosophers of science of his day, such as Herbert Spencer and William Whewell in England and Friedrich Lange in Germany.[33] Behind Engels's deep understanding of the historical development of science in his time, according to Bernal, was a dialectical perception in which the "concept of nature was always as a whole and as a process."[34] In this, Engels had borrowed critically from Hegel, recognizing that behind the latter's idealist presentation of dialectical change in his *Logic* were processes that could be said to inhere objectively in nature, as captured in human cognition.

In addressing the first of the three dialectical "laws" or ontological principles that Engels had drawn from Hegel—how changes in quantity can lead to qualitative transformations and its opposite—Bernal emphasized its essential character for natural scientific thought: "With remarkable insight, Engels says:—'The so-called constants of physics are for the most part nothing but designations of nodal points where quantitative addition or withdrawal of motion calls forth a qualitative change in the state of the body in question.'... We are only now beginning to appreciate the essential justice of these remarks and the significance of such nodal points." In this regard, Bernal stressed Engels's reference to Dmitri Mendeleev's periodic table as exemplary of qualitative transformations arising from continuous quantitative changes, as well as the relation of Engels's basic notions to discoveries associated with the rise of quantum theory.[35] Engels's approach, as the British Marxist mathematician Hyman Levy indicated, pointed to the concept of "phase change" as employed in modern physics.[36]

Today, we know that this dialectical principle holds for biology as well. For example, increasing population density of microorganisms (a quantitative increase) can cause a change in genetic expression, leading to the formation of something new (a qualitative change). As bacterial populations increase, the signals (chemicals) emitted by each organism accumulate to a level that activates genes, leading to the production of a mucilaginous biofilm phase in which the organisms become embedded. Biofilms may be composed of a number of organisms and attach organisms to almost any surface, from water pipes to rocks in streams to teeth to soil roots.[37]

Engels's second law, the interpenetration of opposites, was more difficult to define in an operational sense, but is still of supreme importance for scientific inquiry. In Bernal's explanation, this stood for two related principles: (1) "everything implies its opposite"; and (2) there were "no hard and fast lines in nature." Engels illustrated the latter point by referring to Lankester's famous discovery that the horseshoe crab (*Limulus*) was an arachnid, part of the spider and scorpion family, a revelation that had startled the scientific world and threw previous biological classifications askew.[38] In his application of this dialectical principle to physics and to the question of matter and movement (or energy), Bernal contended, "Engels approached very close to the modern ideas of relativity."[39] Engels's notion of the unity of opposites is often seen in today's Marxian dialectics in terms of the role of internal relations, in which at least one of the *relata* is dependent on the other.[40] As Engels himself observed, the recognition that mechanical relations with "their imagined rigidity and absolute validity have been introduced into nature only by our reflective minds ... is the kernel of the dialectical conception of nature."[41]

The negation of the negation, Engels's third informal dialectical law, which, as Bernal noted, seemed so paradoxical in mere words, was meant to convey that, in the course of its historical development or evolution over time, anything within the objective world is bound to generate something different, a new emergent reality, representing new material relations and emergent levels, often through the action of recessive factors or residual elements, previously overcome, that still inhere in the present. Material existence as a whole could be seen as leading to a hierarchy of organizational levels, while transformative change often meant the shift from one organizational level to another, as in the seed to the plant.[42]

The development of what are called "emergent properties" is now considered a basic biological and ecological concept. In an ecological context, it occurs when communities of species interact in ways that produce new characteristics, mostly unpredicted, arising from the behavior of the individual species in the community.[43] A four-acre farm field with a mixture of four different species (a polyculture) may lead to higher total yield than four acres devoted to only growing each of the individual species separately. This can occur for a variety of reasons: for example, better use of sunlight and water, and decreased insect damage in the polyculture field.

Coevolution of organisms also produces new properties. For example, over evolutionary time, insects feeding on plant leaves lead to the development of numerous defense mechanisms in plants. These include producing chemicals that inhibit the insect's feeding and emitting chemicals that recruit organisms (frequently small wasps) that lay their eggs in the insect, which is then killed as the eggs develop. But the back-and-forth continues. In at least one instance, that of the tomato hornworm caterpillar, the wasp also has to inject a virus that deactivates the caterpillar's immune system to enable the wasp's eggs to develop. Evolution is constantly creating something different, sometimes dramatically, as *organisms* interact. In some cases, this leads to fundamental changes in whole ecosystems and the rise of new dominant species in particular environments. As Engels wrote, emergence, in the sense of "the negation of the negation, *really does take place* in both [plant and animal] kingdoms of the natural world."[44]

As a historian of science, Engels, according to Bernal, was remarkable in his insights into the three great scientific revolutions of the nineteenth century: (1) thermodynamics—the laws of the conservation and interchangeability of forms of energy and of entropy; (2) the analysis of the organic cell and the development of physiology; and (3) Darwin's theory of evolution based on natural selection by innate variation.[45] As Ilya Prigogine, winner of the 1977 Nobel Prize in Chemistry, was later to observe, Engels's great insight was to recognize that these three revolutions in physical science "rejected the mechanistic worldview" and drew "closer to the idea of an historical development of nature."[46]

In Bernal's account, among Engels's concerns was the pursuit of "the synthesis of all the processes affecting life, animal ecology, and [biological] distribution."[47] What made this synthesis possible was his conception of dialectical movement and change, emphasizing the complexity of material interactions and the introduction of new emergent powers, in a process of origin, development, and decline. "The central idea in Dialectical Materialism," Bernal declared, "is that of transformation.... The essential task of the materialist dialectic is the explanation of the qualitatively new," uncovering the conditions governing the emergence of a new "organizational hierarchy."[48]

In this respect, Engels's pioneering achievement was to utilize his dialectical conception of nature to throw light on all four materialist problems of "origin" that remained after Darwin: (1) *the origin of the universe*, which Engels insisted was a self-origin as envisioned in the nebular hypothesis of Immanuel Kant and Pierre-Simon Laplace; (2) *the origin of life*, in which Engels refuted Justus von Liebig's and Hermann Helmholtz's notion of the eternity of life and pointed instead to a chemical origin focusing on the complex of chemicals underlying the protoplasm, particularly proteins; (3) *the origin of human society*, in which Engels went further than any other thinker of his time in explaining the evolution of the hand and tools through labor, and with them the brain and language, anticipating later discoveries in paleoanthropology; and (4) *the origin of the family*, in which he explained the original matrilineal basis of the family and the rise of the patriarchal family with private property.[49]

In this way, Engels, Bernal insisted, had anticipated or prefigured many of the developments in materialist science. "Engels, who welcomed the principle of the conversion of one form of energy into another, would equally have welcomed the transformation of matter into energy. Motion as the mode of existence of matter [Engels's great postulate] would here acquire its final truth."[50] As Bernal noted elsewhere, Engels "saw more clearly than most distinguished physicists of his time the importance of energy and its inseparability from matter. No change in matter, he declared, could occur without a change in energy, and vice versa.... [The] substitution of motion for force which Engels battles for throughout was the starting-point of Einstein's own criticism of mechanics."[51]

Yet it was the broad perspective on ecology emanating from Engels's dialectics that constituted the most critical insight of the *Dialectics of Nature* and is the reason why a return to Engels's way of reasoning remains so important. As Bernal argued, one of Engels's crucial contributions was his critique of notions of the absolute human conquest of nature. Engels had powerfully diagnosed the failure of human society, and particularly the capitalist mode of production, to foresee the ecological consequences of its actions, tracing "the effects of undesired physical consequences of human interference with nature such as cutting down forests and the spreading of deserts."[52]

Other leading British socialist scientists of the 1930s and '40s were equally impressed by Engels's ecological warnings. For the great biochemist and science historian Joseph Needham, Engels could be described as someone for "whom nothing escaped." Engels thus pointed out that, in Needham's words, "A time may some day come when the struggle of mankind against the adverse conditions of life on our planet will have become so severe that further social evolution will become impossible," referring to the eventual extinction of the human species.⁵³ For Needham, such a critical standpoint, which rejected the crude hypothesis of linear progress, also served to illuminate the extraordinary waste and ecological destruction of capitalist society—as in the case of Brazil where coffee was grown to feed locomotive fireboxes. This raised the question of a "thermodynamic interpretation of justice" since the alienation of nature (including the alienation of energy), as Engels had intimated, was "squandering" real human possibilities in the present and future.⁵⁴

Biologist J. B. S. Haldane—one of the two leading British figures along with R. A. Fisher in the neo-Darwinian synthesis reconciling Darwinian biology with the revolution in genetics—saw Engels as "the chief source" of materialist dialectics. Comparing Engels to Charles Dickens in relation to the Industrial Revolution, Haldane emphasized that Engels saw deeper and further. "Dickens had a firsthand knowledge of these conditions [of poverty and pollution]. He described them with burning indignation and in great detail. But his attitude was one of pity rather than hope. Engels saw the misery and the degradation of the workers, but he saw through it. Dickens never suggested that if they were to be saved they must save themselves. Engels saw that this was not only desirable but inevitable."⁵⁵

The recognition of the importance of Engels's dialectics of nature has extended into our own times. Harvard biologists Richard Levins and Richard Lewontin were to dedicate their now classic work *The Dialectical Biologist* to Engels, drawing heavily, if somewhat critically at points, on his analysis.⁵⁶ Levins and Lewontin's Harvard colleague, paleontologist and evolutionary theorist Stephen Jay Gould, was to observe that Engels provided the best nineteenth-century case for gene-culture coevolution—that is, the best explanation of human evolution in Darwin's own lifetime, given that gene-culture coevolution is the form that all coherent theories of human evolution must take.⁵⁷

It was Engels's development of a dialectics of emergence that was ultimately to prove most revolutionary. The significance of this perspective—ontologically, epistemologically, methodologically—was captured by Needham in his own pathbreaking analysis of "integrative levels" (or emergence) in *Time, the Refreshing River* (a title that referred back to the great ancient materialist, Heraclitus):

> Marx and Engels were bold enough to assert that it [the dialectical process] happens actually in evolving nature itself, and that the undoubted fact that it happens in our thought about nature is because we and our thought are part of nature. We cannot consider nature otherwise than as a series of levels of organisation, a series of dialectical syntheses. From the ultimate particle to atom, from atom to molecule, from molecule to colloidal aggregate, from aggregate to living cell, from cell to organ, from organ to body, from animal body to social association, the series of organisational levels is complete. Nothing but energy (as we now call matter and motion) and the levels of organisation (or the stabilised dialectical syntheses) at different levels have been required for the building of our world.[58]

Engels in the Anthropocene

It is widely recognized in contemporary science (though not yet official) that the Holocene Epoch in geological time, extending back almost twelve thousand years, has come to an end, beginning in the 1950s, displaced by the current Anthropocene Epoch. The onset of the Anthropocene was brought about by a Great Acceleration in the anthropogenic impacts on the environment, such that the scale of the human economy has now come to rival the major biogeochemical cycles of the planet itself, resulting in rifts in the planetary boundaries that define the Earth System as a safe home for humanity.[59] The Anthropocene thus stands for what Lankester had earlier called the "Kingdom of Man," in the critical sense in which this was meant; that is, humanity was increasingly the "disturber" of the natural environment on a planetary scale. Hence, society had no choice but to seek the rational application of science, and thus the overturning of a social order in which science has been relegated to a mere means by which "treasure and luxury are opened to capitalists."[60] What this meant, in Engels's (and Marx's) more forceful terms, was that the condition for the rational regulation of the metabolism between humanity and nature, and hence the rational application of science, was the transformation of the mode of production and distribution. Any other course invited the accumulation of catastrophe.[61]

It is in the Anthropocene that Engels's dialectic of ecology can be seen as finally coming into its own. It is here that his emphasis on the interdependence of everything in existence, the unity of opposites, internal relations, discontinuous change, emergent evolution, the reality of ecosystem and climate destruction, and the critique of linear notions of progress can all be seen as essential to the very future of humanity and the earth as we know it. Engels was acutely aware that in modern scientific conceptions "the whole of nature also is now merged in history, and history is only differentiated from natural history as the evolutionary process of *self-conscious* organisms."[62] Insofar as humanity was alienated from its own labor and production process, and therefore from its metabolism with nature, this could only mean the destruction of nature as well as society. The quantitative growth of capital led to a qualitative transformation of the human relation to the earth itself, which only a society of associated producers could rationally address. This was related to the fact that a particular qualitative mode of production (such as capitalism) was associated with a specific matrix of quantitative demands, while a qualitatively transformed mode of production (as in socialism) could lead to a very different quantitative matrix.

Engels argued that capitalism was "squandering" the world's natural resources, including fossil fuels.[63] He indicated that urban pollution, desertification, deforestation, exhaustion of the soil, and (regional) climate change were all the result of unplanned, uncontrolled, destructive forms of production, most evident in the capitalist commodity economy. In line with Marx, and Liebig, he pointed to London's enormous sewage problem as a manifestation of the metabolic rift, which removed the nutrients from the soil and shipped them one-way to the overcrowded cities where they became a source of pollution.[64] He underscored the class basis of the spread of the periodic epidemics of smallpox, cholera, typhus, typhoid, tuberculosis, scarlet fever, whooping cough, and other contagious diseases that were affecting the environmental conditions of the working class, along with poor nutrition, overwork, exposure to toxics at work, and workplace injuries of all kinds. He highlighted, based on the new science of thermodynamics, that historical ecological change was often irreversible and that humanity's own survival was ultimately in question.[65] In terms of the current relations of production and the environment, he wrote of a society faced with *ruin or revolution*. The social murder of workers in urban environments and the famines in colonial Ireland and India were seen as indications of the extreme exploitation, ecological degradation, and even wholesale extermination of populations just below the surface of capitalist society.[66]

On all these bases, Engels, like Marx, argued that the social metabolism with nature should be regulated by the associated producers in conformity to (or in coevolution with) nature's laws as understood by science, while fulfilling individual and collective needs. Such a rational application of science, however, was impossible under capitalism. Nor was development itself controllable under capitalism, since it was based on immediate, individual gain. To implement a comprehensive, rational scientific approach in line with human needs and sustainable environmental conditions required a society in which a system of long-term planning in the interest of the chain of human generations could be put into operation.[67]

Implicit in Engels's analysis from the very beginning was a notion of what we can call the *environmental proletariat*. Thus, while capitalism was concerned with the "political economy of capital," the working class in its most oppressed and also in its most radical phases was concerned with the entirety of existence, always starting from elemental needs. To call the objectives of workers a "political economy of the working class," as Marx once did, may not be wrong, but it would be more correct in today's terminology to say that workers, in their more revolutionary struggles, are primarily striving to create a new *political ecology of the working class*, concerned with their whole environment and basic living conditions, which can only be achieved on a communal basis.[68] It was this that was captured so well in Engels's *Condition of the Working Class in England*, where he systematically exposed the pollution of air and water, the contaminated sewers, the adulterated food, the lack of nutrition, the toxics at work, the frequent injuries, and the high morbidity and mortality of the working class—and saw the struggle for socialism as the only genuine way forward.

Indeed, *The Condition of the Working Class in England* raised issues that are now once again coming to the fore in the Anthropocene. For Marx, Engels's youthful work was to exert an enduring influence leading him to designate "periodical epidemics" as a manifestation of the metabolic rift alongside the destruction of the soil. Many pages of *Capital* were devoted simply to attempting to update Engels's epidemiological analysis decades later.[69] Today, in the context of the pandemic represented by COVID-19, these insights take on a renewed importance as a place from which to begin in the long revolution for an ecosocialist world. Yet, to bring such analyses forward, it is necessary to explore a dialectical science (and art) rooted in a conception of the complex "oneness" of humanity and nature.

All Things Are Sold

Engels admired the poetry of Percy Bysshe Shelley, whom he considered a "genius." He wrote in his youth of "a tenderness and originality in the depiction of nature such as only Shelley can achieve."[70] In the opening stanzas of Shelley's *Mont Blanc*, we find a materialist dialectics of nature and mind not unlike Engels's own:

> The everlasting universe of things
> Flows through the mind, and rolls its rapid waves,
> Now dark—now glittering—now reflecting gloom—
> Now lending splendour, where from secret springs
> The source of human thought its tribute brings
> Of waters—with a sound but half its own[71]

Like Shelley, who in *Queen Mab* wrote of bourgeois society's alienation of nature along with love—"All things are sold: the very light of Heaven / Is venal; earth's unsparing gifts of love"—Engels saw the deep need for the reconciliation of humanity with nature, which only a revolution could bring.[72]

CHAPTER FOURTEEN

Late Soviet Ecology

Soviet ecology presents us with an extraordinary set of historical ironies. On the one hand, the USSR in the 1930s and '40s violently purged many of its leading ecological thinkers and seriously degraded its environment in the quest for rapid industrial expansion. The end result has often been described as a kind of "ecocide," symbolized by the Chernobyl nuclear accident, the assault on Lake Baikal, and the drying up of the Aral Sea, as well as extremely high levels of air and water pollution.[1] On the other hand, the Soviet Union developed some of the world's most dialectical contributions to ecology, revolutionizing science in fields such as climatology, while also introducing pioneering forms of conservation. Aside from its famous *zapovedniki*, nature reserves for scientific research, it sought to preserve and even to expand its forests. As environmental historian Stephen Brain observes, it established "levels of [forest] protection unparalleled anywhere in the world." Beginning in the 1960s the Soviet Union increasingly instituted environmental reforms, and in the 1980s was the site of what has been called an "ecological revolution." A growing recognition of this more complex reality has led scholars in recent years to criticize the "ecocide" description of Soviet environmental history as too simplistic.[2]

From the 1960s on, Soviet ecological thought grew rapidly together with the environmental movement, which was led primarily by scientists. In the 1970s and '80s this evolved into a mass movement, leading to the emergence in the USSR of the largest conservation organization in the world. These developments resulted in substantial changes in the society. For example, between 1980 and 1990 air pollutants from stationary sources fell by over 23 percent.[3]

More significant from today's standpoint was the role the Soviet Union played from the late 1950s on in the development of global ecology. Soviet climatologists discovered and alerted the world to the acceleration of global climate change; developed the major early climate change models; demonstrated the extent to which the melting of polar ice could create a positive feedback, speeding up global warming; pioneered paleoclimatic analysis; constructed a new approach to global ecology as a distinct field based on the analysis of the biosphere; originated the nuclear winter theory; and probably did the most early on in exploring the natural-social dialectic underlying changes in the Earth System.[4]

Soviet ecology can be divided into roughly three periods: (1) early Soviet ecology, characterized by revolutionary ecological theories and key conservation initiatives from the 1917 Revolution up to the mid-1930s; (2) the middle or Stalin period, from the late 1930s to the mid-1950s, dominated by purges, rapid industrialization, the Second World War, the onset of the Cold War, and aggressive reforestation; and (3) late Soviet ecology from the late 1950s to 1991, marked by the development of a dialectical "global ecology," and the emergence of a powerful Soviet environmental movement—responding in particular to the extreme environmental degradation of the decade following Stalin's death in 1953. The end product was a kind of negation of the negation in the ecological realm, but one that was to be superseded finally by the wider forces leading to the USSR's demise.

Although much has been written about the early and middle periods of Soviet ecology, relatively little has been written about late Soviet ecology. Western ecological Marxism emerged largely in ignorance of rapidly developing Soviet ecological science and philosophy. Yet late

Soviet ecology remains of extraordinary importance to us today, representing a valuable legacy that can potentially aid us in our efforts to engage with the present planetary emergency.

Soviet Ecology under Lenin and Stalin

Early Soviet ecology was extraordinarily dynamic. Lenin had strongly embraced ecological values, partly under the influence of Marx and Engels, and was deeply concerned with conservation. He read Vladimir Nikolaevich Sukachev's *Swamps: Their Formation, Development and Properties* and was, Douglas Weiner has speculated, "affected by the holistic, ecological spirit of Sukachev's pioneering text in community ecology." Immediately after the October 1917 Revolution, Lenin supported the creation of the People's Commissariat of Education under the leadership of Anatolii Vasil'evich Lunacharskii, which was given responsibility for conservation. In 1924 the All-Russian Conservation Society (VOOP) was created with an initial membership of around one thousand. The Education Commissariat with Lenin's backing set up the celebrated ecological reserves, known as *zapovedniki*, of relatively pristine nature, set apart for scientific research. By 1933 there were thirty-three zapovedniki encompassing altogether some 2.7 million hectares.[5]

Key Soviet ecological thinkers, besides Sukachev, included Vladimir Vernadsky, who published his epoch-making *The Biosphere* in 1926; Alexander Ivanovich Oparin, who in the early 1920s (simultaneously with J. B. S. Haldane in Britain) developed the main theory of the origins of life; and the brilliant plant geneticist Nikolai Ivanovich Vavilov, who discovered the primary sources of germplasm or genetic reservoirs (known as Vavilov areas) tied to the areas of earliest human cultivation around the world—in locations such as Ethiopia, Turkey, Tibet, Mexico, and Peru. Others, such as leading Marxian theorist and close Lenin associate Nikolai Bukharin and historian of science Y. M. Uranovsky, generalized such discoveries in terms of historical materialism. Bukharin, following Vernadsky, emphasized the human relation to the biosphere and the dialectical interchange between humanity and nature. Zoologist Vladimir Vladimirovich Stanchinskii pioneered the development of energetic analysis of ecological communities (and trophic levels), and was a leading promoter and defender of the zapovedniki. Stanchinskii was the editor of the USSR's first formal ecology journal. Physicist Boris Hessen achieved worldwide fame for reinterpreting the history and sociology of science in historical materialist terms.

However, with Lenin's death and the rise of Stalin, issues of Soviet conservation and genetics were politicized and bureaucratized within a repressive state. This led to the elimination of many leading scientists and intellectuals, particularly those who questioned Trofim Denisovich Lysenko, a dominant figure in Soviet biology for three decades from the mid-1930s to the late 1950s—first through his directorship of the Lenin All-Union Academy of Agricultural Sciences and then the Institute of Genetics of the USSR Academy of Sciences. Noted scientists who resisted Lysenko's often exaggerated claims that by using various techniques, such as vernalization and hybridization, it was possible to speed up plant growth and generate greater productivity in agriculture, were purged. As a result, the USSR in this period lost some of its most creative ecological thinkers. Bukharin, viewed by Stalin as a rival, and Hessen, who was closely associated with Bukharin and Vavilov, were both executed. Vavilov, who had opposed Lysenko on genetics, was imprisoned, where he died a few years later of malnutrition—to be dumped into an unmarked grave.[6]

In 1927, the issue of using the zapovedniki for "acclimatization" research—removal of wild and/or domestic animals and plants from their original habitat and placement in new habitats in an attempt to transform nature—arose in Soviet biology. Sukachev and Stanchinskii strongly defended the zapovedniki against those promoting the acclimatization agenda, arguing that they should remain inviolable. In 1933, Stanchinskii came directly into conflict with Lysenko (and his chief ally Issak Izrailovich Prezent) regarding the zapovedniki and acclimatization, leading to Stanchinskii's 1934 arrest, imprisonment, and torture. He was to die in prison (after a second arrest) in 1942.[7]

The consequences for Soviet ecological science, particularly in areas related to agriculture, were disastrous. Membership in VOOP, which had risen to 15,000 by 1932, declined to around 2,500 in 1940. More and more, the zapovedniki were converted from reserves for the scientific study of pristine nature into a new role as transformation-of-nature centers.[8]

Nevertheless, in two major areas, forestry and climatology, Soviet ecology continued to develop. One of the key intellectual achievements was Sukachev's first introduction in 1941, developed more fully in 1944, of the concept of biogeocoenosis (alternatively biogeocoenose), which was to be extraordinarily influential both in the USSR and in the wider world, and was the main rival to Arthur Tansley's ecosystem category.[9] A botanist and ecologist, Sukachev had been influenced by Georgii Fedorovich Morozov, considered the founder of Russian scientific forestry, who died in 1920. Morozov helped introduce systemic thinking into Russian ecology by making extensive use of the concept of biocoenosis, or biological community, coined by the German zoologist Karl Möbius in 1877.

Sukachev's concept of "biogeocoenosis" was a further development on biocoenosis, intended to incorporate the abiotic environment. It was conceived in dialectical-energetic terms as a more unified and dynamic category than the notion of the ecosystem. The concept of biogeocoenosis grew out of and had an integral connection to Vernadsky's notions of the biosphere and biogeochemical cycles. According to Sukachev, in his landmark 1964 work *Fundamentals of Forest Biogeocoenology*, written with N. Dylis, "The idea of the interaction of all natural phenomena . . . is one of the basic premises of materialistic dialectics, well proved by the founders of the latter, K. Marx and F. Engels."[10] "A Biogeocoenose," as Sukachev famously defined it,

> is a combination on a specific area of the earth's surface of homogeneous natural phenomena (atmosphere, mineral strata, vegetable, animal, and microbiotic life, soil, and water conditions), possessing its own specific type of interaction of these components and a definite type of interchange of their matter

and energy among themselves and with other natural phenomena, and representing an internally contradictory dialectical unity, being in constant movement and development.[11]

In a 1960 article he further explained:

> Since the existence of mutual influences or interaction of the components is the most characteristic feature of the [integrative ecological] concept in question, we believe that "biogeocoenosis" (from the Greek words *koinos* "common" and the prefixes bio "life" and geo "earth," which emphasize the participation in this general unity of living things and inert elements of the earth's surface) is the more accurate and descriptive term [as compared with all alternatives]. . . .
>
> A biogeocoenosis may be defined as any portion of the earth's surface containing a well-defined system of interacting living [biotic] (vegetation, animals, microorganisms) and dead [abiotic] (lithosphere, atmosphere, hydrosphere) natural components, i.e. a system of obtaining and transforming matter and energy and exchanging them with neighboring biogeocoenoses and other natural bodies that remain uniform.
>
> The continuous interaction of all the components among themselves and with surrounding natural objects means that each biogeocoenosis is a dynamic phenomenon, constantly moving, changing, and developing.[12]

Hence, "each organism and each specimen," Sukachev argued, "is in dialectical unity with the environment." Nevertheless, a key aspect of the ecological condition was that multicellular organisms higher on "the evolutionary ladder"—that is, characterized by a wider range of adaptive mechanisms and specialization in relation to their environment—experienced a "growth of relative autonomy." The biogeocoenosis could then be seen as dialectically evolving in complex ways, with organisms actively changing their environments—a reality that demanded specific investigations. "The biogeocoenosis as a whole," he wrote, "develops through the interaction of all its variable components and in accordance with special laws. The very process of interaction among the components constantly disrupts the established relationships, thereby affecting the evolution of the biogeocoenosis as a whole."[13] Like dialectical frameworks in general, Sukachev's biogeocoenosis (even more than its main conceptual rival, ecosystem) emphasized internal dynamics, contradictory changes, and instability in ecological processes.

The dialectical, integrative approach in Soviet ecology promoted by figures like Morozov and Sukachev, which was rooted in detailed empirical research into specific conditions, led to the recognition of the extent to which forest-ecological-system health was essential to hydrology and the control of climate. This broad ecological understanding helped give rise in 1948 to the Great Stalin Plan for the Transformation of Nature, which was conceived as a grand attempt to reverse anthropogenic regional climate change in deforested areas, with an emphasis on the promotion of watersheds. Already in 1936, the Soviet government had created the Main Administration of Forest Protection and Afforestation, which established "water-protective forests" in wide belts across the country. While forests in parts of the Soviet Union were exploited relentlessly for industrial use, the best old-growth forests of the Russian heartland were protected, with ecological concerns given priority, eventually creating a total "forest preserve the size of France, which grew over time to an area the size of Mexico" (roughly two-thirds of the contiguous United States).[14]

The Great Stalin Plan for the Transformation of Nature, introduced in the context of attempts at ecological restoration following the Second World War, was the most ambitious plan of afforestation in all of history up to that point. It sought to create some 6 million hectares (15 million acres) of entirely new forest in the forest-steppe and steppe regions, and constituted "the world's first explicit attempt to reverse human-induced climate change." The trees were planted in shelterbelts along rivers (and roads) and around collective farms, with the goal of staving off the drying influence of winds emanating from

Central Asia, while protecting watersheds and agriculture. Although the plan had not been realized at the time of Stalin's death (when it was discontinued), a million hectares of new forest were planted, with 40 percent surviving.[15] Yet, even while this afforestation plan was being carried out, some 85 percent of the territory of the zapovedniki was formally liquidated in 1951, to be reestablished under the leadership of Sukachev and others during the resurrected conservation movement of the late 1950s.[16]

One reason for the limited success of the Great Stalin Plan was Lysenko's entry into forestry and his battle for control of Soviet afforestation. In 1948, Lysenko had achieved his greatest victory, with the Lenin All-Union Academy of Agricultural Sciences' declaration that Mendelian genetics was a form of bourgeois idealism. With the introduction of the Great Stalin Plan for the Transformation of Nature, Lysenko turned his attention to forestry, taking direct control of the Main Administration for Field-Protective Afforestation. He concocted a "nest method" of planting trees based on the notion that tree seedlings planted in dense formations would collectively defend themselves from other species, reducing the amount of labor required to clear areas for planting. Here, however, Lysenko was opposed at every step by Sukachev, who countermanded his orders on the ground several times and reported to the Ministry of Forest Management in 1951 that 100 percent of the forest seedlings planted in the Ural territorial administration with Lysenko's nest method had died.[17]

From 1951, two years before Stalin's death, and continuing until 1955, Sukachev, as the dean of Soviet botany—director of the Academy of Science's Institute of Forests, head of the Academy Presidium's Commission on Zapovedniki, and editor of the *Botanical Journal*—courageously launched an intellectual war against Lysenko. In article after article he wrote and edited for the *Botanical Journal* and the *Bulletin of the Moscow Society of Naturalists* (the journal of Russia's oldest and most prestigious scientific society) Sukachev, in what Weiner has called a "monumental battle against Lysenko," sharply criticized Lysenko's theories and methods. Later, in 1965, Sukachev was to accuse Lysenko of fraudulent practices. Young biologists viewed Sukachev as a hero and secretly flocked to his banner. In 1955 Sukachev was elected president of the Moscow Society of Naturalists (MOIP), a position he occupied until his death in 1967. This symbolized a dramatic decline in Lysenko's power and a shift in Soviet ecology (although Lysenko's final removal as head of the Institute of Genetics was not until 1965, under Brezhnev). Following Sukachev's election as president of the MOIP, a concerted campaign to reestablish the zapovedniki began. At that point the Soviet conservation movement began to rise out of the ashes. Membership in VOOP grew to 136,000 in 1951, and by 1959 had topped 910,000. The 1960s saw a spectacular rise of student conservation brigades nurtured by the MOIP under Sukachev.[18]

Meanwhile, Soviet climatology had been making extraordinary advances through the work of figures such as E. K. Fedorov (Y. K. Fyodorov), famous for his work on the Arctic, and Mikhail Ivanovich Budyko, who specialized initially in the emerging field of energetics, focusing on exchanges of energy and matter in a global context. Budyko's pathbreaking *Heat Balance of the Earth Surface*, published in 1958, earned him the prestigious Lenin Prize. In this work he developed a method for calculating the various components of the heat balance of the entire Earth System. This was crucial in opening the way to the founding of physical climatology as a field. Appointed in 1954 as director of Leningrad's Main Geophysical Observatory, at age thirty-four, Budyko played a crucial role in delineating multiple aspects of "the global ecological system." He was to be awarded the Blue Planet Prize in 1998 (the same year as David Brower in the United States) for founding physical climatology, early warnings on the acceleration of global warming, developing the nuclear winter theory, and pioneering global ecology. Budyko built his theoretical and empirical analysis on Vernadsky's biosphere concept and saw Sukachev's work on the biogeocoenosis as "essential in developing modern ideas of interrelations between organisms and the environment." Sukachev was to rely in turn on Budyko's energy flow analysis in his own work.[19]

Late Soviet Ecology

One of the tragedies of Soviet ecology is that the USSR's degradation of its environment worsened in the first decade after Stalin's death in 1953, with the discontinuation of the Great Stalin Plan for the Transformation of Nature and the more rapacious exploitation of resources. Six days after Stalin's death the Ministry of Forest Management was abolished and forest conservation was reduced to a much lower priority. (Yet it was not until the post-Soviet era that Vladimir Putin was finally to sign altogether out of existence Stalin's Group I of protected forests—those under the highest level of protection and preservation.)[20]

The USSR obtained high rates of growth through a form of extensive development, drawing constantly on more labor and resources. By the end of the 1950s the weaknesses of this approach, and the need to develop more intensive forms of development that took into account resource limits, were already becoming apparent. However, inertia within the system, and an accelerating Cold War, prevented a transition to a more rational economic development path.[21]

The worst damage was done during the Malenkov and Khrushchev years. Partly as a result, these years saw the rise of what was to be an immense environmental movement growing initially out of the scientific community. Khrushchev's "Virgin Lands" program, beginning in 1954, targeted the plowing up of 33 million hectares of so-called virgin land for the expansion of agriculture. Initial successes were obtained, but these were soon followed by dust bowls. In the late 1950s the Soviet leadership decided for the first time to interfere with the ecology of Lake Baikal, the oldest and deepest freshwater lake in the world. In the early 1960s the Soviet Presidium ordered the diversion of the two main rivers feeding into the Aral Sea, the Amu Darya and the Syr Darya, in order to provide irrigation for cotton farming in Soviet Eurasia. The Aral Sea consequently shrank to a tenth of its original size.[22]

These developments were met with a powerful response from scientists and conservationists. In 1964 Sukachev, as head of the MOIP, sent a letter to Soviet geographers in order to draw them into the fight to save Lake Baikal. Two years later he was one of a group of scientists who signed a collective letter to the media demanding protection of Lake Baikal. Baikal became a symbol of ecological destruction, leading to the extraordinary growth of the Soviet environmental movement. By 1981, VOOP membership had risen to 32 million, and by 1985 to 37 million, constituting the largest nature protection organization in the world. During the Brezhnev to Gorbachev years, the Soviet leadership introduced more and more environmental measures.[23]

Fedorov, one of the leading climatologists, became a member of the Presidium of the Supreme Soviet of the USSR and headed the Institute of Applied Geophysics of the State Committee of the USSR on Hydrometeorology and Control of the Natural Environment. In the early 1960s Fedorov's views with respect to the environment could be described as human exemptionalist, though in 1962 he raised the critical issue of sea-level rise with a melting of the Greenland ice sheet. But a decade later he had clearly shifted in an ecological direction. His 1972 *Man and Nature* presented a Marxian environmental perspective explicitly linked to that of Barry Commoner in the West. Like most Soviet ecologists at the time, Fedorov accepted some aspects of the Club of Rome's 1972 *Limits to Growth* argument, which focused on natural resource limits to economic growth. But he insisted on an approach that more fully accounted for social and historical factors. Moreover, he argued that the authors of *The Limits to Growth* had erred in failing to consider the crucial challenge represented by climate change. Fedorov's arguments relied directly on Marx's theory of socio-ecological metabolism: "The authors of the materialist theory of social development," he wrote, "regarded interaction (metabolism) between people and nature as a vital element in human life and activity and showed that the socialist organization of society would have every possibility to ensure optimal forms of such interaction." With respect to climate, he pointed to Marx and Engels's early discussions of anthropogenic climate change on a regional basis (and the threat of desertification) in relation to the writings of Karl Fraas. Fedorov represented the USSR at the first World Conference on Climate in

Geneva in 1979, where he stressed the urgency of action, declaring that "future climate changes are unavoidable. They will become probably irreversible during the nearest decades"—if an international plan were not soon worked out.[24]

However, the scientific revolutions in climatology and global ecology in the Soviet Union had their main origins in the work of Budyko, who was the acknowledged world leader in the study of the heat balance of the earth. He was also the world's primary analyst of the effect of the polar ice on the climate, and was the first to delineate the ice albedo effect as a global warming feedback mechanism. Budyko was also the first to point to the dangerous acceleration in global average temperature that would result from such positive feedbacks. He went on to pioneer studies of paleoclimatic changes in earth history and to develop "global ecology" as a distinct field, based on a dialectical, biospheric analysis, in the tradition of Vernadsky and Sukachev. Budyko promoted a theory of "critical epochs" in the earth's history, which were characterized by "ecological crises" and "global catastrophes," and he extended this analysis to the growing threat of "anthropogenic ecological crisis."[25]

In 1961 Fedorov and Budyko called the All-Union Conference on the Problem of Climate Modification by Man in Leningrad to address the emerging problem of climate change—the first such conference in the world. That same year Budyko presented his paper "The Heat and Water Balance Theory of the Earth's Surface" to the Third Congress of the Geographical Society of the USSR, in which he arrived at his famous conclusion that anthropogenic climate change was now inevitable under business as usual, and that human energy usage needed to be addressed. In 1962, he published his landmark article "Climate Change and the Means of Its Transformation" in the USSR's *Bulletin of the Academy of Sciences*, in which this conclusion was again advanced, together with the observation that the destruction of ice cover could generate "a significant change in the regime of atmospheric circulation." By 1963 Budyko compiled an atlas of the world's heat balance system. "Budyko energy balance models" soon became the basis of all complex climate modeling. In 1966 he published (together with colleagues) an article on "The Impact of Economic Activity on the Climate," describing the history of anthropogenic climate change. In it he indicated that human beings—through actions such as deforestation, swamp drainage, and city construction—had long affected "the microclimate, i.e. local changes in the meteorological regime of the surface layer of the atmosphere." What was new, however, was that anthropogenic climate change was now occurring over large territories and globally.

However, the discovery of ice-albedo feedback (the melting of white sea ice and the increase in the aborption of solar energy by blue sea water, decreasing the reflectivity of the earth) and its dynamic effect on global warming was to change everything. Budyko had presented his basic analysis on this as early as 1962, in an article on "Polar Ice and Climate." But the extent that the global climate, and not just the climate of the Arctic, would be affected was not yet clear. In his 1969 article, "The Effect of Solar Radiation Variation on the Climate of the Earth," he provided a full and concrete assessment of the polar sea ice-albedo feedback mechanism and its relation to climate change. The observations were startling. Similar results on climate sensitivity pointing to catastrophic global climate change were presented that same year by William Sellers at the University of Arizona. From that point on, climate change moved from being a peripheral concern to an increasingly urgent global issue. Meanwhile, Budyko's explorations of the effects of aerosol loading led him to introduce the possibility of using planes to dump aerosols (sulfur particles) in the stratosphere as a possible geoengineering counter to climate change, given his belief that capitalist economies, especially, would not be able to limit their growth, energy use, or emissions. All of these conclusions were driven home in his 1972 book, *Climate and Life*. Although anthropogenic global warming had first been described by Guy Stewart Callendar as early as 1938, the discovery of significant feedback effects and greater climate sensitivity now posed the question of a potential runaway global ecological crisis in approaching decades.[26]

For Soviet climatologists, such as Fedorov, a Soviet delegate to the Pugwash conferences who also served as Vice President of the World

Council of Peace, and Budyko, the issue of peace was closely related to the environment.[27] It was Soviet climatologists, primarily based on the work of Budyko and G. S. Golitsyn, who first developed the nuclear winter theory in the case of a full-scale nuclear exchange—whereby over a hundred gargantuan firestorms set off by nuclear weapons would increase the aerosol loading in the atmosphere sufficiently to bring temperatures across whole continents down by several degrees and possibly several tens of degrees, thereby leading to the destruction of the biosphere and human extinction. The basis of this analysis was developed by the Soviets a decade before their counterparts in other countries. It would play a big role in the development of the anti-nuclear movement and the eventual backing away from the brink of nuclear holocaust during the later stages of the (first) Cold War.[28]

The enormous range and comprehensiveness of Budyko's ecological contributions were particularly evident in his later work, where he sought to define "global ecology" as a distinct field. He played a foundational role in the development of paleoclimatic analysis, examining the history of "global catastrophes" in earth history, associated with alterations in the climate—using this to develop further insights into the significance of anthropogenic climate change. In describing global ecology as a distinct area of analysis he emphasized that previous ecological work had been directed overwhelmingly at local conditions, or at most an "aggregate of local changes." Global ecology, in contrast, was that area of ecology concerned with the operation of the biosphere as a whole, and had arisen as a result of the sudden increase in the human capacity to alter atmospheric and ocean systems. Here again the emphasis was on the dialectical interaction between organisms and the environment. Budyko stressed Oparin's crucial observation—associated with the theory of life's origins—that organisms had generated the atmosphere as we know it, extrapolating this to a consideration of the human role with respect to the atmosphere. In his various analyses of the evolution of *Homo sapiens*, Budyko invariably went back to Engels's exploration in "The Part Played by Labour in the Transformation from Ape to Man" of what is now known as "gene-culture coevolution." Likewise, Budyko's *Global Ecology* pointed to

Marx's comment in a letter to Engels on the desertification tendencies of civilization. All ecological analysis, Budyko indicated, was modeled on metabolism, the process of material exchange between life and the environment.[29]

Some of Budyko's early heat balance work had been carried out together with leading Soviet geographers A. A. Grigoriev and Innokenti P. Gerasimov. The goal was a more integral dialectical science capable of addressing the evolution of the biosphere. Budyko and Gerasimov postulated that it was paleoclimatic change that had created the dynamic conditions millions of years ago in Africa for the evolution of the early hominins, including the Australopithecines and the genus *Homo*. In *Geography and Ecology*, a collection of his essays from the 1970s, Gerasimov provided an elegant theoretical merger of the notion of the geographic landscape with Sukachev's biogeocoenosis.

Scarcely less important was Budyko's analysis of the social aspects of what he considered to be the approaching "global ecological crisis." Here he emphasized the difficulties posed by the system of capital accumulation. All economic expansion was constrained by the fact that "the stability of the global ecological system is not very great." There was no way out of this dilemma except through economic and ecological planning, namely a "socialist planned economy" aimed at the realization of Vernadsky's "noosphere," or an environment ruled by reason.[30]

Crossing the intellectual boundaries represented by C. P. Snow's "two cultures," Budyko connected his analysis to the ideas of Soviet social and environmental philosophers, specifically those of Ivan T. Frolov, the dynamic editor-in-chief from 1968 to 1977 of the USSR's leading philosophy journal *Problems of Philosophy* (*Voprosy filosofi*). It was largely owing to Frolov's efforts that Soviet social philosophy in the 1970s and '80s began to revive, based on the conscious reintegration of ecological and humanistic values into dialectical materialism. In this new analysis, inspiration was drawn from Marx's deep humanism and naturalism in the *Economic and Philosophic Manuscripts* and the *Grundrisse*, as well as from his later ecological critique in *Capital*.

This emerging Soviet ecological Marxism deliberately circumvented the Frankfurt School in the West with its less materialist emphasis and suspicion of science—though accepting the analysis of Antonio Gramsci. Frolov and others called for the development of a "dialectical integral unity" on materialist-ecological grounds. The resulting critical philosophy and social science was rooted in the whole Soviet tradition of scientific ecology from Vernadsky to Sukachev to Budyko.[31]

Frolov's *Global Problems and the Future of Mankind*, published in 1982, represented an important first attempt in the creation of a new ethic of global ecological humanism. Moreover, a second work published that same year, *Man, Science, Humanism: A New Synthesis*, went still further in developing this new dialectical humanism-naturalism. Although Frolov's vision showed traces of technologism (especially in his treatment of food production), the overall perspective was deeply humanist in its analysis and its values. The human relation to nature, he indicated, quoting from Marx's *Economic and Philosophic Manuscripts*, needed to be governed not simply by the laws of sustainable production, but the "laws of beauty." He argued in these years for "moving away from the illusion of anthropocentrism and rejecting the traditional hegemonistic relationship to nature."[32]

But perhaps the most astonishing product of this revival of Soviet critical ecological thinking was the 1983 collection *Philosophy and the Ecological Problems of Civilisation*, edited by A. D. Ursul.[33] This volume was remarkable in that it brought together leading ecological philosophers like Frolov with such major natural-science figures as Fedorov and Gerasimov. The understanding of Marx and Engels's ecological thought demonstrated here—though still treated in a somewhat fragmented way—was profound. As Gerasimov explained, "Marx characterized labour as a process in which man 'starts, regulates, and controls the material re-actions [metabolism] between himself and nature.'... Man's interaction with nature needs to be subordinated to the general principles of metabolic processes." Similarly, Frolov, in criticizing the historically specific ecological depredations of capitalist society wrote: "The danger of an ecological crisis has become real not because the use of technical mechanisms and devices in the 'metabolism' of man and nature in itself . . . but primarily because this industrial development is realised on the basis of the socioeconomic, spiritual, and practical setups of the capitalist mode of production." It was essential, he argued, for society to focus on "ecodevelopment" or "ecologically justified development," taking into account "the objective dialectic and inner contradictoriness of the interaction of society and nature."[34]

A core aspect of Frolov's stance was his argument that although struggles to create a more ecologically rational world ran the risk of utopianism, since they necessarily got ahead of the development of material-social forces, the severity of the global ecological threat nevertheless demanded a "rational realism" that was utopian-like in character.[35]

The various essays in *Philosophy and the Ecological Problems of Civilisation* displayed signs of the characteristic Soviet faith in progress and technology and the overcoming of ecological constraints. Yet, the "ecological problems of civilization" were nonetheless presented with considerable depth and sophistication—particularly where the more radical and scientific thinkers were concerned. For Fedorov, arguing from the standpoint of climate science, the challenge was that "the scale of society's activity" now made it "necessary to take into account the quantities of all our planet's elements" and the "anthropogenic impact" on them. He illustrated this by reference to global warming, citing the work of Budyko. Turning to "the production of forest biogeocoenosis," philosopher N. M. Mamedov emphasized the need for a restoration ecology that would reestablish the integrity of ecosystems. Ursul pointed out that Vernadsky had long ago taught that humanity was becoming a geological force, and emphasized that "the extension of the scale of the ecological problem from a regional to a global, and even a cosmic one" represented a new challenge to society, and in effect a new geological epoch.[36]

Late Soviet ecological analysis was well ahead of most ecological socialism in the West in understanding the new planetary dynamic, associated with climate change in particular, and in the construction of a distinct global ecology. To be sure, by focusing their critique on the global ecological problem and on capitalism Soviet thinkers often skirted the ecological problems of the USSR itself. Still, Frolov had gained his reputation in the late 1960s through a major critical assessment of the whole sorry history of Lysenkoism, in which he openly contested the very idea of "party science." Gerasimov's *Geography and Ecology* was remarkable in its direct confrontation (in an essay written in 1977) with major Soviet ecological problems. Thus he explicitly, if somewhat schematically, highlighted in the Soviet context: (1) the history of the destruction of the Aral Sea, (2) the controversial diversion of rivers, (3) the causes of desertification, (4) the imperative of protecting Lake Baikal, (5) the need to restore the taiga forests, (6) destructive forms of timber exploitation, (7) irrational, non-scientific mining practices, (8) controlling air pollution in cities, (9) removal of industrial wastes from urban areas, and (10) actions to limit new forms of radioactive and toxic waste. What was needed, he insisted, was "an *ecologization* of modern science." As the preeminent Soviet geographer, Gerasimov took the huge step of arguing that ecology, not economy, should become the focal point of geography as a field.[37]

Soviet economists in this period were engaged in a fierce debate over the proper relation of economic growth calculations to social welfare. P. G. Oldak took a leading role in the 1970s and '80s in arguing for the replacement of the standard economic growth calculations with a new approach focusing on "gross social wealth" as the basis for socioeconomic decisions. Lenin, Oldak pointed out, had made it clear that the goal of socialism should be the free development of each member of the population on the widest possible (that is, not narrowly economistic or mechanistic) basis, taking account of qualitative factors. With this as the justification, Oldak proposed a new accounting that would directly incorporate into the main planning criteria not only accumulated material wealth but also services, the knowledge sector, the condition of natural resources, and the health of the population. Given an "excess of the anthropogenic load on natural systems over their potential for self-regeneration," it might even be rational, he suggested, to choose to curtail production altogether for a time in order to transition to "a new [and more sustainable] production level."[38]

In 1986–87, Frolov became the editor in chief of *Kommunist*, the Communist Party's main theoretical organ; in 1987–89 (after Chernobyl) he was one of Gorbachev's key advisers; and in 1989–91 he was editor-in-chief of *Pravda*. Frolov was responsible for much of the ecological cast that Gorbachev gave to his public pronouncements, which were accompanied by a speeding up of environmental reform measures.

Nevertheless, the much wider shift in power relations in the Soviet state and the destabilization of the society that Gorbachev had introduced with *glasnost* and *perestroika* led to a deepening of Soviet political-economic contradictions, the rapid dismantling of its hegemony in Eastern Europe, splits in the top echelons of the Soviet *nomenklatura*, and a dissolution of the whole power system—leading to the demise of the USSR in 1991.

Soviet Ecology in the Twenty-First Century

The foregoing argument points to a complex historical reality not captured in the hegemonic depiction of the course of Soviet environmental history from the mid-1930s on as one continuous story of extreme ecological degradation, even ecocide. From an ecological perspective, the USSR can be seen as a society that generated some of the worst ecological catastrophes in history but it also gave birth to some of the most profound ecological ideas and practices, based on materialist, dialectical, and socialist intellectual foundations. Characterized by the growth of repressive bureaucratic control and the emergence of new class relations, the USSR by the late 1930s had ceased to be meaningfully socialist in the sense of moving in the direction of a society governed by the associated producers, and instead is best described as a post-revolutionary society of a distinct type, neither capitalist nor socialist.[39] Yet the existence of economic planning and a wide sphere of social property ownership, plus the intellectual legacies of Marxian theory in terms of materialist, dialectical, and socialist thinking, all ran deep. However distorted the development of the Soviet Union became in terms of its original socialist objectives, it did promote alternative forms of socialization. The purges of ecological thinkers and various environmental depredations in the Stalin or middle period gave way in the end to enormous achievements in the development of a distinct global ecology—in a kind of negation of the negation. It was in the Soviet Union, based on the theories of the biosphere and biogeocoenosis, that the analysis of accelerated climate change began, and it was from Moscow and Leningrad, not Washington and New York, that the first warnings of runaway global warming and the theory of nuclear winter first emanated.

The historic turning point in the reemergence of Soviet environmentalism took place in the early 1950s with Sukachev's struggle against Lysenko, the growing role of the Moscow Society of Naturalists, the rise of student conservation brigades, and the eventual emergence of VOOP as the largest conservation organization in the world. In the 1960s, beginning with Brezhnev, significant environmental legislation was passed, but implementation was generally ineffective due to conflict with plant managers, class-economic barriers, failure to disseminate information (cloaked in secrecy), and the still-nascent development of the environmental movement. Joan DeBardeleben's remarkably balanced assessment of "The New Politics in the USSR" contends that, despite important environmental initiatives, "pro-development forces on the whole were considerably stronger than the pro-environmental forces in the Brezhnev period." Still, environmental progress was discernible. Thus, the number of zapovedniki by 1983 had gradually expanded to 143—going beyond the 128 that existed in 1951, before the great bulk were liquidated under Stalin (and well beyond the thirty-three originally established under Lenin).[40]

In the Gorbachev era, beginning in 1985, everything changed. What followed has been characterized by Laurent Coumel and Marc Elie in *The Soviet and Post-Soviet Review* as a "tragic ecological revolution"—the tragedy lying mainly in the demise of the Soviet Union that cut it short, leading to a dramatic decline both in the environmental movement and in the state's responsiveness to ecological issues in the post-Soviet years, as capitalism resumed control.[41]

Following Chernobyl in 1986, the Soviet environmental movement became more powerful. In addition to VOOP, some three hundred major environmental organizations were operating throughout the USSR. "From 1987 to 1990, all across the USSR, plants were closed, planned projects were re-sited or re-tooled for a less polluting type of production, or projects were canceled altogether. The most prominent examples included the cessation of work on the planned river diversion projects, cancellation of the Volga-Chograi canal, closing of biochemical plants, and plans to convert the Baikalsk Pulp and Paper Plant to furniture production." Environmental movement pressure resulted in the closing down of over a thousand large enterprises in these years.[42]

Dramatic results were apparent in relation to carbon dioxide emissions as well. Already in the 1960s the country had begun to shift from coal as its main energy source to natural gas. In 1988 carbon emissions peaked. They fell dramatically in the two years after that, due chiefly to the aggressive switchover from coal to natural gas.[43]

Implying falsely that a critical-scientific Soviet ecology was nonexistent, U.S. historian Paul Josephson observed in 2010 that there were "no Soviet counterparts to Rachel Carson's *Silent Spring* or the Club of Rome's *Limits to Growth*."[44] Yet late Soviet ecology did generate such works as Sukachev's *Fundamentals of Forest Biogeocoenology*, Federov's *Man and Nature*, Budyko's *Climate and Life*, *Global Ecology*, and *The Evolution of the Biosphere*, and Frolov's *Man, Science, Humanism*. Like Carson's later work these were all influenced by Vernadsky's *The Biosphere* and Oparin's theory of the origins of life.[45] These contributions enormously advanced ecological science and thinking, and pointed to the need for a rapid ecological restructuring of human society throughout the globe. In the twenty-first century a recognition of the positive achievements of Soviet ecology is obviously crucial if we are to create the Great Transition now called for by environmentalists worldwide.

Late Soviet ecology, moreover, left a legacy of economic planning (coupled with signs of an emergent ecological planning) that, for all of its weaknesses and false turns, represented in many ways a massive human achievement; one from which we need to learn today if we are to find a way to regulate the human metabolism with nature and to surmount the present global ecological crisis. It began a process of ecological transition that, if carried out fully, could have had immeasurable positive effects.

Writing on "Socialism and Ecology" in 1989, Paul Sweezy argued that unless "the planning system" represented by such societies could somehow be preserved "and adapted to serve the needs of the new situation," and unless the potential of so-called actually existing socialist societies to operate, unlike capitalism, on other bases than the pursuit of economic riches, were somehow harnessed, it might simply be "too late for civilized humanity to restore the necessary conditions for its own survival."[46] This is a specter that haunts us today more than ever. The answer to our present problems requires some sort of convergence with the notion of the planned regulation of the environment in accordance with human needs: the primary message of late Soviet ecology.

CHAPTER FIFTEEN

The Return of Nature and *Marx's Ecology*
Interview by Alejandro Pedregral

John Bellamy Foster writes me before leaving Eugene, Oregon: "We had to evacuate. And we have to travel a long ways. But I will try to send the interview by the morning." The massive fires on the West Coast of the United States had triggered the air quality index up to values of 450, and in some cases over the maximum of 500—an extremely dangerous health situation. Forty thousand people in Oregon had left their homes and another half a million were waiting to flee if the threat grew. "Such is the world of climate change," Foster states. Professor of sociology at the University of Oregon and editor of *Monthly Review*, twenty years ago Foster revolutionized Marxist ecosocialism with *Marx's Ecology*. This book, together with *Marx and Nature* by Paul Burkett, opened Marxism to a second wave of ecosocialist critique that confronted all kinds of entrenched assumptions about Karl Marx in order to elaborate an ecosocialist method and program for our time. The great development of Marxist ecological thought in recent years—which has shown how, despite writing in the nineteenth century, Marx is essential for reflecting on our contemporary ecological degradation—is in part the product of a turn carried out by Foster and others linked to *Monthly Review*. Current researchers in this area, associated with what came to be known as the school of the metabolic rift from the central notion Foster rescued from volume 3 of Marx's *Capital*, have developed numerous ecomaterialist lines of research in the social and natural sciences—from imperialism and the study of the exploitation of the oceans, to social segregation and epidemiology. On the occasion of the release of his latest book, *The Return of Nature*, a monumental genealogy of great ecosocialist thinkers that has taken him twenty years to complete, Foster tells us about the path these key figures traveled, from the death of Marx to the emergence of environmentalism in the 1960s and '70s, as well as about the relationship of his new book to *Marx's Ecology* and the most prominent debates of current Marxist ecological thought. His reflections thus serve to help us rethink the significance of this legacy, in view of the urgent need for a project that transcends the conditions that threaten the existence of our planet today.

Alejandro Pedregal: In *Marx's Ecology*, you refuted some very established assumptions about the relationship between Marx and ecology, both within and outside of Marxism—namely, that the ecological thought in Marx's oeuvre was marginal; that his few ecological insights were mostly if not solely found in his early work; that he held Promethean views on progress; that he saw in technology and the development of the productive forces the solution to the contradictions of society with nature; and that he did not show a genuine scientific interest in the anthropogenic effects on the environment. Your work, along with that of others, disputed these assumptions and shifted many paradigms associated with them. Do you think that these ideas persist in current debates?

John Bellamy Foster: Within socialist and ecological circles in English-speaking countries, and indeed I think in most of the world, these early criticisms of Marx on ecology are all now recognized as disproven. They not only have no basis in fact, but are entirely contradicted by Marx's very powerful ecological treatment, which has been fundamental to the development of ecosocialism and increasingly to all social-scientific treatments of the ecological ruptures generated by capitalism. This is particularly evident in the widespread and growing influence of Marx's theory of the metabolic rift, the understanding of which keeps expanding and which has been applied now to nearly all of our current ecological problems. Outside the English-speaking world, one still occasionally encounters some of the earlier misconceptions, no doubt because the most important works so far have been in English, and much of this has not yet been translated. Nevertheless, I think we can treat those earlier criticisms as now almost universally understood to be invalid, not simply due to my work, but also that of Paul Burkett in *Marx and Nature*, Kohei Saito in *Karl Marx's Ecosocialism*, and many others. Hardly anyone on the left is so simplistic today as to see Marx as a Promethean thinker in the sense of promoting industrialization over all else. There is now a widespread understanding of how science and the materialist conception of nature entered his thought, a perception reinforced by the publication of some of his scientific/ecological extract notebooks in the *Marx-Engels Gesamtausgabe* project. Thus, I don't think the view that Marx's ecological analysis is somehow marginal in his thought is given much credence among socialists in the English-speaking world today, and it is rapidly receding everywhere else. Ecological Marxism is a very big topic in Europe, Latin America, China, South Africa, the Middle East—in fact, nearly everywhere. The only way in which Marx's ecological analysis can be seen as marginal is if one were to adopt an extremely narrow and self-defeating definition of what constitutes ecology. Moreover, in science, it is often the most "marginal" insights of a thinker that prove most revolutionary and cutting edge.

Why were so many convinced earlier on that Marx had neglected ecology? I think the most straightforward answer is that most socialists simply overlooked the ecological analysis present in Marx. Everyone read the same things in Marx in the prescribed manner, skipping over what was then designated as secondary and of little importance. I remember talking to someone years ago who said there were no ecological discussions in Marx. I asked if he had ever read the chapters on agriculture and rent in volume 3 of *Capital*. It turned out he hadn't. I asked: "If you haven't read the parts of *Capital* where Marx examines agriculture and the soil, how can you be so sure that Marx did not deal with ecological questions?" He had no answer. Other problems were due to translation. In the original English translation of *Capital*, Marx's early usage of *Stoffwechsel*, or metabolism, was translated as *material exchange* or *interchange*, which hindered rather than helped understanding. But there were also deeper reasons, such as the tendency to overlook what Marx meant by materialism itself, which encompassed not just the materialist conception of history, but also, more deeply, the materialist conception of nature.

The important thing about Marx's ecological critique is that it is unified with his political-economic critique of capitalism. Indeed, it can be argued that neither makes any sense without the other. Marx's critique of exchange value under capitalism has no significance outside of his critique of use value, which related to natural-material conditions. The materialist conception of history has no meaning unless it is seen in relation to the materialist conception of nature. The alienation of labor cannot be seen apart from the alienation of nature. The exploitation of nature is based on capital's expropriation of the "free gifts of nature." Marx's very definition of human beings as the self-mediating beings of nature, as István Mészáros explained in *Marx's Theory of Alienation*, is based on a conception of the labor process as the *metabolism of human beings and nature*. Science as a means of enhancing the exploitation of labor can't be separated from science conceived as the domination of nature. Marx's notion of social metabolism cannot be divided off from the question of the metabolic rift. And so on. These things were not actually separated in Marx, but were removed from each other by later left thinkers, who generally ignored ecological questions, or who employed idealist, mechanist, or dualist perspectives and thus robbed the critique of political economy of its real material basis.

AP: In regard to Prometheanism, you have shown in your work how

Marx's reflections on Prometheus are to be read in relation to his own scholarly research on Epicurus, as well as to the Roman poet Lucretius, and thus need to be interpreted as linked to the secular knowledge of the Enlightenment, rather than as a blind advocacy for progress. However, the dominant use of the term *Prometheanism* remains quite common in Marxist literature, which gives room to certain accelerationist and techno-fetishist trends that reclaim Marx for their aims. Should this notion be challenged more effectively, at least in relation to Marx and his materialist thought?

JBF: This is a very complicated issue. Everyone knows that Marx praised Prometheus. He was a devotee, of course, of Aeschylus's *Prometheus Bound*, which he reread frequently. In his dissertation he compared Epicurus to Prometheus. And Marx himself was even caricatured as Prometheus in the context of the suppression of the *Rheinische Zeitung* in a famous image that appears in volume 1 of Marx and Fredrick Engels's *Collected Works*. It thus became common for various critics within and without Marxism to characterize Marx's views as Promethean, particularly in such a way as to suggest that he saw extreme productivism as the chief aim of society. Not having any proof that Marx put industrialization before human social (and ecological) relations, his critics simply employed the term *Promethean* as a way of making their point *without evidence*, merely taking advantage of this common association with Marx.

Yet, this was a distortion in quite a number of ways. In the Greek myth, Prometheus, a Titan, defied Zeus by giving fire to humanity. Fire of course has two manifest qualities. One is light, the other is energy or power. In the interpretation of the Greek myth in Lucretius, Epicurus was treated as the bringer of light or knowledge in the sense of Prometheus, and it was from this that Voltaire took the notion of Enlightenment. It was in this same sense that Marx himself praised Epicurus as Prometheus, the giver of light, celebrating him as the Enlightenment figure of antiquity. Moreover, Marx's references to Aeschylus's *Prometheus Bound* always emphasized Prometheus' role as a revolutionary protagonist in defiance of the Olympian gods.

In the age of the Enlightenment itself, the Prometheus myth was seen, not surprisingly, as all about Enlightenment, not about energy or production. Walt Sheasby, a great ecosocialist with whom I worked in the early days of *Capitalism, Nature, Socialism* and while I was editor of *Organization and Environment*, wrote an extraordinary piece for the latter journal in March 1999, establishing conclusively that the notion of Prometheanism and the Promethean myth was used until the nineteenth century primarily in this sense of Enlightenment. I am not sure when the usage changed. But certainly Mary Shelley's *Frankenstein, or, the Modern Prometheus* and Pierre-Joseph Proudhon's *The Philosophy of Poverty* represented a shift where Prometheanism came to mean industrialism and machinery, symbolizing the Industrial Revolution. Here, Prometheus was seen as standing for mechanical power. It is interesting that Marx took on Proudhon's mechanistic Prometheanism directly, attacking all such notions in *The Poverty of Philosophy*. Yet, the Promethean myth became reified as a story of industrialization, something the ancient Greeks themselves could never possibly have imagined, and the common identification of Marx with Prometheus in people's minds became a way therefore of faulting him on ecological grounds. Interestingly, the charge that Marx was Promethean, which you find in such figures such as Leszek Kolakowski, Anthony Giddens, Ted Benton, and Joel Kovel, was directed against Marx exclusively and at no other thinker, which points to its ideological character.

The closest anyone could come to finding evidence that Marx was Promethean in the sense of glorifying industrialization as its own end was in his panegyric to the bourgeoisie in the first part of the *Communist Manifesto*, but this was simply a prelude to his critique of the same bourgeoisie. Thus, he turned around a few pages later, ushering in all the contradictions of the bourgeois order, referring to the sorcerer's apprentice, ecological conditions (town and country), the business cycles, and of course the proletariat as the grave digger of capitalism. In fact, there is nowhere that Marx promotes industrialization as an objective in itself as opposed to free, sustainable human development.

Explaining all of this, though, takes time and, while I have brought up all of these points at various occasions in my work, it is usually sufficient simply to show that Marx was not at all a Promethean thinker, if what is meant by this is the worship of industry, technology, and productivism as ends in themselves, or a belief in an extreme mechanistic approach to the environment. In these concrete terms, setting aside the confusions borne of myth, there can be no doubt.

AP: Twenty years after *Marx's Ecology*, the extensive work of the metabolic rift school has transformed today's debates about Marxism and ecology. What are the continuities and changes between that context and the current one?

JBF: There are several different strands of discussion and debate. One, the most important, as I indicated, is a vast amount of research into the metabolic rift as a way of understanding the current planetary ecological crisis and how to build a revolutionary ecosocialist movement in response. Basically, what has changed things is the spectacular rise of Marxian ecology itself, throwing light on so many different areas, not only in the social sciences, but in the natural sciences as well. For example, Mauricio Betancourt has recently written a marvelous study for *Global Environmental Change* on "The Effect of Cuban Agroecology in Mitigating the Metabolic Rift." Stefano Longo, Rebecca Clausen, and Brett Clark have applied Marx's method to the analysis of the oceanic rift in *The Tragedy of the Commodity*. Hannah Holleman has used it to explore dust bowls past and present in *Dust Bowls of Empire*. A considerable number of works have utilized the metabolic rift conception to understand the problem of climate change, including Brett Clark, Richard York, and myself in our book *The Ecological Rift* and Ian Angus in *Facing the Anthropocene*. These works, as well as contributions by others, such as Andreas Malm, Eamonn Slater, Del Weston, Michael Friedman, Brian Napoletano, and a growing number of scholars and activists too numerous to name, can all be seen basically in this light. An important organization is the Global Ecosocialist Network in which John Molyneux has played a leading role, along with System Change Not Climate Change in the United States. Naomi Klein's work has drawn on the metabolic rift concept. It has played a role in the Landless Workers' Movement (MST) in Brazil and in discussions around the question of ecological civilization in China.

Another issue concerns the relations between Marxian ecology and both Marxist feminist social reproduction theory and the new analyses of racial capitalism. All three of these perspectives have drawn in recent years on Marx's concept of *expropriation* as integral to his overall critique, extending beyond *exploitation*. It is these connections that motivated Brett Clark and myself to write our recent book *The Robbery of Nature* on the relation between *robbery and the rift*, that is, the expropriation of land, use values, and human bodies, and how this is related to the metabolic rift. An important area is the whole realm of ecological imperialism and unequal ecological exchange on which I have worked with Brett Clark and Hannah Holleman.

Today, there are some new criticisms of Marx on ecology aimed at the metabolic rift theory itself, saying it is dualistic rather than dialectical. But this of course is a misconception, since for Marx the social metabolism between humanity and (extra-human) nature through the labor and production process is by definition the *mediation* of nature and society. In the case of capitalism, this manifests itself as an *alienated mediation* in the form of the metabolic rift. Such an approach, focusing on labor/metabolism as the dialectical mediation of totality, could not be more opposed to dualism.

Others have said that if classical Marxism addressed ecological questions, they would have appeared in subsequent socialist analyses after Marx, but did not. That position too is wrong. In fact, that is the question taken up in *The Return of Nature*, which was expressly intended to explore the dialectic of continuity and change in socialist and materialist ecology in the century after the deaths of Charles Darwin and Marx, in 1882 and 1883 respectively.

AP: Indeed, in *Marx's Ecology* you focused on the emergence and formation of Marx's materialism in correlation to that of Darwin's and

Alfred Russell Wallace's theory of evolution, ending precisely with the deaths of the first two. Now, in your new book, you start from this point to trace an intellectual genealogy of key ecosocialist thinkers until the appearance of the ecological movement in the 1960s and '70s. For a long time, some of these stories did not receive enough attention. Why did it take so long to recover them? And how does the rediscovery of these links help us understand the emergence of the ecologist movement differently?

JBF: *The Return of Nature* continues the method of *Marx's Ecology*. This can be seen by comparing the epilogue of the earlier book to the argument of the later one. *Marx's Ecology*, apart from its epilogue, ends with the deaths of Darwin and Marx; *The Return of Nature* begins with their funerals and with the one person who was known to be present at both funerals, E. Ray Lankester, the great British zoologist who was Darwin's and Thomas Huxley's protégé and Marx's close friend. *The Return of Nature* is not directed simply at the development of Marxist ideas, but at the socialists and materialists who developed what we today call ecology as a critical form of analysis. Moreover, we can see how these ideas were passed on in a genealogical-historical fashion.

Like all Marxian historiography, this, then, is a story of origins and of the dialectic of continuity and change. It presents a largely unbroken genealogy that extends, though in complex ways, from Darwin and Marx to the explosion of ecology in the 1960s. Part of my argument is that the socialist tradition in Britain from the late nineteenth to the mid-twentieth century was crucial in this. Not only was this the main period of the development of British socialism, but in the sciences the most creative work was the product of a kind of synthesis of Darwin and Marx along evolutionary ecological lines. The British Marxist scientists were closely connected to those revolutionary Marxist thinkers involved in the early and most dynamic phase of Soviet ecology—nearly all of whom were later purged under Joseph Stalin—but unlike their Soviet counterparts, the British-left scientists were able to survive and develop their ideas, ushering in fundamentally new socioecological and scientific perspectives.

A common criticism of *Marx's Ecology* from the beginning, raised, for example, in the journal *Capitalism, Nature, Socialism* right after the book was published, was that, even if Marx had developed a powerful ecological critique, this had not been carried forward in subsequent socialist thought. There were two answers to this. The first was Rosa Luxemburg's statement that Marx's science had reached far beyond the immediate movement and the issues of the time, and that, as new contradictions and challenges arose, new answers would be found in Marx's scientific legacy. In fact, it is true that Marx's perception of the ecological crisis of capitalism, based in tendencies of his time, was far ahead of the historical development and movement, which in some ways makes his analysis more valuable, not less. But the other answer is that the presumption that there was no socialist ecological analysis was false. Indeed, ecology as a critical field was largely the creation of socialists. I had already tried to explain this in the epilogue to *Marx's Ecology*, but much more was needed. The challenge was to uncover the history of socialist and materialist ecology in the century after Marx. But doing this was a huge undertaking since there was no secondary literature to speak of, except in some respect Helena Sheehan's marvelous *Marxism and the Philosophy of Science*.

I commenced the archival research for *The Return of Nature* in 2000, around the time that *Marx's Ecology* was published. The idea was always to explore further the issues brought up in the epilogue, focusing on the British context. But at the same time, as I began this work, I also took on the position of co-editor (and eventually sole editor) of *Monthly Review*, and that naturally pulled me back to political economy, which governed my work for years. Moreover, when I wrote on ecology in these years, I had to deal first and foremost with the immediate crisis. So I could only work on an intensive project like *The Return of Nature* when the pressure was off, during short vacations from teaching. As a result, the work proceeded slowly over the years with innumerable interruptions. I might never have finished the book except for constant encouragement by a few friends, particularly John Mage, to whom the book is dedicated, and the fact that the ecological problem came to loom so large that, for *Monthly*

Review itself, the ecological critique became as important as the critique of political economy, making the development of systematic historical approach more necessary than ever.

However, the bigger reason the book took so long was that these stories were not known and it required an enormous amount of archival research and pursuit of obscure sources, including works that no one had read for more than half a century. Great works were cast aside and grew moldy in obscure corners. Other writings were not published or had appeared only in hard-to-find places. The role of thinkers such as J. B. S. Haldane, Joseph Needham, J. D. Bernal, Hyman Levy, and Lancelot Hogben in the development of ecological thought was, despite their earlier prominence, then unknown or forgotten, in part a casualty of the internecine struggles within Marxism itself. Also forgotten were the great left classicists such as Benjamin Farrington, George Thomson, and Jack Lindsay. With all of this to deal with, grasping the vast scope of the analyses, placed in their proper historical context, took time.

But the historical linkages, as you say, were definitely there. The story leads in the end to figures like Barry Commoner and Rachel Carson, and also to Stephen Jay Gould, Richard Levins, Richard Lewontin, Steven and Hilary Rose, Lindsay, and E. P. Thompson, who became Britain's leading antinuclear activist—all of whom were immensely impacted, although in different ways, by this intellectual and political inheritance. In answer to your question on how this history can help us in today's struggles, perhaps the most succinct response is the statement of Quentin Skinner, whom I quote in the introduction of *The Return of Nature*, who says that the only purpose of such histories is to demonstrate "how our society places limitations on our imaginations." He adds that "we are all Marxists to this extent."

AP: *Marx's Ecology* mentions how your own internalization of the legacy of Georg Lukács and Antonio Gramsci prevented you from using the dialectical method for the realm of nature. You point out how, due to this common weakness, Western Marxism had partly abandoned the field of nature and the philosophy of science to the dominion of mechanist and positivist variants of thought. However, *The Return of Nature* begins precisely by questioning some assumptions about Lukács central to the departure of Western Marxism from the dialectics of nature. What conditions delayed so many findings of this importance? What were the main effects that these assumptions had on Marxism, particularly in relation to ecology?

JBF: Maybe I can explain this somewhat through my own intellectual development. When I was an undergraduate, I studied the works of Immanuel Kant, G. W. F. Hegel, Arthur Schopenhauer, Marx, Engels, V. I. Lenin, and Max Weber fairly extensively, as well as thinkers such as Herbert Marcuse, Mészáros, Ernst Cassirer, H. Stuart Hughes, and Arnold Hauser. So, when I got to graduate school, I had a pretty good general idea of the boundaries between Kantianism/neo-Kantianism and Hegelianism/Marxism. I was therefore surprised, in participating in courses on critical theory, to find that the very first proposition taught was that the dialectic did not apply to nature, based primarily on the authority of a footnote in Lukács's *History and Class Consciousness*, where he had criticized Engels on the dialectics of nature. Only by rejecting the dialectic of nature, it was argued, could the dialectic be defined in terms of the identical subject-object of the historical process.

Of course, Lukács himself, as he later pointed out, had never totally abandoned the notion of "merely objective dialectics" or the dialectics of nature, which he referred to elsewhere in *History and Class Consciousness*. Indeed, in his famous 1967 preface to *History and Class Consciousness*, Lukács, following Marx, insisted on a dialectical mediation between nature and society via labor as metabolism, and in that sense on a dialectics of nature conception. The same argument was made in his *Conversations with Lukács*, which I read in the early 1980s.

It was in this context that I internalized, to some extent, at a practical level, without ever fully embracing, the Western Marxist philosophical notion that the dialectic was applicable only to the human historical realm and not to nature (or natural science), which was given over to mechanism or positivism. I came to see the historical dialectic in terms of the Vician principle that we can understand history because we have made it, as advanced by Marxist historian E. P. Thompson—even though I recognized that, at a deeper level, this was not entirely satisfactory because human beings do not make history alone, but do so in conjunction with the universal metabolism of nature of which human society is an emergent part. But my interests in the 1980s were mainly geared toward political economy and history, where such issues seldom arose. As far as the human historical realm was concerned, it was easy enough to bracket the question of the dialectics of nature.

It was when I turned more directly to the question of ecology in the late 1980s and '90s that this problem became unavoidable. The dialectics of nature could only consistently be set aside on idealist or mechanical materialist grounds. Still, in writing *Marx's Ecology* I consciously avoided, for the most part, any explicit, detailed consideration of the dialectics of nature in relation to Marx, given the complexity of the issues, which I was not then prepared to address, though clearly Marx's concept of social metabolism took him in that direction. Thus, in the epilogue to *Marx's Ecology*, I simply referred to Marx's reference to the "dialectical method" as the way of dealing with the "free movement in matter," and how this was part of the inheritance he had taken from Epicurus and other earlier materialists, mediated by his studies of Hegel. As an epistemological approach, I indicated, this could be defended as heuristically equivalent to the role that teleology played for human cognition in Kant. But the wider ontological question of "so-called objective dialectics," as this appeared in Engels (and in Lukács), and its relation to Marx, was mostly avoided (left implicit) in my book.

I did not address the dialectics of nature explicitly in any detail until 2008, in a chapter that I wrote for a book on dialectics edited by Bertell Ollman and Tony Smith (later included in *The Ecological Rift*). Here, I was still caught in what I called "the Lukács problem," even if I understood that, for the later Lukács, Marx's metabolism argument offered a broad pathway out of the whole epistemological-ontological dilemma. Whereas another pathway, I argued, was to be found in what Marx had called the "dialectic of sensuous certitude" represented by the materialism of Epicurus, Francis Bacon, and Ludwig Feuerbach, and incorporated into Marx's early work. Yet, my approach there, even if arguably a step forward, was in various ways inadequate. Part of the difficulty, as I came to understand it, lay in the philosophical limitations, and at the same time much greater scientific scope, of a materialist dialectic, which could never be a closed, circular system as in Hegel's idealist philosophy—or a totalizing system consisting exclusively of internal relations and windowless monads. The dialectic for Marx was open, not closed, as was the case for the physical world itself.

The question of the dialectics of nature was to be central to *The Return of Nature*. One element was the study of the later Lukács, particularly *The Young Hegel* and the *Ontology of Social Being*. A key factor here was Lukács's treatment of Hegel's reflection determinations, which helped me understand the way in which Engels's dialectical naturalism had been inspired to a considerable extent by the Doctrine of Essence in Hegel's *Logic*. Another element affecting my views, going back to *Marx's Ecology*, was the critical realism of Roy Bhaskar, especially his *Dialectic: The Pulse of Freedom*. But at the heart of my project in *The Return of Nature* was the close scrutiny of Engels's *Dialectics of Nature* itself—as well as Lenin's philosophical writings—which had untold depth. This allowed me to chart the influence that Engels exerted on subsequent thinkers—most notably, in terms of the dialectics of nature problem itself, on Needham, Christopher Caudwell, and Lindsay. In addition, William Morris in the arts and Haldane, Bernal, Hogben, and Levy in the sciences offered a variety of powerful insights into dialectical and materialist ecology.

AP: Lukács also noted how the division of alienated labor served to increase the disciplinary divisions of knowledge according to the needs of functional specialization of capital. As a philosophy of praxis, Marxism is proposed as a totalizing project, among other things, to recompose the many varied rifts that capitalism had expanded or imposed: nature and society, but also science and art. A central theme of your new book is the existence of parallel approaches to ecology and socialism in science and art. How did these links contribute to materialist ecosocialist thought? And how can they help us rethink this interaction in relation to ecology and the ecosocial crisis we face?

JBF: In writing *The Return of Nature*, Morris's statement in *News from Nowhere* that there were two insurmountable forms of knowledge, the sciences and the arts, was constantly on my mind. All of the Marxist thinkers concerned with ecology crossed these boundaries in various ways, so the parallel developments had to be examined in any genealogical-historical account. Clearly, the analytical development of ecology as a science and its relation to the dialectics of nature evolved mainly through the scientific stream. But it was hardly possible to isolate this from socialist aesthetics.

Thus, Lankester was friends with Morris and the Pre-Raphaelites. Hogben took the main inspiration for his socialism from Morris. In Morris, we find an analysis rooted in the conception that all unalienated work contains art, a notion he drew from John Ruskin, but to which he added depth via Marx. Morris also reproduced independently of Marx the notion of the social character of all art. Caudwell brilliantly captured both the aesthetic and scientific strands of the overall ecological critique. His aesthetics drew on the concept of *mimesis* based in Aristotle and in the radical British classical tradition of the Cambridge ritualists represented by Jane Harrison, which Caudwell then merged with materialist dialectics. Caudwell's powerful approach led to George Thomson's extraordinary analyses of the origins of poetry and drama.

This whole aesthetic-ecological development on the left culminated with the Australian Marxist Jack Lindsay, who due to his enormous range of classical, literary, philosophical, and scientific studies was to bring together notions on the dialectics of nature, drawing on both aesthetics and science. It is no accident that thinkers like Lukács, Mészáros, and Thompson thought so highly of Lindsay, whose work is not sufficiently valued, perhaps because navigating his corpus of 170 volumes, extending from the ancient classics to literature, poetry, history, and the philosophy of science is simply too daunting.

AP: Engels is a key character in your new book. For a long time, within certain Marxisms, Engels was accused of having vulgarized Marx's thought, but you point out the relevance and complexity of Engels's dialectical materialism for a social and ecological critique of capitalism. Although increasingly recognized, you can still find a certain disdain for Engels and for his work's ties to Marx. How did this happen? How do we contest these positions from the standpoint of Marxist ecological thought?

JBF: I remember hearing David McClellan speak in December 1974, not long after he had written his biography of Marx. I was completely taken aback by an extraordinary tirade against Engels, which was the core of his talk. This was my first real introduction to the attacks on Engels that in so many ways came to define the Western Marxist tradition in the days of the Cold War, and which have carried over into the post–Cold War era. All of this was clearly less about Engels as a thinker than it was about the "two Marxisms," as Alvin Gouldner called it. Western Marxism and, to a considerable extent, the academic world claimed Marx as their own, as an urbane thinker, but for the most part rejected Engels as supposedly too "crude," casting him in the role of spoiler, as the person who had created a "Marxism" that had nothing to do with Marx, and who was thus responsible for the economism, determinism, scientism, and vulgar philosophical and political perspectives of the Second International and beyond, all the way to Stalin.

It should not perhaps surprise us, therefore, that while we can find hundreds, even thousands, of books and articles that mention Engels's *Dialectics of Nature*, there is hardly anything to be learned from them because they either treat his views in a doctrinaire way, as in much of the old official Marxism, or, in the case of the Western Marxist philosophical tradition, simply quote a few lines from *Dialectics of Nature*, or sometimes *Anti-Dühring*, so as to establish his vulgarization of Marxism. Others, like Terrell Carver, who has written extensively on Engels, devote themselves not to furthering an understanding of Engels's work, but to the systematic severing of Engels's work from that of Marx.

I remember looking at Karl Padover's *Letters of Karl Marx* and wondering why it felt like such an arid empty work, despite the fact that it was filled entirely with Marx's own words. I realized it was because almost all the letters were to Engels and Engels was left out of the book, so it is a one-sided conversation, as if only Marx counted and was simply talking to himself. The Marx-Engels correspondence is definitely a two-sided conversation, and takes on much of its brilliance as a continual dialogue between these two magisterial thinkers, who together founded historical materialism.

In terms of Marxian ecology, Engels is essential. Because as brilliant as Marx's analysis was in this regard, we cannot afford to ignore the vast contributions of Engels to class-based epidemiology, the main subject of his *Condition of the Working Class in England*, to the dialectics of nature and emergence, to the critique of the conquest of nature, or to the understanding of human evolutionary development. Engels's critical appropriation of Darwin in *Anti-Dühring* was fundamental to the development of evolutionary ecology. The emergentist materialism developed in *Dialectics of Nature* is central to a critical scientific worldview.

AP: *Monthly Review* has always shown great sensitivity to the revolutionary struggles of the third world. Lenin's theory of imperialism, together with that of monopoly capital by Paul Sweezy and Paul Baran, dependency theory (in Ruy Mauro Marini and Samir Amin, among others) and its dialogue with world-systems analysis, or the contributions of Mészáros, among many other influences, have been essential for the elaboration of your specific ecosocialist critique. Unfortunately, and to some extent in connection to the limitations of Western Marxism, the link between ecology and imperialism has often been underestimated in other Marxist and ecological currents.

Some have even considered imperialism an outdated category to deal with global capitalism. Why is it that this separation between geopolitics and ecology remains so strong in certain sectors of the left? Is a different approach to these matters possible?

JBF: In my generation in the United States, impacted by the Vietnam War and the coup in Chile, most of those drawn to Marxism came to it by way of opposing imperialism. It was partly for this reason that I was attracted early on to *Monthly Review*, which, practically from its birth in 1949, has been a major source of the critique of imperialism, including dependency theory and world-system analysis. Harry Magdoff's writings on imperialism, in *The Age of Imperialism* and *Imperialism: From the Colonial Age to the Present*, are central to us, as well as work on imperialism by Paul Baran, Paul Sweezy, Oliver Cromwell Cox, Che Guevara, Andre Gunder Frank, Walter Rodney, Samir Amin, Immanuel Wallerstein, and a host of others. The fact that the most revolutionary perspective in the United States has historically come from the Black movement, which has always been more internationalist and anti-imperialist in its perspective, has been crucial in defining the radical U.S. left. Yet, with all of this, there have always been major social democratic figures in the United States, such as Michael Harrington, who have made their peace with U.S. imperialism. Today, some of the representatives of the new movement for "democratic socialism" regularly turn a blind eye to Washington's ruthless interventions abroad.

Of course, none of this is new. Variants of the conflict over imperialism within the left can be seen as far back as the early socialist movement in England. H. M. Hyndman, the founder of the Social Democratic Federation, and George Bernard Shaw, one of the leading Fabians, both supported the British Empire and "social imperialism." On the other side were figures associated with the Socialist League, such as Morris, Eleanor Marx, and Engels, all of whom were anti-imperialists. It was the issue of imperialism that was most decisively to split the European socialist movement at the time of the First World War, as recounted in Lenin's *Imperialism, the Highest Stage of Capitalism*.

Within the New Left in Britain from the 1960s, imperialism was a major source of contention. Those who identified with the First New Left, such as Thompson, Ralph Miliband, and Raymond Williams, were strongly anti-imperialist, while the Second New Left, associated in particular with the *New Left Review*, either saw imperialism as a progressive force in history, as in the case of Bill Warren, or tended to downplay its significance altogether. The result, particularly with the rise of globalization ideology in this century, was a dramatic decline in studies of imperialism—though accompanied by growing cultural studies of colonialism and postcolonialism—in both Britain and the United States. The logical outcome of this is that a figure as influential today in the left academy as David Harvey has recently pronounced that imperialism has been "reversed," with the West now on the losing end.

All of this takes us to the question of the very weak performance on the left generally in developing a theory of ecological imperialism, or unequal ecological exchange. This is a product of the systematic failure to explore capitalism's ruthless *expropriation* of the resources and ecology of most of the world. This is about use value, not just exchange value. Thus, the famines introduced in India under British colonial rule had to do with how the British forcibly altered the food regime in India, shifting the use values, metabolic relations, and the hydrological infrastructure essential to human survival, while also draining away India's surplus. Although this process of ecological expropriation has long been understood by the left in India, and in much of the rest of the Global South, it is still not fully grasped by Marxists in the Global North. An exception is Mike Davis's excellent *Late Victorian Holocausts*.

Similarly, the massive expropriation of guano from Peru to fertilize European soil, which had been robbed of its nutrients, a manifestation of the metabolic rift, was to have all sorts of long-term negative developmental effects on Peru, and included the importation of Chinese laborers under horrific conditions with no survivors to dig the guano. All of this was tied to what Eduardo Galeano called *The Open Veins of Latin America*.

What this tells us is that the issues of ecology and imperialism have always been intimately related and are becoming more closely intertwined all the time. The *Ecological Threat Register 2020* report from the Institute of Economics and Peace indicates that as many as 1.2 billion people may be displaced from their homes, becoming climate refugees, by 2050. Under such historical conditions, imperialism can no more be analyzed independently from the planetary ecological destruction that it has brought into being than the planetary ecological crisis can be addressed independently from the imperialism in which it is being played out today. This was the message that Brett Clark and I sought to convey in *The Robbery of Nature*, and that the two of us, together with Hannah Holleman, endeavored to explain in our article "Imperialism in the Anthropocene," published in the July–August 2019 issue of *Monthly Review*. In that article, we concluded: "There can be no ecological revolution in the face of the current existential crisis unless it is an anti-imperialist one, drawing its power from the great mass of suffering humanity.... The poor shall inherit the earth or there will be no earth left to inherit."

AP: As we have seen, interest in Marx's ecosocialism has grown greatly in recent decades. But, of course, this goes beyond Marx's historical context. Why is it important for current ecological thought to return to the ideas of Marx? And what are the challenges for Marxist ecological thought today?

JBF: Marx's ecology is a starting point and a set of foundations, not an end point. It is in Marx's thought above all that we find the foundations of the critique of political economy that was also a critique of capitalism's ecological depredations. This was no accident, since Marx dialectically presented the labor process as the social metabolism (the mediation) of nature and society. In Marx, capitalism, in alienating the labor process, also alienated the metabolism between humanity and nature, thereby generating a metabolic rift. Marx took this to its logical conclusions, arguing that no one owns the earth, that not even all the people in all the countries of the world own the earth, that they simply have the responsibility to care for it and, if possible, improve it for the chain of future generations as good heads of the household. He defined socialism as the rational regulation of the metabolism of humanity and nature, so as to conserve as much as possible on energy and promote full human development. There is nothing in conventional or even left green theory—however much capitalism may be questioned in part—that has this unity between ecological and economic critique, or as comprehensive a historical synthesis. Consequently, in our planetary emergency, ecosocialism has come to rest inevitably on Marx's foundational conception. The environmental movement, if it is to matter at all, has to be ecosocialist.

But, of course, I would not have written *The Return of Nature*, which focuses on the century following Marx and Darwin's deaths, if socialist ecology simply began and ended with Marx. It is crucial to understand how socialist dialectical, materialist, and ecological perspectives developed from the late nineteenth to the late twentieth century in order to grasp the historical theory and practice that feeds into today's struggles. Our task now is not simply to linger on the past, but to pull all of this together to engage with the challenges and burdens of our historical time. Marx serves to demonstrate the essential oneness of our political-economic-ecological contradictions and their basis in the present alienated social and ecological order. This helps us unmask the contradictions of the present. But to carry out the necessary change, we need to do so with an eye to how the past informs the present and allows us to envision necessary revolutionary action.

The purpose of Marxian ecological thought is not merely to understand our present social and ecological contradictions, but to transcend them. Given that humanity is facing greater dangers than ever before and is on a runaway capitalist train headed over the cliff, this has to be our chief concern. Facing up to the planetary ecological emergency means we must be more revolutionary than ever before, and not be afraid to raise the question of altering society, as Marx said, "from top to bottom," starting from where we are. The piecemeal and reformist approach of most environmentalism, which puts faith in the market and technology, while making its peace in large part with the prevailing system, with its unceasing, totalizing ecological destruction, will not work, even in the short run. There is now more than a century of socialist critique of the ecological contradictions of capitalism, which has enormous theoretical power and points to a different philosophy of praxis. In our current growing recognition that there is no choice but to leave capitalism's burning house, we need the deeper theoretical understanding of human, social, and ecological possibility, of freedom as necessity, offered by ecological Marxism. As Doris Lessing, who appears briefly in *The Return of Nature*, stated in her introduction to *The Golden Notebook*: "Marxism looks at things as a whole and in relation to each other." This is the revolutionary capacity we most need today.

PART III

The Future of History

CHAPTER SIXTEEN

Capitalism and Degrowth: An Impossibility Theorem

In the opening paragraph to his 2009 book, *Storms of My Grandchildren,* James Hansen, perhaps the world's foremost scientific authority on global warming, declared: "Planet Earth, creation, the world in which civilization developed, the world with climate patterns that we know and stable shorelines, is in imminent peril.... The startling conclusion is that continued exploitation of all fossil fuels on Earth threatens not only the other millions of species on the planet but also the survival of humanity itself—and the timetable is shorter than we thought."[1]

In making this declaration, however, Hansen was only speaking of a *part* of the global environmental crisis currently threatening the planet, namely, climate change. Recently, leading scientists (including Hansen) have proposed nine planetary boundaries, which mark the safe operating space for the planet. Three of these boundaries—climate change, biodiversity, and the nitrogen cycle—have already been crossed, while others, such as freshwater use and ocean acidification, are emerging planetary rifts. In ecological terms, the economy has now grown to a scale and intrusiveness that is both overshooting planetary boundaries and tearing apart the biogeochemical cycles of the planet.[2]

Hence, almost four decades after the Club of Rome raised the issue of "the limits to growth," the economic growth idol of modern society is once again facing a formidable challenge.[3] What is known as "degrowth economics," associated with the work of Serge Latouche in particular, emerged as a major European intellectual movement in 2008 with the historic conference in Paris on "Economic De-Growth for Ecological Sustainability and Social Equity," and has since inspired a revival of radical Green thought, as epitomized by the 2010 "Degrowth Declaration" in Barcelona.

Ironically, the meteoric rise of degrowth (*décroissance* in French) as a concept has coincided over the last three years with the reappearance of economic crisis and stagnation on a scale not seen since the 1930s. The degrowth concept therefore forces us to confront the questions: Is degrowth feasible in a capitalist grow-or-die society—and if not, what does this say about the transition to a new society?

According to the website of the European degrowth project, "Degrowth carries the idea of a voluntary reduction of the size of the economic system which implies a reduction of the GDP."[4] "Voluntary" here points to the emphasis on voluntaristic solutions—though not as individualistic and unplanned in the European conception as the "voluntary simplicity" movement in the United States, where individuals, usually well-to-do, simply choose to opt out of the high-consumption market model. For Latouche, the concept of "degrowth" signifies a major social change: a radical shift from growth as the main objective of the modern economy, toward its opposite (contraction, downshifting).

An underlying premise of this movement is that, in the face of a planetary ecological emergency, the promise of green technology has proven false. This can be attributed to the Jevons Paradox, according to which greater efficiency in the use of energy and resources leads not to conservation but to greater economic growth, and hence more pressure on the environment.[5] The unavoidable conclusion—associated with a wide variety of political-economic and environmental thinkers, not just those connected directly to the European degrowth project—is that there needs to be a drastic alteration in the economic trends operative since the Industrial Revolution. As Marxist economist Paul Sweezy put it more than two decades ago: "Since there is no way to increase the capacity of the environment to bear the [economic and population] burdens placed on it, it follows that the adjustment must come entirely from the other side of the equation. And since the disequilibrium has already reached dangerous proportions, it also follows that what is essential for success is a reversal, not merely a slowing down, of the underlying trends of the last few centuries."[6]

Given that wealthy countries are already characterized by ecological overshoot, it is becoming more and more apparent that there is indeed no alternative, as Sweezy emphasized, but a *reversal* in the demands placed on the environment by the economy. This is consistent with the argument of ecological economist Herman Daly, who has long insisted on the need for a steady-state economy. Daly traces this perspective to John Stuart Mill's famous discussion of the "stationary state" in his *Principles of Political Economy*, which argued that if economic expansion was to level off, as the classical economists expected, the economic goal of society could then shift to the qualitative aspects of existence, rather than mere quantitative expansion.

A century after Mill, Lewis Mumford insisted in his *Condition of Man*, first published in 1944, that not only was a stationary state in Mill's sense ecologically necessary, but that it should also be linked to a concept of "basic communism . . . [that] applies to the whole community the standards of the household," distributing "benefits according to need"—a view that drew upon Marx.

Today this recognition of the need to bring economic growth in overdeveloped economies to a halt, and even to shrink these economies, is seen as rooted theoretically in Nicholas Georgescu-Roegen's *The Entropy Law and the Economic Process*, which established the basis of modern ecological economics.[7]

*De*growth as such is not viewed, even by its proponents, as a stable solution, but one aimed at reducing the size of the economy to a level of output that can be maintained perpetually at a steady state. This might mean shrinking the rich economies by as much as a third from today's levels by a process that would amount to negative investment, since not only would new net investment cease but also only some, not all, worn-out capital stock would be replaced. A steady-state economy, in contrast, would carry out replacement investment but would stop short of new net investment. As Daly defines it, "A steady-state economy" is "an economy with constant stocks of people and artifacts, maintained at some desired, sufficient levels by low rates of maintenance 'throughput,' that is, by the lowest feasible flows of matter and energy."[8]

Needless to say, none of this would come easily, given today's capitalist economy. In particular, Latouche's work, which can be viewed as exemplary of the European degrowth project, is beset with contradictions, resulting not from the concept of degrowth *per se*, but from his attempt to skirt the question of capitalism. This can be seen in his 2006 article, "The Globe Downshifted," where he argues in convoluted form:

> For some on the far left, the stock answer is that capitalism is the problem, leaving us stuck in a rut and powerless to move towards a better society. Is economic contraction compatible with capitalism? This is a key question, but one that it is important to answer without resort to dogma, if the real obstacles are to be understood....
>
> Eco-compatible capitalism is conceivable in theory, but unrealistic in practice. Capitalism would require a high level of regulation to bring about the reduction of our ecological footprint. The market system, dominated by huge multinational corporations, will never set off down the virtuous path of eco-capitalism of its own accord....
>
> Mechanisms for countering power with power, as existed under the Keynes-Fordist regulations of the Social-Democratic era, are conceivable and desirable. But the class struggle seems to have broken down. The problem is: capital won....
>
> A society based on economic contraction cannot exist under

capitalism. But capitalism is a deceptively simple word for a long, complex history. Getting rid of the capitalists and banning wage labour, currency and private ownership of the means of production would plunge society into chaos. It would bring large-scale terrorism.... We need to find another way out of development, economism (a belief in the primacy of economic causes and factors) and growth: one that does not mean forsaking the social institutions that have been annexed by the economy (currency, markets, even wages) but reframes them according to different principles.[9]

In this seemingly pragmatic, non-dogmatic fashion, Latouche tries to draw a distinction between the degrowth project and the socialist critique of capitalism by (1) declaring that "eco-compatible capitalism is conceivable" at least in theory; (2) suggesting that Keynesian and so-called Fordist approaches to regulation, associated with social democracy, could—if still feasible—tame capitalism, pushing it down "the virtuous path of eco-capitalism"; and (3) insisting that degrowth is not aimed at breaking the dialectic of capital-wage labor or interfering with private ownership of the means of production. In other writings, Latouche makes it clear that he sees the degrowth project as compatible with continued valorization (that is, augmentation of capitalist value relations) and that anything approaching substantive equality is considered beyond reach.[10]

What Latouche advocates most explicitly in relation to the environmental problem is the adoption of what he refers to as "reformist measures, whose principles [of welfare economics] were outlined in the early 20th century by the liberal economist Arthur Cecil Pigou [and] would bring about a revolution" by internalizing the environmental externalities of the capitalist economy.[11] Ironically, this stance is identical with that of neoclassical environmental economics— while distinguished from the more radical critique often promoted by ecological economics, where the notion that environmental costs can simply be internalized within the present-day capitalist economy is sharply attacked.[12]

"The ecological crisis itself is mentioned" in the current degrowth project, as Greek philosopher Takis Fotopoulos has critically observed, "in terms of a common problem that 'humanity' faces because of the degradation of the environment, with no mention at all of the differentiated class implications of this crisis, i.e., of the fact that the economic and social implications of the ecological crisis are primarily paid in terms of the destruction of lives and livelihood of the lower social groups—either in Bangladesh or in New Orleans—and much less in terms of those of the elites and the middle classes."[13]

Given that it makes the abstract concept of economic growth its target, rather than the concrete reality of capital accumulation, degrowth theory—in the influential form articulated by Latouche and others—naturally faces difficulty confronting today's reality of economic crisis/stagnation, which has produced unemployment levels and economic devastation greater than at any time since the 1930s. Latouche himself wrote in 2003 that "there would be nothing worse than a growth economy without growth."[14] But, faced with a capitalist economy caught in a deep structural crisis, European degrowth analysts have little to say. The Barcelona Degrowth Declaration simply pronounced: "So-called anti-crisis measures that seek to boost economic growth will worsen inequalities and environmental conditions in the long run."[15] Neither wishing to advocate growth, nor to break with the institutions of capital—nor, indeed, to align themselves with workers, whose greatest need at present is employment—leading degrowth theorists remain strangely silent in the face of the greatest economic crisis since the Great Depression.

To be sure, when faced with "actual degrowth" in the Great Recession of 2008–2009 and the need for a transition to "sustainable degrowth," noted ecological economist Joan Martinez-Alier, who had taken up the degrowth banner, offered the palliative of "a short-run Green Keynesianism or a Green New Deal." The goal, he said, was to promote economic growth and "contain the rise in unemployment" through public investment in green technology and infrastructure. This was viewed as consistent with the degrowth project, as long as such Green Keynesianism did not "become a doctrine of continuous economic growth."[16] Yet how working people were to fit into this largely technological strategy, predicated on ideas of energy efficiency that degrowth analysts generally reject, was left uncertain.

Indeed, rather than dealing with the unemployment problem directly—through a radical program that would give people jobs aimed at the creation of genuine use values in ways compatible with a more sustainable society—degrowth theorists prefer to emphasize shorter working hours, and separate "the right to receive remuneration from the fact of being employed" by means of the promotion of a universal basic income. Such changes are supposed to allow the economic system to shrink and, at the same time, guarantee income to families—all the while keeping the underlying structure of capital accumulation and markets intact.

Yet, looked at from a more critical standpoint, it is hard to see the viability of shorter work hours and basic income guarantees on the scale suggested other than as elements in a transition to a post-capitalist (indeed socialist) society. As Marx said, the rule for capital is: "Accumulate, accumulate! That is Moses and the prophets!"[17] To break with capitalism's institutional basis of the "law of value," or to question the structure underpinning the exploitation of labor, both of which would be threatened by a sharp reduction of working hours and substantial income guarantees, is to raise larger questions of system change—ones that leading degrowth theorists seem unwilling to acknowledge at present. Moreover, a meaningful approach to the creation of a new society would have to provide not merely income and leisure, but it would also need to address the human need for useful, creative, non-alienated work.

Even more problematic is the attitude of much of current degrowth theory toward the global South. "Degrowth," Latouche writes,

> must apply to the South as much as to the North if there is to be any chance to stop Southern societies from rushing up the blind alley of growth economics. Where there is still time, they should aim not for development but for disentanglement— removing the obstacles that prevent them from developing

differently.... Southern countries need to escape their economic and cultural dependence on the North and rediscover their own histories—interrupted by colonialism, development and globalization—to establish distinct indigenous cultural identities.... Insisting on growth in the South, as though it were the only way out of the misery that growth created, can only lead to further Westernization.[18]

Lacking an adequate theory of imperialism, and failing to address the vast chasm of inequality separating the richest from the poorest nations, Latouche thus reduces the whole immense problem of underdevelopment to one of cultural autonomy and subjection to a westernized growth fetish. This can be compared to the much more reasoned response of Herman Daly, who writes:

> It is absolutely a waste of time as well as morally backward to preach steady-state doctrines to underdeveloped countries before the overdeveloped countries have taken any measure to reduce either their own population growth or the growth of their per-capita resource consumption. Therefore, the steady-state paradigm must first be applied in the overdeveloped countries. ... One of the major forces necessary to push the overdeveloped countries toward a ... steady-state paradigm must be Third World outrage at their overconsumption.... The starting point in development economics should be the "impossibility theorem" ... that a U.S.-style high mass consumption economy for a world of 4 billion people is impossible, and even if by some miracle it could be achieved, it would certainly be short-lived.[19]

The notion that degrowth as a concept can be applied in essentially the same way both to the wealthy countries of the center and the poor countries of the periphery represents a category mistake resulting from the crude imposition of an abstraction (degrowth) on a context in which it is essentially meaningless, e.g., Haiti, Mali, or even, in many ways, India. The real problem in the global periphery is overcoming imperial linkages, transforming the existing mode of production, and creating sustainable-egalitarian productive possibilities. It is clear that many countries in the South with very low per capita incomes cannot afford *degrowth* but could use a kind of *sustainable development*, directed at real needs such as access to water, food, health care, education, etc. This requires a radical shift in social structure away from the relations of production of capitalism/imperialism. It is telling that in Latouche's widely circulated articles there is virtually no mention of those countries, such as Cuba, Venezuela, and Bolivia, where concrete struggles are being waged to shift social priorities from profit to social needs. Cuba, as the *Living Planet Report* has indicated, is the only country on Earth with high human development and a sustainable ecological footprint.[20]

It is undeniable today that economic growth is the main driver of planetary ecological degradation. But to pin one's whole analysis on overturning an abstract "growth society" is to lose all historical perspective and discard centuries of social science. As valuable as the degrowth concept is in an ecological sense, it can only take on genuine meaning as part of a critique of capital accumulation and part of the transition to a sustainable, egalitarian, communal order; one in which the associated producers govern the metabolic relation between nature and society in the interest of successive generations and the earth itself (socialism/communism as Marx defined it).[21] What is needed is a "co-revolutionary movement," to adopt David Harvey's pregnant term, which will bring together the traditional working-class critique of capital, the critique of imperialism, the critiques of patriarchy and racism, and the critique of ecologically destructive growth (along with their respective mass movements).[22]

In the generalized crisis of our times, such an overarching, co-revolutionary movement is conceivable. Here, the object would be the creation of a new order in which the valorization of capital would no longer govern society. "Socialism is useful," E. F. Schumacher wrote in *Small Is Beautiful*, precisely because of "the possibility it creates for the overcoming of the religion of economics," that is, "the modern trend towards total quantification at the expense of the appreciation of qualitative differences."[23]

In a sustainable order, people in the wealthier economies (especially those in the upper income strata) would have to learn to live on "less" in commodity terms in order to lower per capita demands on the environment. At the same time, the satisfaction of genuine human needs and the requirements of ecological sustainability could become the constitutive principles of a new, more communal order aimed at human reciprocity, allowing for qualitative improvement, even *plenitude*.[24] Such a strategy—not dominated by blind productivism—is consistent with providing people with worthwhile work. The ecological struggle, understood in these terms, must aim not merely for degrowth in the abstract but more concretely for *deaccumulation*—a transition away from a system geared to the accumulation of capital without end. In its place we need to construct a new co-revolutionary society, dedicated to the common needs of humanity and the earth.

CHAPTER SEVENTEEN

The Ecology of Marxian Political Economy

It is no secret today that we are facing a planetary environmental emergency, endangering most species on the planet, including our own, and that this impending catastrophe has its roots in the capitalist economic system. Nevertheless, the extreme dangers that capitalism inherently poses to the environment are often inadequately understood, giving rise to the belief that it is possible to create a new "natural capitalism" or "climate capitalism" in which the system is turned from being the enemy of the environment into its savior.[1] The chief problem with all such views is that they underestimate the cumulative threat to humanity and the earth arising from the existing relations of production. Indeed, the full enormity of the planetary ecological crisis can only be understood from a standpoint informed by the Marxian critique of capitalism.

A common weakness of radical environmental critiques of capitalism is that they rely on abstract notions of the system based on nineteenth-century conditions. As a result, many of the historically specific underpinnings of environmental crises related to twentieth- and twenty-first-century conditions have been insufficiently analyzed. Marx's own indispensable ecological critique was limited by the historical period in which he wrote, namely, the competitive stage of capitalism, and thus he was unable to capture certain crucial characteristics of environmental destruction that were to emerge with monopoly capitalism. The following analysis, therefore, will discuss not only the ecological critique provided by Marx (and Engels), but also that of later Marxian and radical political economists, including such figures as Thorstein Veblen, Paul Baran, Paul Sweezy, and Allan Schnaiberg.

Marx and the Capitalist *Raubbau*

It is seldom recognized that Marx's very first political economic essay, "Debates on the Law on Theft of Wood," written in 1842 during his editorship of *Rheinische Zeitung*, was focused on ecological issues. A majority of those in jail in Prussia at that time were peasants arrested for picking up dead wood in the forests. In carrying out this act, the peasants were merely exercising what had been a customary right, but one disallowed with the spread of private property. Observing the debates on this issue in the Rhineland Diet (the provincial assembly of the Rhineland), Marx commented that the dispute centered on how best to protect the property rights of landowners, while the customary rights of the population in relation to the land were simply ignored. Impoverished peasants were viewed as the "enemy of wood" because the exercise of their traditional rights to gather wood primarily as fuel for cooking and warming their homes transgressed the ownership rights of private property holders.[2]

It was not long after this that Marx began his systematic research into political economy. It therefore should not surprise us that as early as his *Economic and Philosophic Manuscripts of 1844* he was already focusing on the issue of primary accumulation, that is, the expropriation of the peasantry, who were being removed from the land in the course of capitalist development. It was this separation of workers from the earth as means of production that he was later to refer to in *Capital* as the "historical precondition of the capitalist mode of production" and its "permanent foundation," the basis for the emergence of the modern proletariat.[3] Capitalism began as a system of encroachment on nature and public wealth.

Here it is important to recognize that at the very root of Marx's critique of political economy was the distinction between use value and exchange value. Every commodity, he explained in the opening pages of *Capital*, had both a use value and an exchange value, with the latter increasingly dominating the former. Use value was associated with the requirements of production in general and with the basic human relation to nature, that is, fundamental human needs. Exchange value, in contrast, was oriented to the pursuit of profit. This established a contradiction between capitalist production and production in general, the natural conditions of production.

This contradiction was most evident in Marx's time in terms of what came to be known as the Lauderdale Paradox, named after James Maitland, the Eighth Earl of Lauderdale (1759–1839). Lauderdale was one of the early classical political economists, author of *An Inquiry into the Nature of Public Wealth and into the Means and Causes of Its Increase* (1804). Public wealth, he explained, consisted of use values, which, like water and air, oftentimes existed in abundance, while private riches were based on exchange values, which demanded scarcity. Under such conditions, he charged, the expansion of private riches went hand in hand with the destruction of public wealth. For instance, if water supplies that had previously been freely available were monopolized and a fee placed on wells, then the measured riches of the nation would be increased at the expense of public wealth.

"The common sense of mankind," Lauderdale declared, "would revolt" at any proposal to increase private riches "by creating a scarcity of any commodity generally useful and necessary to man." But the bourgeois society in which he lived, he recognized, was already doing that. Thus, Dutch colonists had in particularly fertile periods burned "spiceries" or paid natives to "collect the young blossoms or green leaves of the nutmeg trees" to kill them off, while planters in Virginia by legal enactment burned a certain share of their crops to maintain the price. "So truly is this principle understood by those whose interest leads them to take advantage of it," he wrote, "that nothing but the impossibility of general combination protects the public wealth against the rapacity of private avarice."[4]

Marx saw the Lauderdale Paradox, arising out of "the inverse ratio of the two kinds of value" (use value and exchange value), as one of the chief contradictions of bourgeois production. The entire pattern of capitalist development was characterized by the wasting away and destruction of the natural wealth of society.[5] "For all its stinginess," he wrote, "capitalist production is thoroughly wasteful with human material, just as its way of distributing its products through trade, and its manner of competition, make it very wasteful of material resources, so that it loses for society [public wealth] what it gains for the individual capitalist [private riches]."[6]

The domination of exchange value over use value in capitalist development and the ecological impact of this can also be seen in Marx's general formula of capital, M-C-M′. Capitalism is commonly described as a system conforming to simple commodity production, C-M-C, in which money is simply an intermediary in a process of production and exchange, beginning and ending with particular use values embodied in concrete commodities. In sharp contrast, Marx explained that capitalist production and exchange takes the form of M-C-M′, in which money capital is advanced for labor and materials with which to produce a commodity, which can then be sold for *more money*, M′ or M + Δm (surplus value), at the end of the process. The crucial difference here is that *the process never really ends*, since money or abstract value is the object. The M′ is reinvested in the following period, resulting in M′-C-M″, which leads to M″-C-M‴ in the period after that, and so on.

In order to maintain a given share of wealth under this system, the capitalist must continually seek to expand it. The law of value therefore constantly whispers to each individual capitalist and to the capitalist class as a whole, "Go on! Go on!" This requires the incessant revolutionization of production to displace labor power and promote profits in the service of ever-greater accumulation. Moreover, as production grows "the consuming circle within circulation" must grow correspondingly. Intrinsic to the capital relation, Marx insisted, was the refusal to accept any absolute boundaries to its advance, which were treated as mere barriers to be surmounted. These propositions, intrinsic to Marx's political economy, constituted the foundations for what Schnaiberg was later to call the "treadmill of production" model.[7]

Marx's most pointed ecological contribution, however, lay in his theory of metabolic rift. Building on the work of the great German chemist Justus von Liebig, Marx argued that in shipping food and fiber hundreds and thousands of miles to the new urban centers of industrial production, where population was increasingly concentrated, capital ended up robbing the soil of its nutrients, such as nitrogen, phosphorus, and potassium, which instead of being returned to the earth created pollution in the cities. Liebig called this *"Raubbau"* or the robbery system. As Ernest Mandel put it in his *Marxist Economic Theory*:

> Serious scientists, notably the German Liebig, had drawn attention to a really disturbing phenomenon, the increased exhaustion of the soil, the *Raubbau*, resulting from greedy capitalist methods of exploitation aimed at getting the highest profit in the shortest time. Whereas agricultural societies like China, Japan, ancient Egypt, etc., had known a rational way of carrying on agriculture which conserved and even increased the fertility of the soil over several thousand years, the capitalist *Raubbau* had been able, in certain parts of the world, to exhaust the fertile layer of soil . . . in half a century.[8]

For Marx this capitalist *Raubbau* took the form of "an irreparable rift" within capitalist society in the metabolism between humanity and the earth—"a metabolism prescribed by the natural laws of life itself"—requiring its "systematic restoration as a regulative law of social production." In the industrialization of agriculture, he suggested, the true nature of "capitalist production" was revealed, which "only develops . . . by simultaneously undermining the original sources of all wealth—the soil and the worker."

In order to understand the significance of this ecological critique for Marx's overall critique of capitalism, it is necessary to recognize that the labor and production process was itself designated, in his analysis, as the metabolic relation between human beings and nature. Marx's primary definition of socialism/communism was therefore that of a society in which "the associated producers govern the human metabolism with nature in a rational way . . . accomplishing it with the least expenditure of energy." Along with this, he developed the most radical conception of sustainability possible, insisting that no one, not even all the countries and peoples of the world taken together, owned the earth; that it was simply held in trust and needed to be maintained in perpetuity in line with the principle of *boni patres familias* (good heads of the household). His overall ecological critique thus required that instead of the open rifts developed under capitalism, there needed to be closed metabolic cycles between humanity and nature. This allowed him to incorporate thermodynamic conceptions into his understanding of economy and society.[9]

The totality of Marx's ecological insights went, of course, beyond the foregoing points. Space does not allow full treatment of them here. Still, it is worth noting that his analysis together with that of Engels also touched on such critical issues as the "squandering" of fossil fuels and other natural resources; desertification; deforestation; and regional climate change—already understood by scientists in Marx's day as resulting in part from the human degradation of the local environment.[10]

Monopoly Capital and the Environment

Elements of Marx's general ecological critique resonated with developments in material science, providing inspiration directly and indirectly for a number of important materialist scientists and philosophers of science in the decades that followed. Things were quite different, however, within Marxian political economy, where Marx's critique of the capitalist *Raubbau* was rarely acknowledged (or drawn upon) between the close of the nineteenth century and the close of the twentieth.[11]

The main discoveries of Marxian and radical political economy in the ecological realm in the twentieth century can be seen as arising out of responses to the changed conditions associated with the monopoly stage of capital, and the altered environmental regime that it brought into being. The earliest theorists of monopoly capitalism were Rudolf Hilferding in Germany and Thorstein Veblen in the United States. Hilferding, although building his analysis directly on Marx's political economy, had surprisingly little to say about environmental conditions. In contrast, Veblen—a socialist economist influenced by Marx but not himself a Marxist—saw the transition from free competition to the age of the monopolistic corporation as having immense implications for the environment, resource use, and economic waste.

In his final, 1923 work, *Absentee Ownership and Business Enterprise in Recent Times*, Veblen stressed that "the American plan" of resource exploitation was one of accumulation by encroachment on both the environment and on the Indigenous population. In line with the Lauderdale Paradox, it took the form of "a settled practice of converting all public wealth to private gain on a plan of legalized seizure." The "custom," he wrote, was "to turn every public need to account as a means of private gain, and to capitalise it as such."

In the stage of free competition, Veblen argued, "staple resources" had been overexploited "by speeding up the output and underbidding on the price," leading to "a rapid exhaustion, with waste, of the natural supply." This set the stage for monopoly capital or what Veblen referred to as "absentee ownership," with its more collusive methods of turning public wealth to private gain by means of the careful regulation of scarcity and monopolistic pricing. This evolution was especially evident in the timber, coal, and oil industries, each of which initially involved prodigious waste, and led to eventual monopoly control by a relatively few absentee owners. As a result of these developments, Veblen noted, the "enterprise of lumbermen during the period since the middle of the nineteenth century has destroyed appreciably more timber than it has utilised."[12]

Veblen's more important ecological insights, however, had to do with the transformation of use value and consumption under the new regime of big business. A characteristic of monopoly capitalism was the virtual elimination of price competition by corporations, which was accompanied by the restriction of output. This allowed for monopolistic (or oligopolistic) pricing, which produced large gains for the giant enterprises. With price warfare effectively banned, "competitive strategy" was primarily "confined to two main lines of endeavour: to reduce the production-cost of a restricted output; and to increase their sales without lowering prices." Veblen pointed out that the very effectiveness of monopoly capital in containing production costs—by holding down wages and thereby, in Marxian terms, increasing the rate of surplus value—meant that at any given price the margin available for increases in sales costs (without cutting into profit margins) expanded. Thus a larger and larger share of the total cost of goods was associated with promotion of sales as opposed to the production of the commodity.[13] The implications of this for the use value structure of the economy were profound. "One result," he stated,

> has been a very substantial and progressive increase of sales-cost; very appreciably larger than an inspection of the books would show. The producers have been giving continually more attention to the saleability of their product, so that much of what appears on the books as production-cost should properly be charged to the production of saleable appearances. The distinction between workmanship and salesmanship has progressively been blurred in this way, until it will doubtless hold true now that the shop-cost of many articles produced for the market is mainly chargeable to the production of saleable appearances.[14]

He saw this as applying especially to the "vogue of 'package goods'":

> The designing and promulgation of saleable containers—that is, to say such containers as will sell the contents on the merits of the visual effect of the container—has become a large and, it is said, a lucrative branch of the business of publicity. It employs a formidable number of artists and "copy writers" as well as of

itinerant spokesmen, demonstrators, interpreters; and more than one psychologist of eminence has been retained by the publicity agencies for consultation and critical advice on the competitive saleability of rival containers and the labels and doctrinal memoranda which embellish them. The cost of all this is very appreciable.... It is presumably safe to say that the containers account for one-half the shop cost of what are properly called "package goods," and for something approaching one-half the price paid by the consumer. In certain lines, doubtless, as, e.g., in cosmetics and household remedies, this proportion is exceeded by a very substantial margin.[15]

The upshot of the infiltration of "salesmanship" into production was the proliferation of economic waste—defined by Veblen in *The Theory of the Leisure Class* as "expenditure" that "does not serve human life or human well-being on the whole." Indeed, much of the initial demand for purchased goods under monopoly capitalism was due to "invidious pecuniary comparison," that is, status distinctions arising from having something beyond the reach of others, as well as the various forms of "conspicuous consumption" and "conspicuous waste" associated with this. The more one could display the ostentatiousness of one's life the higher the social prestige. Corporate advertising encouraged such invidious comparisons first among the rich and then within the middle and working classes, often by instilling in people a fear of loss of social status.[16]

It is crucial to understand that the problem, raised by Veblen, of the transformation of consumption and the distortion of use values under capitalism played no significant role in the earlier work of Marx or his immediate followers (or indeed in that of other nineteenth-century critics of the system). To be sure, Engels wrote that under capitalism "the useful effect" of a commodity "retreats far into the background, and the sole incentive becomes the profit to be made on selling."[17] Implicit in this view was the notion that use values could be subordinated to exchange values and the structure of consumption to the forces of production. Yet, nowhere in *Capital* did Marx provide any analysis of the "interaction of production and consumption resulting from technical change" and the accompanying transformation of the use-value structure of the economy. The reason was that in nineteenth-century competitive capitalism workers' consumption goods (as distinct from capitalist luxury goods) were not yet subjected to the gargantuan "sales effort," which was to arise fully only with monopoly capitalism.[18] While waste was commonplace in competitive capitalism—arising from the irrationality and duplication inherent to competition itself—such waste did not have the same "functional" role for accumulation that it was later to acquire under monopoly capitalism, where the chief problem was no longer efficiency of production, on the supply side, but the generation of markets, on the demand side. For this reason, advertising and marketing in general, along with such factors as product differentiation, played only a minuscule role in the nineteenth century. Analysis of these developments thus had to await their appearance in the early twentieth century. This analysis was accomplished first by Veblen, and then—in a synthesis of Marx and Veblen—in Baran and Sweezy's *Monopoly Capital* in 1966.

For Baran and Sweezy the principal problem under monopoly capitalism was the absorption of the enormous economic surplus resulting from the constantly expanding productivity of the system. This economic surplus could be absorbed in three ways: capitalist consumption, investment, or waste.[19] Capitalist consumption was limited by the drive to accumulate on the part of the capitalist class, while investment itself was constrained by market saturation (due principally to the repression of wage-based consumption and conditions of industrial maturity). Hence, capitalism in its monopoly stage was threatened by a problem of markets and a declining rate of utilization of both productive capacity and employable labor.[20] Under such circumstances, the deepening reliance on economic waste served to keep markets going, becoming a necessary part of the monopoly-capitalist economy.

Baran and Sweezy argued that economic waste took various forms, notably military spending and the sales effort, the latter including:

"advertising, variation of the products' appearance and packaging, 'planned obsolescence,' model changes, credit schemes, and the like." The sales effort preceded capitalism's monopoly stage, but it was only under monopoly capitalism that it assumed "gigantic dimensions."

The most obvious form of the sales effort was of course advertising, which grew by leaps and bounds in the twentieth century. Perhaps the "dominant function" of advertising for the system, Baran and Sweezy observed, was that of "waging, on behalf of the producers and sellers of consumer goods, a relentless war against saving and in favor of consumption."[21] Yet advertising, they recognized, was only the tip of the iceberg where modern marketing was concerned, which today also includes targeting, motivation research, product management, sales promotion, and direct marketing.[22] According to Blackfriars Communications, the United States in 2005 spent over $1 trillion, or around 9 percent of GDP, on various forms of marketing.[23]

However, the main structural impact of the sales effort on the system for Baran and Sweezy, following Veblen, was to be found in "the emergence of a condition in which the sales and production efforts interpenetrate to such an extent as to be virtually indistinguishable." This marked "a profound change in what constitutes socially necessary costs of production as well as in the nature of the social product itself." Under these circumstances, constant model changes, product obsolescence, wasteful packaging, etc., all served to reorder the relations of consumption—altering the use value structure of capitalism and enlarging the waste incorporated within production. They estimated that automobile model changes alone were costing the country some 2.5 percent of its GDP. In comparison to this the expenditures of the automobile manufacturers on advertising were minuscule. "In the case of the automobile industry," they wrote, "and doubtless there are many others that are similar in this respect, by far the greater part of the sales effort is carried out not by obviously unproductive workers such as salesmen and advertising copy writers but by seemingly productive workers: tool and die makers, draftsmen, mechanics, assembly line workers." They concluded, "What is certain is the negative statement which, notwithstanding its negativity, constitutes one of the most important insights to be gained from political economy: an output the volume and composition of which are determined by the profit maximization policies of oligopolistic corporations neither corresponds to human needs nor costs the minimum possible amount of human toil and human suffering."[24]

Adopting a related perspective, Michael Kidron conservatively estimated in his *Capitalism and Theory* that in 1970, 61 percent of U.S. production could be classified as economic waste—resources diverted to the military, advertising, finance and insurance, waste in business, conspicuous luxury consumption, etc.[25] Increasingly, what was being produced under monopoly capitalism were formal or *specifically capitalist use values*, the primary "usefulness" of which lay in the exchange value they generated for corporations.[26]

Rational standards of human welfare and resource use, Baran and Sweezy claimed, required an entirely different approach to production. As early as 1957, in *The Political Economy of Growth*, Baran suggested that the optimum economic surplus in a planned economy would be less than that of maximum-potential economic surplus— requiring a slower rate of economic growth—due, among other reasons, to the need to curtail certain "noxious types of production (coal mining, for example)."[27] Likewise Sweezy argued in the 1970s that the need for every worker to have a car to go to work was not a product of human nature but artificially generated as a result of the whole "automobile-industrial complex" of so-called modernized capitalist society. The system of privatized (but publicly subsidized) transportation "externalized" costs such as air pollution, urban decay, and traffic fatalities onto the rest of society, while generating huge profits for corporations. In contrast, a more rational society would produce social use values: "functional, aesthetically attractive and durable," meeting genuine human needs, utilizing "methods of production compatible with humanized labor processes."[28]

Other thinkers in the same period developed related notions. John Kenneth Galbraith advanced his famous thesis of the "dependence effect" applicable to oligopolistic capitalism in *The Affluent Society* in 1958. He argued that the very process of "production of goods creates the wants that the goods are presumed to satisfy"—a thesis designed to overthrow the neoclassical theory of consumer sovereignty. Joan Robinson in her Richard T. Ely Lecture to the American Economic Association in 1971 (with Galbraith as the chair) raised the issue of the "Second Crisis of Economic Theory." Mistakenly assuming that Keynes had provided the solution to "the first crisis," the level or *quantity* of production, Robinson went on to contend that now was the time to turn to the "second crisis," the *quality* or content of production. Military production, pollution, inequality, and poverty were all being generated, she argued, not in spite of—but *because of*—the strategies adopted to expand capitalist growth. In the same year Barry Commoner, in his *The Closing Circle*, highlighted the ecological dangers associated, in particular, with the petrochemical industry, which he argued was deeply embedded in an increasingly toxic mode of production driven by profit.[29]

Elements of this general ecological critique of monopoly capitalism were drawn together in Allan Schnaiberg's 1980 treatise, *The Environment: From Surplus to Scarcity*, one of the founding works of environmental sociology. Already in the 1970s, environmentalists had begun to speak of environmental impact as a result of three factors: population, affluence (or consumption), and technology—with the last two factors, consumption and technology, standing for the role of the economy.[30] The structure of Schnaiberg's book was clearly derived from this, with chapters 2 through 5 focusing successively on population, technology, consumption, and production. Schnaiberg's brilliance was to draw on Marxian and radical political economy to show that the first three of these were conditioned by the fourth, making what he called "the treadmill of production" the fundamental environmental problem. He wrote of the "monopoly capital treadmill," and insisted: "Both the volume and source of . . . treadmill production is high-energy monopoly-capital industry."

For Schnaiberg, the monopoly stage of capitalism was geared to labor-saving, energy-intensive production. By constantly displacing labor and producing ever-greater economic surplus, which overflowed corporate coffers, the system generated a growing problem of effective demand—which it then attempted to solve by introducing various extraordinary means of expanding consumption. Contemporary consumption, he argued in Galbraithian terms, did not reveal consumer preferences so much as the profitability requirements of corporations—with consumer choices circumscribed by modern marketing and the technology of the treadmill. Schnaiberg's realistic conclusion was that attempts to address the ecological problem by focusing on population, consumption, or technology would inevitably fail—since the real problem was the treadmill of production itself.[31]

The treadmill of production (or of accumulation), as we have seen, can be explained in Marx's terms, using the general formula for capital—or M-C-M′, which in the next period of production, becomes M′-C-M″, and in the period after that M″-C-M‴, *ad infinitum*. For Marx, capital was a system of self-expanding value. It had, as Sweezy was to say, "no braking mechanism other than periodic economic breakdowns."[32] This is the basis of the standard ecological critique directed at capitalism, which emphasizes the scale effect of capitalist growth in relation to the earth's limited carrying capacity. Hence, it is rightly assumed that to solve the ecological problem it is necessary to intervene in order to slow down, stop, reverse, and eventually dismantle the treadmill, particularly at the center of the system. Nevertheless, the standard treadmill perspective, if taken by itself, tends to reduce the ecological problem to a *quantitative* one, deemphasizing the more *qualitative* aspects of the dialectic, represented today by the promotion of specifically capitalist use values and thus economic waste.

Here it is useful to stress that the C in the M-C-M′ relation, standing for the concrete use value aspect of the commodity, has now become transformed under monopoly control into a *specifically capitalist use value*, which we can designate as C^K—to stand for the almost complete subordination of use value to exchange value in the development of the commodity. C^K thus stands for the mediating role of *capital* (in German *Kapital*) within consumption, whereby more and more products derive their "use values" primarily from their role in facilitating the overall process of capital accumulation.

The problem of M-C-M' then becomes one of M-C^K-M', in which the qualitative as well as quantitative problems of accumulation/ecological destruction are asserted through the creation of formal use values. In today's packaged goods, the package, designed to sell the commodity and incorporated into its production costs, is now the larger part of the commodity. Thus Campbell soup marketers commonly refer to the soup as the mere *substrate* of the product. Or to take a more economically significant example, since the 1930s the production cost of the motor vehicle has only been a small part of the final sales price, most of which is related to marketing and distribution. As Stephen Fox stated in his *Mirror Makers: A History of American Advertising*, today's cars are "two-ton packaged goods, varying little beneath the skins of their increasingly outlandish styling." The average automobile sold in the United States today has lower fuel efficiency than the Model T Ford.[33] All of this suggests that use value, C, associated with the conditions of production in general, has increasingly given way under monopoly capitalism, to specifically capitalist use value, C^K—incorporating all sorts of socially unproductive features, with the object of generating higher sales, and hence realizing profit, M'.

It is this relentless reduction of consumption to the needs of capital accumulation by means of the alienation of use value (e.g., making plastic wrapping part of the production price of a loaf of bread) that lies behind the worst aspects of what is mistakenly thought of as "consumerism": the seemingly endless demand for superfluous, even toxic, products associated with today's throwaway society.[34] How else do we explain that, worldwide, upwards of 500 billion and perhaps as many as a trillion plastic shopping bags (given away for free) are consumed every year; that some 300 billion pounds of packaging are disposed of every year in the United States; and that 80 percent of all U.S. goods are used once and then thrown away? Much of this is toxic waste; Americans discard seven billion tons of PVC (polyvinyl chloride) plastic—the most hazardous plastic product—annually. In 2008 the Center for Health, Environment and Justice issued a report indicating that an ordinary new shower curtain, which uses

PVC plastic, released 108 separate volatile compounds in the home environment over twenty-eight days of ordinary usage, creating a level of these compounds that was sixteen times beyond what was recommended by the U.S. Green Building Council.[35]

Quite apart from its toxic nature, the economic and ecological waste embedded in the production and consumption process is enormous. "To say that 'capitalism has been simultaneously the most efficient *and* the most wasteful productive system in history,'" Douglas Dowd wrote in *The Waste of Nations*, "is to point to the contrast between the great efficiency with which a particular factory produces and packages a product, such as toothpaste, and the contrived and massive inefficiency of an economic system that has people pay for toothpaste a price over 90 percent of which is owed to the marketing, not the production, of the dentifrice."[36]

William Morris, who saw the very beginnings of monopoly capitalism, referred to "the mass of things which no sane man could desire, but which our useless toil makes—and sells."[37] Today we have to recognize that many of these superfluous goods carry enormous costs to the environment and human health. Indeed, many of our most common use values, as Barry Commoner explained, are the products of modern chemistry—introducing synthetic chemicals that are carcinogenic, mutagenic, and teratogenic into production, consumption, and the environment. These goods are cheap to produce (being energy- and chemical-intensive, not labor-intensive), they sell, and they generate high profit margins for corporations. The fact that many of them are virtually indestructible (non-biodegradable) and if incinerated—to prevent them from overwhelming landfills—give off dioxin and other deadly toxins, is viewed by the economic system as simply beside the point.[38]

In the face of such contradictions, radical economist Juliet Schor has written of the "materiality paradox," which suggests that people in our society are not too materialistic, but rather are not materialistic enough. We no longer retain, reuse, and repair products, because we have been taught to expect them to break down or fall apart due to product obsolescence, and then quickly discard them. Indeed, as a society, we have become entrapped in a still deeper pattern of psychological obsolescence, promoted by modern marketing, encouraging us to throw away what we have only just bought—as soon as it is no longer "new."[39]

The Meaning of Revolution

The ecological critique generated by twentieth-century monopoly capital theory—the bare outlines of which I have sought to present here—only adds additional force to Marx's classical ecological critique of capitalism. Every day we are destroying more and more public wealth—air, water, land, ecosystems, species—in the pursuit of private riches, which turns consumption into a mere adjunct to accumulation, thereby taking on more distorted and destructive forms.

The metabolic rift in the relation of humanity to the earth that Marx described in the nineteenth century has now evolved into multiple ecological rifts transgressing the boundaries between humanity and the planet. It is not just the *scale* of production but even more the *structure* of production that is at fault in today's version of the capitalist *Raubbau*. "Such is the dialectic of historical process," Baran wrote, "that *within the framework of monopoly capitalism* the most abominable, the most destructive features of the capitalist order become the very foundations of its continuing existence—just as slavery was the *conditio sine qua non* of its emergence."[40]

It is the historic need to combat the absolute destructiveness of the system of capital at this stage—replacing it, as Marx envisioned, with a society of substantive equality and ecological sustainability—which constitutes the essential meaning of revolution in our time.

CHAPTER EIGHTEEN

On Fire This Time

For a year, from 2018 to 2019, the world witnessed what appeared to be the beginnings of an ecological revolution, a new historical moment unlike any that humanity had yet experienced.[1] As Naomi Klein suggested at the time in her book *On Fire*, not only was the planet burning, but a revolutionary climate movement was rising up and was in response.[2] Outwardly, the signs of climate protest melted away due to the COVID-19 pandemic early in 2020. Yet, the struggle will undoubtedly reach new heights as the effects of the pandemic subside, spurred on by even greater levels of environmental concern. The climate revolution on the ground during the late summer and fall of 2018, stretching over the following year, thus remains a defining historical moment. Here is a brief chronology of the events, focusing on climate actions in Europe and North America—though it should be stressed that the whole world was—and in many ways still is—objectively (and subjectively) *on fire this time*.

- **August 2018:** Fifteen-year-old Greta Thunberg begins her school strike outside the Swedish Parliament.
- **October 8, 2018:** The United Nations Intergovernmental Panel

on Climate Change (IPCC) releases *Special Report on Global Warming of 1.5°C* pointing to the need for "systems transitions . . . unprecedented in terms of scale."[3]
- **October 17, 2018:** Extinction Rebellion activists occupy UK Greenpeace headquarters demanding the staging of mass civil disobedience to address the climate emergency.
- **November 6, 2018:** Alexandria Ocasio-Cortez (Democrat) is elected as a Congressional Representative on a platform that includes a Green New Deal.[4]
- **November 13, 2018:** Members of the Sunrise Movement occupy House Speaker Nancy Pelosi's Congressional office; newly elected Representative Ocasio-Cortez joins them.
- **November 17, 2018:** Extinction Rebellion activists block five bridges over the Thames in London.
- **December 10, 2018:** Sunrise Movement activists flood key Democratic Party Congressional offices demanding the creation of a Select Committee for a Green New Deal.
- **December 19, 2018:** Members of Congress in support of a Select Committee for a Green New Deal rises to forty.
- **January 25, 2019:** Thunberg tells World Economic Forum: "Our house is on fire. . . . I want you to act as if our house is on fire. Because it is."[5]
- **February 7, 2019:** Representative Ocasio-Cortez and Senator Edward Markey introduce the Green New Deal Resolution in Congress.[6]
- **March 15, 2019:** Nearly 2,100 youth-led climate strikes occur in 125 countries with 1.6 million participating (100,000 in Milan, 40,000 in Paris, 150,000 in Montreal).[7]
- **April 15–19, 2019:** Extinction Rebellion shuts down large parts of central London.
- **April 23, 2019:** Speaking to both Houses of Parliament, Thunberg states: "Did you hear what I just said? Is my English okay? Is the microphone on? Because I am beginning to wonder."[8]
- **April 25, 2019:** Extinction Rebellion protesters blockade the London Stock Exchange, gluing themselves across its entrances.

- **May 1, 2019:** UK Parliament declares a Climate Emergency shortly after similar declarations by Scotland and Wales.
- **August 22, 2019:** Senator and presidential candidate Bernie Sanders unveils the most comprehensive Green New Deal plan to date (outside the Green Party), proposing a public investment of $16.3 trillion over ten years.[9]
- **September 12, 2019:** The number of Congressional co-sponsors of the Green New Deal Resolution reaches 107.[10]
- **September 20, 2019:** Four million people join the global climate strike, staging more than 2,500 events in 150 countries; 1.4 million protest in Germany alone.[11]
- **September 23, 2019:** Thunberg tells the United Nations: "People are suffering. People are dying. Entire ecosystems are collapsing. We are in the beginning of a mass extinction, and all you can talk about is money and fairy tales of eternal economic growth. How dare you!"[12]
- **September 25, 2019:** IPCC *Special Report on the Ocean and Cryosphere* is released, indicating that many low-lying megacities and small islands, especially in tropical regions, will experience "extreme sea level events" *every year* by 2050.[13]

The outpouring of climate change protests over 2018–19 were largely in response to the IPCC's October 2018 report, which declared that carbon dioxide emissions need to peak in 2020, drop by 45 percent by 2030, and reach zero net emissions by 2050 for the world to have a reasonable chance of avoiding a catastrophic 1.5°C increase in global average temperature.[14] Untold numbers of people have suddenly become aware that, in order to pull back from the edge of the cliff, it is necessary to initiate socioeconomic change on a scale commensurate with that of the Earth System crisis that humanity is facing. This has resulted in System Change Not Climate Change, the name of the leading U.S. ecosocialist movement, becoming the mantra of the entire global grassroots climate movement.[15]

The meteoric rise of Thunberg and the student climate strike movement, the Sunrise Movement, Extinction Rebellion, and the

Green New Deal, all within the brief span of a year, coupled with the actual protests and strikes of millions of climate change activists, the vast majority of them young, meant a massive transformation of the environmental struggle in the advanced capitalist states. Virtually overnight, the struggle shifted from its previous more generic climate action framework toward the more radical climate justice and ecosocialist wings of the movement.[16] The climate action movement has been largely reformist, merely seeking to nudge business-as-usual in a climate-conscious direction. The 400,000-person climate march in New York in 2014, organized by the People's Climate Movement, proceeded to 34th Street and 11th Avenue, a non-destination, rather than to the United Nations where climate negotiators were meeting, with the result that it had more the character of a parade than a protest.[17]

In contrast, climate justice organizations such as Extinction Rebellion, the Sunrise Movement, and the Climate Justice Alliance are known for their direct action. The new movement is younger, bolder, more diverse, and more revolutionary in its outlook.[18] In the present struggle for the planet, there is a growing recognition that the social and ecological relations of production must be transformed. Only a transformation that is revolutionary in terms of scale and tempo can pull humanity out of the trap that capitalism has imposed. As Thunberg told the UN Climate Change Conference on December 15, 2018, "If the solutions within this system are so impossible to find then maybe we should change the system itself."[19]

The Green New Deal: Reform or Revolution?

What has made the struggle for an ecological revolution a seemingly unstoppable force in 2018–19 was the rise of the Green New Deal, a program that represents the coalescence of the movement to arrest climate change with the struggle for economic and social justice, focusing on the effects on workers and frontline communities. However, the Green New Deal was not originally a radical-transformational strategy, but rather a moderate-reformist one.

The phrase *Green New Deal* took hold in 2007 in a meeting between Colin Hines, former head of Greenpeace's International Economics unit, and *Guardian* economics editor Larry Elliott. Faced with growing economic and environmental problems, Hines suggested a dose of Green Keynesian spending, labeling it a Green New Deal after Franklin Roosevelt's New Deal during the Great Depression in the United States. Elliott, Hines, and others, including British entrepreneur Jeremy Leggett, launched the UK Green New Deal Group later that year.[20]

The idea caught on quickly within environmental policy circles. Pro-corporate *New York Times* columnist Thomas Friedman began promoting the term in the United States at about the same time as a new capitalist ecomodernist strategy.[21] Barack Obama was to advance a Green New Deal proposal in his 2008 campaign. However, he dropped the Green New Deal terminology along with what remained of its substance after the midterm elections in 2010.[22] In September 2009, the UN Environment Programme issued a report titled *Global Green New Deal*, consisting of a sustainable growth plan.[23] That same month, the Green European Foundation published *A Green New Deal for Europe*, a Keynesian green capitalist strategy, today known as the European Green New Deal.[24]

All of these proposals, introduced under the mantle of a Green New Deal, were top-down combinations of Green Keynesianism, ecomodernism, and corporatist technocratic planning incorporating a marginal concern for promoting employment and eradicating poverty, while standing for a mildly reformist green capitalism. In this respect, the first Green New Deal proposals had more in common with Franklin Roosevelt's First New Deal, from 1933 to 1935 in the United States, which was corporatist and heavily pro-business in character, than with the Second New Deal from 1935 to 1940, which was animated by the great revolt of industrial labor in the mid-to-late 1930s.[25]

In sharp contrast to these early corporatist proposals, the radical version of the Green New Deal that has gained traction in the last year in the United States has its historical inspiration in the great revolt from below in the Second New Deal. A key force in this metamorphosis was the Climate Justice Alliance that arose in 2013 through the coalescence of various primarily environmental justice organizations. The Climate Justice Alliance currently unites sixty-eight different frontline organizations, representing low-income communities and communities of color, engaged in immediate struggles for environmental justice and supporting a just transition.[26]

The critical concept of a just transition had its origins in the 1980s, in the efforts of ecosocialist Tony Mazzocchi of the Oil, Chemical and Atomic Workers Union to build a radical labor-environmental justice movement, which was later promoted by the United Steel Workers.[27] Directed at overcoming the chasm between economic and ecological struggles, a just transition is now recognized as the main principle, beyond the safeguarding of the climate itself, in the struggle for a people's Green New Deal.

The Green New Deal first metamorphosed into a radical grassroots strategy—or a People's Green New Deal in the terms of Science for the People—during Jill Stein's two successive Green Party presidential campaigns in 2012 and 2016.[28] The Green Party's Green New Deal had four pillars: (1) an Economic Bill of Rights, including the right to employment, workers' rights, the right to health care (Medicare for All), and the right to tuition-free, federally funded higher education; (2) a Green Transition, promoting investment in small businesses, green research, and green jobs; (3) Real Financial Reform, including relieving homeowner and student debt, democratizing monetary policy, breaking up financial corporations, ending government bailouts of banks, and regulating financial derivatives; and (4) a Functioning Democracy, revoking corporate personhood, incorporating a Voter's Bill of Rights, repealing the Patriot Act, and cutting military spending by 50 percent.[29]

There can be no doubt about the radical (and anti-imperialist) nature of the Green Party's original Green New Deal platform. Its designated halving of U.S. military spending was the key to its plan to increase federal spending in other spheres. At the heart of the Green Party's Green New Deal was thus an attack on the economic, financial, and military structure of the U.S. empire, while focusing its economic policy proposals on a Green Transition that would provide up to twenty million new green jobs.[30] The Green Transition part of the program was, ironically, the weakest component of the Green Party's Green New Deal. The innovation of the Green Party, however, was to link vital environmental change to what it conceived as equally necessary social change.

But it was not until the radical Green New Deal burst forth in Congress in November 2018, spearheaded by the newly elected Congressional Representative Ocasio-Cortez, following the midterm U.S. elections, that it suddenly became a major factor in the U.S. political landscape. Ocasio-Cortez had decided to run for office after joining the hard-fought Indigenous-led protest aimed at blocking the Dakota Access Pipeline at Standing Rock in North Dakota in 2016–17. In campaigning in New York's 14th Congressional District (representing the Bronx and part of north-central Queens), she signed the Sunrise Movement's No Fossil Fuel Money pledge, with the result that the Sunrise Movement canvassed for her, contributing to her surprise election victory against ten-term incumbent Representative Joe Crowley.[31] The Sunrise Movement's sit-in in Pelosi's office in support of a Green New Deal a week after the midterm elections was immediately joined by Ocasio-Cortez, who, together with Markey, was to introduce the Green New Deal Resolution in Congress.

Ocasio-Cortez's campaign drew much of its inspiration from Sanders's self-described democratic socialist campaign for president in 2016 that led to the revival of the Democratic Socialists of America (DSA), which Ocasio-Cortez joined prior to her election. From the start, the people's Green New Deal resolution thus took on what was in many ways an ecosocialist character.[32]

In the fourteen-page Green New Deal Resolution presented by Ocasio-Cortez and Markey in February 2019, the reality of the climate emergency is laid out along with the extent of U.S. responsibility. This is juxtaposed to "related crises" manifested in the decline in life expectancy, wage stagnation, diminishing class mobility, soaring inequality, the racial divide in wealth, and the gender earnings gap.

The solution offered is a Green New Deal that would achieve net-zero greenhouse gas emissions through a "just transition," creating "millions of good, high wage jobs" in the process of securing a sustainable environment. It is designed to "promote justice and equity by stopping current, preventing future, and repairing historic oppression of Indigenous peoples, communities of color, migrant communities, deindustrialized communities, depopulated rural communities, the poor, low-income workers, women, the elderly, the unhoused, people with disabilities, and youth"—referred to in this resolution as "frontline and vulnerable communities."

The Green New Deal Resolution is based on a "10-year national mobilization." In this period, the goal is to achieve "100 percent of the power demand in the United States through clean, renewable, and zero-emission energy sources." Other measures include opposing "domestic or international monopolies"; supporting family farming; building a sustainable food system; establishing a zero-emission vehicle infrastructure; promoting public transit; investing in high-speed rail; ensuring the international exchange of climate-related technology; creating partnerships with frontline communities, labor unions, and worker cooperatives; providing job guarantees, training, and higher education to the working population; ensuring high-quality, universal health care for the entire U.S. population; and protecting public lands and waters.[33]

Unlike the Green Party's New Deal, the Democratic Party's Green New Deal Resolution as introduced by Ocasio-Cortez and Markey does not directly oppose financial capital or U.S. spending on the military and empire. Rather, its radical character is confined to linking a massive mobilization to combat climate change to a just transition for frontline communities, including redistributive economic measures. And yet there is no doubt about the radical nature of the demands put forward, which if carried out fully would require a mass mobilization of the entire society aimed at a vast transformation of U.S. capital and the expropriation of the fossil fuel industry.

Sanders's 34-page Green New Deal plan went still further.[34] It required 100 percent renewable energy for electricity and transportation by 2030 (amounting to a 71 percent reduction in U.S. carbon emissions) and complete decarbonization by 2050 at the latest. It planned to accomplish all of this by devoting $16.3 trillion in public investment to the massive mobilization of resources to displace fossil fuels; insisting on a just transition for both workers and frontline communities; declaring a climate change national emergency; reauthorizing the Civilian Conservation Corps of the New Deal; and banning offshore drilling, fracking, and mountaintop removal coal mining. It would offer $200 billion to the Green Climate Fund to support necessary transformations in poor countries with the aim of helping reduce carbon emissions in less industrialized nations by 36 percent by 2030.

To ensure a just transition for workers, Sanders proposed "up to five years of a wage guarantee, job placement assistance, relocation assistance, health care, and a pension based on their previous salary," along with housing assistance, to all workers displaced due to the switch away from fossil fuels. Workers would receive training for different career paths, including fully paid four-year college education. Health care cost would be covered by Medicare for All. The principles of environmental justice would be adhered to in order to protect frontline communities. Funding would be provided to impacted frontline communities, including the Indigenous. Tribal sovereignty will be respected, with the Sanders plan offering $1.12 billion for tribal land access and extension programs. In addition, the government "will set aside $41 billion to help large confined animal feeding operations" convert to "ecologically regenerative practices," to be coupled with support to family farms.

Funding would come from a number of sources including: (1) "massively raising taxes on corporate polluters' and investors' fossil fuel income and wealth" as well as "raising penalties on pollution from fossil fuel energy generation" by corporations; (2) eliminating subsidies to the fossil fuel industry; (3) "generating revenue from the wholesale of energy produced by regional Power Marketing Authorities"—with the added revenue being used to support the Green New Deal to be collected until 2035, after which electricity would be provided virtually free to customers aside from operations and maintenance costs; (4) cutting back on military spending directed at safeguarding global oil supplies; (5) collecting additional tax revenue resulting from the increase in employment; and (6) making corporations and the wealthy pay their "fair share."[35]

The Sanders Green New Deal was thus distinguished from Ocasio-Cortez and Markey's House Resolution in (1) setting a definite timeline for greenhouse gas emission cuts (one more ambitious for the United States, due to its unique responsibilities, than what is required by the world on average under the global carbon budget); (2) its direct confrontation with fossil capital; (3) explicitly basing its just transition on the needs of the working class as a whole, while focusing in particular on frontline communities; (4) specifying, like the earlier Green Party New Deal proposal, the creation of twenty million new jobs; (4) banning offshore drilling, fracking, and mountaintop removal coal mining; (5) confronting the role of the military in safeguarding the global fossil fuel economy; (6) stipulating $16.3 trillion in federal government expenditures on the Green New Deal over ten years; and (7) relying on taxes on polluting corporations to help fund the Green New Deal itself.[36] The Sanders plan, however, backed off from the Green Party's bold proposal to halve military spending.

The People's Green New Deal strategies now being advanced constitute what in socialist theory is called revolutionary reforms, that is, reforms that promise a fundamental restructuring of economic, political, and ecological power, and that point toward rather than away from the transition from capitalism to socialism. The scale of the changes envisaged are far greater, representing a more formidable threat to the power of capital, than those posed by the Second New Deal of the late 1930s. The complete disinvestment in fossil fuels, including fossil fuel reserves, constitutes a kind of abolitionism driven by sheer necessity that finds its closest analogy, in terms of its overall economic-scale effects, in the abolition of slavery in the United States. It has been estimated that in 1860, slaves constituted "the largest single financial asset in the entire U.S. economy, worth more than all manufacturing and railroads combined."[37] Today, taking on the fossil fuel industry and related industries and infrastructure, including the entire financial structure, raises analogous conflicts over wealth and power in terms of the sheer scale involved, and is only conceivable as part of a general ecological and social transformation. Thus, the Inter-American Development Bank declared in 2016 that the world's energy companies were facing the potential loss of $28 trillion as a result of the world's need to keep fossil fuels in the ground.[38]

As capital understood from the start, these changes would threaten the entire political-economic order, since once the population was mobilized for change, the whole metabolism of capitalist production would be challenged.[39] Energy corporations, Klein writes, will "have to leave trillions of dollars' worth of proven fossil fuel reserves [which they count as assets] in the ground."[40] For the climate justice movement to take on fossil capital and the reigning capitalist system as a whole in this way requires social mobilization and class struggle on an enormous scale, with the major transformations in production-energy to be introduced in a mere handful of years.

To be sure, none of the Green New Deal proposals are anywhere near to conceiving, much less tackling, the immensity of the task that the current planetary emergency demands. But they are sufficiently grounded in necessity that they could spark a global revolutionary struggle for freedom and sustainability, since the changes contemplated go against the logic of capital itself and cannot be achieved without a mobilization of the population as a whole on an emergency basis.

Still, there are lingering contradictions to the radical Green New Deal strategies related to their emphasis on economic growth and capital accumulation. The constraints imposed by the need to stabilize the climate are severe, requiring changes in the underlying structure of production. Nevertheless, all of the current Green New Deal proposals largely eschew any mention of direct conservation of resources or cuts in overall consumption—much less emergency measures like rationing as an equitable non-price-related means of reallocating society's scarce resources (a fairly popular measure in the United States in the Second World War).[41] None consider the full level of waste built into the current accumulation system and how that could be turned to ecological advantage. Instead, all of the plans are based on the notion of promoting rapid, exponential economic growth or capital accumulation—despite the fact that this would compound the planetary emergency, and despite the fact that the real successes of the Second New Deal had much less to do with growth than economic and social redistribution.[42] As Klein cautions, a Green New Deal plan would fail dismally both in protecting the planet and in carrying out a just transition if it were to take the path of "climate Keynesianism."[43]

The IPCC and Mitigation Strategies

None of this is to deny that a tectonic shift appears to be underway. The radical Green New Deal strategies now being advocated threaten to blow apart the IPCC-led scientific-policy process with respect to what can and should be done to combat climate change, which has hitherto interdicted all left-social perspectives. In sharp contrast to its careful scientific treatment of the causes and consequences of climate change, which have been relatively free from political intervention, the IPCC's approach to the social actions necessary to mitigate the climate emergency has been dictated in large part by the current political-economic hegemony. Mitigation strategies for reducing carbon dioxide emissions worldwide have thus far been heavily impacted by the near total domination of capitalist relations of accumulation and the hegemony of neoclassical economics. The guidelines built into such mitigation scenarios heavily restrict the parameters of change under consideration via such devices as integrated assessment models, or IAMs, large computer models that integrate energy markets and land use with greenhouse gas projections, and shared socioeconomic pathways, or SSPs, consisting of five different business-as-usual pathways based on largely technological frameworks, with substantial economic growth and the lack of climate policy implementation formally built into all the models.

The result of such deliberately conservative models, which write off all alternatives to business as usual, is the proliferation of unrealistic assessments of what can be done and what has to be done.[44] In general, the mitigation scenarios incorporated into the IPCC process (1) implicitly assume the need to perpetuate the current political-economic hegemony; (2) downplay changes in social relations in favor of technocratic change, much of it based on technologies that do not exist or are unfeasible; (3) stress supply-side, mainly price-related and technological factors, rather than demand-side factors, or direct reductions in ecological consumption so as to lower emissions; (5) rely on so-called negative emissions—capturing carbon dioxide from the atmosphere and somehow sequestering it—so as to allow the overshooting of emission targets; (6) leave the mass of the population out of account, assuming that change will be managed by managerial elites with minimal public participation; and (7) postulate slow responses, leaving out the possibility (indeed necessity) of ecological revolution.[45]

Hence, though the scale of *climate change* and its socioecological impacts are well captured by IPCC models and projections, the scale of the *social change* required to meet this challenge is systematically downgraded in the hundreds of mitigation models utilized by the IPCC. Magic bullets emanating from market-price interventions (such as carbon trading) and futuristic technology involving inventions that are not feasible on the necessary scale and that rely on negative emissions are resorted to instead.[46] Such models point to catastrophic outcomes for which the only defenses are presumed to be so-called market efficiency and nonexistent and/or irrational baroque technology, since these approaches supposedly allow society to proceed with its current productive mode largely unaltered.

Thus, most climate mitigation models incorporate bioenergy with carbon capture and storage (BECCS) technology, which promotes growing plants (principally trees) on a massive scale to be burned to produce energy, while simultaneously capturing the carbon dioxide released into the atmosphere and somehow sequestering or storing it, as in geologic and ocean sequestration. If implemented, the most ambitious scheme would require a quantity of land equal to two Indias and an amount of freshwater approximating that which is currently used by world agriculture despite world water shortages.[47] Nor is the avid promotion of such purely mechanistic approaches an accident. It is deeply embedded in how these reports are constructed and the underlying capitalist order they serve.

In the words of leading climatologist Kevin Anderson of the Tyndall Centre for Climate Change Research in the United Kingdom:

> The problem is that delivering on the 1.5–2°C commitment demands emissions cuts for wealthy nations of more than 10% each year, far beyond rates typically considered possible in the current economic system. It is in seeming to remedy this impasse that IAMs have an important and dangerous role. Behind a veneer of objectivity, the use of these leviathan computer models has professionalized the analysis of climate-change mitigation by substituting messy and contextual politics with non-contextual mathematical formalism. Within these professional boundaries, IAMs synthesize simple climate models, with a belief in how finance works and technologies change, buttressed by an [orthodox] economic interpretation of human behaviour....
>
> Typically, IAMs use models based on free-market axioms. The algorithms embedded in these models assume marginal changes near economic equilibrium, and are heavily reliant on small variations in demand that result from marginal changes in prices. The Paris Climate Agreement, by contrast, sets a mitigation challenge that is far removed from the equilibrium of today's market economy, requiring immediate and radical change across all facets of society.[48]

The reality, Anderson stresses, is that current climate-scenario modeling and projections provided by the IPCC and incorporated into national plans are based on assumptions drawn from the general equilibrium analysis of neoclassical economics, building in notions of gradualist changes, based on the requirements of the profit system. Such stipulations in mitigation scenarios are meaningless in the context of the current climate emergency and dangerous in that they inhibit necessary action—so that nonexistent technology is seen as the only savior. Of the numerous models considered by the IPCC in its 2018 report, *all* require carbon dioxide reduction (CDR) or so-called negative emissions, mostly by technological means but also including afforestation.[49] The truth is that the whole mitigation approach within the IPCC, Anderson explains, has been an "accelerating failure," guiding a process that is radically opposed to its projections, with the result that "annual CO_2 emissions have increased by about 70% since 1990." Since the effects of such emissions are cumulative and nonlinear, with all sorts of positive feedbacks, the "ongoing failure to mitigate emissions has pushed the challenge from a moderate change in the economic system to a revolutionary overhaul of the system. This is not an ideological position; it emerges directly from a scientific and mathematical interpretation of the Paris Climate Agreement."[50]

Recognizing the accelerating climate emergency, the IPCC in its 2018 report departed from its previous reports in mildly encouraging the development of approaches to climate-change mitigation that include demand-side considerations. This means finding ways to reduce consumption, usually through increased efficiency (though typically downplaying the well-known Jevons Paradox, where increasing efficiency under capitalism leads to increased accumulation and consumption).[51] A number of mitigation scenarios have been introduced that demonstrate that demand-side interventions are the fastest way to address climate change—and even, in one model, suggesting that the below-1.5ºC target can be met with only slight overshoot and without reliance on capital-intensive, energy-intensive negative-emission technologies, but rather depending on improved agricultural and forestry practices.[52] These results are achieved, moreover, *within* the extremely restrictive assumptions of the IPCC mitigation models, which formally build in (via IAMs and SSPs) significant rapid economic growth while formally excluding all climate policy (or political) interventions. It has therefore been suggested by some radical critics, such as Jason Hickel and Giorgos Kallis, that a demand-side sociopolitical approach that emphasizes abundance and redistributive policies, while placing limits on profits and growth (which today mainly benefit the .01 percent), is demonstratively far superior in mitigation terms and constitutes the only realistic solution.[53]

A major virtue of the rise of radical or people's Green New Deal strategies, therefore, is that they open up the realm of what is possible in accord with actual necessity, raising the question of transformative change as the only basis of human-civilizational survival: *the freedom of necessity*.[54] Here it is important to recognize that an ecological and social revolution under present historical conditions is likely to pass through two stages that we can call *ecodemocratic* and *ecosocialist*.[55] The self-mobilization of the population will initially take an ecodemocratic form, emphasizing the building of energy alternatives combined with a just transition, but in a context generally lacking any systematic critique of production or consumption. Eventually, the pressure of climate change and the struggle for social and ecological justice, spurred on by the mobilization of diverse communities, can be expected to lead to a more comprehensive ecorevolutionary view, penetrating the veil of the received ideology.

Still, the fact remains that the attempt to construct a radical Green New Deal in a world still dominated by monopoly-finance capital will be constantly threatened by a tendency to revert to Green Keynesianism, where the promise of unlimited jobs, rapid economic growth, and higher consumption militate against any solution to the planetary ecological crisis. As Klein remarks in *On Fire*:

> Any credible Green New Deal needs a concrete plan for ensuring that the salaries from all the good green jobs it creates aren't immediately poured into high-consumer lifestyles that inadvertently end up increasing emissions—a scenario where everyone has a good job and lots of disposable income and it all gets spent on throwaway crap.... What we need are transitions that recognize the hard limits on extraction and that simultaneously create new opportunities for people to improve quality of life and derive pleasure outside the endless consumption cycle.[56]

The path toward ecological and social freedom requires abandoning a mode of production rooted in the exploitation of human labor and the expropriation of nature and peoples, leading to ever more frequent and severe economic and ecological crises. The overaccumulation of capital under the regime of monopoly-finance capital has made waste at every level integral to the preservation of the system, creating a society in which what is rational for capital is irrational for the world's people and the earth.[57] This has led to the wasting away of human lives on unnecessary labor spent on producing useless commodities, requiring the squandering of the world's natural-material resources. Conversely, the extent of this profligate waste of human production and wealth, and of the earth itself, is a measure of the enormous potential that exists today for expanding human freedom and fulfilling individual and collective needs while securing a sustainable environment.[58]

In the current climate crisis, it is the imperialist countries at the center of the system that have produced the bulk of the carbon dioxide emissions now concentrated in the environment. It is these nations that still have the highest per capita emissions. These same states, moreover, monopolize the wealth and technology necessary to reduce global carbon emissions dramatically. It is therefore essential that the wealthy nations take on the larger burden for stabilizing the world's climate, reducing their carbon dioxide emissions at a rate of 10 percent or more a year.[59] It is the recognition of this responsibility on the part of rich nations, together with the underlying global necessity, that has led to the sudden rise of transformative movements like Extinction Rebellion.

Over the longer run, however, the main impetus for worldwide ecological transformation will come from the Global South where the planetary crisis is having its harshest effects—on top of an already imperialist world system and a growing gap between rich and poor countries as a whole. It is in the periphery of the capitalist world that the legacy of revolution is the strongest—and the deepest conceptions of how to carry out such needed change persist. This is especially evident in countries such as Cuba, Venezuela, and Bolivia, which have sought to revolutionize their societies despite the harsh attacks by the imperialist world system and in spite of their historic dependence (in the cases of Venezuela and Bolivia)—itself imposed by the hegemonic structures of the global economy—on energy extraction. In general, we can expect the Global South to be the site of the most rapid growth of an environmental proletariat, arising from the degradation of material conditions of the population in ways that are equally ecological and economic.[60]

The role of China in all of this remains crucial and contradictory. It is one of the most polluted and resource-hungry countries in the world, and its carbon emissions are so massive as to constitute a global-scale problem. Nevertheless, China has done more than any other country thus far to develop alternative-energy technologies geared to the creation of what is officially referred to as an *ecological civilization*. Remarkably, it remains largely self-sufficient in food due to its system of agriculture, in which the land is social property and agricultural production is mainly reliant on small producers with remnants of collective-communal responsibility. What is clear is that the present and future choices of the Chinese state, and even more the Chinese people, with respect to the creation of an ecological civilization are likely to be key in determining the long-term fate of the earth.[61]

Ecological revolution faces the enmity of the entire capitalist system. At a minimum it means going against the logic of capital. In its full development, it means transcending the system. Under these conditions, the reactionary response of the capitalist class backed by its rear guard on the far right will be regressive, destructive, and unrestrained. This could be seen in the numerous attempts by Donald Trump's administration to remove the very possibility of making the changes necessary to combat climate change (seemingly in order to burn the world's ships behind it), beginning with its withdrawal from the Paris Climate Agreement and its acceleration of fossil fuel extraction. Ecological barbarism or ecofascism are palpable threats in the current global political context and are part of the reality with which any mass ecological revolt will need to contend.[62] Only a genuine revolutionary, and not a reformist, struggle will be able to propel itself forward in these circumstances.

An Age of Transformational Change

It is commonplace in the social science literature, representing the reigning liberal ideology, to see society as simply constituted by the actions of the individuals that make it up. Other, more critical thinkers sometimes present the opposite view that individuals are the product of the overall social structure. A third generic model sees individuals as affecting society and society affecting individuals in a kind of back-and-forth motion, viewed as a synthesis of structure and agency.[63]

In contrast to all of these mainstream, mostly liberal, approaches, which leave little room for genuine social transformation, Marxian theory, with its historical-dialectical approach, relies on what the critical-realist philosopher Roy Bhaskar has called a "transformational model of social activity" in which individuals are historically born into and socialized in a given society (mode of production), which sets the initial parameters of their existence.[64] However, these conditions and productive relations change in unpredictable, contingent ways during the course of their lives, leading to unintended consequences, contradictions, and crises. Caught in historical situations not of their choosing, human beings, acting both spontaneously and through organized social movements, reflecting class and other individual and collective identities, seek to alter the existing structures of social reproduction and social transformation, giving rise to critical historical moments consisting of radical breaks and revolutions, and new emergent realities. As Karl Marx wrote, "Men make their own history, but they do not make it just as they please; they do not make it under circumstances chosen by themselves, but under circumstances directly encountered, given and transmitted from the past."[65]

Such a transformational model of social activity supports a theory of human self-emancipation in history. Existing social relations become fetters on general human development, but they also give rise to fundamental contradictions in the labor and production process—or what Marx called the social metabolism of humanity and nature—leading to a period of crisis and transformation, threatening the revolutionary overturning of the social relations of production, or the relations of class, property, and power.[66] Today we are presented with such severe contradictions in the metabolism of nature and society and in the social relations of production, but in a manner for which there is no true historical precedent.

In the Anthropocene, the planetary ecological emergency overlaps with the overaccumulation of capital and an intensified imperialist expropriation, creating an epochal economic and ecological crisis.[67] It is the overaccumulation of capital that accelerates the global ecological crisis by propelling capital to find new ways to stimulate consumption to keep the profits flowing. The result is a state of impending planetary Armageddon, threatening not just socioeconomic stability, but the survival of human civilization and the human species itself. For Klein, the core explanation is simple: noting that "Marx wrote about capitalism's 'irreparable rift' with 'the natural laws of life itself,'" she goes on to underscore that "many on the left have argued that an economic system built on unleashing the voracious appetites of capital would overwhelm the natural systems on which life depended."[68] And this is exactly what has happened in the period since the Second World War, through the great acceleration of economic activity, overconsumption on the part of the wealthy, and the resulting ecological destruction.

Capitalist society has long glorified the domination of nature. William James, the great pragmatist philosopher, famously referred in 1906 to "the moral equivalent of war." It is seldom mentioned, though, that James's moral equivalent was a *war on the earth*, in which he proposed "to form for a certain number of years a part of the army enlisted against *Nature*."[69] Today, we have to reverse this and create a new, more revolutionary moral equivalent of war, one directed not at the enlisting of an army to conquer the earth, but directed at the self-mobilization of the population to save the earth as a place of human habitation. This can only be accomplished through a struggle for ecological sustainability and substantive equality aimed at resurrecting the global commons. In the words of Thunberg, speaking to the United Nations on September 23, 2019, "Right here, right now is where we draw the line. The world is waking up. And change is coming, whether you like it or not." The world is on fire this time.

CHAPTER NINETEEN

COVID-19 and Catastrophe Capitalism

COVID-19 has accentuated as never before the interlinked ecological, epidemiological, and economic vulnerabilities imposed by capitalism. As the world enters the third decade of the twenty-first century, we are seeing the emergence of catastrophe capitalism as the structural crisis of the system takes on planetary dimensions.

Since the late twentieth century, capitalist globalization has increasingly adopted the form of interlinked commodity chains controlled by multinational corporations, connecting various production zones, primarily in the Global South, with the apex of world consumption, finance, and accumulation primarily in the Global North. These commodity chains make up the main material circuits of capital globally that constitute the phenomenon of late imperialism identified with the rise of generalized monopoly-finance capital.[1] In this system, exorbitant imperial rents from the control of global production are obtained not only from *global labor arbitrage*, through which multinational corporations with their headquarters in the center of the system overexploit industrial labor in the periphery, but also increasingly through *global land arbitrage*, in which agribusiness multinationals expropriate cheap land (and labor) in the Global South so as to produce export crops mainly for sale in the Global North.[2]

In addressing these complex circuits of capital in today's global economy, corporate managers refer both to supply chains and value chains, with supply chains representing the movement of the physical product, and value chains directed at the "value added" at each node of production, from raw materials to the final product.[3] This dual emphasis on supply chains and value chains resembles in some ways the more dialectical approach developed in Karl Marx's analysis of the *commodity chains* in production and exchange, encompassing both use values and exchange values. In the first volume of *Capital*, Marx highlighted the dual reality of natural-material use values (the "natural form") and exchange values (the "value form") present in each link of "the general chain of metamorphoses taking place in the world of commodities."[4] Marx's approach was carried forward by Rudolf Hilferding in his *Finance Capital*, where he wrote of the "links in the chain of commodity exchanges."[5]

In the 1980s, world-system theorists Terence Hopkins and Immanuel Wallerstein reintroduced the commodity-chain concept based on these roots within Marxian theory.[6] Nevertheless, what was generally lost in later Marxian (and world-system) analyses of commodity chains, which treated these as exclusively economic/value phenomena, was the material-ecological aspect of use values. Marx, who never lost sight of the natural-material limits in which the circuit of capital took place, had stressed "the negative, i.e. destructive side" of capitalist valorization with respect to the natural conditions of production and the metabolism of human beings and nature as a whole.[7] The "irreparable rift in the interdependent process of social metabolism" (the metabolic rift) that constituted capitalism's destructive relation to the earth, whereby it "exhausted the soil" and "forced the manuring of English fields with guano," was equally evident in "periodical epidemics," resulting from the same organic contradictions of the system.[8]

Such a theoretical framework, focusing on the dual, contradictory forms of commodity chains, which incorporate both use values and exchange values, provides the basis for understanding the combined ecological, epidemiological, and economic crisis tendencies of late imperialism. It allows us to perceive how the circuit of capital under late imperialism is tied to the etiology of disease via agribusiness, and how this has generated the COVID-19 pandemic. This same perspective focusing on commodity chains, moreover, allows us to understand how the disruption of the flow of use values in the form of material goods and the resulting interruption of the flow of value have generated a severe and lasting economic crisis. The result is to push an already stagnant economy to the very edge, threatening the toppling of the financial superstructure of the system. Finally, beyond all of this lies the much greater planetary rift engendered by today's catastrophe capitalism, exhibited in climate change and the crossing of various planetary boundaries, of which the present epidemiological crisis is simply another dramatic manifestation.

Circuits of Capital and Ecological-Epidemiological Crises

Remarkably, during the last decade, a new, more holistic One Health-One World approach to the etiology of disease arose, mainly in response to the appearance of recent zoonotic diseases (or zoonoses) such as SARS, MERS, and H1N1 transmitted to humans from nonhuman animals, wild or domesticated. The One Health model integrates epidemiological analysis on an ecological basis, bringing together ecological scientists, physicians, veterinarians, and public health analysts within an approach that has a global scope. However, the original ecological framework that motivated One Health, representing a new, more comprehensive approach to zoonotic disease, has recently been appropriated and partially negated by such dominant organizations as the World Bank, the World Health Organization, and the Centers for Disease Control and Prevention in the United States. Hence, the multisector approach of One Health has been rapidly converted into a mode of bringing together such varied interests as public health, private medicine, animal health, agribusiness, and big pharma to strengthen the response to what are regarded as episodic epidemics, while signifying the rise of a broad corporatist strategy in which capital, specifically agribusiness, is the dominant element. The result is that the connections between epidemiological crises and the capitalist world economy are systematically downplayed in what purports to be a holistic model.[9]

There thus arose in response a new, revolutionary approach to the etiology of disease, known as Structural One Health, building critically on One Health, but rooted in the broad historical-materialist tradition. For proponents of Structural One Health the key is to ascertain how pandemics in the contemporary global economy are connected to the circuits of capital that are rapidly changing environmental conditions. A team of scientists, including Rodrick Wallace, Luis Fernando Chaves, Luke R. Bergmann, Constância Ayres, Lenny Hogerwerf, Richard Kock, and Robert G. Wallace, have together written a series of works such as *Clear-Cutting Disease Control: Capital-Led Deforestation, Public Health Austerity, and Vector-Borne Infection* and, more recently, "COVID-19 and Circuits of Capital" (by Rob Wallace, Alex Liebman, Luis Fernando Chaves, and Rodrick Wallace) in the May 2020 issue of *Monthly Review*. Structural One Health is defined as "a new field, [which] examines the impacts global circuits of capital and other fundamental contexts, including deep cultural histories, have upon regional agroeconomics and associated disease dynamics across species."[10]

The revolutionary historical-materialist approach represented by Structural One Health departs from the mainstream One Health approach by (1) focusing on commodity chains as drivers of pandemics; (2) discounting the usual "absolute geographies" approach that concentrates on certain locales in which novel viruses emerge while failing to perceive the global economic conduits of transmission; (3) seeing the pandemics not as an episodic problem, or random "black swan" events, but rather as reflecting a general structural crisis of capital, in the sense explicated by István Mészáros in his *Beyond Capital*; (4) adopting the approach of dialectical biology associated with Harvard biologists Richard Levins and Richard Lewontin in *The Dialectical*

Biologist; and (5) insisting on the radical reconstruction of society at large in ways that would promote a sustainable "planetary metabolism."[11] In his *Big Farms Make Big Flu* and other writings, Robert G. (Rob) Wallace draws on Marx's notions of commodity chains and the metabolic rift, as well as the critique of austerity and privatization based in the notion of the Lauderdale Paradox (according to which private riches are enhanced by the destruction of public wealth). Thinkers in this critical tradition thus rely on a dialectical approach to ecological destruction and the etiology of disease.[12]

Naturally, the new historical-materialist epidemiology did not appear out of thin air, but was built on a long tradition of socialist struggles and critical analyses of epidemics, including such historic contributions as (1) Frederick Engels's *Conditions of the Working Class in England*, which explored the class basis of infectious diseases; (2) Marx's own discussions of epidemics and general health conditions in *Capital*; (3) the British zoologist (Charles Darwin and Thomas Huxley's protégé and Marx's friend) E. Ray Lankester's treatment of the anthropogenic sources of disease and their basis in capitalist agriculture, markets, and finance in his *Kingdom of Man*; and (4) Levins's "Is Capitalism a Disease?"[13]

Especially important in the new historical-materialist epidemiology associated with Structural One Health is the explicit recognition of the role of global agribusiness and integration of this with detailed research into every aspect of the etiology of disease, focusing on the new zoonoses. Such diseases, as Rob Wallace stated in *Big Farms Make Big Flu*, were the "inadvertent biotic fallout of efforts aimed at steering animal ontogeny and ecology to multinational profitability," producing new deadly pathogens.[14] Offshore farming consisting of monocultures of genetically similar domestic animals (eliminating immune firebreaks), including massive hog feedlots and vast poultry farms coupled with rapid deforestation and the chaotic mixing of wild birds and other wildlife with industrial animal production—not excluding wet markets—have created the conditions for the spread of new deadly pathogens such as SARS, MERS, Ebola, H1N1, H5N1, and now SARS-CoV-2. Over half a million people globally died of

H1N1 while the deaths from SARS-CoV-2 have already exceeded that by a factor of ten.[15]

"Agribusinesses," Rob Wallace writes, "are moving their companies into the Global South to take advantage of cheap labor and cheap land," and "spreading their entire production line across the world."[16] Avians, hogs, and humans all interact to produce new diseases. "Influenzas," Wallace tells us, "now emerge by way of a globalized network of corporate feedlot production and trade, wherever specific strains first evolve. With flocks and herds whisked from region to region—transforming spatial distance into just-in-time expediency—multiple strains of influenza are continually introduced into localities filled with populations of susceptible animals."[17] Large-scale commercial poultry operations have much higher odds of hosting these virulent zoonoses. Value-chain analysis has been used to trace the etiology of new influenzas such as H5N1 along the poultry production commodity chain.[18] Influenza outbreaks in southern China emerge in the context of "a 'historical present' within which multiple virulent recombinants arise out of a mélange of agroecologies originating at different times by both path dependence and contingency: in this case, ancient (rice), early modern (semi-domesticated ducks), and present-day (poultry intensification)." This analysis has been extended by radical geographers, such as Bergmann, working on "the convergence of biology and economy beyond a single commodity chain and up into the fabric of the global economy."[19]

The interconnected global commodity chains of agribusiness, which provide the bases for the appearance of novel zoonoses, ensure that these pathogens move rapidly from one place to another, exploiting the chains of human connection and globalization, with the human hosts moving in days, even hours, from one part of the globe to the other. Wallace and his colleagues write in "COVID-19 and Circuits of Capital": "Some pathogens emerge right out of centers of production. . . . But many like COVID-19 originate on the frontiers of capital production. Indeed, at least 60 percent of novel human pathogens emerge by spilling over from wild animals to local human communities (before the more successful ones spread to the rest of the world)." As they sum up the conditions of the transmission of these diseases,

> the underlying operative premise is that the cause of COVID-19 and other such pathogens is not found just in the object of any one infectious agent or its clinical course, but also in the field of ecosystemic relations that capital and other structural causes have pinned back to their own advantage. The wide variety of pathogens, representing different taxa, source hosts, modes of transmission, clinical courses, and epidemiological outcomes, have all the earmarks that send us running wild-eyed to our search engines upon each outbreak, and mark different parts and pathways along the same kinds of circuits of land use and value accumulation.[20]

The imperial restructuring of production in the late twentieth and early twenty-first centuries—which we know as globalization—was the result primarily of global labor arbitrage and the overexploitation (and superexploitation) of workers in the Global South, including the purposeful contamination of the local environments, for the benefit primarily of the centers of world capital and finance. But it was also driven in part by global land arbitrage that took place simultaneously through multinational agribusiness corporations. According to Eric Holt-Giménez in *A Foodie's Guide to Capitalism*, "the price of land" in much of the Global South "is so low in relation to its land rent (what it is worth for what it can produce) that the capture of the difference (arbitrage) between low price and high land rent will provide investors with a handsome profit. Any benefits from actually growing crops are secondary to the deal.... Land arbitrage opportunities come about by bringing new land—with an attractive land rent—into the global land market where rents can actually be capitalized."[21] Much of this was fed by what is called the Livestock Revolution, which made livestock into a globalized commodity based on giant feedlots and genetic monocultures.[22]

These conditions have been promoted by various development

banks in the context of what is euphemistically known as "territorial restructuring," which involves removing subsistence farmers and small producers from the land at the behest of multinational corporations, primarily agribusinesses, as well as rapid deforestation and ecosystem destruction. These are also known as twenty-first-century land grabs, accelerated by high prices for basic foods in 2008 and again in 2011, as well as private wealth funds seeking tangible assets in the face of uncertainty after the Great Financial Crisis of 2007–2009. The result is the greatest mass migration in human history, with people being thrown off the land in a global process of depeasantization, altering the agroecology of whole regions, replacing traditional agriculture with monocultures, and pushing populations into urban slums.[23]

Rob Wallace and his colleagues observe that historian and critical-urban theorist Mike Davis and others "have identified how these newly urbanizing landscapes act as both local markets and regional hubs for global agricultural commodities passing through.... As a result, forest disease dynamics, the pathogens' primeval sources, are no longer constrained to the hinterlands alone. Their associated epidemiologies have themselves turned relational, felt across time and space. A SARS can suddenly find itself spilling over into humans in the big city only a few days out of its bat cave."[24]

Commodity Chain Disruption and the Global Bullwhip Effect

The new pathogens generated unintentionally by agribusiness are not themselves natural-material use values, but rather are toxic residues of the capitalist production system, traceable to agribusiness commodity chains as part of a globalized food regime.[25] Yet, in a kind of metaphorical "revenge" of nature as first depicted by Engels and Lankester, the ripple effects of combined ecological and epidemiological disasters introduced by today's global commodity chains and the actions of agribusiness, giving rise to the COVID-19 pandemic, have disrupted the entire system of global production.[26]

The effect of lockdowns and social distancing, shutting down production in key sectors of the globe, has shaken supply/value chains internationally. This has generated a gigantic "bullwhip effect" rippling up from both the supply and demand ends of the global commodity chains.[27] Moreover, the COVID-19 pandemic has occurred in the context of a global regime of neoliberal monopoly-finance capital that has imposed worldwide austerity, including on public health. The universal adoption of just-in-time production and time-based competition in the regulation of global commodity chains has left corporations and facilities such as hospitals with few inventories, a problem compounded by urgent stockpiling of some goods on the part of the population.[28] The result is extraordinary dislocation of the entire global economy.

Today's global commodity chains—or what we call *labor-value chains*—are organized primarily in order to exploit lower unit labor costs (taking into account both wage costs and productivity) in the poorer countries of the Global South where world industrial production is now predominantly located. Unit labor costs in India in 2014 were 37 percent of the U.S. level, while China's and Mexico's were 46 and 43 percent, respectively. Indonesia was higher with unit labor costs at 62 percent of the U.S. level.[29] Much of this is due to the extremely low wages in countries in the South, which are only a small fraction of the wage levels of those in the North. Meanwhile, arm's-length production carried out under multinational corporation specifications, along with advanced technology introduced into the new export platforms in the Global South, generates productivity on levels comparable in many areas to that of the Global North. The result is an integrated global system of exploitation in which the differences in wages between countries in the Global North and the Global South are greater than the difference in productivities, leading to very low unit labor costs in countries in the South and generating enormous gross profit margins (or economic surplus) on the export price of goods from the poorer countries.

The enormous economic surpluses generated in the Global South are logged in gross domestic product accounting as *value added* in the

North. However, they are better understood as *value captured* from the South. This whole new system of international exploitation associated with the globalization of production constitutes the deep structure of late imperialism in the twenty-first century. It is a system of world exploitation/expropriation formed around the global labor arbitrage, resulting in a vast drain of value generated from the poor to the rich countries.

All of this was facilitated by revolutions in transportation and communication. Shipping costs dived as standardized shipping containers proliferated. Communication technologies such as fiber-optic cables, mobile phones, the Internet, broadband, cloud computing, and video conferencing altered global connectivity. Air travel cheapened rapid travel, annually growing by an average of 6.5 percent between 2010 and 2019.[30] Around a third of U.S. exports are intermediate products for final goods produced elsewhere, such as cotton, steel, engines, and semiconductors.[31] It is out of these rapidly changing conditions, generating an increasingly integrated, hierarchical international accumulation structure, that the present global commodity-chain structure arose. The result was the connecting of all parts of the globe within a world-system of oppression, a connectivity that is now showing signs of destabilizing under the impacts of the U.S. trade war against China and the global economic effects of the COVID-19 pandemic.

The COVID-19 pandemic, with its lockdowns and social distancing, is "the first global supply-chain crisis."[32] This has led to losses in economic value, vast unemployment and underemployment, corporate collapse, increased exploitation, and widespread hunger and deprivation. Key to understanding both the complexity and chaos of the present crisis is the fact that no CEO of a multinational corporation anywhere has a complete map of the firm's commodity chain.[33] Usually, the financial centers and procurement officers in corporations know their first-tier suppliers, but not their second-tier (that is, the suppliers of their suppliers), much less the third- or even fourth-tier suppliers. As Elisabeth Braw writes in *Foreign Policy*, "Michael Essig, a professor of supply management at the Bundeswehr University of Munich calculated that a multinational company such as Volkswagen has 5,000 suppliers (the so-called tier-one suppliers), each with an average of 250 tier-two suppliers. That means that the company actually has 1.25 million suppliers—the vast majority of whom it doesn't know." Moreover, this leaves out the third-tier suppliers. When the novel coronavirus outbreak occurred in Wuhan in China, it was discovered that 51,000 companies globally had at least one direct supplier in Wuhan, while five million companies had at least one two-tier supplier there. On February 27, 2020, when the supply chain disruption was still largely centered on China, the World Economic Forum, citing a report by Dun & Bradstreet, declared that more than 90 percent of the Fortune 1000 multinational corporations had a tier-one or tier-two supplier affected by the virus.[34]

The effects of SARS-CoV-2 have made it urgent for corporations to try to map their entire commodity chains. But this is enormously complex. When the Fukushima nuclear disaster occurred, it was discovered that the Fukushima area produced 60 percent of the world's critical auto parts, a large share of world lithium battery chemicals, and 22 percent of the world's .300-millimeter silicon wafers, all crucial to industrial production. Attempts were made at that time by some monopoly-finance corporations to map their supply chains. According to the *Harvard Business Review*, "Executives of a Japanese semiconductor manufacturer told us it took a team of a 100 people more than a year to map the company's supply networks deep into the sub-tiers following the earthquake and tsunami [and the Fukushima nuclear disaster] in 2011."[35]

Faced with commodity chains in which many of the links in the chain are invisible, and where the chains are breaking in numerous places, corporations are faced with interruptions and uncertainties in what Marx called the "chain of metamorphoses" in the production, distribution, and consumption of material products, coupled with erratic changes in overall supply demand. The scale of the coronavirus pandemic and its consequences on world accumulation are unprecedented, with the global economic costs still increasing. At the end of March, some three billion people on the planet were in lockdown or social-distancing mode.[36] Most corporations have no emergency plan for dealing with the multiple breaks in their supply chains.[37] The scale of the problem had manifested itself in the early months of 2020 in tens of thousands of force majeure declarations, beginning first in China and then spreading elsewhere, where various suppliers indicated they were unable to fulfill contracts due to extraordinary external events. This was accompanied by numerous "blank sailings" standing for scheduled voyages of cargo ships that were canceled with the goods being held up due to either failure of supply or demand.[38] In early April, the U.S. National Retail Federation indicated that March 2020 saw a five-year low in the shipment of twenty-foot equivalents (of containers) in ship cargo, with shipments expected to plummet much more rapidly from that point.[39] Airline passenger flights all over the world decreased by around 90 percent, leading the major U.S. airlines to leverage "the bellies and passenger cabins of their aircraft [in order to redirect them] for cargo flights, often removing seats and using the empty tracks to secure cargo."[40]

According to estimates in March 2021 by the Organization of Economic Cooperation and Development (OECD), the economic fallout from the COVID-19 pandemic led to a drop in annual world merchandise trade in 2020 of 8 percent, and a drop in the world service trade that same year of 18 percent. This represented a plummet in world trade in a single year that exceeded that of all other annual declines in the post–Second World War period, including during the 2008–2010 Great Financial Crisis.[41]

The dire effects of the disruption of global supply chains during the pandemic have been particularly evident with respect to medical equipment. Premier, one of the chief general purchasing organizations for hospitals in the United States, indicated that it normally purchases up to 24 million N95 respirators (masks) per year for its member health care providers and organizations, while in January and February 2020 alone its members used 56 million respirators. In late March, Premier was ordering 110 to 150 million respirators, while its member organizations such as hospitals and nursing homes when surveyed indicated they had barely more than a week's supply. The demand for medical masks soared while the global supply froze up.[42] COVID-19 test kits were also in chronically short supply globally until China revved up production in late March.[43]

Not the least of the vulnerabilities exposed is what is known as *supply-chain finance*, which allows corporations to defer payments to suppliers, with the help of bank finance. According to the *Wall Street Journal*, some corporations have supply-chain financing obligations that dwarf their reported net debt. These debts owed to suppliers are sold by other financial interests in the form of short-term notes. Credit Suisse owns notes that are owed by large U.S. corporations such as Kellogg and General Mills. With a general disruption of commodity chains, this intricate chain of finance, which is itself the object of speculation, is inherently placed in a crisis mode itself, creating additional vulnerabilities in an already fragile financial system.[44]

Imperialism, Class, and the Pandemic

SARS-CoV-2, like other dangerous pathogens that have emerged or reemerged in recent years, is closely related to a complex set of factors including: (1) the development of global agribusiness with its expanding genetic monocultures that increase susceptibility to the contraction of zoonotic diseases from wild to domestic animals to humans; (2) destruction of wild habitats and disruption of the activities of wild species; and (3) human beings living in closer proximity. There is little doubt that global commodity chains and the kinds of connectivity they have produced have become vectors for the rapid transmission of disease, throwing this whole globally exploitative pattern of development into question. As Stephen Roach of the Yale School of Management, formerly chief economist of Morgan Stanley and the principal originator of the global labor arbitrage concept, has written in the context of the coronavirus crisis, that what the financial headquarters of corporations wanted was "low-cost goods irrespective of what those cost efficiencies entailed in terms of [the lack of] investing in public health, or I would also say [the lack of] investing in environmental protection and the quality of the climate." The result of such an unsustainable approach to "cost efficiencies" is the contemporary global ecological and epidemiological crises and their financial consequences, further destabilizing a system that was already exhibiting an "excessive surge" characteristic of financial bubbles.[45]

Initially, rich countries are at the epicenter of the COVID-19 pandemic and financial fallout, but the overall crisis, incorporating its economic as well as epidemiological effects, will inevitably hit poor countries harder. How a planetary crisis of this kind is handled is ultimately filtered through the imperial-class system. In March 2020, the COVID-19 Response Team of Imperial College in London issued a report indicating that in a global scenario in which SARS-CoV-2 was unmitigated, with no social distancing or lockdowns, 40 million people in the world would die, with higher mortality rates in the rich countries than in poor countries because of the larger proportions of the population that were sixty-five or older, as compared with poor countries. This analysis ostensibly took into account the greater access to medical care in rich countries. But it left out factors like malnutrition, poverty, and the greater susceptibility to infectious diseases in poor countries. Nevertheless, the Imperial College estimates, based on these assumptions, indicated that in an unmitigated scenario the number of deaths would be in the range of 15 million in East Asia and the Pacific, 7.6 million people in South Asia, 3 million people in Latin America and the Caribbean, 2.5 million people in Sub-Saharan Africa, and 1.7 million in the Middle East and North Africa—as compared with 7.2 million in Europe and Central Asia and around 3 million in North America.[46]

Basing their analysis on the Imperial College's approach, Ahmed Mushfiq Mobarak and Zachary Barnett-Howell at Yale University wrote an article for the establishment journal *Foreign Policy* titled "Poor Countries Need to Think Twice About Social Distancing." In their article, Mobarak and Barnett-Howell were very explicit, arguing that "epidemiological models make clear that the cost of not intervening in rich countries would be in the hundreds of thousands to millions dead, an outcome far worse than the deepest economic recession imaginable. In other words, social distancing interventions and aggressive suppression, even with their associated economic costs, are overwhelmingly justified in high-income societies"—to save lives. However, the same is not true, they suggested, for poor countries, since they have relatively few elderly individuals in their populations as a whole, generating, according to the Imperial College estimates, only around half the mortality rate. This model, they admit, "does not account for the greater prevalence of chronic illnesses, respiratory conditions, pollution, and malnutrition in low-income countries, which could increase the fatality rates from coronavirus outbreaks." But largely ignoring this in their article (and in a related study conducted through the Yale Economics Department), these authors insist that it would be better, given the impoverishment and vast unemployment and underemployment in these countries, for the populations not to practice social distancing or aggressive testing and suppression, and to put their efforts into economic production, presumably keeping intact the global supply chains that primarily start upstream in low-wage countries.[47] No doubt the deaths of millions of people in the Global South is considered by these authors to be a reasonable tradeoff for the continued growth of the empire of capital.

As Mike Davis wrote at the outset of the pandemic, twenty-first-century capitalism points to "a permanent triage of humanity . . . dooming part of the human race to eventual extinction." He asked:

> But what happens when COVID spreads through populations with minimal access to medicine and dramatically higher levels of poor nutrition, untended health problems and damaged immune systems? The age advantage will be worth far less to poor youth in African and South Asian slums.
>
> There's also some possibility that mass infection in slums and poor cities could flip the switch on coronavirus's mode of infection and reshape the nature of the disease. Before SARS emerged in 2003, highly pathogenic coronavirus epidemics were confined to domestic animals, above all pigs. Researchers soon recognized two different routes of infection: fecal-oral, which attacked the stomach and intestinal tissue, and respiratory, which attacked the lungs. In the first case, there was usually very high mortality,

while the second generally resulted in milder cases. A small percentage of current positives, especially the cruise ship cases, report diarrhea and vomiting, and, to quote one report, "the possibility of SARS-CoV-2 transmission via sewage, waste, contaminated water, air conditioning systems and aerosols cannot be underestimated."

The pandemic has now reached the slums of Africa and South Asia, where fecal contamination is everywhere: in the water, in the home-grown vegetables, and as windblown dust. (Yes, shit storms are real.) Will this favor the enteric route? Will, as in the case of animals, this lead to more lethal infections, possibly across all age groups?[48]

Although this prospect did not emerge, Davis's argument made plain the gross immorality of a position that says social distancing and aggressive suppression of the virus in response to the pandemic should take place in rich countries and not poor. Such imperialist epidemiological strategies are all the more vicious in that they take the poverty of the populations of the Global South, the product of imperialism, as the justification for a Malthusian or social Darwinist approach, in which millions would die in order to keep the global economy growing, primarily for the benefit of those at the apex of the system. Contrast this to the approach adopted in socialist-led Venezuela, the country in Latin America with the least number of deaths per capita from COVID-19, where collectively organized social distancing and social provisioning is combined with expanded personalized screening to determine who is most vulnerable, widespread testing, and expansion of hospitals and health care, developing on the Cuban and Chinese models.[49]

Economically, the Global South as a whole, quite apart from the direct effects of the pandemic, is destined to pay the highest cost. The breakdown of global supply chains due to canceled orders in the Global North (as well as social distancing and lockdowns around the globe) and the refashioning of commodity chains that will follow, will leave whole countries and regions devastated.[50]

Here, it is crucial to recognize as well that the COVID-19 pandemic has come in the middle of an economic war for global hegemony unleashed by the Donald Trump administration and directed at China, which has accounted for some 37 percent of all cumulative growth of the world economy since 2008.[51] This is seen by the Trump and Biden administrations as a war by other means. As a result of the tariff war, many U.S. companies had already pulled their supply chains out from China. Levi's, for example, has reduced its manufacturing in China from 16 percent in 2017 to 1–2 percent in 2019. In the face of the tariff war and the COVID-19 pandemic, two-thirds of 160 executives surveyed across industries in the United States have recently indicated that they had already moved, were planning to move, or were considering moving their operations from China to Mexico, where unit labor costs are now comparable and where they would be closer to U.S. markets.[52] Washington's economic war against China is currently so fierce that the Trump administration refused to drop the tariffs on personal protection equipment, essential to medical personnel, until late March.[53] Trump meanwhile appointed Peter Navarro, the economist in charge of his economic war for hegemony with China, as head of the Defense Production Act to deal with the COVID-19 crisis.

In his roles in directing the U.S. trade war against China and as policy coordinator of the Defense Production Act, Navarro accused China of introducing a "trade shock" that lost "over five million manufacturing jobs and 70,000 factories" and "killed tens of thousands of Americans" by destroying jobs, families, and health. He is now declaring that this has been followed by a "China virus shock."[54] On this propagandist basis, Navarro proceeded to integrate U.S. policy with respect to the pandemic around the need to fight the so-called China virus and pull U.S. supply chains out of China. Yet, since about a third of all global intermediate manufacturing products are currently produced in China, most heavily in the high-tech sectors, and since this remains key to the global labor arbitrage, the attempt at such restructuring will be vastly disruptive, to the extent that it is possible at all.[55]

Some multinationals that had moved their production out of China learned the hard way later that the decision did not "free" them from their dependency on it. Samsung, for example, has started flying electronic components from China to its factories in Vietnam—a destination for companies that are eager to escape the trade-war tariffs. But Vietnam was also found to be vulnerable, because they rely heavily on China for materials or intermediate parts.[56] Similar cases happened in neighboring Southeast Asian countries. China is Indonesia's biggest trading partner, and roughly 20 to 50 percent of the country's raw materials for industries come from China. In February, factories in Batam, Indonesia, already had to deal with raw materials from China drying up (which counts for 70 percent of what was produced in that region). Companies there said that they considered getting materials from other countries but "it's not exactly easy." For many factories, the feasible option was to "cease operations completely."[57] Capitalists like Cao Dewang, the Chinese billionaire founder of Fuyao Glass Industry, predicted the weakening of China's role in the global supply chain after the pandemic but concluded that, at least in the short term, "it's hard to find an economy to replace China in the global industry chain"—citing many difficulties from "infrastructure shortcomings" in Southeast Asian countries, higher labor costs in the Global North, and the obstacles that "rich countries" have to face if they want to "rebuild manufacturing at home."[58]

The COVID-19 crisis is not to be treated as the result of an external force or as an unpredictable "black swan" event, but rather belongs to a complex of crisis tendencies that are broadly predictable, though not in terms of actual timing. Today, the center of the capitalist system is confronted with secular stagnation in terms of production and investment, relying for its expansion and amassing of wealth at the top on historically low interest rates, high amounts of debt, the drain of capital from the rest of the world, and financial speculation. Income and wealth inequality are reaching levels for which there is no historical analogue. The rift in world ecology has attained planetary proportions and is creating a planetary environment that no longer constitutes a safe place for humanity. New pandemics are arising on the basis of a system of global monopoly-finance capital that has made itself the main vector of disease. State systems everywhere are regressing toward higher levels of repression, whether under the mantle of neoliberalism or neofascism.

The extraordinarily exploitative and destructive nature of the system is evident in the fact that blue-collar workers everywhere were declared essential critical infrastructure workers (a concept formalized in the United States by the Department of Homeland Security) and were expected to carry out production mostly without protective gear while the more privileged and dispensable classes socially distance themselves.[59] A true lockdown would be much more extensive and would require state provisioning and planning, ensuring that the whole population was protected, rather than focusing on bailing out financial interests. It is precisely because of the class nature of social distancing, as well as access to income, housing, resources, and medical care, that increased mortality from COVID-19 in the United States has been falling primarily on populations of color, where conditions of economic and environmental injustice are most severe.[60]

Social Production and the Planetary Metabolism

Fundamental to Marx's materialist outlook was what he called "the hierarchy of ... needs."[61] This meant that human beings were material beings, part of the natural world, as well as creating their own social world within it. As material beings they had to satisfy their material needs first—eating and drinking, providing food, shelter, clothing, and the basic conditions of healthy existence, before they pursued their higher developmental needs, necessary for the full realization of human potential.[62] Yet, in class societies it was always the case that the vast majority, the real producers, were relegated to conditions in which they were caught in a constant struggle to meet their most basic needs. This has not fundamentally changed. Despite the enormous wealth created over centuries of growth, millions upon millions of people in even the wealthiest capitalist society remain in a precarious condition in relation to such basics as food security, housing, clean water, health care, and transportation—under conditions in which three billionaires in the United States own as much wealth as the bottom half of the population.

Meanwhile, local and regional environments have been put in danger—as have all of the world ecosystems and the Earth System itself as a safe place for humanity. An emphasis on global "cost efficiencies" (a euphemism for cheap labor and cheap land) has led multinational capital to create a complex system of global commodity chains, designed at every point to maximize the over-/superexploitation of labor on a worldwide basis, while also turning the entire world into a real estate market, much of it as a field for operation of agribusiness. The result has been a vast draining of surplus from the periphery of the global system and a plundering of the planetary commons. In the narrow system of value accounting employed by capital, most of material existence, including the entire Earth System and the social conditions of human beings, insofar as these do not enter the market, are considered externalities, to be robbed and despoiled in the interest of capital accumulation. What has mistakenly been characterized as "the tragedy of the commons" is better understood, as Guy Standing pointed out in *Plunder of the Commons*, as "the tragedy of privatizing." Today, the famous Lauderdale Paradox, introduced by the Earl of Lauderdale in the early nineteenth century, in which public wealth is destroyed for the enhancement of private riches, has the entire planet as its field of operation.[63]

The circuits of capital of late imperialism have taken these tendencies to their fullest extent, generating a rapidly developing planetary ecological crisis that threatens to engulf human civilization as we know it—a perfect storm of catastrophe. This comes on top of a system of accumulation that is divorced from any rational ordering of needs for the population independent of the cash nexus.[64] Accumulation and the amassing of wealth in general are increasingly dependent on the proliferation of waste of all kinds. In the midst of this disaster, a New Cold War and a growing likelihood of thermonuclear destruction have emerged, with an increasingly unstable and aggressive United States at the forefront. This has led the *Bulletin of Atomic Scientists* to move its famous doomsday clock to 100 seconds to midnight, the closest to midnight since the clock started in 1947.[65]

The COVID-19 pandemic and the threat of increasing and more deadly pandemics is a product of this same late-imperialist development. Chains of global exploitation and expropriation have destabilized not only ecologies but the relations between species, creating a toxic brew of pathogens. All of this can be seen as arising from the introduction of agribusiness with its genetic monocultures; massive ecosystem destruction involving the uncontrolled mixture of species; and a system of global valorization based on treating land, bodies, species, and ecosystems as so many "free gifts" to be expropriated, irrespective of natural and social limits.

Nor are new viruses the only emerging global health problem. The overuse of antibiotics within agribusiness as well as modern medicine has led to the dangerous growth of bacterial superbugs generating increasing numbers of deaths, which by midcentury could surpass annual cancer deaths, and inducing the World Health Organization to declare a "global health emergency."[66] Since communicable diseases, due to the unequal conditions of capitalist class society, fall heaviest on the working class and the poor, and on populations in the periphery, the system that generates such diseases in the pursuit of quantitative wealth can be charged, as Engels and the Chartists did in the nineteenth century, with social murder. As the revolutionary developments in epidemiology represented by One Health and Structural One Health have suggested, the etiology of the new pandemics can be traced to the overall problem of ecological destruction brought on by capitalism.

Here, the necessity of a "revolutionary reconstitution of society at large" rears its head once again, as it has so many times in the past.[67] The logic of contemporary historical development points to the need for a more communal, commons-based system of social metabolic reproduction, one in which the associated producers rationally regulate their social metabolism with nature, so as to promote free development of each as the basis of the free development of all, while conserving energy and the environment.[68] The future of humanity in the twenty-first century lies not in the direction of increased economic and ecological exploitation/expropriation, imperialism, and war. Rather, what Marx called "freedom in general" and the preservation of a viable "planetary metabolism" are the most pressing necessities today in determining the human present and future, and even human survival.[69]

CHAPTER TWENTY

Ecological Catastrophe or Ecological Civilization

It is common today to argue that the Holocene Epoch in geological history has given way to a new geological Epoch of the Anthropocene, in which human beings are increasingly the main geological force affecting the Earth System, overshadowing all other factors. The Anthropocene marks a Great Climacteric or *age of epochal transition* in human history. From the perspective of historical social science, this presents us with a fundamental problem: a crisis of civilization.[1] Not only does the growing rift in the Earth System threaten, with the continuation of capitalist business as usual, the entire realm of human civilization, in the sense of an advanced, ordered society; it could potentially undermine the conditions of human life itself, as well as that of innumerable other species.[2] In this way, the Anthropocene represents an unprecedented challenge.

Civilization—the rise of which was preceded by settled agriculture and the growth of an economic surplus, and which is commonly associated with the development of writing and class-based urban society—had its origins in the geological epoch of the Holocene. Although certain regional civilizations have collapsed in the past, partly due to ecological factors, overall the Holocene was conducive to the rise of highly ordered societies. In contrast, the Anthropocene Epoch, arising as a result of capitalist development, raises the question of a crisis on a planetary scale and of the necessary forging of an *ecological civilization,* that is, the shift to a society of substantive equality and ecological sustainability. It is no accident that this has been recognized most fundamentally in China, which, despite its own deep ecological contradictions, draws on the theoretical legacy of historical materialism—adding to its long civilizational-cultural dynamic a materialist revolutionary outlook. Nor should it surprise us that the question of ecological civilization has become pivotal to Marxian theory in the twenty-first century, building on its classical roots, and is now engendering a worldwide ecosocialist movement.

To address the historical specificity of the crisis of contemporary civilization and the challenge of creating an ecological civilization, it is necessary to begin with the historical conditions marked by today's Great Capitalist Climacteric. Only then can we address the necessary epochal transition before us. Moreover, it is important to recognize that this is linked to the question of capitalism and socialism, that is, the organization of production: the metabolism of nature and society.

Approaching the question of ecological civilization from a Marxian perspective, moreover, requires, first and foremost, a critical outlook on the concept of civilization itself, recognizing its historical class-based character and how this is related to our present ecological crisis. To speak of the making of an ecological civilization is also to evoke the long transition from capitalism to socialism.

The Great Capitalist Climacteric

It is commonplace today to refer to a global ecological crisis. Yet, the concept of crisis, though indispensable, is inadequate by itself to express the full extent of the ecological challenge that humanity faces in the Anthropocene Epoch.[3] The world is facing a prolonged Earth System emergency, requiring a radical transition in the social metabolism of humanity and nature. The Anthropocene crisis of today has been defined as an "anthropogenic rift" in the Earth System endangering civilization and innumerable species, including our own, associated with the crossing of planetary boundaries.[4] From a historical materialist perspective, this raises the question of a social-environmental transition that must accelerate in a revolutionary manner if the challenge of the Anthropocene is to be met. Such an ecological and social revolution would necessarily constitute a protracted process, occurring in stages with all sorts of advances and retreats.[5] The material roots of this transition in the social-environmental relation have long been in the making and more generally have their basis in the development of capitalism and class-based civilization.

Here it is useful to turn to the notion of "The Great Climacteric" introduced by environmental geographers Ian Burton and Robert Kates in the 1980s to refer to the social aspects of the changing human relation to the environment beginning with the Industrial Revolution in the late eighteenth century.[6] Burton and Kates figuratively used the date of Thomas Malthus's 1798 *An Essay on the Principle of Population* as the starting point for the Great Climacteric and saw this as extending to the year 2048, 250 years later.[7] A Great Climacteric in this sense represents a long period of crisis of a whole society or civilization associated with changing relations of production and environmental relations.

As Burton and Kates wrote: "Applied to population, resources, and environment throughout the world," the notion of a Great Climacteric "captures the idea of a period that is critical and where serious change for the worse may occur. It is a time of unusual danger."[8] It also raises the issue of the search for "a new global equilibrium" between human beings and the planet. Such a new, dynamic equilibrium state, involving evolutionary change, along with relatively stable, resilient relations, they argued, would need to be reached by the mid-twenty-first century. The idea of a Great Climacteric in society, viewed in this way (but without the Malthusian framework), is consistent with the emergence of the Anthropocene—seen as having its seeds in the Industrial Revolution with the rise of fossil fuels, but coming into being only with the Great Acceleration in the post–Second World War era. The most dramatic stratigraphic traces of the Great Acceleration marking the Anthropocene are to be found in radionuclides from aboveground nuclear weapons testing in the late 1940s and 1950s.[9] The Great Climacteric thus stands both for a planetary emergency and for the necessity of a *social-historical transition* to transform the human relation to the Earth System so as to conform to the requirements of sustainable human development.[10]

Given that the world economy has now reached a scale where its normal operations, such as current forms of energy use, threaten to disrupt the biogeochemical cycles of the entire Earth System,[11] it is clear that some kind of adjustment will be necessary between what Karl Marx called the "social metabolism" of production and the "universal metabolism of nature"—in the direction of a more sustainable society.[12] Otherwise the Earth System will necessarily impose its own limits on human society, leading to the demise of civilization, with untold costs to our own and other species.

As Marxian economist Paul Sweezy wrote back in 1989, "The general nature of the [environmental] crisis," can be seen

> as a radical (and growing) disjunction between on the one hand the demands placed on the environment by the modern global economy, and on the other hand the capacity of the natural forces embedded in the environment to meet these demands.

He depicted the capitalist system as a juggernaut aimed at ever-greater accumulation of capital as an end in itself. Individual capitals were "checked," in this expansive drive,

> only... by the impersonal forces of the market and in the longer run, when the market fails, by devastating crises. Implicit in the very concept of this system are interlocked and enormously powerful drives to both creation and destruction. On the plus side, the creative drive relates to what humankind can get out of nature for its own uses; on the negative side, the destructive drive bears most heavily on nature's capacity to respond to the demands placed upon it. Sooner or later, of course, these two

drives are contradictory and incompatible. And since . . . the adjustment must come from the side of the demands imposed on nature rather than from the side of nature's capacity to respond to those demands, we have to ask whether there is anything about capitalism as it has developed over recent centuries to cause us to believe that the system could curb its destructive drive and at the same time transform its creative drive into a benign environmental force. The answer, unfortunately, is that there is absolutely nothing in the historic record to encourage such a belief.[13]

Sweezy saw these same creative/destructive drives as applying not simply to capitalism but also to the "really existing socialism" of his day. Here he argued that post-revolutionary societies—rising out of conditions of underdevelopment and having to survive in a larger and hostile capitalist world economy—were under pressure to emulate and catch up with the more advanced capitalist economies. Due to these historical conditions, the impact of post-revolutionary societies on the environment had thus far been scarcely distinguishable from that of their capitalist counterparts. Nevertheless, post-revolutionary societies did not have the same *inner* drive to environmental destruction—since they were not inherently ruled by the capital accumulation process.[14] The existence of planning, moreover, made it more likely that such countries could effectively address environmental problems once these were brought to the forefront of their societal agendas.[15]

This argument points to what we might call—going beyond Burton and Kates' formulation—the Great Capitalist Climacteric, which requires the reestablishment of a kind of equilibrium between production and the planet, transcending capitalism and the alienation of labor and nature. However, rather than a mere static equilibrium, this can be conceived as a dynamic one of coevolution or creative sustainability. A failure on the part of society to push beyond the status quo in order to construct such a sustainable order can only lead to *cumulative catastrophe*.[16] The threatened catastrophe is one of civilization and potentially of the human species, along with innumerable other living species. The way out at present can only be by means of a transitional ecological civilization, which while still carrying the vestiges of class society, moves toward socialism and sustainable human development.

The Critique of Civilization

The question of the crisis of civilization is continually being raised by today's science, confronted as it is with the reality of climate change. As Kevin Anderson of the Tyndall Centre for Climate Change in the UK stated in "Climate Change Going Beyond Dangerous—Brutal Numbers and Tenuous Hope," an increase in global average temperatures by 2°C is "beyond dangerous," while an increase by 4°C—the world to which we are presently heading under business as usual—threatens global civilization itself. "It is fair to say," he writes,

> based on many (and ongoing) discussions with climate change colleagues, that there is a widespread view that a 4°C future is *incompatible with any reasonable characterization of an organized, equitable, and civilized global community*. A 4°C future is also beyond what many people think we can reasonably adapt to. Besides the global society, such a future will also be devastating for many if not the majority of ecosystems. Beyond this, and perhaps more alarmingly, there is a possibility that a 4°C world would not be stable, and that it might lead to a range of "natural" feedbacks, pushing the temperatures still higher.[17]

In socialist terms, the ecological crisis of civilization, that the real threat of a 4°C (or even higher) future represents is most usefully viewed through the lens of the historical *critique of civilization*—a critique introduced into socialist theory by French Utopian socialist Charles Fourier, and later developed further in the work of thinkers such as Marx, Frederick Engels, and William Morris. The concept of "civilization" first appeared in eighteenth-century France, and soon after in England and Germany, and was closely related to the concept of culture.[18] It took its principal meaning from a contrast with barbarism, and in that sense grew out of the Latin distinction between "modes of life that are civil" identified with the city-state or *civitatis,* and the ways of life of barbarians—as that was articulated, for example, by the Roman geographer Strabo (ca. 64 BCE–AD 24).[19] In contrast, in today's social sciences and humanities, civilization "in the singular"—French historian Fernand Braudel wrote in his *A History of Civilizations*—is often taken to "denote something which all civilizations share, however unequally," namely cultural assets that are a product of development, such as "writing, mathematics, the cultivation of plants, and the domestication of animals."[20] The historical precondition for civilization, in this sense, was settled agriculture and the production of an economic surplus. Hence civilization, as a generic stage of human development, is commonly associated with the rise of written language, the urban revolution, the state, class divisions, and private property.[21]

Fourier was best known for his critique of civilization whereby he questioned the forms of property, production/overproduction, division of labor, wage slavery, poverty, and patriarchy associated especially with the latest phase of civilization, now known as capitalism. Civilization, in his view, represented both a higher stage of cultural and economic development (in comparison to what he called the stages of savagery and barbarism), and, at the same time in many ways a step backward or barbarism at a higher level—in that it stood for a more intense, while anarchic, form of production and exploitation.

Criticizing bourgeois civilization, Fourier wrote, "The vicious circle of industry has been so clearly recognized, that people on all sides are beginning to suspect it, and feel astonished *that, in civilization, poverty should be the offspring of abundance.*"[22] Jonathan Beecher wrote, "Fourier's critique of civilization"

> was nothing if not comprehensive. When the occasion demanded, he was able to list up to 144 "permanent vices" of civilization running from the slavery of the wage system to the "excitation of hurricanes and all sorts of climatic excesses."[23]

Hence, the term civilization had a complex, often pejorative, meaning in nineteenth-century socialist thought. Marx and Engels too used the term in this way—although frequently employing it more broadly to refer to pre-capitalist as well as capitalist class formations. As Sven Beckert states in *The Empire of Cotton*:

> Whether celebrating the material advances generated from slavery or calling for slavery's abolition, many contemporaries agreed by the 1850s that global economic development required physical coercion. Karl Marx sharpened the arguments made all around him by concluding in 1853 that "bourgeois civilization" and "barbarity" were joined at the hip.[24]

"Civilization," as far back as antiquity, Engels wrote, "was defined first and foremost by its sharpening of the opposition between town and country."[25] It was precisely this that created the rift in the social metabolism between civilization and the environment, which was intensified with the exploitation of the urban proletariat under capitalism.[26]

Other early Marxian thinkers, notably William Morris, were to expand on this critique of civilization. For Morris, it was the entire culture of class society, of which capitalism was only a manifestation, that was the target. "Revolution" was both "the intelligence of civilization" and its negation.[27] The countries within "the ring of civilization" for Morris were characterized by "organized misery"; they "were glutted with the abortions of the market, and force and fraud were used unsparingly to 'open up' countries *outside* that pale.... When the civilized World-Market coveted a country not yet in its clutches, some transparent pretext" was invented to justify an invasion.[28] Without denying the significance of civilization in the sense of the general advancement of culture, Morris compared it unfavorably in many ways to "non-civilisation," as this was cynically referred to in the West. For Morris, all of this reflected the class-based, imperialist nature of capitalism which saw itself as the epitome of culture and civilization, while perpetrating a greater barbarism.

When addressing ecological problems, Marx often saw them as reflecting the contradictions of civilization in the broader historical sense, as well as bourgeois society more specifically. Thus in writing in *Capital,* Volume 2, on deforestation, he stated: "The development of civilisation and industry in general has always shown itself so active in the destruction of forests that everything that has been done for their conservation and production is completely insignificant in comparison."[29] Likewise, in writing about land cover change and desertification, as this had appeared since ancient times, Marx famously observed, in relation to the work of the German agronomist Carl Fraas: "Climate and flora change in *historical* times," that is, in the period of civilization or written history. Indeed,

> with cultivation—depending on its degree—the "moisture" so beloved by the peasants gets lost (hence also the plants migrate from south to north).... The first effect of cultivation is useful, but finally devastating through deforestation, etc.... The conclusion is that cultivation—when it proceeds in natural growth and is not *consciously controlled* . . . leave deserts behind it, Persia, Mesopotamia, etc., Greece. So once again an unconscious socialist tendency![30]

Commenting on this passage by Marx on Fraas, leading Soviet climatologist Fedorov wrote:

> This quote can well apply to many present-day Western researchers of the problem of interaction between society and the natural environment. Just like Fraas they [Marx and Engels] feel that the spontaneously developing culture leads to a crisis in the relationship between society and nature, and their calculations provide a fairly good illustration of just how this may happen. And, just like Fraas, they display (possibly unconsciously) certain "socialist tendencies."[31]

Here Federov had in mind the implicit criticism of capitalism and contemporary civilization that pervaded Western ecological thought in the early 1970s. What was emerging was the recognition of the deep contradiction in the metabolism with nature as a whole, highlighted in the nineteenth century by Marx. Indeed, Marx and Engels, Federov emphasized, "regarded interaction (metabolism) between people and nature as a vital element of human life and activity."

Marx saw the "unconscious socialist tendency," arising from ecological degradation, as evident in the whole history of developing civilization, though manifesting itself fully only under capitalism.[32] His theory of metabolic rift was developed specifically to address the disruption in the relation between human beings and the soil that developed as a result of industrialized agriculture and the extreme division between town and country, as "the urban population . . . achieve[s] an ever-growing preponderance." Large-scale capitalist agriculture, he argued, progressively "disturbs the metabolic interaction between man and the earth."[33] It thus creates a "rift" in the soil nutrient cycle, thereby "robbing the soil," and "ruining the more long-lasting sources of that fertility."[34] By creating a break in the social metabolism between human beings and the earth, undermining the universal metabolism of nature, it disrupts the eternal-natural conditions of soil fertility. In this way

> it produces conditions that provoke an irreparable rift in the interdependent process of social metabolism, a metabolism prescribed by the natural laws of life itself. The result of this is a squandering of the vitality of the soil, which is carried by trade far beyond the bounds of a single country.[35]

The significance of this emerging ecological climacteric, for Marx, was that it disrupted the conditions of production that had allowed for the development of civilization up to that point. The answer to this civilizational crisis, coming to a head under capitalism, was the necessary creation of a socialist society in which the associated producers rationally regulated the metabolism between humanity and nature—doing so in a way that conserved their energy, and that fulfilled their own specific human-species needs. No individual, Marx observed, owns the earth, not even all the people on the globe own the earth; they are simply "its possessors, its beneficiaries" and are responsible for maintaining it for future generations as "boni patres familias" (good heads of the household).[36]

This sense of ecological crisis as a crisis not simply of capitalism but also representing a broader threat encompassing the entire range of human culture and civilization took a concrete form in Marx and Engels's writings on *Ireland and the Irish Question*. Here Marx gave material expression to the ecological destruction that the colonial-capitalist system was forcing upon workers and the dispossessed. This is evident in the "Record of a Speech on the Irish Question Delivered by Karl Marx to the German Workers' Educational Association in London on December 16, 1867." In examining the Great Irish Famine of 1845–46, Marx depicted it as a product in part of the despoliation of the soil resulting from destructive colonial intensification of the metabolic rift that he had already described in relation to English agriculture. "Fertilizers [soil nutrients] were exported with the produce and the rent, and the soil was exhausted. Famines often set in here and there, and owing to the potato blight there was a general famine in 1846. A million people died of starvation. The potato blight resulted from the exhaustion of the soil, it was a product of English rule."[37]

Marx, however, was not principally concerned here with the Great Irish Famine. Rather, his analysis focused on the subsequent transformation of Irish agriculture in the mid-nineteenth century, during which people were being replaced on the land by cattle and sheep under the colonial rationale that the land was unsuited for crops for people. In his notes to an earlier undelivered lecture on the Irish question, he referred to "the gradual deterioration and exhaustion of the source of national life, the soil."[38] Commenting on the decrease in yield per acre from 1847 to 1865, he pointed to a drop in the production of oats by 16 percent, flax by 48 percent, turnips by 36 percent, and potatoes by 50 percent.[39] With the exhaustion of the soil, the population deteriorated physically. There was "an *absolute increase* in the number of deaf-mutes, blind, insane, idiotic, and decrepit inhabitants" in the decreasing population.[40]

"The Irish question," Marx declared, "is therefore not simply a question of nationality but a question of land and existence. *Ruin or revolution is the watchword.*"[41] Here the question of revolution stemmed not simply from the nationality question, or colonialism, nor merely from labor exploitation; instead it raised the larger issue of the *ecological ruin* of the peasantry and agricultural laborers as an unconscious tendency to revolution, that is, the objective conditions forcing transformative change. It followed that not only class and nationality, but also ecological ruin, a general devastation threatening the entire culture, could constitute the basis for revolution.

Exterminism or Ecological Civilization

Similar considerations to those raised by Marx, borne of the historic critique of class-based civilization, coupled with a growing perception of global environmental destruction, led twentieth-century ecological and social critics like Lewis Mumford, E. P. Thompson, and Rudolf Bahro to refer to the growing ecological threat to civilization as a whole. For Mumford in *The City in History*: as a result of modern industrial development, particularly capitalism, "the very survival of civilization, or indeed of any large and unmutilated portion of the human race is now in doubt."[42] Earlier in *The Condition of Man*, he had observed: "What happened to Greece, Rome, China, or India has no parallel in the world today: when those civilizations collapsed they were surrounded by neighbors that had reached nearly equal levels of culture, whereas if Western civilization should continue its downward course it will spread ruin to every part of the planet."[43]

Mumford's view of civilization, although not strictly Marxist in form, was complex and radical, derived from the historical critique long embedded in socialist thought whereby civilization itself was regarded as a transitional, class-based cultural formation—both a form of advance and a new barbarism. As he wrote in 1966 in *The Myth of the Machine*:

I use the term "civilization" in quotation marks . . . to denote the group of institutions that first took form under kingship. Its chief features, constant in varying proportions throughout history, are the centralization of political power, the separation of classes, the lifetime division of labor, the mechanization of production, the magnification of military power, the economic exploitation of the weak, and the universal introduction of slavery and forced labor for industrial and military purposes. These institutions would have completely discredited both the primal myth of divine kingship and the derivative myth of the machine had they not been accompanied by another set of collective traits that deservedly claim admiration: the invention and keeping of the written record, the growth of visual and musical arts, the effort to widen the circle of communication and economic intercourse far beyond the range of any local community; ultimately the purpose to make available to all men the discoveries and inventions and creations, the works of art and thought, the values and purposes that any single group has discovered.

The negative institutions of "civilization," which have besmirched and bloodied every page of history, would never have endured so long but for the fact that its positive goods, even though they were arrogated to the use of a dominant minority, were ultimately of service to the whole human community.[44]

It was this complex understanding of civilization, which carried with it both the collective cultural legacies of the past, as well the alienated heritage of barbarism, exploitation, and conflagration, that informed the work of the most acute socialist ecological analysts faced with the more universal dangers of the late twentieth century.

Protesting against the in-built tendency toward a nuclear conflagration, as well as environmental destruction, with the reigniting of Cold War hostilities under Reagan, Thompson penned "Notes on Exterminism, the Last Stage of Civilization."[45] The term "exterminism" was applied to "those characteristics of a society—expressed, in differing degrees, within its economy, its polity and its ideology—which thrust it in a direction whose outcome must be the extermination of multitudes."[46] Thompson was particularly concerned with the dangers of nuclear holocaust, but gave the concept a wider scope that also pointed to environmental destruction.

These grave concerns were carried forward by Bahro, who observed:

> In order to furnish a basis for resistance to rearmament plans, the visionary British historian E. P. Thompson wrote an essay in 1980 about *exterminism*, as the last stage of civilisation. Exterminism doesn't just refer to military overkill, or to the neutron bomb—it refers to *industrial civilization as a whole.* . . . Thompson's statements about the "increasing determination of the extermination process," about the "last dysfunction of humanity, its total self-destruction," characterize the situation as a whole. . . . As an inseparable consequence of military and economic progress we are in the act of destroying the biosphere which gave birth to us.
>
> To express the extermination-thesis in Marxian terms, one could say that the relationship between productive and destructive forces is turned upside down. Like others who looked at civilization as a whole, Marx had seen the trail of blood running through it, and that "civilisation leaves deserts behind it." In ancient Mesopotamia it took 1500 years for the land to grow salty, and this was only noticed at a very late stage, because the process was slow. Ever since we began carrying on a productive material exchange with nature, there has been this destructive side. And today we are forced to think apocalyptically, not because of culture-pessimism, but because this destructive side is gaining the upper hand.[47]

Today warnings of the potential collapse of civilization, now on a global, not just a regional scale, have reemerged, reflecting the growing recognition of the anthropogenic rift in the Earth System. Such warnings today are coming first and foremost from natural scientists (and historians of science), in works like Martin Rees's *Our Final Hour,* Jared Diamond's *Collapse,* James Lovelock's *The Revenge of Gaia,*

James Hansen's *The Storms of My Grandchildren*, and Naomi Oreskes and Erik Conway's *The Collapse of Western Civilization*.[48] What such mainstream analyses typically lack, however, is a social-historical critique of capitalism and of class-based civilization in general, together with a vision of ecological civilization.

In Diamond's *Collapse*, it is recognized that modern society is accelerating down a "non-sustainable course." Under these circumstances, "the only question," he writes, "is whether the world's environmental problems . . . will become resolved in pleasant ways of our own choice, or in unpleasant ways not of our choice, such as warfare, genocide, starvation, disease epidemics, and collapse of societies."[49] Indeed, some societies in the "Third World" are already, he argues in an imperialistic vein, collapsing, while the real question is whether this will be extended to "First World societies."[50] Yet Diamond's reified analysis of societies is curiously devoid of social and historical categories, much less class analysis. The whole question of environmentally induced civilizational collapse throughout history is treated in his book largely in terms of individualistic, behavioristic, demographic, and technological categories. In this kind of flattened analysis, issues such as class, the division of labor, mode of production, capitalism, the state, and inequality are notable in their absence. The mediation between society and the environment is addressed largely in technocratic or scientific terms. Nowhere is this lack of social and historical framework more evident than in Diamond's strident defense of "big business" on the environment, including major oil companies like Chevron. This is accompanied by a long disquisition—hidden in "further readings" to the final chapter— in which he argues that whether there is an environmental collapse of present-day society simply rests on the behavior and values of individuals, acting as voters and as consumers.[51] The form of civilization he wants to preserve, and that he identifies with civilization as a whole, is neoliberal civilization.

Standard liberal analyses of this kind can be contrasted to the classical Marxian view, with its critique of class-based civilization and capitalism, and its advocacy of a revolutionary transition to socialism.

From a historical materialist perspective, civilization—itself a historical product—is something to be both critically defended in certain respects and opposed in others, with the goal being its historical transcendence. This reflects civilization's dual role as the repository of historical cultures, along with its destructive, exploitative, imperialist, and frequently barbaric character. The critical defense of and at the same time opposition to civilization was a crucial part of the whole revolutionary argument. In referring in the opening pages of the *Communist Manifesto* to the choice between the "revolutionary reconstitution of society at large" or the "common ruin of the contending classes" Marx and Engels had in mind the downfall of the Roman Empire where there had been a civilizational collapse.[52] The answer for the founders of historical materialism was always a revolutionary transcendence (*Aufhebung*), not the collapse of civilization but its transformation. The Marxian theory of change focuses on revolution, and not on breakdown, either of the economy or the ecology.

Nothing could be more opposed to this historical materialist conception, therefore, than the view that a *collapse* of civilization is actually to be welcomed today. As one world-ecology-system theorist has opined: "Is the 'collapse' of a civilization that plunges nearly half its population into malnutrition to be feared? The Fall of Rome after the fifth century and the collapse of feudal power in Western Europe ushered in golden ages in living standards for the vast majority."[53]

Certain self-styled anarchist "anticivilization" thinkers like Jensen have gone so far as to promote the idea of a vanguard dedicated to the immediate "taking down" of civilization itself by way of violence, destroying dams and electrical grids.[54] All hope, Jensen contends, lies in the elimination of a life based on cities.

Nevertheless, any emphasis on the *positive aspects of civilizational collapse*, akin to the fall of Rome, but in today's circumstances (notably, climate change) necessarily raising the specter of the catastrophic demise of global civilization, portending the death and dislocation of hundreds of millions, perhaps billions, of people, is grossly irresponsible from a historical materialist standpoint.[55] This is all the more the case since the greatest immediate impact will be on the world's poor, particularly populations in the periphery who have been subjected to centuries of imperialism. Although socialists, looking back historically, may have understandably sympathized with the barbarian invaders against imperial Rome (as in William Morris's romances), it is in the nature of historical materialism always to identify with the radical transformation and transcendence of societies from within and the development of a higher society. "Socialism or barbarism," as raised most notably by Rosa Luxemburg, cannot be twisted into meaning that we should conceive barbarism or catastrophic civilizational decline as a viable, indirect path to socialism.[56]

It is here that the vital question of ecological civilization asserts itself. With the brief, contradiction-laden renaissance of Soviet ecological thought in the late 1970s and early 1980s, which sought to challenge the then hegemonic view in the USSR, the ecological problem came to be characterized—in line with classical historical materialist thought—as a *general problem of civilization*. This was evident in the important collection, edited by A. D. Ursul, *Philosophy and the Ecological Problems of Civilisation*, to which some of the Soviet Union's leading ecological scientists and philosophers contributed. This led to the concept of "ecological civilization," with a systematic discussion of "Ecological Civilization" appearing in Soviet works in 1983–84.[57] The same notion entered almost immediately into Chinese Marxism as well, where it was to become a central category of analysis—and where today it has taken on a very prominent role in ongoing discussions of China's developmental path.[58]

Ecological civilization in the Marxian sense means the struggle to transcend the logic of all previous class-based civilizations, and particularly capitalism, namely, the interconnections between the domination/alienation of nature and the domination/alienation of humanity. This view provided the framework for *Philosophy and the Ecological Problems of Civilization*.[59] The preface to that book raised the danger of "the extinction of the biosphere."[60] The opening chapter by P. N. Fedoseev, vice president of the USSR Academy of Sciences, delved into the issue of "rejection of the gains of civilization," implicit in many Green attempts to address the ecological problem, that advocated historically disembodied utopias (either backward-looking or technocratic).[61] Leading ecological philosopher Ivan Frolov, following Marx, emphasized that the human metabolism with nature was mediated by the labor and production processes and by science and thus depended on the mode of production.[62] Philosopher Yu. P. Trusov wrote dialectically of "the principle of the exploitation and defense of nature," and of "the unity of the use and reproduction of natural resources."[63] Notably, philosopher V. A. Los explored how "culture is becoming an antagonist . . . of nature" and referred to the need to construct a new "ecological culture" or civilization, reconstructing on more sustainable grounds the role of science and technology in relation to the environment. As he explained: "It is in the course of shaping an ecological culture that we can expect not only a theoretical solution of the acute contradictions existing in the relations between man and his habitat under contemporary civilization, but also their practical tackling."[64]

Hence, from a historical-materialist standpoint the emerging global ecological crisis pointed to the objective necessity for an ecological revolution and ecological civilization, as a crucial step in the transition to an ecological socialism.[65] This was in line with the long history of ecological analysis within Marxism. Marx and Engels had dealt with ecological contradictions beyond simply the degradation of the soil and the division between town and country, encompassing such issues as industrial pollution, the depletion of coal reserves, the destruction of forests, the degradation of food through adulteration and additives, etc. Engels remarked in a letter to Marx that humanity was "a squanderer *of past* solar heat," exhausting coal supplies, as well as other natural resources.[66] In recent years, Marx's theory of metabolic rift has been extended by Marxian theorists to address numerous problems, encompassing the manifold rifts in the carbon metabolism (climate change), ocean metabolism, land cover, forestry, fire management, agriculture, food systems, fertilizers, animal husbandry, freshwater supplies, mining, and urban agriculture.[67] It has also been used to develop analyses of unequal ecological exchange, ecological imperialism, and environmental justice. One of the principal concerns has been the emergence of rifts in planetary boundaries in the Anthropocene. Significantly, the Anthropocene crisis, as indicated above, has been described within science as an "anthropogenic rift" in the Earth System.[68]

We are thus moving toward a more unified understanding of both the global metabolic rift, and the recognition of the necessity of a transitional ecological civilization. An ecological civilization cannot be seen as a simple technological or modernizing response to the immense ecological challenges of a Great Climacteric associated with the Anthropocene. Rather, it requires changes in the forces and relations of production and in the state and society: a massive shift, but necessarily occurring in stages, toward realized socialism/communism, that is, a social formation aimed at substantive equality and ecological sustainability, emphasizing sustainable human development—one that involves collective action and planning. It requires that cultural resources, the long heritage of humanity in its many social formations, be brought to bear on the need to create a bridge to a sustainable future.

Ecological Civilization, East and West, North and South

Civilization should never of course be seen as a monopoly of, or as emerging primarily in, the West. In 1974, the great British scientist, Marxist, and sinologist Joseph Needham, one of the foremost synthesizing thinkers of the twentieth century, gave a talk on "An Eastern Perspective on Western Anti-Science." Here Needham addressed the environmental problem, and its relation to the misuse of science under capitalism. Having recently read William Leiss's *The Domination of Nature,* Needham commented on how the domination of nature by mechanistic science was connected in Western capitalist culture to the domination of humanity.[69] This had led irrationally—as its dialectical opposite—in the 1970s to a growing Western tradition of anti-science. In response, Needham pointed to revolutionary China as the locus of an alternative, less alienated, more coherent tradition. What was needed, he explained, in Marxian terms, was "a stage in human consciousness so advanced that intelligence can regulate its relationship to Nature, minimizing the self-destructive aspects of human desires, and maximizing the freedom of the human individual within a classless and egalitarian society."[70] He pointed to Herbert Marcuse's recognition that the distinctive aspects of non-Western cultures—in those places where capitalism had not triumphed—when combined with the critical viewpoint of historical materialism, could be, in Needham's words, the key to "avoiding the repressive and destructive uses of advanced technologies."[71] As he had already indicated in his *Science and Civilization in China,* a distinctive feature of Chinese science, despite its backwardness in some respects, was "an organic philosophy of Nature . . . closely resembling that which modern science has been forced to adopt after three centuries of mechanical materialism."[72] Chinese civilization and the Chinese Revolution, in his view, thus offered resources of hope for a quite different future.

China has of course changed tremendously in the four decades since Needham made these observations. It would be wrong to downplay the deep ecological and social challenges that China faces in its current developmental path and its complex contradictory relation to the capitalist road. Beijing is known through the world not only as a great cultural and political center, but also today for having some of the worst urban air pollution on the entire planet. And China faces other horrendous environmental problems. Yet Western scientists, such as James Hansen and Michael E. Mann, disturbed by the failure of Western economies to address climate change, are increasingly turning to China as a source not so much of optimism as of hope.[73]

In her 2018 book *Will China Save the Planet?* Barbara Finamore, senior atrategic director for Asia at the Natural Resources Defense Council, laid out China's extraordinary environmental advances in the last decade and a half in its attempt to transform itself into an ecological civilization. Similar observations were made by John Cobb in *China and Ecological Civilization* in 2019.[74]

This emphasis of some Western scientists on China as a potential beacon of hope with respect to the climate, in the face of the default of the neoliberal West, was dramatically illustrated in the widely read environmental book *The Collapse of Western Civilization: A View from the Future* by leading science historians Oreskes and Conway.[75] Set in the year 2393, the book is a science-fiction history in which an unknown Chinese historian of the late twenty-fourth century looks back at how climate change led to untold disaster around the world and the final collapse of Western civilization and its capitalist society. Meant as a serious warning, most of the book is actually a discussion of historically documented events in the late twentieth and early twenty-first century. It focuses on how the anarchic world-capitalist system, centered in the West—in what is referred to as the "Penumbral Age" of neoliberalism—failed to address climate change, resulting, in the end, in its own collapse. However, what is most telling in this story, related by an anonymous Chinese historian three centuries in the future, is how late twenty-first-century China, unlike any other society, managed to respond in a planned and coordinated manner, including moving its population inward in response to sea level rise, saving its people and culture.[76]

This emphasis on planning and coordination offers us a clue as to why scientists and historians of science in the West are so ready to see China, despite its own serious environmental problems, as a potential ray of hope in the necessary ecological transition in the Anthropocene. This can be analyzed further by turning to an article that Sweezy wrote on "Socialism and Ecology" in 1989, in the midst of the fall of the Soviet bloc. Referring to the socialist planned economies, he stated:

> The lesson of this experience [the history of the past seven decades] is not that socialist planning is necessarily environmentally destructive but that up to now it has never transcended the capitalist context from which it emerged. If and when the time comes that a socialist country is able to reorder its priorities from catching up and defense [with respect to capitalism] to protection and preservation of the environment—seen as the life and death questions they are rapidly becoming—it cannot be

excluded in advance that the planning system can be adapted to serve the needs of the new situation.

This is the reason it is so important that actually existing socialism should survive its present crisis. There are no guarantees, but at least it is a system with a potential that capitalism totally lacks. If all the existing socialist countries take the capitalist road in the present conjuncture, so much time will have been lost that it may be too late for civilized humanity to restore the necessary conditions for its own survival.[77]

It is precisely for this reason, in the increasingly desperate conditions represented by the Anthropocene and the Great Capitalist Climacteric, that so much hope—much of it of course Utopian, but nonetheless completely rational in today's desperate circumstances—is now being directed at China. Some of Sweezy's worst fears were realized, and the planned economies did generally move down the capitalist road, in the majority of cases abandoning socialism entirely.[78] China, however, in its socialism with Chinese characteristics never completely renounced its socialist goals, nor gave up on the planning system entirely. It remained in important respects still a post-revolutionary society, if deeply affected by the capitalist world market, with all the contradictory characteristics and manifold possibilities that entailed.

Bearing all of this in mind, the question arises as to whether China, propelled from below, might once again make a great change. Could China, as some scholars and activists suggest, initiate an ecological revolution based on rural reconstruction and an abandonment of its current hyper-industrialist path?[79] Could it play a role of global leadership in promoting ecological civilization in the Anthropocene—a role that the United States as the hegemonic power has currently abdicated (a fact punctuated by Trump's rise to the presidency)? Or is China too immersed in the capitalist road, too characterized by extremes of inequality, too unable to draw on social forces at its roots, to make this switch? These are key questions that cannot be answered at present. Yet, there is no doubt that in its New Era since 2009 and particularly since 2012 China has been moving in a direction of ecological civilization.[80] The response of the Chinese people themselves to these challenges will be crucial in determining the future course of the Chinese revolution and the fate of humanity on a global level.

Recognizing the importance of China's role, and that of other countries of the Global South, in any conceivable path toward a new ecological civilization, still leaves enormous uncertainty as to what will actually happen. But it nonetheless points to where the needed ecological revolution might conceivably take hold and under what possible conditions. It should not surprise us, given its complex and distinctive history, that while China's primary goal has been catching up with the economic development of the West—thereby promoting very high rates of growth with the attendant horrendous environmental problems—it has also, looking to the future, raised the issue of "ecological civilization" and has taken huge steps at shifting resources and technology toward environmental amelioration.[81] China stands today paradoxically at a kind of turning point of its own, which will have an enormous impact on the world as a whole: it is known worldwide for some of the most serious forms of environmental damage on earth, but at the same time no country seems to be accelerating so rapidly into the new world of alternative energy.[82]

The question is not so much whether China can successfully accomplish a transition to an ecological civilization in terms of its present productive relations; rather it is a matter of whether China might be the site, or one of a number of sites, possibly stretching across the Global South and in certain locales in the Global North—one thinks here of the courageous battle of Indigenous-led Water Protectors at Standing Rock in North Dakota in opposition to the Dakota Access Pipeline—in which a world ecological revolution might be launched.[83] With all of its environmental contradictions, China has forcefully raised the issue of the forging of an "ecological civilization" as a project—something that is still lacking among the leading capitalist powers within the hegemonic core of the world economy.

What is certain is that human history is at a turning point. Never before has it faced such a challenge. As Noam Chomsky has stated,

"It seems to me unlikely that civilization can survive really existing capitalism and the sharply attenuated democracy that goes along with it."[84] On this basis, he has stated that based on current trajectories "the era of human civilization . . . may now be approaching its inglorious end."[85] Hence, in Chomsky's view, there is no alternative but a revolt against capitalism, including the entire capitalist world market. The dire facts constituting today's Earth System emergency are stubborn things, and the world's options are clearly limited. What is needed in the end across the globe as a whole, in order to create the new, essential ecological civilization, is nothing less than a worldwide ecological and social revolution against the capitalist mode of production—a revolution that is most likely to emerge first in the Global South, given the depth of the economic and ecological crises there and the struggle against economic and ecological imperialism. In the Great Capitalist Climacteric, the future depends on the rise globally of a *new environmental proletariat,* representing the greater part of today's endangered humanity, and providing the revolutionary impetus for a more substantively equal and ecologically sustainable world.[86]

CHAPTER TWENTY-ONE

The Capitalinian: The First Geological Age of the Anthropocene

The geologic time scale, dividing the 4.6 billion years of Earth history into nested eons, eras, periods, epochs, and ages, is one of the great scientific achievements of the last two centuries. Each division is directed at environmental change on an Earth System scale based on stratigraphic evidence, such as rocks or ice cores. At present, the earth is officially situated in the Phanerozoic Eon, Cenozoic Era, Quaternary Period, Holocene Epoch (beginning 11,700 years ago), and Meghalayan Age (the last of the Holocene ages beginning 4,200 years ago). The current argument that the planet has entered into a new geological epoch, the Anthropocene, is based on the recognition that Earth System change as represented in the stratigraphic record is now primarily due to anthropogenic forces. This understanding has now been widely accepted in science, but nevertheless has not yet been formally adopted by the International Commission on Stratigraphy of the International Union of Geological Sciences, which would mean its official adoption throughout science.

Under the assumption that the Anthropocene will soon be officially designated as the earth's current epoch, there remains the question of the geological age with which the Anthropocene begins, following the last Holocene age, the Meghalayan. Adopting the standard nomenclature for the naming of geological ages, we propose, in our role as professional environmental sociologists, the term *Capitalinian* as the most appropriate name for the new geological age, based on the stratigraphic record, and conforming to the historical period that environmental historians see as commencing around 1950, in the wake of the Second World War, the rise of multinational corporations, and the unleashing of the process of decolonization and global development.[1]

In the Anthropocene Epoch, it is clear that any designation of ages, while necessarily finding traces in the stratigraphic record, has to be seen, in part, in terms of human socioeconomic organization, not purely geologically. The most widely accepted social-scientific designation for the predominant world economic system over the last few centuries is capitalism. The capitalist system has passed through various stages or phases, the most recent of which, arising after the Second World War under U.S. hegemony, is often characterized as global monopoly capitalism.[2] Beginning with the first nuclear detonation in 1945, humanity emerged as a force capable of massively affecting the entire Earth System on a geological scale of millions (or perhaps tens of millions) of years. The 1950s are known for having ushered in "the synthetic age," not only because of the advent of the nuclear age, but also due to the massive proliferation of plastics and other petrochemicals associated with the global growth and consolidation of monopoly capitalism.[3]

The designation of the first geological age of the Anthropocene as the Capitalinian is, we believe, crucial, because it also raises the question of a possible second geological age of the Anthropocene Epoch. The Anthropocene stands for a period in which humanity, at a specific point in its history, namely the rise of advanced industrial capitalism following the Second World War, became the principal geological force affecting Earth System change—which is not to deny the importance of numerous other geological forces, which are not all affected by human action, such as plate tectonics, volcanism, erosion, and weathering of rocks in shaping the Earth System's future. If capitalism in the coming century were to create such a deep anthropogenic rift in the Earth System through the crossing of planetary boundaries that it led to the collapse of industrial civilization and a vast die-down of human species ensued—a distinct possibility under business as usual according to today's science—then the Anthropocene Epoch and no doubt the entire Quaternary Period would come to an end, leading to a new epoch or period in geological history, with a drastically diminished human role.[4] Barring such an end-Anthropocene and even end-Quaternary extinction event, the socioeconomic conditions defining the Capitalinian will have to give rise to a radically transformed set of socioeconomic relations, and indeed a new mode of sustainable human production, based on a more communal relation of human beings with each other and the earth.

Such an environmental climacteric would mean pulling back from the current crossing of planetary boundaries, rooted in capital's creative destruction of conditions of life on the planet. This reversal of direction, reflecting the necessity of maintaining the earth as a safe home for humanity and for innumerable other species that live on it, is impossible under a system geared to the exponential accumulation of capital. Such a climatic shift would simply require for human survival the creation of a radically new material-environmental relation with Earth. We propose that this necessary (but not inevitable) future geological age to succeed the Capitalinian by means of ecological and social revolution be named the *Communian*, derived from communal, community, commons.

The Anthropocene versus Capitalocene Controversy

The word *Anthropocene* first appeared in the English language in 1973 in an article by Soviet geologist E. V. Shantser on "The Anthropogenic System (Period)" in *The Great Soviet Encyclopedia*. Here, Shantser referred to the Russian geologist A. P. Pavlov's introduction in the 1920s of the notion of the "'Anthropogenic system (period),' or 'Anthropocene.'"[5] During the much of the twentieth century, Soviet science played a leading role in numerous fields, including climatology, geology, and ecology, forcing scientific circles in the West to pay close attention to its findings. As a result, the Shantser article, describing developments in geochronology, would likely have been fairly well known to specialists, having appeared in such a prominent source.[6]

Pavlov's introduction of the term *Anthropocene* was closely related to the Soviet geochemist Vladimir I. Vernadsky's landmark book *The Biosphere* in 1926, which provided an early proto-Earth System outlook, revolutionizing how the relationship between humans and the planet was understood.[7] Pavlov used the concept of the Anthropocene (or Anthropogene) to refer to a new geological period in which humanity was emerging as the main driver of planetary ecological change. In this way, Pavlov and subsequent Soviet geologists provided an alternative geochronology, one that substituted the Anthropocene (Anthropogenic) Period for the entire Quaternary. Most important, Pavlov and Vernadsky strongly emphasized that anthropogenic factors had come to dominate the biosphere in the late Holocene. As Vernadsky observed in 1945, "Proceeding from the notion of the geological role of man, the geologist A. P. Pavlov [1854–1929] in the last years of his life used to speak of the *anthropogenic era*, in which we now live.... He rightfully emphasized that man, under our very eyes, is becoming a mighty and ever-growing geological force.... In the 20th Century, man for the first time in the history of the Earth knew and embraced the whole biosphere, completed the geological map of the planet Earth, and colonized its whole surface. *Mankind became a single totality in the life of the Earth*."[8]

The current usage of *Anthropocene*, however, derives from atmospheric chemist Paul J. Crutzen's recoining of the term in February 2000, during a meeting of the International Geosphere-Biosphere Program in Cuernavaca, Mexico, where he declared, "We're not in the Holocene any more. We're in the ... Anthropocene!"[9] Crutzen's use of the term *Anthropocene* was not based on stratigraphic research but on a direct understanding of the changing Earth System rooted principally in perceptions of anthropogenic climate change and the anthropogenic thinning of the ozone layer (research for which he was awarded the Nobel Prize in chemistry in 1995). Crutzen's designation of the Anthropocene as a new geological epoch thus reflected, from the beginning, a sense of crisis and transformation in the human relation to the earth.[10] As Crutzen, geologist Will Steffen, and environmental historian John McNeill declared a few years later: "The term Anthropocene . . . suggests that the Earth has now left its natural geological Epoch, the present interglacial state called the Holocene. Human activities have become so pervasive and profound that they rival the great forces of Nature and are pushing the Earth into planetary *terra incognita*. The Earth is rapidly moving into a less biologically diverse, less forested, much warmer, and probably wetter and stormier state."[11] Similar views on the effect of anthropogenic changes on the Earth System were presented by one of us in the early 1990s: "In the period after 1945 the world entered a new stage of planetary crisis in which human activities began to affect in entirely new ways the basic conditions of life on earth. . . . As the world economy continued to grow, the scale of human economic processes began to rival the ecological cycles of the planet, opening up as never before the possibility of planetary-wide ecological disaster. Today, few doubt that the [capitalist] system has crossed critical thresholds of sustainability."[12]

Perhaps the best way of understanding the overall threat brought about by the Anthropocene Epoch, as depicted by science, is in terms of an "anthropogenic rift," in which the socioeconomic effects of human production—today largely in the form of capitalism—have created a series of ruptures in the biogeochemical processes of the Earth System by crossing critical ecological thresholds and planetary boundaries, with the result that all of Earth's existing ecosystems and industrial civilization itself are now imperiled.[13] By pointing to the Anthropocene Epoch, natural scientists have underscored a new climacteric in Earth history and a planetary crisis that needs to be addressed to preserve Earth as a safe home for humanity.

It should be mentioned that the widespread notion that the Anthropocene Epoch stands for "the age of man," frequently presented in the popular literature, is entirely opposed to the actual scientific analysis of the new geological epoch. Logically, to refer to anthropogenic causes of Earth System change does not thereby ignore social structures and inequality, nor does it imply that humanity has somehow triumphed over the earth. Rather, the Anthropocene Epoch, as conceptualized within science, not only incorporates social inequality as a crucial part of the problem, but also views the Anthropocene as standing, at present, for a planetary ecological crisis arising from the forces of production at a distinct phase of human historical development.[14]

Yet, despite the crucial importance of the designation of the Anthropocene Epoch in promoting an understanding not only of the current phase of the Earth System but also of the present ecological emergency, the notion of the Anthropocene has come under heavy attack within the social sciences and humanities. Many of those outside the natural sciences are not invested in or informed about the natural-scientific aspects of Earth System change. They therefore react to the designation of the Anthropocene within geochronology in purely cultural and literary terms divorced from the major scientific issues, reflecting the famous problem of the "two cultures," dividing the humanities (and frequently the social sciences) off from natural science.[15] In this view, the prefix *anthro* is often interpreted as simply having a human-biological dimension while lacking a socioeconomic and cultural one. As one posthumanist critic has charged, not only the notion of the Anthropocene, but even "the phrase *anthropogenic climate change* is a special brand of blaming the victims of exploitation, violence, and poverty."[16]

Today, the most prominent alternative name offered for the Anthropocene is that of the *Capitalocene*, conceived as a substitute designation for the geochronological epoch of the Earth System following the Holocene. Leading environmental historian and historical-materialist ecological theorist Andreas Malm argues that the Anthropocene, as the name of a new epoch in the geologic time scale, is an "indefensible abstraction" since it does not directly address the social reality of *fossil capital*. Thus, he proposes substituting the *Capitalocene* for the Anthropocene, shifting the discussion from a geology of humankind to a geology of capital accumulation.[17] In practical as well as scientific terms, however, this runs into several problems. The term *Anthropocene* is already deeply embedded in natural science, and it represents the recognition of a fundamental change in human and geological history that is critical to understanding our period of planetary ecological crisis.

Although it is true that the Anthropocene was generated by capitalism at a certain phase of its development, the substitution of the name *Capitalocene* for the *Anthropocene* would abandon an essential critical view embodied in the latter. The notion of the Anthropocene as demarcated in natural science stands for an irreversible change in humanity's relation to Earth. There can be no conceivable industrial civilization on Earth from this time forward where humanity, if it is to continue to exist at all, is no longer the primary geological force conditioning the Earth System. This is the critical meaning of the Anthropocene. To substitute the term *Capitalocene* for *Anthropocene* would be to obliterate this fundamental scientific understanding. That is, even if capitalism is surmounted, through a "Great Climacteric," representing the transition to a more sustainable world order, this fundamental boundary will remain.[18] Humanity will continue to operate on a level in which the scale of human production rivals the biogeochemical cycles of the planet, and hence the choice is between unsustainable human development and sustainable human development. There is no going back—except through a civilizational crash and a massive die-down—to a time in which human history had little or no effect on the Earth System.

If a true mass extinction and planetary civilizational collapse were to occur, this would be an end-Anthropocene or even end-Quaternary extinction event, not a continuation of the Anthropocene. As the great British zoologist E. Ray Lankester (Charles Darwin and Thomas Huxley's protégé and Karl Marx's close friend) remarked in 1911 in *The Kingdom of Man*, given its massive and growing disruption of the ecological conditions of human existence, humanity's "only hope is to control . . . the sources of these dangers and disasters."[19]

The enormous historical, geological, and environmental challenges now facing humanity demand, we believe, a shifting of the terrain of analysis to the question of *ages* rather than *epochs* in the geologic time scale. If the world entered the Anthropocene Epoch around 1950, we can also say that the Capitalinian Age began at the same time. The Capitalinian in this conception is not coterminous with historical capitalism, given that capitalism had its origins as a world system in the fifteenth and sixteenth centuries. Rather, the Capitalinian Age was a product of global monopoly capitalism in the wake of the Second World War. In order to understand the historical and environmental significance of the emergence of the Capitalinian and to put it in the context of the geologic time scale, it is first necessary to address the question of the changeover from one geological age to another, stretching from the late Holocene Epoch to the early Anthropocene Epoch.

From the Meghalayan to the Capitalinian

The Holocene Epoch (*Holocene* means "entirely recent") was first proposed as a division of geologic time by the French paleontologist Paul Gervais in 1867 and formally adopted by the International Geographic Congress in 1885. It dates back to the end of the last Ice Age and thus refers to the warmer, relatively mild Earth-environmental conditions extending from roughly 11,700 years ago to the present, covering the time during which glaciers receded and human civilizations arose.[20] It was not until around a century and a half after it was first proposed that the Holocene Epoch was formally divided into geological ages. This occurred with the modification of the geologic time scale by the International Commission on Stratigraphy in June 2018, dividing the Holocene into three ages: (1) the Greenlandian, beginning 11,700 years ago, with the end of the Pleistocene Epoch and the beginning of the Holocene; (2) the Northgrippian, beginning 8,300 years ago; and (3) the Meghalayan, extending from 4,200 years ago to the present.

Dividing the Holocene into ages represented a more difficult problem than in other epochs of the Quaternary, given the relatively calm environmental-climatic character of the Holocene.[21] The first division of the Holocene, the Greenlandian, posed no problems because it corresponded to the criteria giving rise to the Holocene Epoch. The Northgrippian came to be designated in terms of an outburst of freshwater from naturally dammed glacial lakes that poured into the North Atlantic, altering the conveyor belt of ocean currents, leading to global cooling. The demarcation of the third division was not as straightforward. There were archaeological reports beginning in the 1970s of a megadrought 4,200 years ago (ca. 2200 BCE) lasting several centuries, which was thought to have led to the demise of some early civilizations in Mesopotamia, Egypt, and elsewhere.

In 2012, paleoclimatologists discovered a stalagmite in Mawmluh cave in the Meghalaya state in northeast India that pointed to a centuries-long drought. This was then taken as the geological exemplar or "golden spike" for the Meghalayan Age. In their original July 15, 2018, press release on the Meghalayan, titled "Collapse of Civilizations Worldwide Defines Youngest Unit of the Geologic Time Scale," the International Commission on Stratigraphy went so far as to declare that a civilizational collapse had occurred around 2200 BCE: "Agricultural-based societies that developed in several regions after the end of the last Ice Age were impacted severely by the 200-year climatic event that resulted in the collapse of civilizations and human migrations in Egypt, Greece, Syria, Palestine, Mesopotamia, the Indus Valley, and the Yangtze River Valley. Evidence of the 4.2 kiloyear climatic event has been found on all seven continents."[22]

This resulted in sharp rebuttals by archaeologists, who argued that the evidence for the sudden collapse of civilizations due to climate change around 2200 BCE does not in actuality exist. Although civilizations did decline, it was most likely over longer periods of time, and there were reasons to believe that an array of social factors played a more significant role than the megadrought.[23] As archaeologist Guy D. Middleton wrote in *Science* magazine: "Current evidence . . . casts doubt on the utility of 2200 BCE as a meaningful beginning to a new age in human terms, whether there was a megadrought or not. . . .

Climate change never inevitably results in societal collapse, though it can pose serious challenges, as it does today. From an archaeological perspective, the new Late Holocene Meghalayan Age seems to have started with a whimper rather than a bang."[24]

The Meghalayan controversy, whatever the final outcome, highlights a number of essential facts. First, as early as 4,200 years ago, geologic time became intertwined in complex ways with historical time. In the case of the Meghalayan, the geological demarcation drew much of its salience from a seeming correspondence to the historical-archaeological record. Second, although the International Stratigraphic Committee moved away from its original reference to the collapse of civilizations and sought instead to define the Meghalayan simply in terms of geologic-stratigraphic criteria, the question of social conditions associated with a geological age can no longer be avoided. Third, during the Holocene, from the earliest civilizations to the present, the issues of environmental change and civilizational collapse recur, on an ever more expanding global scale.

If the Meghalayan Age did in fact come into being in the context of a megadrought, the end-event signaling the passing of the Meghalayan (and the Holocene) happened around 1950, leading to the start of what the Anthropocene Working Group posits as the Anthropocene Epoch and what we are proposing as the accompanying Capitalinian Age.[25] This transition in geologic time, which is deeply intertwined with distinct sociohistorical relations, is associated with the Great Acceleration of global monopoly capitalism in the 1950s, resulting in an age of planetary ecological crisis. This has involved a move away from an environmentally "highly stable epoch" to one "in which a number of key planetary boundary conditions, notably associated with the carbon, nitrogen and phosphorus cycles, are clearly outside the range of natural variability observed in the Holocene."[26] Here, megadroughts, megastorms, rising sea levels, out-of-control wildfires, deforestation, species extinction, and other planetary threats are emerging in fast order—not simply as external forces, but as the product of capitalism's anthropogenic rift in the Earth System.

The Capitalinian Age

The "golden spike" in geologic time determining the end of the Holocene Epoch and the Meghalayan Age—as well as the corresponding emergence of the Anthropocene Epoch and what we are proposing as the Capitalinian Age—has not yet been determined, although a number of candidates are being pursued by the Anthropocene Working Group of the International Commission on Stratigraphy. The two most prominent of these are radionuclides, the result of nuclear testing, and plastics, the creation of the petrochemical industry—both of which are products of the synthetic age and represent the emergence of a qualitative transformation in the human relation to the earth.[27] Although the "Anthropocene strata may be commonly thin," they "reflect a major Earth System perturbation" in the mid-twentieth century, "are laterally extensive, and can include rich stratigraphic detail," in which distinct "signatures" of a new epoch and age are evident.[28]

Anthropogenically sourced radionuclides stem primarily from the fallout from numerous aboveground nuclear tests (and two atomic bombings in war) commencing with the U.S. Trinity detonation at 5:29 a.m. on July 16, 1945, at Alamogordo, New Mexico.[29] The first thermonuclear detonation was the Ivy Mike test on Enewetak Atoll on November 1, 1952. This was followed by the disastrous Castle Bravo test at Bikini Atoll on March 1, 1954, the explosion of which was two and a half times what had been projected, raining down fallout on sailors in a Japanese fishing boat, the *Lucky Dragon*, and on residents of the Marshall Islands, who ended up with radiation sickness. The United States conducted over two hundred atmospheric and underwater tests (and others were carried out in the 1950s and '60s by the Soviet Union, United Kingdom, France, and China), introducing radioactive fallout in the form of Iodine-131, Caesium-137, Carbon-14, and Strontium-90. This nuclear fallout, especially the gaseous and particulate forms, which entered the stratosphere, was dispersed throughout the biosphere, generating widespread global environmental concern, connecting the entire world's population, to some extent, in a common environmental fate.[30]

Radionuclides primarily from nuclear weapons tests are thus the most obvious basis for demarcating the beginning of the Anthropocene Epoch and the Capitalinian Age. They have left a permanent record throughout the planet in sediments, soil, and glacial ice, serving as "robust independent stratigraphic markers" that will be detectable for millennia.[31] The effects of nuclear weapons, beginning with the U.S. bombings of Hiroshima and Nagasaki at the end of the Second World War, stand for a qualitative change in the human relation to the earth, such that it is now possible to destroy life on such a scale that it would take perhaps as much as tens of millions years for it to recover.[32] Indeed, the theory of nuclear winter developed by climatologists suggests that a massive global thermonuclear exchange, generating megafires in a hundred or more major cities, could lead to planetary climate change, more abruptly and in the opposite direction from global warming, through the injection of soot into the stratosphere, causing global or at least hemispheric temperatures to drop several degrees or even "several tens of degrees" Celsius in a matter of a month.[33]

The advent of nuclear weapons technology thus stands for the enormous change in the human relation to the earth around the 1950s, marking the Anthropocene, leaving a distinct signature in the stratigraphic record; it also serves as a moment when specific radioactive elements were introduced into the body composition of all life.[34] Nuclear weapons technology is of course not entirely separable from nuclear energy use, which also presents dangers of global radioactive contamination as in the nuclear accidents at Three Mile Island, Chernobyl, and Fukushima.

Plastics, which emerged as a major element of the economy in the 1950s, were the result of developments in organic chemistry, associated with the Scientific and Technical Revolution and the Second World War. They are a product of the petrochemical industry, thus standing for the further development of fossil capital, which dates back to the Industrial Revolution.[35] As of 2017, over "8,300 million metric tons... of virgin plastics have been produced," exceeding that of almost all other human-made materials.[36] Plastic waste is so pervasive that it is found dispersed throughout the entire world. In fact, "molten plastics... have fused basalt clasts and coral fragments... to form an assortment of novel beach lithologies," and deep ocean mud deposits include microplastics.[37] The majority of plastics, made from hydrocarbon-derived monomers, is not biodegradable, resulting in an "uncontrolled experiment on a global scale, in which billions of metric tons of material will accumulate across all major terrestrial and aquatic ecosystems on the planet."[38] Due to these conditions, plastic is seen as another potential stratigraphic indicator of the Anthropocene.[39]

The production of plastics and petrochemicals in general, like nuclear weapons testing, represents a qualitative shift in the human relationship with the earth. It has resulted in the spread of a host of mutagenic, carcinogenic, and teratogenic (birth-defect causing) chemicals, particularly harmful to life because they are not the product of evolutionary development over millions of years. Like radionuclides, many of these harmful chemicals are characterized by bioaccumulation (concentration in individual organisms) and biomagnification (concentration at higher levels in the food chain/food web) representing increasingly pervasive threats to life. Microplastics actively absorb carcinogenic persistent organic pollutants within the larger environment, making them more potent and toxic.[40] Plastics are durable and resistant to degradation, properties that "make these materials difficult or impossible for nature to assimilate."[41] The omnipresent character of plastics in the Capitalinian is evident in the massive plastic gyres in the ocean and by the existence of microplastic particles in nearly all organic life.

Ecological scientists, such as Barry Commoner, Rachel Carson, Howard Odum, and others, singled out both radionuclides and plastics/petrochemicals/pesticides as embodying the synthetic age that emerged in the 1950s. They provided detailed accounts of the transformation in the relationship between humans and the earth, which today are reflected in contemporary charts on the Great

Acceleration, presenting such Earth System trends as the dramatic increase in the atmospheric concentration of carbon dioxide, ocean acidification, marine fish capture, land use change, and loss of biodiversity. The epicenter for such global environmental disruption has been the United States as the hegemonic power of the capitalist world economy, dominating and characterizing this entire period. In our analysis, the economic and social system of the United States thus epitomizes the Capitalinian, as no other nation has played a bigger historical role in the promotion of the "poverty of power" represented by fossil capital.[42]

At the start of what we are calling the Capitalinian, global monopoly capital, rooted within the United States, entered a period of massive expansion, fueled by the rebuilding of Europe and Japan, the petrochemical revolution, the growth of the automobile complex, suburbanization, the creation of new household commodities, militarization and military technologies, the sales effort (that is, the entire realm of marketing including its peneatration into the production process), and the growth of international trade. With the endless quest for profit spurring the accumulation of capital, production and the material throughputs to support the economic system's operations have greatly expanded, placing more demands on ecosystems and generating more pollution.[43]

Since plastics and other synthetic materials associated with the expansion of the petrochemical industry were readily incorporated into industrial operations, agricultural production, and everyday commodities, new ecological problems inevitably emerged. As Commoner explained in *The Closing Circle*, "The artificial introduction of an organic compound that does not occur in nature, but is man-made and is nevertheless active in a living system, is very likely to be harmful."[44] Such materials do not readily decompose or break down in a meaningful human-historical time frame and thus end up accumulating, presenting an increasing threat to ecosystems and living beings. Pesticides and plastics that have these characteristics are therefore a violation of the informal laws of ecology.

Given the operations of monopoly capitalism and its technological apparatus, the largely uncontrolled development of synthetic materials results in a particularly dangerous situation, often referred to as "the risk society."[45] In the words of Peter Haff, a professor of environmental engineering at Duke University, a capitalist technostructure "has emerged possessing no global mechanism of metabolic regulation. Regulation of metabolism introduces the possibility of a new timescale into system dynamics—a lifetime—the time over which the system exists in a stable metabolic state. But without an intrinsic lifetime, i.e., lacking enforced setpoint values for energy use," this system "acts only in the moment, without regard to the more distant future, necessarily biased towards increasing consumption of energy and materials," racing ahead "without much concern for its own longevity," much less the continuance of what is external to it.[46]

The uncontrollable, alienated social metabolism of global monopoly capitalism, coinciding with the introduction of radionuclides from nuclear testing, proliferation of plastics and petrochemicals, and carbon emissions from fossil capital—along with innumerable other ecological problems resulting from the crossing of critical thresholds—is manifested in the Capitalinian Age, associated with the present planetary crisis. Capitalism's relentless drive to accumulate capital is its defining characteristic, ensuring anthropogenic rifts and ecological destruction as it systematically undermines the overall conditions of life.

Today the moment of truth looms large. We currently reside within a "Great Climacteric"—first identified in the 1980s by geographers Ian Burton and Robert Kates—a long period of crisis and transition in which human society will either generate a stable relation to the Earth System or experience a civilizational collapse, as part of a great die-down of life on earth, or *sixth extinction*.[47]

The future of civilization, viewed in the widest sense, demands that humanity collectively engage in an ecological and social revolution, radically transforming productive relations, in order to forge a path toward sustainable human development. This entails regulating the social metabolism between humanity and the earth, ensuring that it operates within the planetary boundaries or the universal metabolism of nature. Viewed in these terms, there is an objective historical necessity for what we are calling the prospective second geological age of the Anthropocene: the Communian.

THE DAWN OF ANOTHER AGE: THE COMMUNIAN

In a remarkable intellectual development in the closing decade of the Soviet Union, leading Soviet geologists, climatologists, geographers, philosophers, cultural theorists, and others came together to describe the global ecological crisis as a *civilizational crisis* requiring a whole new *ecological civilization*, rooted in historical-materialist principles.[48] This viewpoint was immediately taken up by Chinese environmentalists and has been further developed and applied in China today.[49] If historic humanity is to survive, today's capitalist civilization devoted to the single-minded pursuit of profits as its own end, resulting in an anthropogenic rift in the Earth System, must necessarily give way to an ecological civilization rooted in communal use values. This is the real meaning of today's widely referred to planetary "existential crisis."[50]

In this Great Climacteric, it is not only essential to bring to an end the destructive trends that are ruining the earth as a safe home for humanity, but also, beyond that, it is vital to engineer an actual "reversal" of these trends.[51] For example, carbon concentration in the atmosphere is nearing 420 parts per million (ppm), peaking in May 2021 at 419 ppm, and is headed rapidly toward 450 ppm, which would break the planetary carbon budget. Science tells us that it will be necessary, if global climate catastrophe is to be avoided, to return to 350 ppm and stabilize the atmospheric carbon dioxide at that level.[52] This in itself can be seen as standing for the necessity of a new ecological civilization and the anthropogenic generation of a new Communian Age within the Anthropocene. This eco-revolutionary transition obviously cannot occur through the unbridled pursuit of acquisitive ends, based on the naive belief that this will automatically lead to the greater good—sometimes called "Adam's Fallacy," after the classical economist Adam Smith.[53] Rather, the necessary reversal of existing trends and the stabilization of the human relation to the earth in accord with a path of sustainable human development can only occur through social, economic, and ecological planning, grounded in a new system of social metabolic reproduction.[54]

To create such an ecological civilization in the contemporary world would require a radical (in the sense of *root*) impetus emanating from the bottom of society—outside the realm of the vested interests.[55] This overturning of the dominant social relations of production requires a long revolution emanating from the mass movement of humanity. Today's realities are therefore giving rise to a nascent *environmental proletariat*, defined by its struggle against oppressive environmental as well as economic conditions, and leading to a revolutionary path of sustainable human development. Broad environmental-proletarian movements in this sense are already evident in our time—from the Landless Workers' Movement (MST) in Brazil, the international peasants' movement La Vía Campesina, the Bolivarian communes in Venezuela, and the farmers' movement in India, to the struggles for a People's Green New Deal, environmental justice, and a just transition in the developed countries, to the Red New Deal of the North American First Nations.[56]

The advent of the Communian, or the geological age of the Anthropocene to succeed the Capitalinian, barring an end-Anthropocene extinction event, necessitates an ecological, social, and cultural revolution; one aimed at the creation of collective relations within humanity as a whole as a basis for a wider community with the earth. It thus requires a society geared to both substantive equality and ecological sustainability. The conditions for this new relation to the earth were eloquently expressed by Marx, writing in the nineteenth century, in what is perhaps the most radical conception of sustainability ever developed: "From the standpoint of a higher socio-economic formation [socialism], the private property of particular individuals in the earth will appear just as absurd as the private property of one man in other men [slavery]. Even an entire society, a nation, or all simultaneously existing societies taken together, are not the owners of the earth. They are simply its possessors, its beneficiaries, and have to bequeath it in an improved state to succeeding generations, as *boni patres familias* [good heads of the household]."[57] In the view of the ancient Greek materialist Epicurus, "The world is my *friend*."[58]

The revolutionary reconstitution of the human relation to the earth envisioned here is not to be dismissed as a mere utopian conception, but rather is one of historical struggle arising out of objective (and subjective) necessity related to human survival. In the poetic words of Phil Ochs, the great radical protest singer and songwriter, in his song "Another Age":

> *The soldiers have their sorrow*
> *The wretched have their rage*
> *Pray for the aged*
> *It's the dawn of another age.*[59]

In the twenty-first century, it will be essential for the great mass of humanity, the "wretched of the earth," to reaffirm, at a higher level, its communal relations with the earth: the dawn of another age.[60]

CHAPTER TWENTY-TWO

CONCLUSION
Ecology and the Future of History

The subject of historical knowledge is the struggling, oppressed class itself.
—WALTER BENJAMIN

Nothing so clearly demonstrates the inherent limits of capitalist ideology as its innate denial of the future of history.[1] The capitalist metaphysic, as Jean-Paul Sartre critically observed, is one of a "barred future"; there is "no exit" from the system and its burning house.[2] Even in the context of the present planetary emergency brought on by capital accumulation, Margaret Thatcher's well-known mantra that "there is no alternative" to the regime of capital—a view she repeated so frequently that she was nicknamed with the acronym Tina—continues to exercise its frozen grip on society.[3]

The notion of bourgeois society as "absolutely the end of History," intrinsic to liberal thought, found its most powerful concrete expression in the early nineteenth-century writings of Georg Wilhelm Friedrich Hegel.[4] In recent years, credit for the questionable notion that capitalism marks the termination of the historical process has often been accorded to Francis Fukuyama, based on his 1992 book

The End of History and the Last Man. In advancing the thesis of "a universal and directional history leading up to liberal democracy," Fukuyama, who served as deputy director of policy planning and as deputy director of European political-military affairs in the U.S. State Department during the George H. W. Bush administration, was merely repackaging long-standing claims of liberal ideology in the context of the demise of the Soviet Union, which he took as representing the final defeat of socialism and the ultimate victory of capitalism, closing off history in any meaningful sense. Humanity, according to this hegemonic view widely circulated in the 1990s, had reached its political-economic-ideological apex: there was no future beyond capitalism and liberalism.[5]

Yet, a mere quarter of century after the celebration of the end of history in the permanence of the liberal order, humanity is confronted with a chain of catastrophic threats extending beyond anything it has experienced in the long course of its development—all arising from the laws of motion of capitalism. In the present epochal crisis, there are multiple dire threats to the world as a whole and to "the wretched of the earth" in particular—from economic stagnation in the capitalist core, to the planetary ecological rift, to the epidemiological threat represented by COVID-19, to the renewed imperialism directed at the Global South and the New Cold War on China.[6] All rational responses to this age of threatened catastrophe point to the need for a global transformation aimed at surmounting capitalism's laws of motion and promoting a world of sustainable human development, that is, socialism and ecology. As Karl Marx indicated in the nineteenth century, in those cases where capitalism leads to the ecological destruction of entire social formations and the extermination of the material basis of human existence, the choice left to working populations and their communities inevitably becomes one of "ruin or revolution."[7]

Historically, revolutions have appeared globally in waves.[8] The first stirrings of what can be conceived as a new revolutionary wave, different than the ones that came before but emanating primarily from the Global South, are now emerging in response to capitalism in the Anthropocene. This will likely expand rapidly with the decline of U.S.

world hegemony, related to the rise of China. Twenty-first-century revolutionary praxis necessarily operates within a wider field combining the struggles for socialism and ecology. It represents a new materiality of hope, rooted in the movements of hundreds of millions, potentially billions, of people, seeking to transcend the oppressions of class, race, gender, environmental injustice, and imperialism emanating from the empire of capital. These struggles necessarily entail new revolutionary vernaculars arising in specific historical and cultural contexts, embodying environmental as well as economic realities. In this sense, there is not a single model of proletarian revolution. Rather, today's movements toward socialism and ecology encompass peasant and Indigenous struggles while converging in complex ways with the struggles of a still expanding industrial (and post-industrial) working class confronting a rapidly changing environment engendered by capital's creative destruction.

In all such instances, it is the combined materiality of the economy and the environment that now determines the terrain of resistance and revolt. Struggles that begin from an ecological basis, the most inclusive expressions of the material conditions shaping people's lives, are as vital as economic struggles, and as crucial in the end in defining the class structure of society. Genuine revolutionary movements necessarily combine the two, shaping the nature and culture of social agency in our time. Today the catastrophes unleased by capitalism embrace not only the economy but the entire environment of the planet, leading to the emergence everywhere of what can be called an *environmental proletariat*.

Capitalism as the Barrier to the Future of History

In the *Grundrisse*, written in 1857–58, Marx famously described capital as a "limitless drive" to accumulate that accepted *no boundaries outside itself*. Drawing on Hegel's dialectic of barriers and boundaries, in which *barriers* were understood as something to be surmounted, in contrast to *boundaries*, which represented actual limits, Marx declared:

> Capital is the endless and limitless drive to go beyond its limiting barrier. Every boundary is and has to be a barrier for it. Else it would cease to be capital—money as self-reproductive. If ever it perceived a certain boundary not as a barrier, but became comfortable within it as a boundary, it would itself have declined from exchange value to use value, from the general [abstract] form of wealth to a specific, substantial mode of the same.... The quantitative boundary of the surplus value appears to it as a mere natural barrier, as a necessity which it constantly tries to violate and beyond which it constantly seeks to go....
>
> Capital drives beyond national barriers and prejudices as much as beyond nature worship, as well [as] all traditional, confined, complacent, encrusted satisfactions of present needs, and reproductions of old ways of life. It is destructive toward all of this, and constantly revolutionizes it, tearing down all barriers which hem in the development of the forces of production, the expansion of needs, the all-sided development of the forces of production, and the exploitation and exchange of natural and mental forces. But from the fact that capital posits every such limit as a barrier and hence gets *ideally* beyond it, it does not by any means follow that it has *really* overcome it, and since every such barrier contradicts its character, its production moves in contradictions which are constantly overcome but just as constantly posited.[9]

The constant positing of contradictions that are only ideally surmounted, but which nonetheless remain and accumulate over the course of capitalism, to the point that more potentially catastrophic crises emerge, has to do with the fact that capital's creative destruction revolutionizes the world in ways limited by its own essential conditions of existence.[10] The one boundary that is permanent, which can never be transgressed, from the standpoint of capital, is the social relation of class-based accumulation itself, and thus it is to this artificially imposed boundary that all the contradictions of the system can ultimately be traced. "The *true barrier* [boundary] to capitalist production," Marx wrote, "is *capital itself.*"[11]

The concrete result of this central contradiction of the capitalist system is that all transformations carried out by capital as part of its process of creative destruction are necessarily associated with fetters on sustainable human development, in the form of alienated second-order mediations, leading to ever more contradictory and catastrophic results.[12] The path to a world of sustainable human development is blocked at every point. It is this limit, determined by the very nature of the system, that now constitutes the fundamental basis of the planetary ecological and economic crisis engulfing the entire world, seemingly closing off the future as history. The more serious the social, economic, and ecological contradictions become the more the ideological response is to seal capitalism off from history, defining it as an immutable reality and denying all other possibilities.

The universalization of the present in such a way as to portray as insurmountable the ruling ideas of society, which are at the same time both the ideas of the ruling class and the ideological bases of its rule, is common to all ruling classes, whether in the form of divine right of kings or the invisible hand of capital.[13] Such universalization, however, becomes more complex in those societies in which historical development is recognized. Here what is above all required is the denial of the future through the "decapitation" of history, as Sartre called it.[14] This decapitation of history is evident in the ubiquitous attempts of both mainstream modernist and postmodernist ideology to deny the historical specificity and thus transitory character of capitalist social relations.

Just as any future beyond capitalism is denied, so is capitalism's genesis presented in the conventional wisdom as predetermined, a mere coming to be of forces that were always present and simply waiting to be set free. The result is the systemic denial of any coherent theory of the historical origins of capitalism, which would contradict its assumed innate character. As Marxian political theorist Ellen Meiksins Wood observed, "Accounts of the origin of capitalism" are "fundamentally circular," assuming "the prior existence of capitalism in order to explain its coming into being. . . . Capitalism seems always to *be* there, somewhere; and it only needs to be released from its chains—for instance from the fetters of feudalism—to be allowed to grow and mature."[15]

The notion that capitalism is a natural and universal, and thus somehow ever-present, only waiting for obstacles to be cleared away to its advance for it to emerge full bloom, can be traced back to the liberal possessive-individualist view of human nature, associated with thinkers from Thomas Hobbes to Adam Smith, the latter stipulating as the basis of his economic vision, an inherent tendency of human beings to "truck, barter, and exchange."[16] In this view, which remains dominant in present-day ideology, capitalism is simply bourgeois human nature, parading as human nature in general, writ large.

Max Weber in the twentieth century was to expand on this fundamental liberal outlook by presenting capitalism as the "most fateful force in our modern life," constituting the highest development of the formally rational, instrumentalist culture that was uniquely identified, in Weber's Eurocentric perspective, with the West. "In Western civilization, and in Western civilization only," he wrote, were to be found "cultural phenomena which (as we like to think) lie in a line of development having *universal* significance and value."[17]

This naturalization of fundamental capitalist relations of production is deeply embedded within neoclassical economics, where historical elements hardly enter at all. In the prevailing reductionist view in the dismal science, the same abstract factors of production associated with capital are seen as common to absolutely all societies. As Thorstein Veblen critically observed in 1908, "A gang of Aleutian Islanders slushing about in the wrack and surf with rakes and magical incantations for the capture of shell-fish are held, in point of taxonomic reality, to be engaged on a feat of hedonistic equilibration in rent, wages, and interest. . . . All situations are, in point of economic theory, substantially alike."[18] Society is seen by conventional economists primarily in a positivistic mode in terms of invariant laws, of which the market in capitalism is the supreme expression.[19] In this view all historical laws associated with particular social systems as historically specific, emergent forms of organization with their own properties, are deemed false. All developments are in effect predetermined by universal, innate, unchanging properties, with capitalist modernity implicitly representing the ultimate working out of these fundamental principles.[20]

In line with this general loss of historical perspective, technology is often treated today as if it were innately capitalist, based on Joseph Schumpeter's famous notion of "creative destruction," which was derived from Marx's conception of capitalism as a revolutionary technological force.[21] The effect of this in current discussions has been to reinforce the belief in the immutability of capitalism with widespread notions of technological determinism, designating all progress as somehow uniquely capitalist and predestined.[22] In the face of climate change, it is generally assumed in the prevailing outlook that all solutions to the most pressing social problems are technological and all technologies that might conceivably address the dire challenges we face are compatible with capitalism.

Central to the denial of historicity of both past and present, related to prevailing notions of economic and technological determinism, is the almost complete identification of capitalism with modernity. As sociologist Peter L. Berger put it in his article "Capitalism and the Disorders of Modernity": "Capitalism is a thoroughly modern phenomenon, perhaps even the most modern phenomenon of all."[23] The main alternative to capitalism in terms of modernity were Soviet-type economies, but with their demise, and the triumph of capitalism, there was seemingly no alternative to capitalism in the context of modernity.[24] Indeed, many leftists, who themselves came to accept the end of history, began to see capitalism itself in terms of a postmodernity in which the future had been decapitated, emphasizing how capital and technological imperatives had annihilated all grand, meta-historical projects.[25]

For cultural critic Leo Marx, "The pessimistic tenor of postmodernism follows from this inevitably diminished sense of human agency."[26] Here the battle with capitalist modernity is reduced to a shadowy postmodern exercise in the cultural interstices of the system, rather than a genuine emancipatory project. This perspective thus becomes one of disenchantment and de-Enlightenment, a stance of perpetual, if somewhat detached and ironic, defeat.[27] As Wood wrote, "In the final analysis, 'postmodernity' for postmodernist intellectuals seems to be not a historical moment but the human condition itself, from which there is no escape."[28] In the words of cultural theorist Keti Chukhrov, "The capitalist undercurrent of these emancipatory and critical theories functions not as a program to exit from capitalism, but rather as the radicalization of the impossibility of this exit."[29]

The cumulative effect of these various interconnected notions of capitalism as the end of history has been to enshrine capitalism as a permanent reality, more phenomenally real and of greater seeming importance to people's lives than the physical universe itself. Capitalism, in fact, is often presented not only as the end of history but as the end of natural history, based in the conquest of nature that is often presented as its greatest achievement. Even the advent of climate change has not quite shaken this hegemonic belief.[30]

Indeed, the notion that capitalism constitutes the ultimate boundary to human existence is so embedded in today's dominant ideology that, as Derrick Jensen and Aric McBay wrote in *What We Leave Behind*, it gives rise to a cultural outlook in which there is an "inversion of what is real and not real," where "dying oceans and dioxin in every mother's breast milk" are considered less real than "industrial capitalism." Hence, we are constantly led to believe that "the end of the world is less to be feared than the end of industrial capitalism. . . . When most people in this culture ask, 'How can we stop global warming?' That's not really what they are asking. They're asking, 'How can we stop global warming without significantly changing this lifestyle that is causing global warming in the first place?' The answer is that you can't. It's a stupid, absurd, and insane question."[31] It is this same ruling ideological view that Fredric Jameson was to capture in his famous aside: "Someone once said that it is easier to imagine the end of the world than the end of capitalism."[32] Nothing indeed so clearly captures the capitalist universalism, parading as realism, that dominates contemporary ideology, closing off the future as history.[33]

A New Eco-Revolutionary Wave

Confronted with the received ideology of a "barred future," which denied the continuing role of revolution in human history, Sartre passionately declared, even "a barred future is still a future." This adamant refusal to accept capitalism as a boundary that could never be crossed drew its essential meaning not simply from an abstract conception of human agency, but also from the fact that we live, as he said, in "a time of incredible revolutions."[34]

The "incredible revolutions" emerging in our time are, as in previous historical eras, aimed at the ever wider social control of the means of production. Yet, unlike some previous class struggles and revolutionary movements, this is no longer conceived today mainly in narrow *economic* terms but also increasingly in *ecological* terms, reflecting the fact that it is the social metabolism between human beings and nature that constitutes the most ineluctable basis of human history. The agent of revolution is increasingly a working class that is not to be conceived in its usual sense as a purely economic force but as an environmental (and cultural) force: an *environmental proletariat*.

From a historical-materialist perspective this should hardly surprise us. Most of the major class struggles and revolutionary movements over the centuries of capitalist expansion have been animated in part by what could be called ecological imperatives—such as struggles over land, food, and environmental conditions—going beyond narrower political-economic objectives. The English Revolution and French Revolution of the seventeenth and eighteenth centuries, respectively, involved intense struggles over land ownership, represented by the Diggers and the Levellers in the former, and the Great Peasant Revolt in the latter.[35] E. P. Thompson concluded his great work *The Making of the English Working Class* by indicating that no one else after William Blake (perhaps with the exception of William Morris) was fully at home in *cultures of resistance* against "Acquisitive Man," both that of the Romantic criticism of Utilitarianism rooted in struggles over the land, aesthetics, and environment, and that of the industrial workers fighting capital. It was the separation of these two great movements, he suggested, that led in the end to a working-class struggle that gravitated toward a mere "resistance movement" rather than a "revolutionary challenge" to capitalism.[36]

Yet, it would be wrong to see this separation as ever being absolute. If the Romantics started with the struggle over the land and nature, they nevertheless, through radical figures like Percy Bysshe Shelley, John Ruskin, and Morris, provided devastating critiques of bourgeois political economy, often overlapping with the working-class struggle.[37] The English proletariat in the nineteenth century fought an environmental struggle that was no less serious due to capitalism's total separation of the workers from the land and the annihilation of a livable environment for those laboring in the industrial cities. Frederick Engels's account of "social murder" in Manchester and other English factory towns in 1844 focused especially on the *environmental conditions* of the working class.[38] Marx, partly inspired by Engels, wrote in 1844:

> Even the need for fresh air ceases to be a need for the worker. Man reverts once more to living in a cave, but the cave is now polluted by the mephitic and pestilential breath of civilization. Moreover, the worker has no more than a precarious right to live in it, for it is for him an alien power that can be daily withdrawn and from which, should he fail to pay, he can be evicted at any time. He actually has to *pay* for this mortuary. A dwelling in the *light*, which Prometheus describes in Aeschylus as one of the great gifts through which he transformed savages into men, ceases to exist for the worker. Light, air, etc.—the simplest *animal* cleanliness—ceases to be a need for man. *Dirt*—this pollution and putrefaction of man, the *sewage* (this word is to be understood in its literal sense) of civilization—becomes an *element of life* for him. *Universal neglect*, putrefied nature, becomes an *element of life* for him.[39]

The proletariat was conceived by Marx as stripped of all direct connections to the means of production, notably the land and natural resources (as well as tools, factories, machinery), on which all human existence depended. It was thereby forced into struggles over capitalism's one-sided destruction of the conditions of life, and the environment, and compelled ultimately to enter into a battle over the entirety of the human social metabolism with nature. "The living conditions of the proletariat," Marx and Engels wrote in *The Holy Family*, "represent the focal point of all inhuman conditions in contemporary society.... It cannot emancipate itself without abolishing the conditions which give it life, and it cannot abolish these conditions without abolishing *all* those inhuman conditions of social life today which are summed up in its own situation."[40]

The question of *materialism* for classical historical materialism was therefore both about what Marx called "the universal metabolism of nature" and about the mode of production (or social metabolism) in a given historical case—the latter viewed as an emergent form of nature with its own properties. In this way, the materialist conception of nature developed by natural science and the materialist conception of history of scientific socialism were seen as dialectically connected. In Marx's analysis, the labor-and-production process was itself defined as the "social metabolism" of humanity and nature. Production was thus both a social relation between human beings and a social-ecological relation between human beings and nature. If economic crises under capitalism were breaks in the accumulation of capital, ecological crises took the form of ruptures in the social metabolism, such that the "the eternal natural condition[s]" of this metabolism were undermined—as explained in Marx's famous theory of the metabolic rift.[41]

In such a perspective, militant class struggles and revolutionary movements were engendered by contradictions that arose in the social metabolism of humanity and nature in both of its material aspects: political-economic and natural-environmental. Revolutionary movements did not simply emerge because of fetters on the expansion of production—what could be seen as more economic causes—but also as a result of the destruction of people's actual living conditions and of the natural conditions of production of themselves. If in the former case, the potential of human development was undermined, in the latter, at least in the more dire instances, as in Ireland, in the mid-nineteenth century, it became a case of "ruin or revolution."[42]

It is this complex understanding of the struggle for the land/nature/environment, which was crucial to classical historical materialism, that explains why Marx and Engels, while emphasizing the role of the proletariat as the leading revolutionary force in developed capitalist economies, never denied either the past or present significance of peasant revolts in the struggle against bourgeois society—an approach that also extended to their growing support from the late 1850s on for all Indigenous struggles against colonialism. Thus, classical historical materialism, as distinct from some socialist tendencies, never portrayed the peasantry as simply a reactionary class. The very issue of proletarianization in the age of "so-called primitive accumulation" (or the age of original expropriation) was connected to the enclosure of the commons and the overthrow of the customary rights of the workers. For Marx, this could not be explained in terms of some kind of economic determinism or the superior productivity of capitalism, but rather was a product of "the opportunity that makes the thief."[43] The populace was fully justified in defending their rights to the commons, that is, their communal property rights. Indeed, the proletarian struggle itself pointed ultimately toward what Marx called "the negation of the negation," the expropriation of the "expropriators."[44]

In the classical historical-materialist view, few things were more important than the abolition of the big land monopolies that divorced the majority of humankind from a direct relation to nature, the land as a means of production, and a communal relation to the earth. Marx delighted in quoting Herbert Spencer's chapter from his *Social Statics* (1851) on "The Right to the Use of the Earth," where Spencer stated: "Equity . . . does not permit property in land, or the rest would live on the earth in sufferance only. . . . It is impossible to discover any mode in which land can become private property. . . . A claim to the exclusive possession of the soil involves land-owning despotism."

Land, Spencer declared, and Marx underscored, properly belongs to "the great corporate body—society." Human beings were "co-heirs" to the earth.[45]

The recognition that struggles over the land and peasant wars were integral to resistance to capitalism can be seen in Marx's statement, in an 1856 letter to Engels, that "the whole thing in Germany will depend on whether it is possible *to back the proletarian revolution by some second edition of the Peasants War*"—that is, through a struggle in which the urban proletariat and the rural peasantry (agricultural laborers) were both engaged, constituting a battle for both the cities and the land.[46] In this Marx was building on the implications of Engels's 1850 *The Peasant War in Germany*. In the context of the rise of revolutionary movements in Russia in the 1870s and '80s, Marx at the end of his life placed heavy emphasis on the archaic Russian commune and sided with the revolutionary Russian populists in seeing the peasantry, who were concerned above all with defending their customary collective relations to the land, as playing a crucial role in the coming Russian Revolution.[47]

It is this same perspective, focusing on the need of all direct producers throughout the globe for the collective control of their own means of production, thus opposing the expropriation of lands and bodies, that led to Marx and Engels's strong attacks, beginning in the late 1850s, on colonialism, along with their defense of the revolts of Indigenous peoples throughout the world. In particular, they supported Indigenous revolts against expropriation and extermination, in Ireland, India, China, Algeria, South Africa, and the Americas.[48] With respect to the East Indies, Marx wrote: "Everyone but Sir Henry Maine and others of his ilk realises that the suppression of communal landownership out there was nothing but an act of English vandalism, pushing the native people not forwards but backwards."[49] Likewise criticizing the destruction by the British of the irrigation system of India and the famines leading to the deaths of millions of people, Marx pointed directly to the devastating effects of Western ecological imperialism.[50] Such a viewpoint anticipated the numerous peasant and proletarian-led peasant wars of the twentieth century, most of these Marxist-inspired revolutions, including those of Mexico, Russia, China, Viet Nam, Algeria, and Cuba—all of which arose in the context of resistance to imperialism, and all involved intense struggles over the land and the environment.[51]

In general, Third World liberation movements have been aimed at both the environment and economy and have been struggles in which peasants and Indigenous peoples have played central roles, together with nascent proletarian and petty bourgeois forces. Often these wars of resistance and revolution have been waged by alliances between a proletariat and peasantry jointly resisting imperialism, fighting for peace, bread, and land.[52] For the great African Marxist liberation fighter Amilcar Cabral, the basis of revolutionary action in a colonial encounter required a "return to the source" of Indigenous culture associated with a given population's historical relations to its material environment.[53]

If capitalism begins with the *extensive, external expropriation* of lands and bodies, it then uses this as the basis from which it constructs a system of *intensive, internal exploitation* of human labor. In this dual process of expropriation and exploitation capitalist private property exhausts the environmental conditions of production and life, seeking to externalize this destruction onto the wider social and ecological realms on a global basis. It follows that as capitalism proceeds with its accumulation on an increasingly global basis, its destruction simply knows no barriers, extending to the world environment as a whole. In *The German Ideology*, Marx and Engels captured this increasingly one-sided, yet all-encompassing destructive character of capitalist production:

> In the development of productive forces there comes a stage when productive forces and means of intercourse are brought into being which, under the existing relations, only cause mischief, and are no longer productive but destructive forces. . . . These productive forces receive under the system of private property a one-sided development only, and for the majority they become destructive forces; moreover a great many of these

forces can find no application at all within the system of private property.... [Labor and production] now diverge to such an extent that material life appears as the end, and what produces this material life, labor ... as the means. Thus things have now come to such a pass that the individuals must appropriate the existing totality of productive forces, not only to achieve self-activity, but, also, merely to safeguard their own existence.[54]

It was, in fact, the perception of the "negative, i.e. destructive side" of capitalist production that Marx sought to capture in his theory of the metabolic rift.[55] His analysis here focused initially on the rift in the soil metabolism associated with the export of soil nutrients with the food and fiber sent to the new densely populated urban areas. This contributed to the pollution of the cities together with the loss of soil fertility in rural areas. Similar rifts or ruptures in the social metabolism between humanity and nature, Marx recognized, were common to capitalism's entire expropriation of nature, and materialized in innumerable ways, not least of all, as he pointed out, in periodic epidemics.[56]

Engels's *Condition of the Working Class in England*, which provided the original materialist understanding of the proletariat that was to be the basis of historical materialism, was concerned with the growth of the industrial working class in the new manufacturing towns and introduced the concept of the industrial reserve army of the unemployed. But most of Engels's analysis in the book was devoted to the social epidemiology of working-class life and the etiology of disease. The combination of the critique of political economy with the critique of environmental and epidemiological conditions and their relation to the reproduction of the laboring class under capitalism helps us to understand the enormous radicalism of that time just couple of years after the 1842 General Strike or Plug Plot Riots, in which factory workers were struggling simultaneously against the economic and environmental degradations created by capitalism.[57] The movements for economic justice in the nineteenth century and into the twentieth century were accompanied by struggles for environmental justice. Socialists, and particularly Marxists, in the early twentieth century were to pioneer in the development of an ecological critique side by side and dialectically interconnected with historical materialism's economic critique.[58]

Today, faced with a planetary ecological crisis, environmental hazards are everywhere, extending from climate change to ocean acidification, to the sixth extinction, to the disruption of the nitrogen and phosphorus cycles, to deforestation and loss of ground cover, to desertification, to ubiquitous pollution by synthetic chemical and radioactive wastes, to pandemics, to the destruction of the soil metabolism. These destructive influences are now part of our daily lives: from heat waves to megastorms to rising sea levels to COVID-19 and other pandemics.

Marx's original notion (based primarily on the work of the great German chemist Justus von Liebig) of the degradation of the soil through the loss of soil nutrients has now given way to concerns about the loss of soil organic matter or soil carbon, a factor contributing to climate change.[59] Everywhere we are confronted with the reality that capitalism has now generated the Anthropocene Epoch in geological time (and what has been referred to here in this book as the the first geological age of the Anthropocene, the Capitalinian Age). The human economy is now the main driver of Earth System change, disrupting planetary boundaries to the point that changes that previously would have only taken place over millions of years are now occurring in decades. All material struggles are now environmental-class as well as economic-class struggles, with the separation between the two fading. More and more it is becoming clear to humanity as a whole that the needed revolutionary break with the system is not simply a question of removing capitalism's fetters on human advance, but, beyond that, and more importantly, countering its systemic destruction of the earth as a place of human habitation (and the habitat of innumerable other species)—a question of *ruin or revolution.*

THE EMERGING ENVIRONMENTAL PROLETARIAT

The objective consequence of the changing social and eoclogical environment, the product of uncontrolled capitalist globalization and accumulation, arising from forces at the center of the system, is inevitably to create a more globally interconnected revolutionary struggle: a new eco-revolutionary wave emanating primarily from the Global South, but with rapidly developing transnational alliances, reflecting the undermining of the material conditions for the "chain of human generations" throughout the planet.[60] In this emerging global conflict, economic struggles are only meaningful if they are also environmental struggles, while environmental movements must equally be economic ones. Ultimately it requires, as Cabral stated, a *return to the source*, drawing vital insights from historic customary-communal-collective cultures, which have to be reinvented, their principles enlarged, under the conditions imposed by capitalism in the Anthropocene. The best way to understand these multiple challenges is in terms of the objectively conditioned role of an emerging environmental proletariat, engaged with promoting a new, more unified social materiality aimed at a world of sustainable human development. All conscious action has the future as its object, which cannot realistically be conceived today apart from ecological revolution.[61]

The prospect of a new eco-revolutionary wave, is foreshadowed by various movements and struggles throughout the world, including (1) the Landless Workers' Movement (MST) in Brazil; (2) the international peasants alliance La Via Campesina; (3) Venezuela's nascent, if besieged, communal state; (4) Cuba's revolutionary ecology and epidemiology; (5) the natural resource nationalist, anti-extractivist, and postcolonial movements in Africa; (6) the Farmer's Revolt in India; (7) China's goal of a socialist-based ecological civilization; (8) the student-led climate strikes in Europe; (9) the Green New Deal, Red New Deal, just transition, environmental justice, and Black Lives Matter struggles in the United States and Canada; and (10) the revival on every inhabited continent of Indigenous environmental struggles.[62] Everywhere these radical movements, occurring at multiple levels, are finding ways to unite with more traditional workers' struggles and calls for a New International of workers and peoples.[63]

Almost unlooked for, Indigenous resistance around the world

has come to play a leading role in the development of what could be called a broad-based environmental-proletarian revolt. In his book, *Our History Is the Future: Standing Rock versus the Dakota Access Pipeline, and the Long Tradition of Indigenous Resistance* (2019), Nick Estes writes:

> Indigenous peoples must lead the way. Our history and long traditions of Indigenous resistance provide possibilities for futures premised on justice. After all, Indigenous resistance is animated by our ancestors' refusal to be forgotten, and it is our resolute refusal to forget our ancestors and our history that animates our vision for liberation. Indigenous revolutionaries are the ancestors from the before and before and the already forthcoming. There is a capaciousness to Indigenous kinship that goes beyond the human.... Whereas past revolutionary struggles have strived for the emancipation of labor from capital, we are challenged not just to imagine, but to demand the emancipation of the earth from capital. For the earth to live, capitalism must die.[64]

In the dire conditions of the Anthropocene Epoch, there is no answer for the human world that does not address the triple threats of capitalism, colonialism, and imperialism. In this sense, history, rather than having come to an end, as claimed by the received ideology, is today entering its most decisive phase. Hundreds of millions of people have now entered actively into the struggle for a world of substantive equality and ecological sustainability, constituting the fundamental meaning of socialism and the future of history in our time. Yet, the planetary revolt of humanity in the twenty-first century will prove "irresistible and irreversible," and thus succeed against all odds, only if it takes the form of a more unified, revolutionary human subject, emanating from "the wretched of the earth," an environmental proletariat.[65] It is time to exit the burning house.

Notes

Preface
1. Georg Lukács, *Tactics and Ethics* (London: New Left Books, 1972), 8.

Introduction
1. Epigraph: John Ruskin, *Deucalion* (Boston: Aldine Publishing, n.d.; written in 1879), 201.
2. See Fredric Jameson, "Future of the City," *New Left Review* 21 (second series) (May–June 2003), 76; Mark Fisher, *Capitalist Realism* (Winchester, UK: Zero Books, 2009), 2.
3. Jason W. Moore, *Capitalism and the Web of Life* (London: Verso, 2016). "The Web of Life is woven: & the tender sinews of life created / And the Three Classes of Men aged by Los's hammer." William Blake, *Collected Poetry and Prose* (New York: Random House, 1982), 100.
4. See chapter 21, "The Capitalinian: The First Geological Age of the Anthropocene," in this book.
5. "Capital," "Capitalism," and "Capitalist," *Oxford English Dictionary: The Compact Edition* (Oxford: Oxford University Press, 1971), vol. 1, 334.
6. Adam Smith, *The Wealth of Nations* (New York: Random House, 1937), 314.
7. Raymond Williams, *Keywords* (Oxford: Oxford University Press, 1983), 50–51.
8. Thomas Hodgskin, *Labour Defended Against the Claims of Capital* (London: Labour Publishing Co., 1922), 71, 95–96.
9. Williams, *Keywords*, 50.

10. "Capitalism," *Oxford English* Dictionary, vol. 1, 334.
11. William Makepeace Thackeray, *The Newcomes* (London: Penguin, 1996), 488; see also 42–44, 110, 422, 521.
12. Karl Marx, *Capital*, vol. 1 (London: Penguin, 1976).
13. Marx, *Capital*, vol. 1, 92, 742.
14. Marx, *Capital*, vol. 1, 247, 875–76. The notion of the "long sixteenth century" was first presented by Fernand Braudel in his work *The Mediterranean* and then was advanced by Immanuel Wallerstein in his analysis of the modern world system. See Fernand Braudel, *The Mediterranean and the Mediterranean World in the Age of Philip II*, vol. 1 (Berkeley: University of California Press, 1996), 326; Immanuel Wallerstein, *The Modern World System I: Capitalist Agriculture and the Origins of the European World-Economy in the Sixteenth Century* (London: Academic Press, 1974), 67–68.
15. Fernand Braudel, *The Perspective of the World* (New York: Harper and Row, 1979), 177–276; Pepijn Brandon, "Marx and the Dutch East India Company," *Historical Materialism*, June 13, 2019, https://www.historicalmaterialism.org/blog/marx-and-dutch-east-india-company.
16. Andreas Malm, *Fossil Capital* (London: Verso, 2016).
17. The broad framework of this analysis was provided in Marx's treatment of "the general law of accumulations." See esp. Marx, *Capital*, vol. 1, 776–80.
18. V. I. Lenin, *Imperialism: The Highest Stage of Capitalism* (New York: International Publishers, 1939); Paul M. Sweezy, "Monopoly Capitalism," in *The New Palgrave Dictionary of Economics* (London: Macmillan, 1987), 541–44; Paul A. Baran and Paul M. Sweezy, *Monopoly Capital* (New York: Monthly Review Press, 1966).
19. Milton Friedman, *Capitalism and Freedom* (Chicago: University of Chicago Press, 1968).
20. See Daniel Singer, *Whose Millennium?: Theirs or Ours* (New York: Monthly Review Press, 1999).
21. Robert Heilbroner, *The Nature and Logic of Capitalism* (New York: W.W. Norton, 1985), 33, 52.
22. Heilbroner, *The Nature and Logic of Capital*, 53–77; Harry Magdoff and Paul M. Sweezy, *Stagnation and the Financial Explosion* (New York: Monthly Review Press, 1987), 153–62; Marx, *Capital*, vol. 1, 247–80.
23. Karl Marx and Frederick Engels, *Collected Works* (New York: International Publishers, 1975), vol. 37, 732–33.
24. Ian Angus, "Anthropocene Working Group: Yes, a New Epoch Has Begun," Climate and Capitalism, January 9, 2016, https://climateandcapitalism.com/2016/01/09/anthropocene-working-group-yes-a-new-epoch-has-begun/.

25. On the synthetic age see John Bellamy Foster, *The Vulnerable Planet* (New York: Monthly Review Press, 1999), 112–24.
26. Will Steffen, "Mid-20th Century 'Great Acceleration,'" in *The Anthropocene as a Geological Time Unit*, ed. Jan Zalasiewicz, Colin N. Waters, Mark Williams, and Colin P. Summerhayes (Cambridge: Cambridge University Press, 2019), 254–60; John R. McNeill and Peter Engelke, *The Great Acceleration* (Cambridge, MA: Harvard University Press, 2014); Ian Angus, *Facing the Anthropocene* (New York; Monthly Review Press, 2016); Stephen A. Marglin and Juliet B. Schor, *The Golden Age of Capitalism* (Oxford: Oxford University Press, 1990). The Industrial Revolution, which marks the beginning of fossil capitalism, is often seen as the origin of the Anthropocene. However, scientists have increasingly seen it as a "precursor" with the stratigraphic markers of the Anthropocene associated with the mid-century Great Acceleration. See John R. McNeill, "The Industrial Revolution and the Anthropocene," in Zalasiewicz, ed., *The Anthropocene as a Geological Time Unit*, 253–54.
27. Barry Commoner, *Science and Survival* (New York: Viking Press, 1966). Signficantly, Commoner also referred in this work to global warming as a potential threat to humanity, 10–11, 125–26.
28. Barry Commoner, *The Closing Circle* (New York: Bantam, 1971); John Bellamy Foster, *The Return of Nature* (New York: Monthly Review Press, 2020), 502–19.
29. Foster, *The Vulnerable Planet*, 112–18.
30. Spencer R. Weart, *The Discovery of Global Warming* (Cambridge, MA: Harvard University Press, 2003), 154–56.
31. E. Ray Lankester, *The Kingdom of Man* (New York: Henry Holt and Co., 1911), 159, 184; Foster, *The Return of Nature*, 61–64.
32. E.V. Shantser, "Anthropogenic System (Period)," in *The Great Soviet Encyclopedia*, vol. 2 (New York: Macmillan, 1973), 140. Attempts have been made to downplay the earlier use of the term "Anthropocene" in English to refer to anthropogenic changes in the biosphere, by referring to it as simply "transcribed" from the Russian. Jacques Grinevald, John McNeill, Naomi Oreskes, Will Steffen, Colin P. Summerhayes and Jan Zalasiewicz, "History of the Anthropocene Concept," in Zalasiewicz, ed., *The Anthropocene as a Geological Time Unit*, 6. It was, however, a *translation* (or one form of the translation) of a definite *scientific concept* used (along with Anthropogene) by Soviet geologists and geochemists, beginning with Pavlov and appearing in English in the 1970s.
33. Vladimir I. Vernadsky, *The Biosphere* (New York: Springer-Verlag, 1998).
34. Vladimir I. Vernadsky, "Some Words About the Noösphere," in *150*

Years of Vernadsky, vol. 2: *The Noösphere* (Washington, D.C.: 21st-Century Science Associates, 2014), 82. There were of course many precursors to Earth System analysis. This included not only Vernadsky, but in many ways even more important, the modern scientific theory of the origins of life and how life's transformation of the atmosphere was key to the development of life itself, originating with the simultaneous discoveries of J. B. S. Haldane in Britain and N. I. Vavilov in the Soviet Union. See J. D. Bernal, *The Origin of Life* (New World Publishing, 1967).

35. G. Evelyn Hutchinson, "The Biosphere," *Scientific American* 233/3 (1970): 45–53
36. Donella H. Meadows, Dennis L. Meadows, Jørgen Randers, and William W. Behrens III, *The Limits to Growth* (Washington, D.C.: Potomac Associates, 1972).
37. James E. Lovelock and Lynn Margulis, "Atmospheric Homeostasis by and for the Biosphere," *Tellus* 26/1–2 (1974): 1–9; James E. Lovelock, *Gaia: A New Look at Life on Earth* (Oxford: Oxford University Press, 2000).
38. Tim Lenton, *Earth System Science: A Very Short Introduction* (Oxford: Oxford University Press, 2016), 1–8.
39. Johan Rockström et al., "A Safe Operating Space for Humanity," *Nature* 461/24 (2009): 472–75; Will Steffen et al., "Planetary Boundaries," *Science* 347/6223 (2015): 736–46; Richard E. Leakey and Roger Lewin, *The Sixth Extinction: Biodiversity and Its Survival* (New York: Anchor, 1996).
40. Will Steffen, "Commentary," in *The Future of Nature: Documents of Global Change*, ed. Libby Robbin, Sverker Sörlin, and Paul Warde (New Haven: Yale University Press, 2013), 486; Paul J. Crutzen, "The Geology of Mankind," *Nature* 415 (2002): 23.
41. Clive Hamilton and Jacque Grinevald, "Was the Anthropocene Anticipated?" *Anthropocene Review* 2/1 (2015): 67.
42. Steffen et al., "Planetary Boundaries"; Rob Wallace, *Dead Epidemiologists* (New York: Monthly Review Press, 2020).
43. Robert Heilbroner, *21st-Century Capitalism* (New York: W.W. Norton, 1993), 151.
44. Thomas More, *Utopia* (Mineola, NY: Dover Books, 1997).
45. Elizabeth Kolbert, *The Sixth Extinction: An Unnatural History* (New York: Picador, 2015); Giovanni Strona and Corey J. A. Bradshaw, "Co-Extinctions Annihilate Planetary Life During Extreme Environmental Change," *Scientific Reports* 8/16274 (2018).
46. United Nations, Intergovernmental Panel on Climate Change (IPCC), *Sixth Assessment Report: Part I: The Physical Science Basis; Summary for Policy Makers* (August 9, 2021), 9.

47. IPCC, *Sixth Assessment Report: Part 1, Summary for Policy Makers*, 10–12, 15–18, 21–23, 34, 41, https://www.ipcc.ch/report/ar6/wg1/downloads/report/IPCC_AR6_WGI_SPM.pdf/; IPCC, *Sixth Assessment Report: Part 3; Summary for Policy Makers* (August 2021) (leaked draft), https://mronline.org/wp-content/uploads/2021/08/summary_draft1.pdf/, C1.3.
48. IPCC, *Sixth Assessment Report: Part I, Summary for Policy Makers*, 18; Climate Central, *Global Weirdness* (New York: Pantheon, 2012).
49. IPCC, *Sixth Assessment Report: Part I, Summary for Policy Makers*, 18.
50. Ibid.
51. IPCC, *Sixth Assessment Report: Part 3; Summary for Policy Makers* (leaked draft); "The Leaked IPCC Reports: Notes from the Editors," *Monthly Review* (October 2021), https://monthlyreview.org/2021/09/08/mr-073-05-2021-09_0/; Scientist Rebellion, "We Leaked the Upcoming IPCC Report" (August 2021), https://scientistrebellion.com/we-leaked-the-upcoming-ipcc-report/.
52. IPCC, *Sixth Assessment Report: Part 3; Summary for Policy Makers* (leaked draft), C1, D.2.2.
53. IPCC,, *Sixth Assessment Report: Part 3; Chapter One* (leaked draft), 1.4, https://mronline.org/wp-content/uploads/2021/08/chapter01.pdf/; "The Leaked IPCC Reports: Notes from the Editors," *Monthly Review* (October 2021); Juan Bordera et al., "Leaked Report of the IPCC Reveals that the Growth Model of Capitalism Is Unsustainable," August 22, 2021, https://mronline.org/2021/08/23/leaked-report-of-the-ipcc-reveals-that-the-growth-model-of-capitalism-is-unsustainable.
54. "Secretary General's Statement on the IPCC Working Group 1 Report on the Physical Science Basis of the Sixth Assessment," United Nations Secretary General, August 9, 2021.
55. "Call for Emergency Action to Limit Global Temperature Increases, Restore Biodiversity, and Protect Health," *British Medical Journal* 374 (September 6, 2021), editorial published simultaneously with over 200 medical and health journals, https://www.bmj.com/content/374/bmj.n1734.
56. John Bellamy Foster, "Marx's Theory of Metabolic Rift," *American Journal of Sociology* 105/2 (September 1999): 366–405.
57. On Roland Daniels see Saito, *Karl Marx's Ecosocialism*, 72–78; Roland Daniels, *Mikrokosmos* (Frankfurt am Main: Lang, 1988).
58. Marx and Engels, *Collected Works*, vol. 30 (New York: International Publishers, 1988), 54–66.
59. Marx, *Capital*, vol. 1, 637. In an analysis of rifts in soil nutrient cycles, building on Marx's classical analysis, Fred Magdoff and Chris Wililams, writing from a contemporary science perspective, describe

three separate rifts in the nturient cycling associated with today's economy. See Fred Magdoff and Chris Williams, *Creating an Ecological Society* (New York: Monthly Review Press, 2017), 8–86. More recently, Magdoff has writen on the crisis of soil organic matter and its relation to the metabolic rift: Fred Magdoff, "Repairing the Soil Carbon Rift," *Monthly Review* 72/11 (April 2021): 1–13.

60. Justus von Liebig, "1862 Preface to *Agricultural Chemistry*," *Monthly Review* 70/3 (July–August 2018): 146–50; Marx, *Capital*, vol. 1, 637; Karl Marx, *Capital*, vol. 3 (London: Penguin, 1981), 949.
61. On this debate with regard to Marx's metabolism concept and the importance of dialectical mediation see chapters 11 and 12 of this book.
62. Marx, *Capital*, vol. 1, 905–7, 914–26.
63. Capitalism can thus be viewed as constituting a particular "ecohistorical period" or regime associated with a particular socioeconomic mode of production, rooted in the alienation of both nature and human labor. Foster, *The Vulnerable Planet*, 34–35.
64. Brett Clark and Richard York, "Carbon Metabolism: Global Capitalism, Climate Change, and the Biospheric Rift," *Theory and Society* 34/4 (2005): 391–428.
65. See Ryan Wishart and R. Jamil Jonna, "Metabolic Rift: A Selected Bibliography," July 6, 2020, https://monthlyreview.org/commentary/metabolic-rift/; Andreas Malm, *The Progress of This Storm* (London: Verso, 2018), 177–80.
66. Rosa Luxemburg, *Rosa Luxemburg Speaks* (New York: Pathfinder Press, 1970), 111.
67. On Marx's later contributions to the metabolic rift analysis, particularly in his notebooks, as well as the centrality of this analysis to overall critique, see Kohei Saito, *Karl Marx's Ecosocialism* (New York: Monthly Review Press, 2017).
68. Frederick Engels, *The Condition of the Working Class in England*, in Marx and Engels, *Collected Works*, vol. 4, 295–583.
69. Marx and Engels, *Collected Works*, vol. 25, 5–309.
70. See Sebastiano Timpanaro, *On Materialism* (London: Verso, 1975).
71. Georg Lukács, *History and Class Consciousness* (Cambridge, MA: MIT Press, 1971), 24.
72. Lukács, *History and Class Consciousness*, 207; Andrew Feenberg, *Lukács, Marx, and the Sources of Critical Theory* ((Totowa, NJ: Rowman and Littlefield, 1981); Foster, *The Return of Nature*, 16–20.
73. Marx and Engels. *Collected Works*, vol. 25, 492; Georg Lukács, *Hegel's False and His Genuine Ontology* (London: Merlin Press, 1978), 31.
74. Georg Lukács, *A Defence of "History and Class Consciousness": Tailism and the Dialectic* (London: Verso, 2000), 102–7.

75. Russell Jacoby, "Western Marxism," *A Dictionary of Marxist Thought*, ed. Tom Bottomore (Oxford: Blackwell, 1983), 523–26; Fredric Jameson, *Valences of the Dialectic* (London: Verso, 2009), 6–7.
76. See, for example, Neil Smith, *Uneven Development* (Athens: University of Georgia Press, 2008), 243–51, and "Nature as Accumulation Strategy," in *Socialist Register 2007* (New York: Monthly Review Press, 2006), 21–31. For a critical analysis of this tendency see Brian Napoletano, John Bellamy Foster, Brett Clark, Pedro S. Urquijo, Michael K. McCall, and Jaime Paneque Gálvez, "Making Space in Critical Environmental Geography for the Metabolic Rift," *Annals of the American Association of Geographers* 109/6 (March 2019): 1811–28.
77. Alfred Schmidt, *The Concept of Nature in Marx* (London: New Left Books, 1971). On the critique of Schmidt see John Bellamy Foster and Brett Clark, *The Robbery of Nature* (New York: Monthly Review Press, 2020), 191–98.
78. Marx and Engels, *Collected Works*, vol. 25, 511–12; Schmidt, *The Concept of Nature in Marx*, 186: Marx and Engels, *Collected Works*, vol. 5, 87.
79. Schmidt, *The Concept of Nature in Marx*, 154–62.
80. Ibid., 188.
81. Ibid., 169.
82. Ibid., 186; Marx and Engels, *Collected Works*, vol. 26, 387; Marx and Engels, *Collected Works*, vol. 25, 460–62.
83. Marx, *Capital*, vol. 1, 638.
84. Rachel Carson, *Silent Spring* (Boston: Houghton Mifflin, 1962).
85. Schmidt, *The Concept of Nature in Marx*, 165.
86. Foster, *The Return of Nature*, 12, 358–68. See also chapter 14 in this book.
87. Foster, *The Return of Nature*, 348–69, 358–501; Helena Sheehan, *Marxism and the Philosophy of Science* (Atlantic Highlands, NJ: Humanities Press, 1985).
88. Brian Napoletano, Brett Clark, Pedro D. Urquijo, and John Bellamy Foster, "Sustainability and Metabolic Revolution in the Work of Henri Lefebvre," *World* 1/1 (December 2020): 300–317.
89. Foster, *The Return of Nature*, 502–26; John Bellamy Foster, Brett Clark, and Hannah Holleman, "Capital and the Ecology of Disease," *Monthly Review* 73/2 (June 2021): 18–21; Rob Wallace, *Big Farms Make Big Flu* (New York: Monthly Review Press, 2016); John Bellamy Foster, Brett Clark, and Hannah Holleman, "Capital and the Ecology of Disease," *Monthly Review* 73/2 (June 2021): 1–23; Wishart and Jonna, "Metabolic Rift: A Selected Bibliography."
90. Joseph Needham, "Foreword," in Marcel Prenant, *Biology and Marxism* (New York: International Publishers, 1943), x.

91. See Kaan Kangal, *Friedrich Engels and the "Dialectics of Nature"* (London: Palgrave Macmillan, 2020); John Bellamy Foster, "Engels's *Dialectics of Nature* in the Anthropocene," *Monthly Review* 72/6 (November 2020): 1–17.
92. See John Bellamy Foster and Paul Burkett, *Marx and the Earth* (Chicago: Haymarket, 2016), 1–56.
93. On how COVID-19 is interlocked with both the metabolic rift and racism see Wallace, *Dead Epidemiologists*, 64–77, 98–105, 168–75.
94. Nancy Fraser, "Behind Marx's Hidden Abode: For an Expanded Conception of Capitalism," *New Left Review* 86 (2014): 63.
95. On expropriation as the key to capitalism's active relation to the environment and its connection to imperialism, racism and exploitative social reproduction, see John Bellamy Foster, Brett Clark, and Richard York, *The Ecological Rift* (New York: Monthly Review Press, 2010), 62; Foster and Clark, *The Robbery of Nature*, 35–53, 78–103; and Michael D. Yates, *Can the Working Class Change the World?* (New York: Monthly Review Press, 2018), 55–56. On *"Raubbau,"* see chapter 9 in this book.
96. See Michael C. Dawson, "Hidden in Plain Sight: A Note on Legitimation Crises and the Racial Order," *Critical Historical Studies* 3/1 (2016): 143–61; Sven Beckert, *Empire of Cotton* (New York: Vintage, 2014), xv–xvi; Nancy Fraser, "Roepke Lecture in Economic Geography—From Exploitation to Expropriation: Historic Geographies of Racialized Capitalism," *Economic Geography* 94/1 ((2018): 10.
97. Marx and Engels, *Collected Works*, vol. 5, 87.
98. On the concept of "ecological civilization" see chapter 20 of this book.
99. On the environmental proletariat see especially chapter 22.
100. Karl Marx and Friedrich Engels, *The Communist Manifesto* (New York: Monthly Review Press, 1964), 2.
101. For analyses on these lines see István Mészáros, *Beyond Leviathan* (New York: Monthly Review Press, 2022); John Bellamy Foster, "Chávez and the Communal State," *Monthly Review* 66/11 (April 2015): 1–17; Marx and Engels, *Collected Works*, vol. 5, 87.
102. Lukács, *History and Class Consciousness*, 86. On the concept of "third nature" see chapter 8 in this book; Karl Marx, *Capital*, vol. 3, 959.

1. Marx and the Rift in the Universal Metabolism of Nature

This chapter is adapted and revised for this book from the following article: John Bellamy Foster, "Marx and the Rift in the Universal Metabolism of Nature," *Monthly Review* 65/7 (December 2003): 1–19.

1. Jason W. Moore, "Transcending the Metabolic Rift," *Journal of Peasant Studies* 38/1 (January 2011): 1–2, 8, 11; Mindi Schneider and Philip M. McMichael, "Deepening, and Repairing, the Metabolic Rift," *Journal*

of *Peasant Studies* 37/3 (July 2010): 478, 482; Alexander M. Stoner, "Sociobiophysicality and the Necessity of Critical Theory," *Critical Sociology* 40/4 (2014): 621–42.
2. Schneider and McMichael, "Deepening, and Repairing, the Metabolic Rift," 481–82. See also Maarten de Kadt and Salvatore Engel-Di Mauro, "Failed Promise," *Capitalism, Nature, Socialism* 12/2 (2001): 50–56.
3. Georg Lukács, *History and Class Consciousness* (London: Merlin Press, 1968), 24.
4. The term "Western Marxism" was first introduced by Maurice Merleau-Ponty in *Adventures of the Dialectic* (Evanston, IL: Northwestern University Press, 1973). It was seen as deriving from the work of Lukács (*History and Class Consciousness*), Karl Korsch, the Frankfurt School, and Antonio Gramsci, and extending to most Western philosophical Marxists. It drew its principal inspiration from the rejection of what were seen as positivistic influences in Marxism, and the concept of the dialectics of nature in particular. See Russell Jacoby, "Western Marxism," in *A Dictionary of Marxist Thought*, ed. Tom Bottomore (Oxford: Blackwell, 1983), 523–26.
5. For an important defense of Engels in this respect see John L. Stanley, *Mainlining Marx* (New Brunswick, NJ: Transaction Publishers, 2002), 1–61. In the dedication to their landmark book *The Dialectical Biologist*, Levins and Lewontin write: "To Frederick Engels, who got it wrong a lot of the time but who got it right where it counted." Richard Levins and Richard Lewontin, *The Dialectical Biologist* (Cambridge, MA: Harvard University Press, 1985), v.
6. Lucio Colletti, *Marxism and Hegel* (London: Verso, 1973), 191–93; Jacoby, "Western Marxism," 524. See also Merleau-Ponty, *Adventures of the Dialectic*, 32; Jean-Paul Sartre, *Critique of Dialectical Reason*, vol. 1 (London: Verso, 2004), 32; Herbert Marcuse, *Reason and Revolution* (Boston: Beacon Press, 1960), 314; Alfred Schmidt, *The Concept of Nature in Marx* (London: New Left Books, 1971), 59–61; Steven Vogel, *Against Nature* (Albany: State University of New York Press, 1996), 14–19.
7. Gramsci explicitly argued that a complete rejection of the dialectics of nature would lead to "idealism" or "dualism" and the destruction of a materialist outlook, voicing this in a discussion of Lukács's *History and Class Consciousness*. Antonio Gramsci, *Selections from the Prison Notebooks* (London: Merlin Press, 1971), 448. For a sharp criticism of Western philosophical Marxism for its move away from materialism and any consideration of natural conditions see Sebastiano Timpanaro, *On Materialism* (London: Verso, 1975).
8. Lukács, *History and Class Consciousness*, 207.

9. Ibid., xvii.
10. Georg Lukács, *Conversations with Lukács* (Cambridge, MA: MIT Press, 1974), 43. Lukács added a clarification on the social aspect in the same paragraph: "Since the metabolism between society and nature is also a social process, it is always possible for concepts obtained from it to react on the class struggle in history."
11. Schmidt, *The Concept of Nature in Marx*, 78–79.
12. Georg Lukács, *A Defence of "History and Class Consciousness": "Tailism and the Dialectic"* (London: Verso, 2003), 96, 106, 113–14, 130–31. The later Lukács recognized, like Marx, that the more contemplative materialism associated with Epicurus, Bacon, Feuerbach, and modern science could generate genuine discoveries in science through processes of sense perception and rational abstraction, particularly when accompanied (as Engels had emphasized) by experimentation. Ultimately, all of this was related to the development of the relations of production, which constantly transformed human metabolic interaction with nature as well as social relations. See Lukács, *History and Class Consciousness*, xix–xx, and *A Defence of "History and Class Consciousness,"* 130–32; John Bellamy Foster, Brett Clark, and Richard York, *The Ecological Rift* (New York: Monthly Review Press, 2010), 229–31.
13. István Mészáros, *Marx's Theory of Alienation* (London: Merlin Press, 1970), 99–119, 162–65, 195–200, and *Beyond Capital* (New York: Monthly Review Press, 1995), 170–77, 872–97. Mészáros used "I" for industry rather than production in *Marx's Theory of Alienation* in his depiction of Marx's conceptual structure, to avoid confusing it with "P" for property. But industry obviously means production.
14. Karl Marx, *Capital*, vol. 3 (London: Penguin, 1981), 949.
15. See John Bellamy Foster, *Marx's Ecology* (New York: Monthly Review Press, 2000), 149–54.
16. Liebig quoted in K. William Kapp, *The Social Costs of Private Enterprise* (New York: Shocken Books, 1971), 35.
17. Karl Marx and Frederick Engels, *Collected Works*, vol. 42 (New York: International Publishers, 1975), 227.
18. Foster, *Marx's Ecology*, 155–62.
19. Karl Marx, *Capital*, vol. 1 (London: Penguin, 1976), 637–38.
20. Marx, *Capital*, vol. 1, 860; Brett Clark and John Bellamy Foster, "Guano: The Global Metabolic Rift and the Fertilizer Trade," in *Ecology and Power*, ed. Alf Hornborg, Brett Clark, and Kenneth Hermele (London: Routledge, 2012), 68–82.
21. Marx, *Capital*, vol. 3, 754, 959.
22. See Ryan Wishart, "The Metabolic Rift: A Selected Bibliography," October 16, 2013, https://monthlyreview.org/ commentary/metabolic-

rift; Foster, Clark, and York, *The Ecological Rift*; Paul Burkett, *Marxism and Ecological Economics* (Boston: Brill, 2006).
23. Moore, "Transcending the Metabolic Rift," 1–2, 8, 11.
24. Stoner, "Sociobiophysicality and the Necessity of Critical Theory," 7. It should be noted that Stoner aims his criticisms of the metabolic rift for its "non-reflexivity" at the present author rather than Marx directly. He does so based on the contention: "We must be careful about ascribing the theory of metabolic rift to Marx, since he did not use this terminology, and was not driven to develop a theory based on such terminology." However, Stoner neglects to provide any explanation (other than a specious reference to Adorno) as to why he thinks all of Marx's statements on the metabolism of nature and society and the rift in the social-ecological metabolism (from the *Grundrisse* in 1857–58 up through *Notes on Adolph Wagner* in 1879–1880) are actually nonexistent or have been falsely attributed to him. Indeed, since Stoner's essay was written Kohei Saito's work has extended still further our knowledge of Marx's treatment of the metabolic rift in his later ecological notebooks. See Kohei Saito, *Karl Marx's Ecosocialism* (New York: Monthly Review Press, 2017).
25. Schneider and McMichael, "Deepening, and Repairing, the Metabolic Rift," 478–82. Schneider and McMichael argue that the rift in the metabolism between nature and society generates an "epistemic rift" in which nature and society become separated within thought, creating various dualisms that depart from a dialectical perspective. Remarkably, they carry this analysis into a partial criticism of Marx's theory. In his value analysis, they suggest, Marx continually "risks a one-sided representation of the society-nature relationship," himself falling prey at times to such methodological dualism, since "the abstraction of value and of nature discount ecological relations in capital theory." Here they fail to recognize that Marx in the treatment of value relations was engaged in *critique*—of the value structure of capital itself. In his conception, capital fails to ground its value abstractions in ecological relations, and this is inherent in its character as an alienated mode of production. Marx makes this clear by sharply distinguishing *value* under capitalism from *wealth*—with the latter, as opposed to the former, having its source in both labor and the earth. See Karl Marx, *Critique of the Gotha Programme* (New York: International Publishers, 1938), 3.
26. Karl Marx and Frederick Engels, *Collected Works*, vol. 30, 54–66.
27. Such an analysis needs to be integrated with Marx's value-theory-based critique. This was accomplished in Paul Burkett, *Marx and Nature* (New York: St. Martin's Press, 1999).
28. Of course society, since it is materially produced, is also objective—a

historical manifestation of the metabolism between nature and humanity. See Lukács, *A Defence of "History and Class Consciousness,"* 100–101, 115.

29. On the role of "isolation" as the key to abstraction in a dialectical approach to science and knowledge see Hyman Levy, *The Universe of Science* (New York: Century Company, 1933), 31–81, and *A Philosophy for a Modern Man* (New York: Alfred A. Knopf, 1938), 30–36; Bertell Ollman, *Dialectical Investigations* (New York: Routledge, 1993), 24–27; Paul Paolucci, *Marx's Scientific Dialectics* (Chicago: Haymarket Books, 2007), 118–23, 136–42; and Richard Lewontin and Richard Levins, *Biology Under the Influence* (New York: Monthly Review Press, 2007), 149–66.

30. See István Mészáros, *Lukács' Concept of Dialectic* (London: Merlin Press, 1972), 61–91.

31. David Harvey, "History versus Theory: A Commentary on Marx's Method in Capital," *Historical Materialism* 20/2 (2012): 12–14, 36.

32. Karl Marx and Friedrich Engels, *MEGA* IV, 26 (Berlin: Akademie Verlag, 2011), 214–19. See also Joseph Beete Jukes, *The Student's Manual of Geology*, 3rd ed. (Edinburgh: Adam and Charles Black, 1872), 476–512; James Hansen, *Storms of My Grandchildren* (New York: Bloomsbury, 2009), 146–47.

33. Michael Hulme, "On the Origin of 'The Greenhouse Effect': John Tyndall's 1859 Interrogation of Nature," *Weather* 64/5 (May 2009): 121–23; Daniel Yergin, *The Quest* (New York: Penguin, 2011), 425–28; Friedrich Lessner, "Before 1848 and After," in Institute for Marxism-Leninism, ed., *Reminiscences of Marx and Engels* (Moscow: Foreign Languages Publishing House, n.d.), 161; Y. M. Uranovsky, "Marxism and Natural Science," in Nikolai Bukharin et al., *Marxism and Modern Thought* (New York: Harcourt, Brace and Co., 1935), 140; Spencer R. Weart, *The Discovery of Global Warming* (Cambridge, MA: Harvard University Press, 2003), 3–4; W. O. Henderson, *The Life of Friedrich Engels*, vol. 1 (London: Frank Cass, 1976), 262.

34. It is interesting to note in this regard that Marx's friend Lankester was to emerge as the most virulent early twentieth-century critic of the catastrophic human destruction of species throughout the globe, particularly in his essay "The Effacement of Nature by Man." See E. Ray Lankester, *Science From an Easy Chair* (New York: Henry Holt, 1913), 373–79.

35. Schneider and McMichael, "Deepening, and Repairing, the Metabolic Rift," 481–82. Others have been even more critical, claiming that Marx's analysis cannot be considered ecological because he did not use the word "ecology" (coined by Haeckel in 1866 but not in general use in Marx and Engels's lifetime—according to the *Oxford*

English Dictionary the first reference to the term in English, outside of translations of Haeckel's work, was in 1893) and because he could not have known about "the development of chemical sciences that produced PCBs, CFCs, and DDT." De Kadt and Engel Di-Mauro, "Failed Promise," 52–54.

36. The Earth System notions of biogeochemical cycles and of the biosphere had their origins in the work of the Soviet scientist V. I. Vernadsky in the 1920s, and reflected the extraordinary development of dialectical ecology in the USSR in this period—prior to the purges, directed at ecologists in particular in the 1930s. See Foster, *Marx's Ecology*, 240–44.

37. See "Œcology," *Oxford English Dictionary*, vol. 2 (Oxford: Oxford University Press, 1971), 1975; "Ecology," *Oxford English Dictionary Online*; Ernst Haeckel, *The History of Creation*, vol. 2, translation supervised and revised by E. Ray Lankester (New York: D. Appleton and Co., 1880), 354; E. Ray Lankester, *The Advancement of Science* (New York: Macmillan, 1890), 287–387; Arthur G. Tansley, "The Use and Abuse of Vegetational Concepts Terms," *Ecology* 16 (1935): 284–307; Foster, Clark, and York, *The Ecological Rift*, 324–34; Peter Ayres, *Shaping Ecology: The Life of Arthur Tansley* (Oxford: Wiley-Blackwell, 2012), 42–44.

38. Eugene P. Odum, "The Strategy of Ecosystem Development," *Science* 164 (1969): 262–70; Frank Benjamin Golley, *A History of the Ecosystem Concept in Ecology* (New Haven: Yale University Press, 1993), 70; Howard T. Odum and David Scienceman, "An Energy Systems View of Marx's Concepts of Production and Labor Value," in *Emergy Synthesis 3: Theory and Applications of the Emergy Methodology, Proceedings from the Third Biennial Emergy Conference, Gainesville, January 2004* (Gainesville, FL: Center for Environmental Policy, 2005): 17–43; Howard T. Odum, *Environment, Power, and Society* (New York: Columbia University Press, 2007), 303, 276; John Bellamy Foster and Hannah Holleman, "A Theory of Unequal Ecological Exchange: A Marx-Odum Dialectic," chapter 10 in this book.

39. I owe this description of the viewpoint of modern soil science and the effects of the changing human metabolism on the nutrient cycle to Fred Magdoff. See Fred Magdoff and Harold Van Es, *Better Soils for Better Crops* (Waldford, MD: Sustainable Agricultural Research and Education Program, 2009).

40. Frederick Engels, *On Marx's Capital* (Moscow: Progress Publishers, 1956), 95.

41. Frederick Engels, *The Housing Question* (Moscow: Progress Publishers, 1975), 92; see also Marx and Engels, *Collected Works*, vol. 25, 460–62.

42. For a reasoned account of the Lysenko controversy see Levins and Lewontin, *The Dialectical Biologist*, 163–96.

43. See John Bellamy Foster, "Marx's Ecology and Its Historical Significance," in Michael R. Redclift and Graham Woodgate, eds., *International Handbook of Environmental Sociology*, 2nd ed. (Northampton, MA: Edward Elgar, 2010), 106–20.
44. See Barry Commoner, *The Poverty of Power* (New York: Bantam, 1976), 236–44; Levins and Lewontin, *The Dialectical Biologist*, and *Biology Under the Influence*; Richard York and Brett Clark, *The Science and Humanism of Stephen Jay Gould* (New York: Monthly Review Press, 2011).
45. Erich Fromm, *The Crisis of Psychoanalysis* (Greenwich, CT: Fawcett Publications, 1970), 153–54. It is noteworthy that in his 1932 article, "The Method and Function of an Analytic Social Psychology," which played such a crucial formative role in the development of the Frankfurt School, Fromm emphasized the need to deal with the nature-society dialectic and pointed to the importance of Nikolai Bukharin's *Historical Materialism*, saying that it "underlines the natural factor in a clear way." Fromm could only have meant Bukharin's use in this work of Marx's concept of metabolism. The Frankfurt School, however, did not follow this path, which would have required a radical reconsideration of the whole, difficult question of the dialectics of nature. Consequently, thinkers such as Fromm, Horkheimer, Adorno, and Marcuse were later to make various broad, critical-philosophical observations on the domination of nature, which all too often lacked substantive, materialist reference points with respect to ecosystem analysis, ecological science, and ecological crises themselves. Although the critical apparatus that they were able to employ allowed them to perceive the general conflict between capitalist society and the environment, the separation that had occurred between Western Marxism and natural science hindered further development in a field that demanded a critical or dialectical naturalism/realism and the recognition of nature's own dynamics. On this general problem see Roy Bhaskhar, *The Possibility of Naturalism* (Atlantic Highlands, NJ: Humanities Press, 1979). On Adorno's limited recognition of the importance of Marx's concept of social metabolism see Deborah Cook, *Adorno on Nature* (Durham, UK: Acumen, 2011), 24–26, 103–4.
46. Marina Fischer-Kowalski, "Society's Metabolism," in Michael Redclift and Graham Woodgate, eds., *International Handbook of Environmental Sociology* (Northampton, MA: Edward Elgar, 1997), 122.
47. Helmut Haberl, Marina Fischer-Kowalski, Fridolin Krausmann, Joan Martinez-Alier, and Verena Winiwarter, "A Socio-Metabolic Transition Towards Sustainability?: Challenges for Another Great Transformation," *Sustainable Development* 19 (2011): 1–14. The

authors of this article avoid attributing the origin of the concept of "social metabolism" to Marx, preferring to cite R. U. Ayres and U. E. Simonis as their first instance of the use of the concept, due to Ayres and Simonis's use of the category of "industrial metabolism" in a 1994 edited volume. Nevertheless, both Fischer-Kowalski and Martinez-Alier were clear in their earlier writings that the concept of "social metabolism" had its origin in Marx. Their failure to note that here may be related to the fact that this article seeks to avoid the question of capitalism altogether, tracing the contemporary ecological problem simply to "industrial society," contradicting in that respect earlier work by at least some of these authors.

48. Wishart, "Metabolic Rift: A Selected Bibliography."
49. Johan Rockström et al., "A Safe Operating Space for Humanity," *Nature* 461 (September 24, 2009): 472–75; Foster, Clark, and York, *The Ecological Rift*, 13–18.
50. "NASA Satellite Measures Earth's Carbon Metabolism," *NASA Earth Observatory*, April 22, 2003, http://earthobservatory.nasa.gov.
51. J. G. Canadell et al., "Carbon Metabolism of the Terrestrial Biosphere," *Ecosystems* (2000) 3: 115–30.
52. James Hansen, "An Old Story But Useful Lessons," September 26, 2013, http://columbia.edu/~jeh1/.
53. Marx, *Capital*, vol. 1, 637–38.
54. "Real labour," Marx wrote, "is purposeful activity aimed at the creation of a use value, at the appropriation of natural material in a manner which corresponds to particular needs." Marx and Engels, *Collected Works*, vol. 30, 55. Obviously the more the labor process is alienated and thus estranged from these essential natural and social conditions, the more it takes on an artificial, unreal form.
55. This is not to say that Marx was completely unaware of the problem of specifically capitalist use values and the wasted labor associated with it. On this see John Bellamy Foster, "James Hansen and the Climate-Change Exit Strategy," *Monthly Review* 64/9 (February 2013): 14.
56. On the role of specifically capitalist use values in today's phase of monopoly-finance capital see John Bellamy Foster, "The Epochal Crisis," *Monthly Review* 65/5 (October 2013): 1–12.
57. See May Morris, *William Morris: Artist, Writer, Socialist*, vol. 2 (Cambridge: Cambridge University Press, 1936), 469–82, and *Collected Works*, vol. 23 (New York: Longman Green, 1915), 98–120, 238–54. Morris's stance was closely related to the general ecological tenor of his socialism evident in his 1890 utopian novel, *News From Nowhere*. See also Harry Magdoff, "The Meaning of Work," *Monthly Review* 34/5 (October 1982): 1–15.

58. Morris, *William Morris: Artist, Writer, Socialist*, 479. Arthur Balfour was the head of the Conservative Party in England at the time Morris was writing, later to become prime minister. The ellipsis before the word "refusing" in the first paragraph of this quote replaces the word "not," which was clearly a typographical error in the preparation of the text.
59. Thorstein Veblen, *Absentee Ownership and Business Enterprise in Recent Times* (New York: Augustus M. Kelley, 1923); Paul A. Baran and Paul M. Sweezy, *Monopoly Capital* (New York: Monthly Review Press, 1966), and "The Last Letters," *Monthly Review* 64/3 (July–August 2012): 68, 73.
60. John Bellamy Foster, Hannah Holleman, and Robert W. McChesney, "The U.S. Imperial Triangle and Military Spending," *Monthly Review* 60/5 (October 2008): 10; "U.S. Marketing Spending Exceeded $1 Trillion in 2005," *Metrics 2.0*, January 26, 2006, http://metrics2.com; U.S. Bureau of Economic Analysis, National Income and Product Accounts, "Government Consumption Expenditures and Investment by Function," Table 3.15.5, http://bea.gov; "U.S. Remains World's Largest Luxury Goods Market in 2012," *Modern Wearing*, October 22, 2012, http://modernwearing.com; "Groundbreaking Study Finds U.S. Security Industry to Be $350 Billion Industry," *ASIS Online*, August 12, 2013, http://asisonline.org.
61. On this see Foster, "James Hansen and the Climate-Change Exit Strategy," 16–18, and "The Epochal Crisis," 9–10.

2. The Great Capitalist Climacteric

This chapter is adapted and revised for this book from the following article: John Bellamy Foster, "The Great Capitalist Climacteric," *Monthly Review* 67/6 (November 2018): 1–18.

1. The term "the Great Climacteric" was used in 1975 by François Bédarida to refer to the debate over changes that occurred in Britain in the Edwardian period and after, marked by hegemonic decline in Britain's position in the capitalist world system, once England was no longer "the workplace of the world." See François Bédarida, *A Social History of England, 1851–1975* (London: Methuen, 1979), 99–103. In a famous June 22, 1941, speech, Winston Churchill referred to Hitler's invasion of Russia as the "fourth climacteric" of the Second World War in Europe. See Winston Churchill, "Alliance with Russia," http://winstonchurchill.org.
2. Ian Burton and Robert W. Kates, "The Great Climacteric, 1798–2048: The Transition to a Just and Sustainable Human Environment," in *Geography, Resources and Environment*, ed. Robert W. Kates and Ian

Burton (Chicago: University of Chicago Press, 1986), vol. 2, 393.
3. Johan Rockström et al., "A Safe Operating Space for Humanity," *Nature* 461/24 (September 2009): 472–75.
4. See Ian Angus, "When Did the Anthropocene Begin . . . And Why Does It Matter?," *Monthly Review* 67/4 (September 2015): 1–11.
5. James E. Hansen and Makiko Sato, "Climate Sensitivity Estimated from Earth's Climate History," draft paper, 2012, http://columbia.edu.
6. "Carbon Clock Countdown," *The Guardian*, January 19, 2017, https://www.theguardian.com/environment/datablog/2017/jan/19/carbon-countdown-clock-how-much-of-the-worlds-carbon-budget-have-we-spent; Remaining Carbon Budget, "Mercator Research Institute on Global Commons and Climate Change," https://www.mcc-berlin.net/en/research/co2-budget.html; http://trillionthtonne.org; Helena Gray, "The IPCC's Red Alert," Carbon Tracker, August 4, 2021, https://mronline.org/2021/09/20/the-ipccs-red-alert/.
7. United Nations Environment Program, "Cut Global Emissions by 7.6 Percent Every Year for Next Decade to Meet 1.5°C Target," November 26, 2019, www.unep.org. Kevin Anderson, "Why Carbon Prices Can't Deliver the 2°C Target," August 13, 2013, http://kevinanderson. "Avoiding Dangerous Climate Change," November 25, 2013, http://kevinanderson.info, "Climate Change Going Beyond Dangerous: Brutal Numbers and Tenuous Hope," *Development Dialogue* (September 2012): 35, http://whatnext.org. See also Dawn Stover, "Two Degrees of Climate Change May Be Too Much," *Bulletin of the Atomic Scientists* (September 4, 2015), http://thebulletin.org; http://trillionthtonne.org.
8. The World Bank (working with the Potsdam Institute for Climate Change) in its *Turn Down the Heat* reports argue that 1.5°C warming is "locked in" in the sense that staying below 2°C and getting back to 1.5° C is the most that is economically and technically feasible today. It goes on to suggest that the 2°C boundary will likely be exceeded and that a 4°C world needs to be avoided—in that way subtly changing the debate. World Bank, *4°—Turn Down the Heat: Confronting the New Climate Normal*, November 23, 2014, xvii, 5, *Turn Down the Heat: Why a 4° World Must Be Avoided*, 2012, http://documents.worldbank.org, xiii. See also Oliver Geden, "Climate Advisers Must Maintain Integrity," *Nature* (May 7, 2015): 27–28, http://nature.com.
9. Hansen and Sato, "Climate Sensitivity Estimated from Earth's Climate History," 14; Fred Pearce, "What Is the Carbon Limit?," *Environment 360*, November 6, 2014, http://e360.yale.edu.
10. James Hansen et al., "Ice Melt, Sea Level Rise, and Superstorms," *Atmospheric Chemistry and Physics Discussions* 15 (2015):

20061–63, 20114–22, http://atmos-chem-phys-discuss.net. On the issue of "unequivocal" and "irreversible damage" associated with the 2°C boundary and the enormous dangers that this implies see Heidi Cullen, *The Weather of the Future* (New York: Harper, 2011), 261–71.

11. James Hansen is the main proponent today of the idea of a possible runaway global warming, first raised in the late 1960s. This is associated with the Venus syndrome or the notion that the heating up of the world's oceans can so alter the atmosphere that Earth comes to resemble Venus. Long before such a scenario could develop, however, humanity would have lost the ability to control climate change through its own actions. It is this then that becomes the real issue. See Hansen and Sato, "Climate Sensitivity"; James Hansen, *Storms of My Grandchildren* (New York: Bloomsbury, 2009), 226–36.

12. Hansen et al., "Ice Melt, Sea Level Rise, and Superstorms": United Nations Climate Change, Cut Global Emissions by 7.6 Percent Every Year for Next Decade to Meet 1.5°C Target."

13. Will Steffen et al., "Planetary Boundaries: Guiding Human Development on a Changing Planet," *Science* 347/6223 (January 15, 2015), https://sciencemag.org. Biomagnification is the magnification of toxins up the food chain, bioaccumulation is the concentration within an individual organism.

14. John Kenneth Galbraith, *The Economics of Innocent Fraud* (Boston: Houghton Mifflin, 2004), 3–9.

15. Ellen Meiksins Wood, *Democracy Against Capitalism* (Cambridge: Cambridge University Press, 1995), 146–53; Francis Fukuyama, *The End of History and the Last Man* (New York: Free Press, 2006).

16. Naomi Klein, *This Changes Everything: Capitalism vs. the Climate* (New York: Simon and Schuster, 2014), 31–63.

17. This phrase is taken from Russell Jacoby, *The Dialectic of Defeat* (Cambridge: Cambridge University Press, 1981). On the rejection of history in postmodernist discourse see John Bellamy Foster, "In Defense of History," in Ellen Meiksins Wood and John Bellamy Foster, eds., *In Defense of History* (New York: Monthly Review Press, 1997), 184–93.

18. For accounts of Green theory and ecologism see Andrew Dobson, *Green Political Thought* (London: Routledge, 1995); Mark J. Smith, *Ecologism* (Minneapolis: University of Minnesota Press, 1998).

19. Paul A. Baran, *The Longer View* (New York: Monthly Review Press, 1969), 32.

20. Russell Jacoby, "Western Marxism," in *A Dictionary of Marxist Thought*, ed. Tom Bottomore (Oxford: Blackwell, 1983), 523–26.

NOTES TO PAGES 68–70   511

21. Alfred Schmidt, *The Concept of Nature in Marx* (London: New Left Books, 1971), 9–10, 155–62; Max Horkheimer and Theodor Adorno, *Dialectic of Enlightenment* (New York: Continuum, 2001; originally 1944).
22. On the first, second, and third stages of ecosocialist discourse see John Bellamy Foster, Foreword, in Paul Burkett, *Marx and Nature* (Chicago: Haymarket, 2014), vii–xiii.
23. Ted Benton, "Marxism and Natural Limits," *New Left Review* 178 (1989): 55, 60, 64; Andre Gorz, *Capitalism, Socialism, Ecology* (London: Verso, 1994). For a detailed assessment of Malthus's theory and Marx's critique of Malthus see John Bellamy Foster, *Marx's Ecology* (New York: Monthly Review Press, 2000), 81–104, 141–49.
24. Ted Benton, ed., *The Greening of Marxism* (New York: Guilford Press, 1996).
25. Raymond Williams, *Resources for Hope* (London: Verso, 1989), 210.
26. Burkett, *Marx and Nature*, 79–98.
27. 'I. I. Rubin, *Essays on Marx's Theory of Value* (Detroit: Black and Red, 1972), 71–75, 107–23; Paul M. Sweezy, *The Theory of Capitalist Development* (New York: Monthly Review Press, 1970; originally 1942), 23–40. The understanding of the relation of the qualitative value (or value-form) analysis in Marx to ecological issues marked the work of Japanese Marxist Shigeto Tsuru, especially, who became one of the leading environmental thinkers in Japan and the world as a whole in the 1960s to 1980s. See Shigeto Tsuru, *Towards a New Political Economy* (Tokyo: Kodansha Ltd., 1976).
28. Elmar Altvater's *The Future of the Market* provided an important reinterpretation of Marx's ecological analysis, preceding Burkett's *Marx's and Nature*. See Elmar Altvater, *The Future of the Market* (London: Verso, 1993).
29. On "production in general" see Karl Marx, *Grundrisse* (London: Penguin, 1973), 85–88.
30. In Marx's theory the value-analytic in its most general form encompasses both use value and exchange value. In this sense Marx saw nature (apart from labor) as contributing to *use value*, i.e., in the material aspects underlying every commodity. But it is more usual to see value-analytic in *exchange value* (or value form) terms, and in this sense nature does not enter directly into value calculations of the system or the constitution of capital. For a detailed discussion see Paul Burkett, "Nature's 'Free Gifts' and the Ecological Significance of Value," *Capital and Class* 23 (1999): 89–110.
31. A property of capitalist income accounting is to exclude domestic labor since it does not contribute directly to profits and accumulation,

and thus is left out of GDP. Not only are domestic workers, primarily women, robbed in such a situation, but the family becomes a way in which capitalism externalizes its costs. The similarity to the exclusion of nature from value has been strongly emphasized by ecofeminist thinkers. See, in particular, Marilyn Waring, *Counting for Nothing* (Toronto: University of Toronto Press, 1999).

32. As Marx wrote, quoting the full sentence: "Natural elements entering as agents into production, which cost nothing, no matter what role they play in production, do not enter as components of capital, but as a free gift of Nature to capital, that is, as a free gift of Nature's productive power to labour, which, however, appears as the productive power of capital, as all other productivity under the capitalist mode of production." Karl Marx and Frederick Engels, *Collected Works* (New York: International Publishers, 1975), vol. 37, 732–33.

33. A somewhat analogous situation exists with respect to labor power. In Marx's theory, capitalist production generally requires that capital pay the worker the value of labor power, that is, its cost of reproduction. But this payment of the cost of reproduction merely allows capital to appropriate that labor power for a given amount of time, exploiting its power to produce, beyond the cost of its own reproduction. With respect to nature, though, capital is under less obligation to cover the costs of reproduction, and outright robbery of nature (the natural conditions of production) is the norm.

34. K. William Kapp, *The Social Costs of Private Enterprise* (Cambridge, MA: Harvard University Press, 1971), 231.

35. Roland Daniels, *Mikrokosmos* (Frankfurt am Main: Verlag Peter Lang, 1988; original ms. 1851), 49. Daniels's work followed the development of the first law of thermodynamics and the application of the principle of the conservation of energy to metabolism in the work of Julius Robert Mayer, one of the co-discoverers of the conservation of energy. See Julius Robert Mayer, "The Motions of Organisms and Their Relation to Metabolism," in R. Bruce Lindsay, ed., *Energy: Historical Development of the Concept* (Stroudsburg, PA: John Wiley and Sons, 1975), 284–307.

36. Karl Marx, *Capital*, vol. 1 (London: Penguin, 1976), 637–38; *Capital*, vol. 3, 949.

37. See Brett Clark and John Bellamy Foster, "Guano: The Global Metabolic Rift in the Fertilizer Trade," in Alf Hornborg, Brett Clark, and Kenneth Hermele, eds., *Ecology and Power* (London; Routledge, 2012), 68–82; John Bellamy Foster and Hannah Holleman, "The Theory of Unequal Ecological Exchange," *Journal of Peasant Studies* 41/1–2 (March 2014): 199–233.

38. Karl Marx and Frederick Engels, *Collected Works*, vol. 30, 54–66; Marx, *Capital*, vol. 3, 949.
39. Marx, *Capital*, vol. 1, 638.
40. Karl Marx, *Capital*, vol. 3 (London: Penguin, 1981), 911.
41. Marx, *Capital*, vol. 3, 959.
42. Burkett, *Marx and Nature*, xx.
43. The classic neo-Marxian theory of ecological crisis within environmental sociology is known as "the treadmill of production" perspective and had its origins in Allan Schnaiberg, *The Environment* (Oxford: Oxford University Press, 1980). Schnaiberg's analysis was heavily influenced by analyses of economic crises and the environment in *Monthly Review*.
44. Marx's comments on the capitalist commodification of animals were largely in response to the French agriculturalist Léonce de Lavergne, whose ideas Marx addressed in *Capital* and in his excerpt notebooks. See Kohei Saito, "Marx's Ecological Notebooks," *Monthly Review* 67/9 (February 2016), 25–41; John Bellamy Foster and Paul Burkett, *Marx and the Earth* (Boston: Brill, 2016); Léonce de Lavergne, *The Rural Economy of England, Scotland and Ireland* (London: William Blackwood and Sons, 1855).
45. Kenneth M. Stokes, *Man and the Biosphere* (Armonk, NY: M. E. Sharpe, 1992), 35–37; Marx and Engels, *Collected Works*, vol. 46, 411.
46. John Bellamy Foster, "The Absolute General Law of Environmental Degradation Under Capitalism," *Capitalism Nature Socialism* 3/3 (1992): 77–81.
47. On the significance of economic and ecological waste to the Marxian critique of capitalism's ecological and social destruction see John Bellamy Foster, "The Ecology of Marxian Political Economy," chapter 7, below.
48. Baran, *The Longer View*, 30.
49. For each additional dollar made by the bottom 90 percent of the population in the United States from 1990–2002 the top 0.01 percent (some 14,000 households) made an additional $18,000. Correspondents of the *New York Times, Class Matters* (New York: Times Books, 2005), 186.
50. On the contemporary phenomena of overaccumulation, stagnation, and financialization see John Bellamy Foster and Robert W. McChesney, *The Endless Crisis* (New York: Monthly Review Press, 2012).
51. Lewis Mumford, *The Condition of Man* (New York: Harcourt Brace Jovanovich, 1973), 411.
52. "Nothing is enough for someone for whom enough is little." Epicurus, *The Epicurus Reader* (Indianapolis: Hackett 1994), 39.
53. Paul Burkett, "Marx's Vision of Sustainable Human Development," *Monthly Review* 57/5 (October 2005): 34–62.

54. Marx, *Capital*, vol. 1, 638; Stefano B. Longo, Rebecca Clausen, and Brett Clark, *The Tragedy of the Commodity* (New York: Rutgers University Press, 2015), 175–203.
55. Del Weston, *The Political Economy of Global Warming* (London: Routledge, 2014), 170–71; Marx, *Capital*, vol. 1, 637–38.
56. Foster, Clark, and York, *The Ecological Rift*, 47, 398, 440.
57. See Weston, *The Political Economy of Global Warming*, 113–52.
58. David Harvey, *The Enigma of Capital* (Oxford: Oxford University Press, 2010), 228–35.
59. Klein, *This Changes Everything*, 293–336.
60. On the two stages of ecological revolution see Fred Magdoff and John Bellamy Foster, *What Every Environmentalist Needs to Know About Capitalism* (New York: Monthly Review Press, 2011), 123–44.
61. Karl Marx, *Critique of the Gotha Programme* (New York: International Publishers, 1938), 10.
62. On the classic conception of democracy as the rule of society by the *demos* (the poor) see Ellen Meiksins Wood and Neal Wood, *Class Ideology and Ancient Political Theory* (Oxford: Blackwell, 1978).
63. Karl Marx and Frederick Engels, *Collected Works* (New York: International Publishers, 1975), vol. 42, 558–59.
64. E. P. Thompson, *Beyond the Cold War* (New York: Pantheon, 1982), 41–80; Rudolf Bahro, *Avoiding Social and Ecological Disaster* (Bath: Gateway Books, 1994), 19. This and the following paragraph draw on John Bellamy Foster, *The Ecological Revolution* (New York: Monthly Review Press, 2009), 27–28.
65. Mumford, *The Condition of Man*, 348, 412.

3. The Anthropocene Crisis

This chapter is adapted and revised for this book from the Foreword by John Bellamy Foster in Ian Angus, *Facing the Anthropocene* (New York: Monthly Review Press, 2016), 9–17.

1. Epigraph: Bertolt Brecht, *Brecht on Theatre* (New York: Hill and Wang, 1964), 275. Clive Hamilton and Jacques Grinevald, "Was the Anthropocene Anticipated?" *Anthropocene Review* 2/1 (2015): 67.
2. Paul J. Crutzen and Eugene F. Stoermer, "The Anthropocene," *Global Change Newsletter*, May 1, 2000, 17; Paul J. Crutzen, "Geology of Mankind," *Nature* 415/6867 (2002): 23; Colin N. Waters et al., "The Anthropocene Is Functionally and Stratigraphically Distinct from the Holocene," *Science* 351/6269 (2016): 137.
3. On the concept of the Capitalinian Age see chapter 21 in this book.
4. Spencer Weart, "Interview with M. I. Budyko: Oral History Transcript," March 25, 1990, http://aip.org; M. I. Budyko, "Polar Ice and Climate,"

in J. O. Fletcher, B. Keller, and S. M. Olenicoff, eds., *Soviet Data on the Arctic Heat Budget and Its Climatic Influence* (Santa Monica, CA: Rand Corporation, 1966), 9–23; William D. Sellars, "A Global Climatic Model Based on the Energy Balance of the Earth Atmosphere System," *Journal of Applied Meteorology* 8/3 (1969): 392–400; M. I. Budyko, "Comments," *Journal of Applied Meteorology* 9, no. 2 (1970): 310.
5. István Mészáros, *The Power of Ideology* (New York: New York University Press, 1989), 128.
6. Karl Marx and Frederick Engels, *Collected Works*, vol. 5 (New York: International Publishers, 1976), 40.
7. George P. Marsh, *Man and Nature* (Cambridge, MA: Harvard University Press, 1965); Frank Benjamin Golley, *A History of the Ecosystem Concept in Ecology* (New Haven: Yale University Press, 1993), 2, 207; Karl Marx, *Capital*, vol. 1 (London: Penguin, 1976), 636–39; *Capital*, vol. 3 (London: Penguin, 1981), 949.
8. Lynn Margulis and Dorion Sagan, *What Is Life?* (New York: Simon and Schuster, 1995), 47; Vladimir I. Vernadsky, *The Biosphere* (New York: Springer, 1998). The concept of the biosphere was originally introduced by the French geologist Edward Suess in 1875, but was developed much further by Vernadsky, and came to be associated primarily with him.
9. Vladimir I. Vernadsky, "Some Words About the Noösphere," in Jason Ross, ed., *150 Years of Vernadsky*, vol. 2 (Washington, D.C.: 21st Century Science Associates, 2014), 82; E. V. Shantser, "The Anthropogenic System (Period)," in *The Great Soviet Encyclopedia*, vol. 2 (New York: Macmillan, 1973), 140. Shantser's article introduced the word *Anthropocene* in English.
10. Richard Levins and Richard Lewontin, *The Dialectical Biologist* (Cambridge, MA: Harvard University Press, 1985), 277; A. I. Oparin, "The Origin of Life," in J. D. Bernal, *The Origin of Life* (New York: World Publishing, 1967), 199–234; and J. B. S. Haldane, "The Origin of Life," in Bernal, *The Origin of Life*, 242–49.
11. Rachel Carson, *Lost Woods* (Boston: Beacon, 1998), 230–31.
12. G. Evelyn Hutchinson, "The Biosphere," *Scientific American* 233/3 (1970): 45–53.
13. Barry Commoner, *The Closing Circle: Nature, Man, and Technology* (New York: Knopf, 1971), 45–62, 138–75, 280.
14. E. Fedorov quoted in Virginia Brodine, *Green Shoots, Red Roots* (New York: International Publishers, 2007), 14, 29. See also E. Fedorov, *Man and Nature* (New York: International Publishers, 1972), 29–30; John Bellamy Foster, "Late Soviet Ecology and the Planetary Crisis," *Monthly Review* 67/2 (June 2015): 9; M. I. Budyko, *The Evolution of the*

Biosphere (Boston: Reidel, 1986), 406. Calls by prominent figures like Fedorov for a more rapid and radical response to environmental problems went largely unheeded by the Soviet state, with tragic results.
15. Fedorov, *Man and Nature*, 146.
16. Hamilton and Grinevald, "Was the Anthropocene Anticipated?," 64.
17. Howard T. Odum, *Environment, Power, and Society for the Twenty-First Century* (New York: Columbia University Press, 2007), 3.
18. Odum, *Environment, Power, and Society*, 263.
19. E. P. Thompson, *Beyond the Cold War* (New York: Pantheon, 1982) 41–80; Rudolf Bahro, *Avoiding Social and Ecological Disaster* (Bath, UK: Gateway, 1994), 19; Odum, *Environment, Power, and Society*, 276–78.
20. Ian Angus, *Facing the Anthropocene* (New York: Monthly Review Press, 2016); Rolf Edburg and Alexei Yablokov, *Tomorrow Will Be Too Late* (Tucson: University of Arizona Press, 1991).
21. Bertolt Brecht, *Tales from the Calendar* (London: Methuen, 1961), 31–32.

4. Crossing the River of Fire

This chapter has been adapted for this book from John Bellamy Foster and Brett Clark, "Crossing the River of Fire," *Monthly Review* 66/9 (February 2015): 1–17.
1. Naomi Klein, *This Changes Everything: Capitalism vs. the Climate* (New York: Simon and Schuster, 2014); "'A Feeling It's Gonna Be Huge': Naomi Klein on People's Climate Eve," interview, *Common Dreams*, September 21, 2014, http://commondreams.org.
2. On this, see Adam Morris, "The 'System Change' Doctrine," *Los Angeles Review of Books*, October 21, 2014, http://lareviewofbooks.org; System Change Not Climate Change, http://systemchangenotclimatechange.org; Klein, *This Changes Everything*, 87–89.
3. William Morris, *Collected Works* (London: Longmans Green, 1914), vol. 22, 131–32; E. P. Thompson, *William Morris: Romantic to Revolutionary* (New York: Pantheon Books, 1976), 244; Naomi Klein, *No Logo* (New York: Picador, 2002), and *The Shock Doctrine* (New York: Henry Holt, 2007).
4. Klein, *This Changes Everything*, 342, 444–47.
5. Ibid., 55.
6. "Carbon Clock Countdown," *The Guardian*, https://www.theguardian.com/environment/datablog/2017/jan/19/carbon-countdown-clock-how-much-of-the-worlds-carbon-budget-have-we-spent; Mercator Research Institute on Global Commons and Climate Change, "Remaining Carbon Budget," https://www.mcc-berlin.net/en/research/co2-budget.html; Intergovernmental Panel on Climate Change

(IPCC), "Carbon Budget Message of IPCC Report Reveals Daunting Challenge," *Huffington Post*, October 4, 2013, http://huffingtonpost.com; Myles Allen et al., "The Exit Strategy," *Nature Reports Climate Change*, April 30, 2009, http://nature.com, 56–58.
7. Klein, *This Changes Everything*, 13, 21, 56, 87; Kevin Anderson, "Why Carbon Prices Can't Deliver the 2° Target," August 15, 2013, http://kevinanderson.info.
8. Klein, *This Changes Everything*, 19, 56. The fact that neoliberal globalization and the creation of the WTO had permanently derailed the movement associated with the Earth Summit in Rio in 1993, including the attempt to prevent climate change, was stressed by one of us more than a dozen years ago at the World Summit for Sustainable Development in Johannesburg 2002, when Klein was present. See John Bellamy Foster, "A Planetary Defeat: The Failure of Global Environmental Reform," *Monthly Review* 54/8 (January 2003): 1–9, based on several talks delivered in Johannesburg, August 2002.
9. Klein, *This Changes Everything*, 21–24.
10. Paul M. Sweezy, *The Theory of Capitalist Development* (New York: Oxford University Press, 1942), 349.
11. Klein, *This Changes Everything*, 179; John Kenneth Galbraith, *The Affluent Society* (New York: New American Library, 1984), 121–28. As the author of *No Logo*, Klein is of course aware of the contradictions of consumption under capitalist commodity production.
12. Joseph A. Schumpeter, *Essays* (Cambridge: Addison-Wesley, 1951), 293.
13. Klein, *This Changes Everything*, 57–58, 115, 479–80.
14. Ibid., 10, 16–17, 115–16, 454; Adolfo Gilly, "Inside the Cuban Revolution," *Monthly Review* 16/6 (October 1964): 69; John Bellamy Foster, "James Hansen and the Climate-Change Exit Strategy," *Monthly Review* 64/9 (February 2013): 13.
15. "U.S. Marketing Spending Exceeded $1 Trillion in 2005," *Metrics Business and Marketing Intelligence*, June 26, 2006, http://metrics2.com; Michael Dawson, *The Consumer Trap* (Urbana: University of Illinois Press, 2005), 1.
16. Klein, *This Changes Everything*, 91–94.
17. Ibid., 381–82, 408–13.
18. Ibid., 176–87; CommonDreams.org, "'A Feeling It's Gonna Be Huge.'"
19. For historical materialist analyses of the extractivism problem in Bolivia and the difficult problem of overcoming it, see Álvaro García Linera, *Geopolitics of the Amazon*, 2012, http://climateandcapitalism.com; Frederico Fuentes, "'The Dangerous Myths of 'Anti-Extractivism,'" May 19, 2014, http://climateandcapitalism.com. As the

author of *The Shock Doctrine*, Klein is cognizant of imperialism but it does not enter in her analysis much here, partly because she is making a point of being balanced by criticizing the left as well as the right.
20. Klein, *This Changes Everything*, 458–60.
21. Ibid., 43, 58–63.
22. Edward S. Herman and Noam Chomsky, *Manufacturing Consent: Noam Chomsky and the Media* (New York: Black Rose Books, 1994), 58. On the "off limits" notion see Robert W. McChesney and John Bellamy Foster, "Capitalism: The Absurd System," *Monthly Review* 62/2 (June 2010): 2.
23. Thompson, *William Morris*, 270–71; Morris, *Collected Works*, vol. 23, 172.
24. Rob Nixon, "Naomi Klein's 'This Changes Everything,'" *New York Times*, November 6, 2014, http://nytimes.com.
25. Dave Pruett, "A Line in the Tar Sands: Naomi Klein on the Climate," *Huffington Post*, November 26, 2014, http://huffingtonpost.com.
26. Elizabeth Kolbert, "Can Climate Change Cure Capitalism?," *New York Review of Books*, December 4, 2014, http://nybooks.com; Naomi Klein and Elizabeth Kolbert, "Can Climate Change Cure Capitalism?: An Exchange," *New York Review of Books*, January 8, 2015, http://nybooks.com.
27. David L. Ulin, "In 'This Changes Everything,' Naomi Klein Sounds Climate Alarm," *Los Angeles Times*, September 12, 2014, http://touch.latimes.com.
28. Michael Signer, "Naomi Klein's 'This Changes Everything' Will Change Nothing," *Daily Beast*, November 17, 2014, http://thedailybeast.com.
29. Mark Jaccard, "I Wish This Changed Everything," *Literary Review of Canada*, November 2014, http://reviewcanada.ca; "Despite California Climate Law, Carbon Emissions May be a Shell Game," *Los Angeles Times*, October 25, 2014, http://latimes.com.
30. Jaccard, "I Wish This Changed Everything"; Paul Krugman, "Errors and Emissions," *New York Times*, September 8, 2014, http://nytimes.com.
31. Will Boisvert, "The Left vs. the Climate: Why Progressives Should Reject Naomi Klein's Pastoral Fantasy—and Embrace Our High-Energy Planet," *The Breakthrough*, September 18, 2014, http://thebreakthrough.org; Bruno Latour, "Love Your Monsters," *The Breakthrough* no. 2, Fall 2011, http://thebreakthrough. Klein situates the Breakthrough Institute within her criticism of the right, questioning its claim to progressive values. Klein, *This Changes Everything*, 57.
32. Erle Ellis, "The Planet of No Return," *The Breakthrough* no. 2, Fall 2011, http://thebreakthrough.org; Boisvert, "The Left vs. the Climate."
33. Klein, *This Changes Everything*, 89.

34. See the important analysis in Richard Smith, "Climate Crisis, the Deindustrialization Imperative and the Jobs vs. Environment Dilemma," *Truthout*, November 12, 2014, http://truth-out.org.
35. David Harvey, *The Engima of Capital* (New York: Oxford University Press, 2010), 228–35.
36. William Morris, *Three Works* (London: Lawrence and Wishart, 1986).
37. Klein, *This Changes Everything*, 466.
38. Simón Bólivar, "Message to the Congress of Bolivia, May 25, 1826," *Selected Works*, vol. 2 (New York: The Colonial Press, 1951), 603.

5. The Fossil Fuels War

This chapter has been adapted and revised for this book from the following article: John Bellamy Foster, "The Fossil Fuels War," *Monthly Review* 65/4 (September 2013): 1–14.

1. See John Bellamy Foster, *The Ecological Revolution* (New York: Monthly Review Press, 2009), 85–105; International Energy Agency, *World Energy Outlook 2010* (OECD/IEA, 2010), 125–26; Ramez Naam, *The Infinite Resource* (Lebanon, NH: University Press of New England, 2013), 47.
2. Pushker A. Kharecha and James E. Hansen, "Implications of 'Peak Oil' for Atmosphere CO_2 and Climate," *Global Biogeochemical Cycles* 22 (2008): 1–10. The term "unconventional fossil fuels" is commonly used to refer to fossil-fuel feedstocks that have not been intensively exploited up to the present, usually because they are of inferior grade and/or require additional technology and added costs for extraction and processing, such as heavy oils, oil sands, shale gas, tight oil, tight gas, oil shale, methane hydrates, and oil from deepwater drilling. "Unconventional Fossil Fuels," *Juice: Alternative Fuels World*, http://alternatefuelsworld.com; International Energy Agency, Glossary of Terms, "unconventional gas" and "unconventional oil," http://iea.org.
3. Charles C. Mann, "What If We Never Run Out of Oil," *Atlantic* 311/ 4, May 2013, 54, 63.
4. Michael T. Klare, *The Race for What's Left* (New York: Henry Holt, 2012), 106.
5. Bill McKibben, "The Fossil Fuel Resistance," *Rolling Stone*, April 25, 2013, 42.
6. The Keystone XL Pipeline is actually part of the larger Keystone pipeline system. The first two phases of this were already completed at the time of the protests and the third phase, the southern leg, was soon finished, with Alberta tar sand oil flowing to the Gulf. But the completion of the critical northern line (phase 4) would have provided a more direct route carrying about twice the oil. As Candice Bernd wrote: "James Hansen called the [Keystone XL] project 'the fuse to the largest carbon

bomb on the planet.'... The northern, cross-border expansion of the project would make that fuse burn faster, doubling the Keystone pipeline system's carrying capacity to more than 800,000 barrels a day." Candice Bernd, "Tar Sands Will Be Piped to the Gulf Coast, With or Without the Northern Segment of Keystone XL," *Truthout*, April 29, 2013, http://truth-out.org.

7. "Keystone XL: Project Overview," Keystonexl.com, https://www.keystonexl.com/kxl-101/.
8. Courtney Lindwall, "The Unlikely Takedown of Keystone XL," Natural Resources Defense Council, June 29, 2021, https://www.nrdc.org/stories/unlikely-takedown-keystone-xl.
9. James Hansen, "Game Over for the Climate," *New York Times*, May 9, 2012, http://nytimes.com, and "Keystone XL: The Pipeline to Disaster," *Los Angeles Times*, April 4, 2013, http://articles.latimes.com.
10. Glenn Gilchrist, "Transportation—Alberta Achilles Heel," April 5, 2013, http://world.350.org; Reid McKay, "Canada Losing Massive Wealth on Oil Price Differential," *CEO.CA*, February 13, 2013, http://ceo.ca; David Biello, "Greenhouse Goo," *Scientific American* 309/1 (July 2013): 61.
11. David Sassoon, "Crude, Dirty and Dangerous," *New York Times*, August 20, 2012, http://nytimes.com; David Biello, "Does Tar Sand Oil Increase the Risk of Pipeline Spills?," *Scientific American*, April 4, 2013, http://scientificamerican.com.
12. McKibben, "The Fossil Fuel Resistance," 40; Michael Levi, *The Power Surge* (New York: Oxford, 2013), 81; "What's Next in the Ongoing Keystone XL Saga," *U.S. News & World Report*, April 5, 2013, http://usnews.com.
13. See Jacob Devaney, "Idle No More: Hints of a Global Super-Movement," *Common Dreams*, January 3, 2013, http://commondreams.org; "First Nations Group Calls for B.C. to Reject Northern Gateway Pipeline Work Permits," *Vancouver Sun*, June 27, 2013, http://vancouversun.com; Brooke Jarvis, "Idle No More: Native-Led Protest Movement Takes on Canadian Government," *Rolling Stone*, February 4, 2013.
14. Ohio Environmental Council, "What Is Fracking?," http://theoec.org; "Baffled About Fracking? You're Not Alone," *New York Times*, May 13, 2011, http://nytimes.com; Levi, *The Power Surge*, 41–49.
15. "Rise in U.S. Gas Production Fuels an Unexpected Plunge in Emissions," *Wall Street Journal*, April 18, 2013, http://online.wsj.com. Such figures are of course misleading in terms of the overall climate problem, since the coal industry has responded to the increased competition of natural gas by increasing coal exports to China and

elsewhere. Indeed, a study by John Broderick and Kevin Anderson of the Tyndall Climate Change Research Institute has indicated that "more than half of the emissions avoided in the U.S. power sector [in 2008–2011] may have been exported as coal." Thus they conclude that "without a meaningful cap on global carbon emissions, the exploitation of shale gas reserves is likely to increase total [global] emissions." John Broderick and Kevin Anderson, *Has US Shale Gas Reduced CO_2 Emissions?*, Tyndall Manchester Climate Change Research, October 2012, http://tyndall.ac.uk, 2. U.S. coal exports help fuel Chinese industry, which then sells a larger part of their output back to the United States. It has been estimated that the United States imported 400 million tons of embedded carbon in Chinese goods in 2008 alone. Bill Chameides, "On U.S. Greenhouse Gas Emissions and Cognitive Dissonance," *The Green Grok*, November 14, 2012, http://blogs.nicholas.duke.edu.
16. Jeff Tollefson, "Methane Leaks Erode Green Credentials of Natural Gas," *Nature*, January 2, 2013, http://nature.com.
17. Matthew Phillips, "More Evidence Shows Drilling Causes Earthquakes," *Bloomberg Businessweek*, April 1, 2013, http://businessweek.com.
18. "Quebec's Lac Mégantic Oil Train Disaster Not Just Tragedy, But Corporate Crime," *The Guardian*, July 11, 2013, http://guardian.co.uk; Jonathan Flanders, "'Pipeline on Rails' Plans for the Railroads Explode in Quebec," *CounterPunch*, July 11, 2013, http://counterpunch.org; "Quebec Train Death Toll at 50," *New York Post*, July 11, 2013, http://nypost.com; "Canadian Tanker Train Crash Raises Fresh Questions on Oil Transportation," *The Guardian*, July 16, 2013, http://guardian.co.uk.
19. Frances Beinecke, "3 Years Later: Act on the Lessons of BP Gulf Oil Spill," The Energy Collective, April 18, 2013, http://theenergycollective.com; Klare, *The Race for What's Left*, 42–49.
20. Klare, *The Race for What's Left*.
21. Susan Solomon et al., "Irreversible Climate Change Due to Carbon Dioxide Emissions," *Proceedings of the National Academy of Sciences* 106/6 (February 10, 2009): 1704–9; Heidi Cullen, *The Weather of the Future* (New York: Harper, 2010), 261–71; James Hansen, "Tipping Point," in Eva Fearn and Kent H. Redford, eds., *State of the Wild 2008* (Washington, D.C.: Island Press, 2008), 7–8; Biello, "Greenhouse Goo," 58–59.
22. See https://www.trillionthtonne.org; Carbon Tracker and the Grantham Research Institute, London School of Economics, *Unburnable Carbon 2012: Wasted Capital and Stranded Assets* (2013), http://carbontracker.org; Myles Allen et al., "The Exit Strategy," *Nature Reports*, April 30,

2009; http://nature.com; and "Warming Caused by Cumulative Carbon Emissions Towards the Trillionth Tonne," *Nature* 458 (April 30, 2009): 1163–66; Malte Meinshausen et al., "Greenhouse Gas Emission Targets for Limiting Global Warming to 2°C," *Nature* 458 (April 20, 2009): 1158–62; Carbon Clock Countdown," *The Guardian*, January 19, 2017, https://www.theguardian.com/environment/datablog/2017/jan/19/carbon-countdown-clock-how-much-of-the-worlds-carbon-budget-have-we-spent.

23. "Ending Its Summer Melt, Arctic Sea Ice Sets a New Low that Leads to Warnings," *New York Times*, September 19, 2012, http://nytimes.com; Andrew Freedman, "A Closer Look at Arctic Sea Ice Melt and Extreme Weather," *Climate Central*, September 19, 2012, http://climatecentral.org; John Vidal and Adam Vaughan, "Arctic Sea Ice Shrinks to Smallest Extent Ever Recorded," *The Guardian*, September 14, 2012, http://guardian.co.uk.

24. Australian Climate Commission, *The Angry Summer* (2013), http://climatecommission.gov.au; Climate Central, *Global Weirding* (New York: Pantheon Press, 2012).

25. "Greenland and Antarctica 'Have Lost Four Trillion Tonnes of Ice' in 20 Years," *The Guardian*, November 29, 2012, http://guardian.co.uk.

26. Levi, *The Power Surge*, 148–49; Naam, *The Infinite Resource*, 161–62.

27. Bruno Burger, "Electricity Production from Solar and Wind in Germany in 2012," Fraunhofer Institute for Solar Energy Systems, February 8, 2013, http://ise.fraunhofer.de; and "Crossing the 20 Percent Mark: Green Energy Use Jumps in Germany," *Spiegel Online International*, August 30, 2011, http://spiegel.de; Levi, *The Power Surge*, 144–45; Naam, *The Infinite Resource*, 163.

28. Mason Inman, "The True Cost of Fossil Fuels," *Scientific American* (April 2013): 58–61; Charles A. S. Hall and Kent A. Klitgaard, *Energy and the Wealth of Nations* (New York: Springer, 2012); Steve Hallett, *The Efficiency Trap* (Amherst, NY: Prometheus Books, 2013), 77; Eric Zencey, "Energy as a Master Resource," in Worldwatch Institute, *State of the World 2013* (Washington, D.C.: Island Press, 2013), 79; Levi, *The Power Surge*, 151–52.

29. Jeremy Leggett, *The Carbon War* (New York: Routledge, 2001), 332.

30. See Horace Campbell, *Global NATO and the Catastrophic Failure in Libya* (New York: Monthly Review Press, 2013); John Bellamy Foster, *Naked Imperialism* (New York: Monthly Review Press, 2006).

31. See Matt McDermott, "Why Japan's Methane Hydrate Exploitation Would Be Game Over for the Planet," *Motherboard*, March 2013, http://motherboard.vice.com; Mann, "We Will Never Run Out of Oil."

32. Executive Office of the President, *The President's Climate Action

NOTES TO PAGES 118-123 523

Plan (June 2013), http://whitehouse.gov, 19; Levi, *The Power Surge*, 99–101, 171–72.
33. Richard York, "Do Alternative Energy Sources Displace Fossil Fuels?," *Nature Climate Change* 2 (2012): 441–43.
34. "Charlie Rose Talks to ExxonMobil's Rex Tillerson," *Bloomberg Businessweek*, March 7, 2013, http://businessweek.com.
35. Juliet Eilperin, "The White House's 'All of the Above' Energy Strategy Goes Global," *Washington Post*, April 24, 2013, http://washingtonpost.com; David Biello, "All-of-the-Above Energy Strategy Trumps Climate Action," *Scientific American*, November 16, 2012, http://blogs.scientificamerican.com.
36. John Bellamy Foster, Hannah Holleman, and Brett Clark, "Imperialism in the Anthropocene," *Monthly Review* 71/3 (July-August 2013), 70–88.
37. Curtis White, *The Barbaric Heart: Faith, Money, and the Crisis of Nature* (Sausalito, CA: PolipointPress, 2009).
38. Lindwall, "The Unlikely Takedown of Keystone XL."
39. Nick Estes, *Our History Is the Future: Standing Rock versus the Dakota Access Pipeline, and the Long Tradition of Indigenous Resistance* (London: Verso, 2019).
40. See, for example, Jeremy Leggett, *The Solar Century* (London: GreenProfile, 2009). This technocratic approach is a product of Leggett's whole history, first as a geologist consulting for the oil industry, then as a Greenpeace leader, then as CEO for Britain's first solar power corporation, and finally as the founder of Carbon Tracker.
41. Bill McKibben, "Global Warming's Terrifying New Math," *Rolling Stone*, July 19, 2012.
42. Zencey, "Energy as a Master Resource," 80–82.
43. "An Interview with Kevin Anderson," Transition Culture, November 2, 2012, http://transitionculture.org; Nicholas Stern, *The Economics of Climate Change: The Stern Review* (Cambridge: Cambridge University Press, 2007), 232. A similar criticism (to that of Anderson) of the *Stern Review*'s contention that carbon emission reductions of more than 1 percent per annum were detrimental to the capitalist economy and thus had to be off limits was made in John Bellamy Foster, Brett Clark, and Richard York, *The Ecological Rift* (New York: Monthly Review Press, 2010), 154–56.
44. "An Interview with Kevin Anderson"; Kevin Anderson and Alice Bows, "Beyond 'Dangerous Climate Change': Emission Scenarios for a New World," *Philosophical Transactions of the Royal Society* 369 (2011): 40–41. It should be added that what would make a vast reduction in consumption possible and at the same time improve the conditions of most of the population is the massive amount of waste built into

monopoly capitalist society, associated with the prodigious expansion of superfluous goods, and vast marketing expenses. Such socially unnecessary expenditures, as Thorstein Veblen explained at the outset of the twentieth century, are built into the production of commodities themselves. See John Bellamy Foster and Brett Clark, "The Planetary Emergency," *Monthly Review* 64/7 (December 2012): 7–16; Thorstein Veblen, *Absentee Ownership and Business Enterprise in Recent Times* (New York: Augustus M. Kelley, 1964), 284–325.

45. The Royal Society, *People and the Planet* (London: Royal Society, April 2012), 9.
46. Johan Rockström et al., "A Safe Operating Space for Humanity," *Nature* 461 (2009): 472–75, http://pubs.giss.nasa.gov; Carl Folke, "Respecting Planetary Boundaries and Reconnecting to the Biosphere," in Worldwatch Institute, *State of the World 2013*, 19–27.
47. Karl Marx and Frederick Engels, *The Communist Manifesto* (New York: Monthly Review Press, 1964), 2. See István Mészáros, "Substantive Equality: The Absolute Condition of Sustainability," in Mészáros, *The Challenge and Burden of Historical Time* (New York: Monthly Review Press, 2008), 258–64.
48. Karl Marx, *Capital*, vol. 1 (London: Penguin, 1976): 637–38.
49. Frederick Engels, *The Condition of the Working Class in England* (Chicago: Academy Chicago Publishers, 1984).
50. The notion of an environmental proletariat is advanced in Foster, Clark, and York, *The Ecological Rift*, 440. See also Fred Magdoff and John Bellamy Foster, *What Every Environmentalist Needs to Know About Capitalism* (New York: Monthly Review Press, 2011), 143–44. An example of the growth of a broad alliance of working people is the Idle No More movement in Canada, in which popular environmental groups, the National Farmers Union, and, increasingly, unionists are allying themselves with a movement led by First Nations, and organized around their treaty rights—in opposition to the Canadian government's rapacious extractivist policies. Much of the struggle has centered on resistance to tar sands oil extraction/production, focusing on native land and water rights. See Gene McGuckin, "Why Unionists Must Build the Climate Change Fight," *Climate & Capitalism*, May 2, 2013, http://climateandcapitalism.com; "Farmers Union: Why We Support Idle No More," *Climate & Capitalism*, April 3, 2013, http://climateandcapitalism.com.
51. James Hansen, "Making Things Clearer: Exaggeration, Jumping the Gun, and the Venus Syndrome," April 15, 2013, http://columbia.edu; Barack Obama, "Remarks by the President on Climate Change," June 25, 2013, https://obamawhitehouse.archives.gov/the-press-office/2013/06/25/remarks-president-climate-change

52. Mészáros, *The Challenge and Burden of Our Historical Time.*

6. Making War on the Planet

This chapter is adapted for this book from the following article: John Bellamy Foster, "Making War on the Planet: Geoengineering and Capitalism's Creative Destruction of the Earth," *Science for the People*, September 2015, https://magazine.scienceforthepeople.org/geoengineering/making-war-on-the-planet/, co-published in *Monthly Review* 70/4 (September 2018): 1–10.

1. On the carbon budget see http://trillionthtonne.org. (Note that the trillionth metric ton refers to cumulative carbon, not carbon dioxide.) Also "Carbon Clock Countdown," *The Guardian*, January 19, 2017, https://www.theguardian.com/environment/datablog/2017/jan/19/carbon-countdown-clock-how-much-of-the-worlds-carbon-budget-have-we-spent.
2. *Jacobin* 26, Earth, Wind, and Fire issue (Summer 2017).
3. James Hansen et al., "Young People's Burden: Requirements of Negative CO_2 Emissions," *Earth System Dynamics* 8 (2017): 577–616; James Hansen et al., "Young People's Burden: Requirements of Negative CO_2 Emissions," July 18, 2017, http://columbia.edu.
4. See John Bellamy Foster, "The Long Ecological Revolution," *Monthly Review* 696 (November 2017): 1–16.
5. Spencer Weart: "Interview with M. I. Budyko: Oral History Transcript," March 25, 1990, http://aip.org; *The Discovery of Global Warming* (Cambridge, MA: Harvard University Press, 2003): 85–88; *Climate and Life* (New York: Academic, 1974), 485. M. I. Budyko and Y. A. Izrael, eds., *Anthropogenic Climate Change* (Tucson: University of Arizona Press, 1991), 1–6; Blue Planet Prize, "The Laureates: Mikhail I. Budyko (1998)," http://af-info.or.jp.
6. M. I. Budyko, *Climatic Changes* (Washington, D.C.: American Geophysical Union, 1977), 235–36, 239–46.
7. Oliver Morton, *The Planet Remade* (Princeton: Princeton University Press, 2016), 137–38.
8. Alan Robock, Luke Oman, and Georgiy L. Stenchikov, "Regional Climate Responses to Geoengineering with Tropical and Arctic SO_2 Injections," *Journal of Geophysical Research* 113 (2008): D16101; Alan Robock, "20 Reasons Why Geoengineering May Be a Bad Idea," *Bulletin of Atomic Scientists* 64/2 (2008): 15; Clive Hamilton, *Earthmasters* (New Haven: Yale University Press, 2003), 64.
9. Robock, "20 Reasons Why Geoengineering May Be a Bad Idea," 16.
10. Ibid.
11. Michael E. Mann and Tom Toles, *The Madhouse Effect* (New York:

Columbia University Press, 2016), 123; Robock, "20 Reasons Why Geoengineering May Be a Bad Idea," 16.

12. Hamilton, *Earthmasters*, 65–67; Robock, "20 Reasons Why Geoengineering May Be a Bad Idea," 17; Daisy Dunne, "Six Ideas to Limit Global Warming with Solar Geoengineering," Carbon Brief, May 9, 2018, http://carbonbrief.org.
13. Hamilton, *Earthmasters*, 52–55; Carbon Brief, "Six Ideas."
14. Hugh Powell, "Fertilizing the Ocean with Iron," *Oceanus* 46/1 (2008), http://whoi.edu; Hamilton, *Earthmasters*, 27–35.
15. Powell, "Fertilizing the Ocean with Iron"; Hamilton, *Earthmasters*, 35.
16. Abby Rabinowitz and Amanda Simson, "The Dirty Secret of the World's Plan to Avert Climate Disaster," *Wired*, December 10, 2017.
17. Rabinowitz and Simson, "The Dirty Secret of the World's Plan to Avert Climate Disaster."
18. Julia Rosen, "Vast Bioenergy Plantations Could Stave Off Climate Change—and Radically Reshape the Planet," *Science*, February 15, 2018; Rabinowitz and Simson, "The Dirty Secret of the World's Plan to Avert Climate Disaster"; ETC Group, Biofuel Watch, Heinrich Böll Stiftung, *The Big Bad Fix: The Case Against Climate Geoengineering* (2017), 22, https://www.boell.de/en/2017/12/01/big-bad-fix-case-against-geoengineering/.
19. Hansen et al., "Young People's Burden."
20. ETC Group, Biofuel Watch, Stiftung, *The Big Bad Fix*, 20–22; Michael Friedman, "Why Geoengineering Is Not a Remedy for the Climate Crisis," MR Online, May 22, 2018, http://mronline.org.
21. Friedman, "Why Geoengineering Is Not a Remedy for the Climate Crisis."
22. Hamilton, *Earthmasters*, 47–50.
23. Vaclav Smil, "Global Energy: The Latest Infatuations," *American Scientist* 99 (2011), http:// americanscientist.org. See also Jeff Goodell, "Coal's New Technology," Yale Environment 360, July 14, 2008, https://e360.yale.edu/features/coals_new_technology_panacea_or_risky_gamble/.
24. Andy Skuce, "'We'd Have to Finish One New Facility Every Working Day for the Next 70 Years'—Why Carbon Capture Is No Panacea," *Bulletin of the Atomic Scientists*, October 4, 2016, https://thebulletin.org/2016/10/wed-have-to-finish-one-new-facility-every-working-day-for-the-next-70-years-why-carbon-capture-is-no-panacea/.
25. Tillerson quoted in Michael Babad, "Exxon Mobil CEO: 'What Good Is It to Save the Planet If Humanity Suffers?'" *Globe and Mail*, May 30, 2017.
26. Karl Marx and Frederick Engels, *Collected Works*, vol. 25 (New York: International Publishers, 1987), 460–61.

27. Paul Burkett, "On Eco-Revolutionary Prudence: Capitalism, Communism, and the Precautionary Principle," *Socialism and Democracy* 30/2 (2016): 87.
28. Barry Commoner, *Making Peace with the Planet* (New York: New Press, 1992), ix.
29. See ETC Group, Biofuel Watch, Stiftung, *The Big Bad Fix*, 10.

7. Nature

This chapter has been adapted for this book from the following essay: John Bellamy Foster, "Nature," in *Keywords for Radicals: The Contested Vocabulary of Late Capitalist Struggle*, ed. Kelly Fritsch, Clare O'Connor, and A. K. Thompson (Chico, CA: AK Press, 2016), 279–86.

1. Raymond Williams, *Keywords* (Oxford: Oxford University Press, 1983), 219.
2. Max Weber, *Critique of Stammler* (New York: Free Press, 1977), 96.
3. Alfred North Whitehead, *The Concept of Nature* (Cambridge: Cambridge University Press, 1920).
4. For criticisms of the concept of natural beauty within aesthetics, see G. W. F. Hegel, *The Philosophy of Nature*, vol. 1 (London: Allen and Unwin, 1970), 3; Theodor Adorno, *Aesthetic Theory* (Minneapolis: University of Minnesota Press, 1977), 68–76.
5. Roy Bhaskar, *A Realist Theory of Science* (London: Verso, 1975), and *The Possibility of Naturalism* (Atlantic Highlands, NJ: Humanities Press), 1979.
6. Keith Tester, *Animals and Society* (New York: Routledge, 1991).
7. On realism and ecology, see David R. Keller and Frank B. Golley, Introduction, in *The Philosophy of Ecology*, ed. Keller and Golley (Athens: University of Georgia Press, 2000), 1–4; John Bellamy Foster, Brett Clark, and Richard York, *The Ecological Rift* (New York: Monthly Review Press, 2010), 289–300.
8. Rachel Carson, *Silent Spring* (Boston: Houghton Mifflin, 1962); Barry Commoner, *The Closing Circle* (New York: Knopf, 1971); William Leiss, *The Domination of Nature* (Boston: Beacon, 1972); Carolyn Merchant, *The Death of Nature* (New York: Harper and Row, 1980); John Bellamy Foster, *The Vulnerable Planet* (New York: Monthly Review Press, 1994); Bill McKibben, *The End of Nature* (New York: Random House, 1989); Richard E. Leakey and Roger Lewin, *The Sixth Extinction* (New York: Doubleday, 1995); Naomi Klein, *This Changes Everything* (New York: Simon and Schuster, 2014).
9. E. O. Wilson, *On Human Nature* (Cambridge, MA: Harvard University Press, 1978), x.
10. On the possessive individualism of capitalist society, affecting its

conception of natural-social relations, see C. B. Macpherson, *The Political Theory of Possessive Individualism* (Oxford: Oxford University Press, 1962).
11. Francis Bacon, "The Masculine Birth of Time," in Benjamin Farrington, *The Philosophy of Francis Bacon* (Chicago: University of Chicago Press, 1964), 59–72.
12. Francis Bacon, *The New Atlantis and The Great Instauration* (London: Wiley-Blackwell, 1991).
13. Francis Bacon, *Novum Organum* (Chicago: Open Court, 1994), 29, 43.
14. Hegel, *The Philosophy of Nature*, 421–23.
15. Karl Marx, *Grundrisse* (London: Penguin, 1973), 409–10.
16. Marx, "Concerning Feuerbach," in *Early Writings* (London: Penguin, 1974), 421–23.
17. Marx, *The Poverty of Philosophy* (New York: International Publishers, 1963), 147.
18. Marx, *Capital*, vol. 3 (London: Penguin, 1981), 949.
19. Marx, *Capital*, vol. 1 (London: Penguin, 1976), 637–38; *Capital*, vol. 3, 959.
20. Marx, *Capital*, vol. 3, 911.
21. Representative works include Andrew Dobson, *Green Political Thought* (New York: Routledge, 1995); Robyn Eckersley, *Environmentalism and Political Theory* (Albany: State University of New York Press, 1992); Mark J. Smith, *Ecologism* (Minneapolis: University of Minnesota Press, 1998); Bill Devall and George Sessions, *Deep Ecology* (Layton, UT: Gibbs Smith, 1985).
22. See, for example, Murray Bookchin, *The Philosophy of Social Ecology* (Montreal: Black Rose, 1995); Paul Burkett, *Marx and Nature* (Chicago: Haymarket, 2014); Stefano Longo, Rebecca Clausen, and Brett Clark, *The Tragedy of the Commodity* (New York: Routledge, 2015); Ariel Salleh, Introduction, in Salleh, ed., *Eco-Sufficiency and Global Justice* (London: Pluto Press, 2009), 1–40; Klein, *This Changes Everything*.
23. Marx, *Capital*, vol. 3, 949.
24. Frederick Engels, *Dialectics of Nature* (Moscow: Progress Publishers, 1934), 180.

8. Third Nature

This chapter has been adapted for this book from the following essay: John Bellamy Foster, "Third Nature," in *Will the Flower Slip Through the Asphalt*, ed. Vijay Prashad (New Delhi: LeftWord Books, 2017), 50–57.
1. Said borrowed the notion of "contrapuntal reading" from music. He used it especially in relation to Jane Austen's *Mansfield Park*, where a colonial-slave sugar plantation is the basis of the family's wealth and

structures the plot without actually entering directly into the narrative. As Said explained, "In practical terms, 'contrapuntal reading' as I have called it means reading a text with an understanding of what is involved when an author shows, for instance, that a colonial sugar plantation is seen as important to the process of maintaining a particular style of life in England. . . . The point is that contrapuntal reading must take account of both processes, that of imperialism and that of resistance to it, which can be done by extending our reading of the texts." One can say that this phenomenon of the veiled reality of imperialism shows up in decolonizing as well as colonizing literatures. Said's own work brings out the violence and physical dislocation of peoples due to imperialism. Klein shows that a contrapuntal reading of Said's work also reveals the deep ecological recesses of resistance in his thought. See Edward Said, *Culture and Imperialism* (New York: Vintage, 1993), 66.
2. Ibid., 225.
3. Ibid.; Neil Smith, *Uneven Development* (Athens: University of Georgia Press, 2008).
4. On Mackinder, see John Bellamy Foster, "The New Geopolitics of Empire," *Monthly Review* 57/8 (January 2006), 1–18.
5. Said, *Culture and Imperialism*, 225–26; Mahmoud Darwish, *Splinters of Bone* (Greenfield Center, NY: Greenfield Review Press, 1974), 23.
6. Said, *Culture and Imperialism*, 67, 226, 239.
7. Karl Marx, *Capital* (London: Penguin, 1976), 877–85; and *Early Writings* (London: Penguin, 1974), 318–19; Said, *Culture and Imperialism*, 33, 225.
8. Said, *Culture and Imperialism*, 156–67; Georg Lukács, *The Theory of the Novel* (Cambridge, MA: MIT Press, 1971), 35ff.
9. Said, *Culture and Imperialism*, 330.
10. On Marx's ecological conception of socialism, see John Bellamy Foster, "Marxism and Ecology," *Monthly Review* 67/7 (December 2015), 4.
11. Said, *Culture and Imperialism*, 336; compare Karl Marx, *The Eighteenth Brumaire of Louis Bonaparte* (New York: International Publishers, 1991), 15.

9. Weber and the Environment

This chapter is adapted and revised for this book from John Bellamy Foster and Hannah Holleman, "Weber and the Environment: Classical Foundations for a Postexemptionalist Sociology," *American Journal of Sociology* 117/6 (May 2012): 1625–73.
1. Frederick Buttel, "Environmental Sociology and the Classical Sociological Tradition," in *Sociological Theory and the Environment: Classical Foundations and Contemporary Insights*, ed. Riley E. Dunlap,

Frederick H. Buttel, Peter Dickens, and August Gijswijt (New York: Rowman & Littlefield, 2002), 35–50.
2. William R. Catton and Riley E. Dunlap, "Environmental Sociology: A New Paradigm," *American Sociologist* 13 (1978): 41–49; "A New Ecological Paradigm for Post-Exuberant Sociology," *American Behavioral Scientist* 24 (1980): 15–47; William R. Catton, "Has the Durkheim Legacy Misled Sociology?," in Buttel et al., *Sociological Theory and the Environment* (New York: Rowman & Littlefield, 2002), 90–115; Frederick H. Buttel, Peter Dickens, Riley E. Dunlap, and August Gijswijt, "Sociological Theory and the Environment: An Overview and Introduction," in Buttel et al., *Sociological Theory and the Environment*, 3–32.
3. In this article the term "environment," unless otherwise indicated, refers to that domain of reality which consists of or is directly related to the natural environment. The term "nature," when used in this general sense, likewise refers to the realm of biophysical existence. The complexity of these terms guarantees that their meanings are somewhat fluid and change with the given context. There is no pure "nature," since nature or the environment, as we know it, is everywhere affected by human actions. "Environmental sociology" is meant to designate the subfield of sociology concerned with the interrelation between environment and society. See Robert J. Antonio, "Climate Change, the Resource Crunch, and the Global Growth Imperative," *Current Perspectives in Social Theory* 26 (2009): 33.
4. Buttel "Environmental Sociology and the Classical Sociological Tradition," 35–50; Emile Durkheim, *The Rules of Sociological Method* (New York: Free Press, 1982); Max Weber, *Economy and Society*, 40 (Berkeley: University of California Press, 1968), and *The Methodology of the Social Sciences* (New York: Free Press, 1949), 25–26, 86.
5. Karl Marx and Frederick Engels, *Marx and Engels on Malthus* (New York: International Publishers, 1954).
6. Buttel et al., "Sociological Theory and the Environment: An Overview," 13–15. See also John Bellamy Foster, "Marx's Theory of Metabolic Rift: Classical Foundations for Environmental Sociology," *American Journal of Sociology* 105/ (1999): 366–401; Richard York, Eugene A. Rosa, and Thomas Dietz, "Footprints on the Earth: The Environmental Consequences of Modernity," *American Sociological Review* 68/2 (2003): 279–300; Gregory Hooks and Chad L. Smith, "The Treadmill of Destruction: National Sacrifice Areas and Native Americans," *American Sociological Review* 69 (2004): 558–75; Andrew K. Jorgenson and Brett Clark, "The Economy, Military and Ecologically Unequal Relationships in Comparative Perspective: A Panel Study of the Ecological Footprints

of Nations, 1975-2000," *Social Problems* 56 (2009): 621-46; Thomas Rudel, "How Do People Transform Landscapes?," *American Journal of Sociology* 115/1 (2009): 129-54; Don Grant, Mary Nell Trautner, Liam Downey, and Lisa Thiebaud, "Bringing the Polluters Back In: Environmental Inequality and the Organization of Chemical Production," *American Sociological Review* 75 (2010): 479-504.

7. James O'Connor, *Natural Causes* (New York: Guilford, 1998); Paul Burkett, *Marx and Nature* (New York: St. Martin's Press, 1999); Foster, "Marx's Theory of Metabolic Rift"; Jason W. Moore, "Environmental Crises and the Metabolic Rift in World-Historical Perspective," *Organization and Environment* 13/2 (2000.): 23-57; Peter Dickens, *Society and Nature* (Cambridge: Polity, 2004); John Bellamy Foster, Brett Clark, and Richard York, *The Ecological Rift* (New York: Monthly Review Press, 2010).

8. Timo Järvikoski, "The Relation of Nature and Society in Marx and Durkheim," *Acta Sociologica* 39/1 (1996): 73-86; Buttel, "Environmental Sociology and the Classical Sociological Tradition"; Eugene A. Rosa, and Lauren Richter, "Durkheim on the Environment: Ex Libris or Ex Cathedra?," *Organization and Environment* 21 (2008): 182-87.

9. Patrick C. West, "Social Structure and Environment: A Weberian Approach to Human Ecological Analysis" (PhD diss., Yale University, 1975).

10. Patrick C. West, "Max Weber's Human Ecology of Historical Societies," in *Theory of Liberty, Legitimacy and Power: New Directions in the Intellectual and Scientific Legacy of Max Weber* (Boston: Routledge & Kegan Paul, 1985), 216-43.

11. Rare citations to West, "Max Weber's Human Ecology of Historical Societies" appear in Buttel, "Environmental Sociology and the Classical Sociological Tradition"; and Murphy, "Ecological Materialism and the Sociology of Max Weber." Whereas there are no citations at all to West's dissertation by environmental sociologists, other than West himself, publishing in English up to the present. Using Web of Science and GoogleScholar we were able to ascertain that West's dissertation has been cited as of 2010 in English by someone other than himself only once and in an article unrelated to environmental sociology. West's book chapter based on his dissertation has been cited (beyond his own work) in a total of four books and four articles, with only three citations (one in an article) occurring prior to 2000. The reason for the relative obscurity of West's work on Weber's environmental thought undoubtedly had to do with the fact that it preceded the organization of environmental sociology as a field. West's dissertation included two main chapters on Weber's ecological contributions, focusing on his

historical-comparative works on religion (*Ancient Judaism, The Religion of China,* and *The Religion of India*) and on *The General Economic History* and also taking into account some crucial methodological issues. Our own analysis, although influenced by West's dissertation, attempts to approach these issues with more breadth and depth, relying on a much broader range of Weber's contributions, and aims at a larger synthesis.

12. Raymond Murphy, *Rationality and Nature* (Boulder, CO: Westview Press, 1994), and *Sociology and Nature* (Boulder, CO: Westview Press, 1997), and "Ecological Materialism and the Sociology of Max Weber," in *Sociological Theory and the Environment: Classical Foundations and Contemporary Insights*, ed. Riley E. Dunlap, Frederick H. Buttel, Peter Dickens, and August Gijswijt (New York: Rowman & Littlefield, 2002), 73–89.

13. Raymond Murphy, *Rationality and Nature*, x. In a more recent essay, Murphy has modified this earlier position, now claiming that embedded within Weber is an "ecological materialism." In support of this he cites West's statement that "Weber's ecological analysis emphasized the interactive role of geography, climate, natural resources, and the material aspects of technology in the structure and change in historical social structures." Murphy does not follow up on this, however, and refers later in the same chapter to Weber's "oversimplified view" of the relation between nature and mind as "characteristic of sociology, leading it to neglect the role of nature." Murphy, "Ecological Materialism and the Sociology of Max Weber," 80.

14. Frederick H. Buttel, Peter Dickens, Riley E. Dunlap, and August Gijswijt, "Sociological Theory and the Environment: An Overview and Introduction," in *Sociological Theory and the* Environment, ed. Dunlap, Buttel, Dickens, and Gijswijt (New York: Rowman and Littlefield, 2002), 8.

15. J. M. Blaut, *The Colonizer's Model of the World* (New York: Guilford, 1993), 83.

16. Buttel, "Sociology and the Environment," 342.

17. David Goldblatt, *Social Theory and the Environment* (Boulder, CO: Westview Press, 1996), 3.

18. Ted Benton, "Biology and Social Science," *Sociology* 25/1 (1991): 12.

19. Ted Benton and Michael Redclift, "Introduction," in *Social Theory and the Global Environment*, ed. Michael Redclift and Ted Benton (New York: Routledge, 1994), 5.

20. Harvey Choldin, "Social Life and the Physical Environment," in *Handbook of Contemporary Urban Life*, ed. D. Street (San Francisco: Jossey-Bass, 1978), 353.

21. Riley E. Dunlap, "Paradigms, Theories, and Environmental Sociology," in *Sociological Theory and the Environment: Classical Foundations and Contemporary Insights,* ed. Riley E. Dunlap, Frederick H. Buttel, Peter Dickens, and August Gijswijt (New York: Rowman & Littlefield, 2002), 332–34 and 341. In making such statements, however, Dunlap has professed himself agnostic on whether such criticisms are directly applicable to Weber and Durkheim themselves, though certainly pertaining to the traditions to which they gave rise.
22. Patrick C. West, "Max Weber's Human Ecology of Historical Societies," in *Theory of Liberty, Legitimacy and Power: New Directions in the Intellectual and Scientific Legacy of Max Weber* (Boston: Routledge & Kegan Paul, 1985), 216.
23. Robert J Antonio, "Climate Change, the Resource Crunch, and the Global Growth Imperative," *Current Perspectives in Social Theory* 26/3 (2009), 4.
24. Joachim Radkau, *Max Weber: A Biography* (Cambridge: Polity, 2009), 443.
25. Weber, *The Protestant Ethic and the Spirit of Capitalism* (London: George Allen & Unwin, 1930), 181, translation slightly altered; Michael Mayerfeld Bell., *An Invitation to Environmental Sociology* (Thousand Oaks, CA: Sage Publications, 1998), 150–51. Although employing the Parsons 1930 translation of *The Protestant Ethic* here and throughout this article, we have altered this passage slightly in conformity with Kalberg's 2009 translation to refer, as Weber did, to "fossil fuel" (Kalberg) as opposed to "fossilized coal" (Parsons).
26. Weber, *From Max Weber* (New York: Oxford University Press, 1946), 364–66.
27. Weber, "German-Agriculture and Forestry," *Kölner Zeitschrift für Soziologie und Sozialpsychologie* 57/1 (2005): 147.
28. Weber, "'Energetic' Theories of Culture," *Mid-American Review of Sociology* 9/2 (1984): 50.
29. Joan Martinez-Alier, *Ecological Economics* (Oxford: Basil Blackwell, 1987), 183–92. For an exception, see Foster, "Marx's Theory of Metabolic Rift," 370.
30. Weber, *Economy and Society,* 70.
31. Albrow, *Max Weber's Construction of Social Theory* (New York: St. Martin's Press, 1990), 146.
32. Martin Albrow, "The Application of the Weberian Concept of Rationalization to Contemporary Conditions," in *Max Weber, Rationality and Modernity,* ed. Scott Lash and Sam Whimster (London: Allen & Unwin, 1987, 182).
33. Ibid.

34. Stephen Kalberg, *Max Weber's Comparative-Historical Sociology* (Chicago: University of Chicago Press, 1994), 69–70, 81, 148–49. An underlying assumption of this article is that Weber's interpretive sociology as encompassed in his concept of *Verstehen* has to be extended to encompass the more complex analysis of causal analytics revealed in his substantive works if the significance of his environmental contributions is to be understood. As Kalberg states: "In [Weber's] substantive texts, causal explanations are not provided alone by the central notion of *Verstehen*," 81. Thus it is in Weber's comparative-historical works that one is most likely to discover the complex interaction between the ideal-type as hypothesis-forming generalization and the manifold causalities revealed in the historical process. Fritz Ringer. *Max Weber's Methodology* (Cambridge, MA: Harvard University Press, 1997), 72–80.
35. Weber, "Some Categories of Interpretive Sociology," *Sociological Quarterly* 22 (Spring (1981): 151–80.
36. Max Weber, *Roscher and Knies* (New York: Free Press, 1975), 107–8.
37. Ibid., 105.
38. Ibid., 107–18. Weber developed some of his key ideas in this respect in relation to Wundt's psychology. But he rejected what he called the "metaphysical belief" and "apologetic" that led Wundt to promote a "belief in 'progress'" in which "the culture of humanity" was seen as positively advancing "into the indefinite future." See Wilhelm Wundt, *Elements of Folk Psychology* (New York: Macmillan, 1916). On Wundt's ideas in relation to sociology, see Don Martindale, *The Nature and Types of Sociological Theory* (Boston: Houghton Mifflin, 1960), 294–97.
39. Ibid., 107–8, 141–42.
40. Weber, *Roscher and Knies*, 107–8, 157; *The Theory of Social and Economic Organization*, 93–94; *Economy and Society*, 7; Audrey M. Lambert, *The Making of the Dutch Landscape* (New York: Seminar Press, 1971), 84–87; Mark Elvin, "Why China Failed to Create an Endogenous Industrial Capitalism: A Critique of Max Weber's Explanation," *Theory and Society* 13/3 (1984): 380. It is perhaps illustrative of the relative neglect by sociologists of Weber's environmental observations that though he refers to the incursion of the Dollard (the flooding of the Ems, the overwhelming of the dikes, and the expansion of the Dollard basin) in a number of his works, including *Economy and Society*, the editorial treatment of this in published versions of his work is confused. The Parsons edition of Weber refers to the incursion as occurring "at the beginning of the twelfth century," while in the first complete English edition of *Economy and Society* the date 1277 is added. Although the floods appeared over the late medieval and early modern periods, with

one in 1287 leading to the loss of 50,000 lives (sometimes thought of as the date of the incursion of the Dollard) the storm flood probably most responsible for forming the Dollard basin occurred in the fifteenth century. The Dollard reached its fullest extent in the early sixteenth century. See Lambert, *The Making of the Dutch Landscape*, 84–86; Adriaan Haartsen and Dre van Marrewijk, "The Dutch Wadden Sea Region," in *Wadden Sea Ecosystem*, no. 12, Lancewad Report, http://waddensea-secretariat.org (2001), 225–56.

41. Weber, *The Methodology of the Social Sciences*, 94; see also Weber, "Marginal Utility Theory and 'The Fundamental Law of Psychophysics,'" 142.

42. See Weber, *The Religion of India* (New York: Free Press, 1958), 337; and West, "Social Structure and Environment," 19–20. Here, Weber wrote that the "drive for gain" in India was "lacking in precisely that which was decisive for the economics of the Occident: the *refraction* and rational immersion of the drive character of economic striving and its accompaniments in a system of rational, inner-worldly ethic of behavior, e.g., the 'inner-worldly' asceticism of Protestantism in the West" (italics added).

43. R. Stephen Warner, "The Role of Religious Ideas and the Use of Models in Max Weber's Comparative Studies on Non-Capitalist Societies," *Journal of Economic History* 30/1 (1970): 81–82, 85–86 See also Neil J. Smelser and R. Stephen Warner, *Sociological Theory* (Middletown, NJ: General Learning Press, 1976), 107, 133; West, "Social Structure and Environment: A Weberian Approach to Human Ecological Analysis, " 19–20.

44. Smelser and Warner, *Sociological Theory*, 133.

45. Weber, *From Max Weber*, 280.

46. Max Weber, *Ancient Judaism* (New York: Free Press, 1952), 80; *Economy and Society* (Berkeley: University of California Press, 1968), 40; *The Religion of India* (New York: Free Press, 1958), 337; *The Methodology of the Social Sciences* (New York: Free Press, 1949), 187; see also Fritz Ringer, *Max Weber's Methodology* , 68–74.

47. Weber, *Economy and Society*, 1178.

48. Weber, "Sociology and Biology," in *Weber: Selections in Translation*, ed. W. G. Runciman (Cambridge: Cambridge University Press, 1978), 390.

49. Stephen Kalberg, *Max Weber's Comparative-Historical Sociology* (Chicago: University of Chicago Press. 1994), 69–70.

50. Weber, *The Methodology of the Social Sciences*, 72, 110.

51. Weber, *Critique of Stammler* (New York: Free Press, 1977), 96.

52. Ibid., 97.

53. Ibid., 95–96.

54. Weber, *The Religion of India*, 340.

55. On neo-Kantianism and Weber's complex relation to it, see Martindale, *The Nature and Types of Sociological Theory*, 220–66, 376–83.
56. Weber, *Critique of Stammler*, 91.
57. Marx Weber, *Roscher and Knies*, 31.
58. Weber, *The Agrarian Sociology of Ancient Civilizations* (London: Verso, 1976), 84.
59. Weber, *Critique of Stammler*, 100–104, 110–11; Ringer, *Max Weber's Methodology*, 99.
60. Kalberg, *Max Weber's Comparative-Historical Sociology*, 148.
61. Weber, *Economy and Society*, 607; Weber, *From Max Weber*, 346.
62. Weber, *Economy and Society*, 468, 607.
63. Weber, *From Max Weber*, 364–66.
64. Radkau, *Nature and Power*, 443.
65. Weber, *Ancient Judaism*, 5–8.
66. Ibid., 8.
67. Ibid., 8–10. See also Weber, *The Agrarian Sociology of Ancient Civilizations*, 134–35; Reinhard Bendix, *Max Weber: An Intellectual Portrait* (Garden City, NY: Doubleday, 1960), 219–22; and West, "Social Structure and Environment," 77–79.
68. Weber, *Ancient Judaism*, 54.
69. Ibid., 8.
70. Ibid., 8–13. See also Bendix, *Max Weber: An Intellectual Portrait*, 219.
71. Ibid., 154–55.
72. Ibid., 124, 128–33. See also Weber, *Economy and Society*, 449; Weber, *The Religion of China* (New York: Free Press, 1951), 21, 23; Bendix, *Max Weber: An Intellectual Portrait*, 229–30.
73. Joachim Radkau, *Nature and Power*, 441-42.
74. Marx's concept of the Asian mode of production was derived primarily from earlier views developed by classical political economists associated with colonial policy such as Adam Smith, James Mill, John Stuart Mill, and Richard Jones. The two Mills were employees of the British East India Company. Jones was Malthus's successor as professor of political economy at the East India College. See John Stuart Mill, *Principles of Political Economy* (New York: Longmans, Green, 1904), 105–6, 255; Karl A. Wittfogel, "Geopolitics, Geographical Materialism and Marxism," *Antipode* 17/1 (1985): 21–72; Donald Winch, *Classical Political Economy and the Colonies* (Cambridge, MA: Harvard University Press, 1965); West. "Social Structure and Environment," 163–64; Perry Anderson, *Lineages of the Absolutist State* (London: New Left Books, 1974), 464–72; and Krader, *The Asiatic Mode of Production*, 5–7, 183.
75. Marx and Engels, *On Colonialism* (New York: International Publishers, 1972), 37.

76. Karl Marx, *A Contribution to a Critique of Political Economy* (Moscow: Progress Publishers, 1970), 21; Brendan O'Leary, *The Asiatic Mode of Production* (Oxford: Basil Blackwell, 1989), 82, 104.
77. Marx, *Capital*, vol. 1 (London: Penguin, 1976), 649–50.
78. Frederick Engels, *Anti-Dühring* (Moscow: Progress Publishers, 1969), 215; see also Anderson. *Lineages of the Absolutist State* (1974), 482.
79. Krader, *The Asiatic Mode of Production*, 286–96.
80. Umberto Melotti, *Marx and the Third World* (London: Macmillan, 1977), 8–21. Weber's argument in *The Religion of India* also focuses on the village community rather than hydraulics, showing some similarity to the later Marx in this respect.
81. Krader, *The Asiatic Mode of Production*, 115.
82. On contemporary criticisms of the Asiatic mode and the hydraulic civilization hypotheses, see Anderson, *Lineages of the Absolutist State*, 548; B. Chandra, "Karl Marx, His Theories of Asian Society, and Colonial Rule," *Review* 5/1: 14, 47; Michael Mann, *The Sources of Social Power* (Cambridge: Cambridge University Press, 1986), vol. 1, 94–98; and J. M. Blaut, *The Colonizer's Model of the World* (New York: Guilford Press, 1993), 80–90. Nevertheless, strong traces of such views, particularly with respect to the hydraulics argument, can still be found in the literature, e.g., Eric L. Jones, *The European Miracle* (Cambridge: Cambridge University Press, 1987), 8.
83. Weber, *Economy and Society*, 198; *General Economic History* (Mineola, NY: Dover, 2003).
84. John Love, "Max Weber's Orient," in *The Cambridge Companion to Weber* (Cambridge: Cambridge University Press. 2000), 175.
85. Weber, *General Economic History*, 97–98.
86. Weber, *The Agrarian Sociology of Ancient Civilizations*, 106; *Economy and Society*, 971–72, 1091; *The Religion of China*, 20–21.
87. Ibid., 20.
88. Weber, *Economy and Society*, 1261; italics added. In relation to China, Elvin indicates that Weber was wrong in his notion of a hydraulic state. "Except for some important large-scale operations that mostly appeared rather late, the greater part of irrigation and flood defense was maintained by collectivities as opposed to supervision and the adjudication of disputes. Doubts about Weber's position here are questions of balance and nuance." See Mark Elvin, "Why China Failed to Create an Endogenous Industrial Capitalism: A Critique of Max Weber's Explanation," *Theory and Society* 13/3 (1984): 386.
89. Weber, *Economy and Society*, 449; italics added.
90. Weber, *The Religion of China*, 64; italics added.
91. Weber, *The Religion of India*, 257; italics added.

92. Radkau, *Max Weber: A Biography*, 82.
93. J. M. Blaut, *Eight Eurocentric Historians* (New York: Guilford Press, 2000), 21–24.
94. Weber, *The Agrarian Sociology of Ancient Civilizations*, 84.
95. Weber, *General Economic History*, 354.
96. Weber, *Economy and Society*, 1091.
97. John Locke, *The Second Treatise of Government* (Indianapolis: Bobbs-Merrill, 1952), 17.
98. Weber, *Economy and Society*, 132.
99. Weber, *General Economic History*, 56.
100. Weber, *From Max Weber*, 379.
101. Weber, *General Economic History*, 9; see also Engels, "The Mark," in *Socialism: Utopian and Scientific* (New York: International Publishers, 1978), 77–93.
102. Weber, *General Economic History*, 66, 71.
103. Ibid., 72.
104. Ibid., 79.
105. Ibid., 111.
106. Bryan S. Turner, "Preface to the New Edition," in Max Weber, *From Max Weber* (London: Routledge, 1991), xxiv.
107. Weber, *From Max Weber*, 368.
108. Ibid., 367.
109. Weber, *General Economic History*, 304.
110. T. S. Ashton, *Iron and Steel in the Industrial Revolution* (Manchester, UK: Manchester University Press, 1951), 17.
111. Radkau, *Nature and Power*, 149.
112. Ibid., 149.
113. Marx and Engels, *Collected Works* (New York: International Publishers, 1975), 3:484. Another sign of the failure of Weber scholars to take the environmental aspects of his analysis seriously is the following sentence, containing a major error, in his *General Economic History*: "The smelting of iron with coal instead of charcoal first begins to be typical in the 16th century, thus establishing the fateful union of iron and coal." The sentence should clearly have said "begins to be typical in the late 18th century." The process of smelting iron with coal was not invented by Andrew Darby until 1709, although historians still debate whether it was first developed by Dud Dudley in the seventeenth century, and then the method was lost. It did not become typical until late in the eighteenth century (in 1788 the number of charcoal furnaces in England and Wales had finally fallen to 24, as compared to coal furnaces which had by then reached 53. W. K. V. Gale, *Iron and Steel* (London: Longmans, Green, 1969), 29; see also John Lord, *Capital and Steam Power, 1750–1800* (London: Frank Cass, 1966),

23–24. Indeed, not only was Weber well aware that the smelting of iron with coal was only introduced in the eighteenth century (he provides 1740 as the date of its first introduction), but he also made this a central part of his argument elsewhere in his *General Economic History,* as indicated in the text above. This curious error might be attributable to the fact that his *General Economic History* was compiled from very scattered notes of his lectures, left behind by Weber and kept by his students. But it is also an indication of the general neglect by sociologists of the environmental aspect of his thought that this contradiction in the text as it has come down to us has apparently gone unnoticed.

114. Vaclav Smil, *Energy in Nature and Society* (Cambridge, MA: MIT Press, 2008), 191.
115. Weber, *General Economic History,* 304.
116. Ibid., 191, 304–5.
117. Tamara L. Whited, Jens F. Engels, Richard C. Hoffmann, Hilde Ibsen, and Wybren Verstegen, *Northern Europe: An Environmental History* (Santa Barbara, CA: ABC-CLIO, 2005), 94.
118. Roland Bechmann, *Trees and Man* (New York: Paragon House, 1990), 154.
119. Weber, *The Protestant Ethic and the Spirit of Capitalism,* 17–22.
120. Randall Collins, *Weberian Sociological Theory* (Cambridge: Cambridge University Press, 1986), 20.
121. Weber, *General Economic History,* 190–91.
122. Weber, *From Max Weber,* 364, 368.
123. Wilhelm Ostwald, "On the Theory of Science," in *Congress of Arts and Science, Universal Exposition, St. Louis, 1904,* vol. 1 (Boston: Houghton Mifflin, 1906), 512; Kenneth Stokes. *Paradigm Lost* (Armonk, NY: M. E. Sharpe, 1995), 136.
124. Lewis Mumford, *Technics and Civilization* (New York: Harcourt, Brace, 1934).
125. Randall Collins, *Weberian Sociological Theory,* 78.
126. J. R. McNeill, *Something New Under the Sun* (New York: W. W. Norton, 2000), 14.
127. Weber, *General Economic History,* 306.
128. Jevons, *The Coal Question,* 138–39; E. J. Hobsbawm, *Industry and Empire* (London: Penguin, 1969), 70–71.
129. Weber, *General Economic History,* 297, 304–6.
130. Weber, *The Religion of China,* 199; Joseph W. H. Lough, *Weber and the Persistence of Religion* (London: Routledge, 2006), 81; Radkau, *Nature and Power,* 106–7.
131. Weber, *General Economic History,* 305.
132. Ibid., 305–6; see also Weber, "'Energetic' Theories of Culture," *Mid-American Review of Sociology* 9/2 (1984): 39.

133. Ibid., 92, 96.
134. Ibid., 302.
135. Ibid., 352.
136. Ibid., 352.
137. Weber, *From Max Weber*, 369.
138. Paul Honigsheim, *The Unknown Max Weber* (New Brunswick, NJ: Transaction Publishers, 2000); see also Foster, "Marx's Theory of Metabolic Rift."
139. Max Weber, *The Russian Revolutions* (Ithaca, NY: Cornell University Press, 1995), 84.
140. Weber, *Economy and Society*, 872.
141. William H. Brock, *Justus von Liebig* (Cambridge: Cambridge University Press, 1997), 177–78.
142. Weber, *The Russian Revolutions*, 84; Marcel Mazoyer and Laurence Roudart, *A History of World Agriculture* (New York: Monthly Review Press, 2006), 366–67.
143. Weber, "German-Agriculture and Forestry," *Kölner Zeitschrift für Soziologie und Sozialpsychologie* 57/1 (2005): 143, 147.
144. Weber, *General Economic History*, 82–83.
145. Weber, *Roscher and Knies*, 296.
146. Radkau, *Max Weber: A Biography*, 94.
147. Weber, "German-Agriculture and Forestry," 139–47, 139.
148. Weber, "German Industries," *Kölner Zeitschrift für Soziologie und Sozialpsychologie* 57/1 (2005): 148–56.
149. W. M. Harper Davis, "The International Congress of Arts and Science," *Popular Science Monthly* 60 (1904): 5–32.
150. Marianne Weber, *Max Weber: A Biography* (New York: John Wiley & Sons, 1975), 290–91; Max Weber, "The Relations of the Rural Community to Other Branches of Social Science," in *Congress of Arts and Science, Universal Exposition, St. Louis, 1904*, vol. 7 (Boston: Houghton Mifflin, 1906), 725–46; Weber, *From Max Weber*, 363–85; Radkau, *Nature and Power*, 60–66. Sociologists generally have ignored the ecological implications of Weber's 1904 presentation in St. Louis. But the same is not true of ecological economists. See Nicholas Georgescu-Roegen, *The Entropy Law and the Economic Process* (Cambridge, MA: Harvard University Press, 1971), 313.
151. Although Turner also presented a version of his thesis in St. Louis at the Universal Exposition in Chicago in 1904, there is no evidence that Weber attended or that the two scholars ever met. Nor is there any record of any direct influence of Turner's ideas on Weber in "Remnants of Romanticism: Max Weber in Oklahoma," *Journal of Classical*

Sociology 5/1 (2005): 54. See also Lawrence A. Scaff, *Max Weber in America* (Princeton: Princeton University Press, 2011), 54.
152. Weber, *From Max Weber*, 372, 383; Scaff, *Max Weber in America*, 60–66.
153. Frederick Jackson Turner, *The Frontier in American History* (New York: Henry Holt, 1921), 13.
154. Weber, *From Max Weber*, 366.
155. Ibid., 364–85.
156. Weber, *Roscher and Knies*, 284–87; Scaff, *Max Weber in America*, 40–43.
157. Scaff, "Remnants of Romanticism," 55.
158. Weber, *Roscher and Knies*, 291; Scaff, *Max Weber in America*, 55.
159. Ibid., 65.
160. Weber, "A Letter from Indian Territory," *Free Inquiry in Creative Sociology* 16/2 (1988): 133–36.
161. Ibid.
162. Ibid., 134–35; Scaff, *Max Weber in America*, 73–97.
163. Allan Schnaiberg, *The Environment: From Surplus to Scarcity* (New York: Oxford University Press, 1980), 7–31.
164. Wilhelm Ostwald, "On the Theory of Science," in *Congress of Arts and Science, Universal Exposition, St. Louis, 1904*, vol. 1 (Boston: Houghton Mifflin, 1906), 339–40.
165. *"Energetische Grundlagen der Kulturwissenschaft"* (Energetic Foundations of a Science of Culture) (Leipzig: Dr. Werner Klinkhardt Verlag, 1909), 44–50. Translation of Ostwald in this and the following paragraphs in this article is by Joseph Fracchia.
166. Ibid., 44.
167. Ibid., 44–50.
168. Ibid.
169. Eugene A. Rosa, Gary E. Machlis, and Kenneth M. Keating, "Energy and Society," *American Review of Sociology* 14 (1988): 151.
170. Radkau, *Max Weber: A Biography*, 73.
171. Ringer, *Max Weber's Methodology*, 53–56.
172. Martinez-Alier, *Ecological Economics*, 183–92; Kenneth Stokes. *Paradigm Lost*, 138.
173. Weber, *The Methodology of the Social Sciences*, 64–65; Weber, "'Energetic' Theories of Culture," 37.
174. Martinez-Alier, *Ecological Economics*, 185.
175. Weber, "'Energetic' Theories of Culture," 38.
176. Ibid., 41; Martinez-Alier, *Ecological Economics*, 190–91.
177. Weber, "'Energetic' Theories of Culture," 38.

178. Ibid., 38–40.
179. Martinez-Alier, *Ecological Economics*, 187.
180. Weber, "'Energetic' Theories of Culture," 48. For a similar critique of an energetic theory of value to that of Weber's, see Engels in Marx and Engels, *Collected Works*, 25:586–87.
181. Weber, "Remarks on Technology and Culture," *Theory, Culture and Society* 22/4 (2005): 31.
182. Weber, "'Energetic' Theories of Culture," 38; Boris Hessen, "The Social and Economic Roots of Newton's 'Principia,'" in Nicholas Bukharin et al., *Science at the Crossroads* (London: Frank Cass, 1971), 197; Martinez-Alier, *Ecological Economics*, 186.
183. Weber, "'Energetic' Theories of Culture," 56.
184. Weber, "Remarks on Technology and Culture," 31.
185. Gerth and Mills, "Introduction," 51; Lawrence A. Scaff, *Fleeing the Iron Cage* (Berkeley: University of California Press, 1989), 224; Ralph Schroeder, "Disenchantment and Its Discontents," *Sociological Review* 43/2 (1995): 227–28.
186. Georg Lukács, "Max Weber and German Sociology," in *Max Weber: Critical Assessments*, vol. 1, ed. Peter Hamilton (New York: Routledge, 1991), 112.
187. Max Horkheimer and Theodor Adorno, *The Dialectic of Enlightenment* (New York: Continuum, 1972), 3–8.
188. Raymond Murphy, *Rationality and Nature* (Boulder, CO: Westview Press, 1994), 32; see also Morris Berman, *The Reenchantment of Nature* (Ithaca, NY: Cornell University Press, 1981); and Murray Bookchin, *Re-Enchanting Humanity* (New York: Cassell, 1995).
189. George G. Iggers, "The Idea of Progress in Historiography and Social Thought Since the Enlightenment," in *Progress and Its Discontents*, ed. Gabriel A. Almond, Marvin Chodorow, and Roy Harvey Pearce (Berkeley and Los Angeles: University of California Press, 1982), 60; see also Gibson, *A Reenchanted World*, 15–16.
190. Ira J. Cohen, "Introduction," in Weber, *General Economic History*, xxvi; Don Martindale and Johanne Riedel, "Introduction," in Weber, *The Rational and Social Foundations of Music*, ed. Martindale and Riedel (Carbondale: Southern Illinois University Press, 1958), xxi; Andrew M. Koch, *Romance and Reason: Ontological and Social Sources of Alienation in the Writings of Max Weber* (Lanham, MD: Lexington Books, 2006), 121–23.
191. Weber, "Some Categories of Interpretive Sociology," 155; Wolfgang Schluchter. *Rationalism, Religion, and Domination* (Berkeley and Los Angeles: University of California Press), 1989, 417.
192. On the addition of the concept of the "disenchantment of the world" to the final, 1920 edition of Weber's original 1905 work, compare the

last edition of Weber's treatise and Talcott Parsons's note to this on 221–22, to the first edition of Weber's treatise translated by Peter Baehr and Gordon C. Wells. See Weber, *The Protestant Ethic and the Spirit of Capitalism*, 105; and Weber, *The Protestant Ethic and the "Spirit" of Capitalism and Other Writings*, 74.
193. Weber, *The Protestant Ethic and the Spirit of Capitalism*, 105.
194. Friedrich Schiller, *The Poems of Schiller* (New York: Henry Holt, 1902), 75; translation according to Taylor, *A Secular Age* (Cambridge, MA: Harvard University Press, 2007).
195. Weber, *The Protestant Ethic and the Spirit of Capitalism*, 78.
196. It is ironic that Kalberg, in opposition to many other Weber scholars, denied the historical and dialectical complexity of Weber's concept of the disenchantment of the world in his classic article on the polymorphous nature of Weber's concept of rationalization. In Kalberg's view, the use of the term "disenchantment" as opposed to "demagification" in English translations was simply an error that conjured "up images of the romanticist's yearning for the *Gemeinschaft*" of "an earlier 'simpler world'" and thus "has not the slightest relationship to Weber's usage of *Entzauberung*." Stephen Kalberg, "Max Weber's Types of Rationality," *American Journal of Sociology* 85/5 (1980): 1146. This is clearly contradicted, however, by the close connection between Weber's term and Schiller's and by Weber's critical use of the concept with respect to modernity. Thus he explicitly raised the issue of disenchantment in works like "Science as a Vocation" as representing "the fate of our times." Weber, *From Max Weber*, 155.
197. Guenther Roth, "Duration and Rationalization: Fernand Braudel and Max Weber," in *Max Weber's Vision of History* (Berkeley and Los Angeles: University of California Press, 1979), 192–93.
198. Weber, *The Protestant Ethic and the Spirit of Capitalism*, 75.
199. Weber, *Economy and Society*, 1156.
200. Sombart quoted in Scaff, *Fleeing the Iron Cage*, 205.
201. Weber, *Economy and Society*, 1178.
202. Buttel et al., "Sociological Theory and the Environment: An Overview and Introduction," 8.
203. Riley E. Dunlap, "Paradigms, Theories, and Environmental Sociology," in *Sociological Theory and the Environment: Classical Foundations and Contemporary Insights*, 332–34.
204. Weber, *The Protestant Ethic and the Spirit of Capitalism*, 181.
205. Goldblatt, *Social Theory and the Environment*, 3.
206. Guenther Roth, "Duration and Rationalization: Fernand Braudel and Max Weber," in *Max Weber's Vision of History*, ed. Roth and Wolfgang Schluchter (Berkeley: University of California Press, 1979), 173.

207. Hettner was best known for his methodological writings on geography, which overlapped with Weber's general perspective. "Both nature and man," he wrote, "are intrinsic to the particular character of the [geographical] areas, and indeed in such intimate union that they cannot be separated from each other." Quoted in Richard Hartshorne, *Perspectives on the Nature of Geography* (Chicago: Rand McNally, 1959). Sociologist Robert E. Park, who was to develop the human ecology approach to urban sociology in the United States, completed his PhD dissertation at Heidelberg under Hettner and Wilhelm Wideband.
208. Guenther Roth, "Duration and Rationalization: Fernand Braudel and Max Weber," 174.
209. Weber, *Economy and Society*, 70. In the nineteenth century it was common to describe desertification processes in terms of "inroads of sand." Thus in *The Book of Nature* John Mason Good wrote: "The most extraordinary inroads of sand storms and sand floods are, perhaps, those which have taken place in the Libyan Desert and in Lower Egypt. M. Denon informs us, in his travels over this part of the world, that the summits of the ruins of ancient cities buried under mountains of drifted sands still appear externally." Good, *The Book of Nature* (New York: J. & J. Harper, 1831), 72. Since Weber in the sentence quoted in the text is referring to environmental factors that had extraordinary effects on civilization there is little doubt that he is describing the desertification process with this region of the world in mind.
210. Weber, *Ancient Judaism*, 10.
211. For example, see Ulrich Beck, "The Reinvention of Politics," in Beck, Anthony Giddens, and Scott Lash, *Reflexive Modernization* (Cambridge: Polity, 1994), 6–7; Frederick Buttel, "Ecological Modernization as Social Theory," *Geoforum* 31 (2000): 63–64; Murray J. Cohen, "Ecological Modernisation, Environmental Knowledge, and National Character," in *Ecological Modernisation Around the World*, ed. Arthur P. J. Mol and David A. Sonnenfeld (London: Frank Cass, 2000), 100; Arthur P. J. Mol and David A. Sonnenfeld, eds., *Ecological Modernisation Around the World*, 21–22).
212. Weber, "'Energetic' Theories of Culture," 56.
213. Weber, "A Letter from Indian Territory," *Free Inquiry in Creative Sociology* 16/2 (1988): 134–35.
214. Weber, *From Max Weber*, 364–85.
215. Wundt, *Elements of Folk Psychology*, 10, 510–12.
216. Weber, *Roscher and Knies*, 118.

10. The Theory of Unequal Ecological Exchange

This chapter is adapted and revised for this book from John Bellamy

Foster and Hannah Holleman, "The Theory of Unequal Ecological Exchange: A Marx-Odum Dialectic," *Journal of Peasant Studies* 41/1–2 (March 2014): 199–233.
1. Neither the metabolic rift, which is associated with contradictions in the human metabolism with nature, nor unequal ecological exchange, which arises from the disparities between relatively urban/industrialized and rural/underdeveloped regions, is exclusively concerned with North-South imperialism, since the effects are internal to given nations/regions as well. Here, however, we will be looking at these theories specifically in relation to imperialism, that is, in terms of the global metabolic rift, and global unequal ecological exchange.
2. Mindi Schneider and Philip McMichael, "Deepening, and Repairing, the Metabolic Rift," *Journal of Peasant Studies* 37/3 (2010): 461.
3. John Bellamy Foster, "Marx's Theory of Metabolic Rift," *American Journal of Sociology* 105/2 (1999): 366–405; and *Marx's Ecology* (New York: Monthly Review Press, 2000); J. W. Moore, "Environmental Crises and the Metabolic Rift in World-Historical Perspective," *Organization and Environment* 13/2 (2000): 123–58; and "Transcending the Metabolic Rift," *Journal of Peasant Studies*, 38/1 (2011): 1–46; Paul Burkett, *Marxism and Ecological Economics* (Boston: Brill, 2006); Rebecca Clausen, "Healing the Rift," *Monthly Review* 59/1 (2007): 40–52; Hannah Wittman. "Reworking the Metabolic Rift," *Journal of Peasant Studies* 36/4 (2009): 805–26; John Bellamy Foster, Brett Clark, and Richard York, *The Ecological Rift* (New York: Monthly Review Press, 2010); Schneider and McMichael, "Deepening, and Repairing, the Metabolic Rift," 461–84; Ryan Gunderson, "The Metabolic Rift of Livestock Agribusiness," *Organization and Environment* 24/4 (2011): 404–22; and Ricardo Dobrovolski, "Marx's Ecology and the Understanding of Land Cover Change," *Monthly Review* 64/1 (2012): 31–39.
4. Alf Hornborg, "Ecosystems and World-Systems," in C. Chase-Dunn and S. J. Babones, eds., *Global Social Change* (Baltimore: Johns Hopkins University Press, 2006), 161–75; and *Global Ecology and Unequal Exchange* (London: Routledge, 2011); Andrew K. Jorgenson, "Unequal Ecological Exchange and Environmental Degradation," *Rural Sociology* 71 (2006): 685–712; Andrew K. Jorgenson and James Rice, "Unequal Exchange and Consumption-Based Environmental Impacts: A Cross-National Comparison," in A. Hornborg, J. R. McNeill, and J. Martinez-Alier, eds., *Rethinking Environmental History: World-System History and Global Environmental Change* (New York: AltaMira, 2007), 273–88; Kirk S. Lawrence, "The Thermodynamics of Unequal Exchange: Energy Use, CO_2 Emissions, and GDP in the World-System, 1975–

2005," *International Journal of Comparative Sociology* 50/3–4 (2009): 335–59; Andrew K. Jorgenson and Brett Clark. "The Economy, Military, and Ecologically Unequal Exchange Relationships in Comparative Perspective: A Panel Study of the Ecological Footprints of Nations, 1975–2000," *Social Problems* 56/4 (2009), 621–46; and Andrew K. Jorgenson and Brett Clark, "Footprints: The Division of Nations and Nature," in A. Hornborg, B. Clark, and K. Hermele, eds., *Ecology and Power: Struggles Over Land and Material Resources in the Past, Present, and Future* (London: Routledge, 2012), 155–67; Brett Clark and John Bellamy Foster, "Ecological Imperialism and the Global Metabolic Rift: Unequal Exchange and the Guano/Nitrates Trade," *International Journal of Comparative Sociology* 50/3–4 (2009): 311–34; Andrew K. Jorgenson., Kelly Austin, and Christopher Dick, "Ecologically Unequal Exchange and the Resource Consumption/Environmental Degradation Paradox," *International Journal of Comparative Sociology* 50/3–4 (2009): 263–84.
5. Our goal here is merely to *open the door* to what we hope will eventually be a comprehensive theory, one that would need to be integrated with issues of history, geography and coevolutionary development, encompassing the whole formation of the world-capitalist system, including its historical logic and crises.
6. Howard T. Odum, "Self-Organization, Transformity, and Information," *Science* 242/4882 (1988): 1132–38; Howard T. Odum, *Environmental Accounting: Emergy and Environmental Decision-Making* (New York: John Wiley and Sons, 1996); Howard T. Odum, *Environment, Power, and Society for the Twenty-first Century: The Hierarchy of Energy* (New York: Columbia University Press, 2007); Howard T. Odum and J. E. Arding, *Emergy Analysis of Shrimp Mariculture in Ecuador* (Narragansett, RI: Coastal Resources Center, University of Rhode Island, 1991); Howard T. Odum and Elisabeth C. Odum, *A Prosperous Way Down* (Boulder: University Press of Colorado, 2001).
7. Odum, *Environment, Power, and Society for the Twenty-first Century*, 276–77.
8. Burkett, *Marxism and Ecological Economics*, 35.
9. Richard Levins, "Dialectics and Systems Theory," in Bertell Ollman and Tony Smith, eds., *Dialectics for the New Century* (New York: Palgrave-Macmillan, 2008), 26–49.
10. Frank B. Golley. A *History of the Ecosystem Concept in Ecology* (New Haven: Yale University Press, 1993), 33.
11. Howard T. Odum, *Systems Ecology* (New York: John Wiley and Sons, 1983), 252.
12. Ibid., 265; David M. Scienceman, "Energy and Emergy," in G. Pillet

and T. Murota, eds., *Environmental Economics: The Analysis of a Major Interface* (Geneva: Roland Leimgruber, 1987), 257–76: David M. Scienceman, "The Emergence of Economics," in *Proceedings, International Society for the System Sciences*, vol. 3 (Edinburgh, 1989): vol. 3, 62–68; David M. Scienceman, "EM VALUE AND LA VALUE," Paper presented to the Annual Meeting of the International Society for the Systems Sciences, University of Denver, July 12–17, 1992, 27–37; Howard T. Odum and David M. Scienceman. "An Energy Systems View of Karl Marx's Concepts of Production and Labor Value," in *Emergy Synthesis 3: Theory and Applications of the Emergy Methodology, Proceedings from the Third Biennial Emergy Conference*, January 2004 (Gainesville, FL: Center for Environmental Policy, 2005), 17–43.

13. Foster, "Marx's Theory of Metabolic Rift," and *Marx's Ecology*; Burkett, *Marxism and Ecological Economics*; Foster et al., *The Ecological Rift*; Mindi Schneider and Philip McMichael, "Deepening, and Repairing, the Metabolic Rift."

14. In what follows we take it for granted that all of the major Marxian theories of imperialism, e.g., those of Marx, Luxemburg, Bukharin, Lenin, Amin, Emmanuel, Baran, Magdoff, and Harvey, are integrally related to theories of unequal exchange. However, our treatment of imperialism is necessarily confined only to those points where unequal-exchange theory (both economic and ecological) directly overlaps with the larger Marxian theory of imperialism. Although a wider synthesis of unequal-exchange analysis with imperialism theory as a whole is, in our view, essential, such an overall synthesis is beyond the limits of our present paper, which may serve, however, to lay some of the crucial foundations for a more unified theory.

15. Donald Winch, *Classical Political Economy and Colonies* (Cambridge, MA: Harvard University Press, 1965).

16. David Ricardo, *On the Principles of Political Economy and Taxation* (Cambridge: Cambridge University Press, 1951), 128–49.

17. E. K. Hunt and Mark Lautzenheiser, *History of Economic Thought* (Armonk, NY: M. E. Sharpe, 2011), 119–20.

18. Samir Amin, *Imperialism and Unequal Development* (Hassocks, Sussex: Harvester Press, 1977), 184.

19. Ricardo, *On the Principles of Political Economy and Taxation*, 135–36.

20. Samir Amin, *Unequal Development* (New York: Monthly Review Press, 1976), 134–35.

21. See Roman Rosdolsky, *The Making of Marx's "Capital"* (London: Pluto Press, 1977), 12; and Michael A. Lebowitz, *Beyond Capital* (New York: St. Martin's Press, 1992).

22. Karl Marx, *Theories of Surplus Value, Part 3* (Moscow: Progress Publishers, 1971), 105–6.
23. Karl Marx, *Capital*, vol. 3 (London: Penguin, 1981), 345.
24. Karl Marx, *Grundrisse* (London: Penguin, 1973), 872.
25. Marx, *Capital*, vol. 3, 345.
26. Karl Marx, "On the Question of Free Trade," in K. Marx, ed., *The Poverty of Philosophy* (New York: International Publishers, 1963), 223; see also Maurice Dobb. *Political Economy and Capitalism* (New York: International Publishers, 1945), 226–27; Rosdolsky, *The Making of Marx's "Capital,"* 307–12.
27. Otto Bauer, *The Question of Nationalities and Social Democracy* (Minneapolis: University of Minnesota Press, 2000), 200; Arghiri Emmanuel, *Unequal Exchange: A Study of the Imperialism of Trade* (New York: Monthly Review Press, 1972), 175.
28. Emmanuel, *Unequal Exchange: A Study of the Imperialism of Trade*, 167; Guglielmo Carchedi, *Frontiers of Political Economy* (London: Verso, 1991), 222–25.
29. Henryk Grossman. *The Law of Accumulation* (London: Pluto Press, 1992), 170.
30. Yevgeni Preobrazhensky, *The New Economics* (Oxford: Oxford University Press, 1965), 5, 227, 262.
31. Emmanuel, *Unequal Exchange: A Study of the Imperialism of Trade*, 167.
32. Ibid., xxxiii–iv, 91, 160–61, 367, 381–83.
33. Ibid., 93.
34. John Brolin, "The Bias of the World," *Lund Studies in Human Ecology* 9 (Lund: Lund University, 2006), 70–71.
35. Paul A Baran and Paul M. Sweezy, "Notes on the Theory of Imperialism," *Monthly Review* 17/10 (1966): 15.
36. Che Guevara, *Che Guevara Reader* (Melbourne: Ocean Press, 1997), 291, 302–3.
37. Wallerstein's initial position seems to have been closer to Emmanuel's. See Immanuel Wallerstein, *World-Systems Analysis* (Durham, NC: Duke University Press, 2004), 28.
38. See, for example, C. Chase-Dunn, *Global Formation* (Lanham, MD: Rowman and Littlefield, 1998), 59.
39. Samir Amin, *Unequal Development; Imperialism and Unequal Development* (Hassocks, Sussex: Harvester Press, 1977); and *The Law of Worldwide Value* (New York: Monthly Review Press, 2010).
40. Ernest Mandel, *Late Capitalism* (London: Verso, 1975), 343–76; Gernot Köhler, *Surplus Value and Transfer Value*, World Systems Archive (1999), available at http://wsarch.ucr.edu/archive/papers/

kohler/svtv.htm; Köhler, "Time Series of Unequal Exchange," in G. Köhler and E. J. Chaves, eds., *Globalization: Critical Perspectives* (New York: Nova Science Publishers, 2003), 373–386; Amin, *The Law of Worldwide Value* and "The Surplus in Monopoly Capitalism and the Imperial Rent," *Monthly Review* 64/3 (2012): 78–85.
41. John Smith, "The GDP Illusion: Value Added Versus Value Capture," *Monthly Review* 64/3 (2012): 86–102.
42. Samir Amin, "The Class Structure of the Contemporary Imperialist System," *Monthly Review* 31/8 (1980): 9–26; and "The Surplus in Monopoly Capitalism and the Imperial Rent"; Köhler, "Time Series of Unequal Exchange," 373–86.
43. Amin, "Self-Reliance and the New International Economic Order," 6; Amin, "The Surplus in Monopoly Capitalism and the Imperial Rent"; and Smith, "The GDP Illusion: Value Added Versus Value Capture," *Monthly Review*, 86–102.
44. William H. Brock, *Justus von Liebig* (Cambridge: Cambridge University Press, 1997), 177–78.
45. Liebig quoted in Erland Mårald, "Everything Circulates: Agricultural Chemistry and Recycling Theories in the Second Half of the Nineteenth Century," *Environment and History* 8 (2002): 74.
46. Marx, *Capital*, vol. 1, 283, 290, 636–39; and *Capital*, vol. 3, 949.
47. Marx *Capital*, vol. 1, 283, 290. Martinez-Alier refers to the unequal exchange concept as "building on earlier notions such as *Raubwirtschaft* or 'plunder economy.'" See Joan Martinez-Alier, *The Environmentalism of the Poor* (Northampton, MA: Edward Elgar, 2002), 214. It is important to note that Weber too, in his *General Economic Theory*, followed Liebig and Marx in raising the question of *Raubbau* as it related to the robbing of the earth by capitalist industry. See "Weber and the Environment," in chapter 9 of this book.
48. Karl Marx and Frederick Engels, *Ireland and the Irish Question* (New York: International, 1972), 290–92. See also Marx, *Capital*, vol. 1, 579–80, 860; *Capital*, vol. 3, 753, 949; Brett Clark and John Bellamy Foster, "Guano: The Global Metabolic Rift and the Fertilizer Trade," in *Ecology and Power*, ed. Al Hornborg, Brett Clark, and Kenneth Hermele (London: Routledge, 2009), 70.
49. Marx, *Capital*, vol. 1, 860. Marx's argument here contradicts Hornborg's contention that "Marx was probably too focused on the exploitation of labor to see that unequal exchange could also take the form of draining another society's natural resources." See Alf Hornborg, "Ecosystems and World-Systems," in C. Chase-Dunn and S. J. Babones, eds., *Global Social Change* (Baltimore: Johns Hopkins University Press, 2006), 169.

50. Brett Clark and John Bellamy Foster, "Guano: The Global Metabolic Rift and the Fertilizer Trade,"
51. Marx, *Texts on Method* (Oxford: Basil Blackwell, 1974), 209.
52. Marx, *Capital*, vol. 1, 287.
53. Marx, *Texts on Method*, 208–10.
54. Sweezy, *The Theory of Capitalist Development* (New York: Monthly Review Press, 1942), 23–55.
55. Marx's value analysis is often seen too narrowly in terms of the quantitative-value problem, and thus related simply to exchange value, or to "value" viewed simply in its quantitative aspect. However, no less crucial to Marx's entire value-theoretic framework is use value (related to production in general and to *real wealth*). Wealth, in Marx's analysis, is derived from both nature and labor as distinct from "value" under capitalism which comes only from labor. Thus, in order to fully grasp Marx's value-theoretic framework, it is necessary to incorporate the qualitative-value problem and the contradictions between (a) exchange value and use value; and (b) "value" and real wealth. See Michael A. Lebowitz, *Following Marx* (Boston: Brill, 2009), 163–66; and Foster et al., *The Ecological Rift*, 61–64. The brilliance of Odum's (and Scienceman's) analysis is that it grasped this larger value-theoretic dialectic of Marx's political economy, leading to an argument on the contradiction between real wealth and capitalist value relations that in many ways paralleled Marx's own—and which becomes the key to a dialectical analysis of unequal ecological exchange.
56. Marx, *Capital*, vol. 1, 290.
57. Marx, *Critique of the Gotha Programme* (Moscow: Progress Publishers, 1938), 3; Marx, *Capital*, vol. 1, 134; Foster et al., *The Ecological Rift*, 61–64. As Heinrich explains, "Things that are not products of labor," in Marx's theory, "do not possess a 'value.' If they are exchanged, they have an exchange value or price, but no value, and this exchange value has to then be explained separately." This sets up a situation where capitalism, as an economic system based on labor values, systematically robs nature, in the sense that no value is accorded to what are referred to in economics as nature's "free gifts" and hence its reproduction is not provided for. See Michael Heinrich, *An Introduction to the Three Volumes of Karl Marx's "Capital"* (New York: Monthly Review Press, 2012), 42.
58. Marx, *Capital*, vol. 1, 134, 636–39.
59. Paul Burkett and John Bellamy Foster, "Metabolism, Energy, and Entropy in Marx's Critique of Political Economy," *Theory and Society* 35/1 (2006): 109–56; Paul Burkett and John Bellamy Foster, "The Podolinsky Myth: An Obituary," *Historical Materialism* 16 (2008):

115–61; Amy E. Wendling, *Karl Marx on Technology and Alienation* (New York: Palgrave Macmillan, 2009). However, Marx and Engels's very extensive studies of thermodynamics were not generally known until recently, even to those studying ecological economics. Thus less than a decade ago Hornborg wrote that Marx was "ignorant ... of thermodynamics." Alf Hornborg, "Ecosystems and World-Systems," in Chase-Dunn and Babones, *Global Social Change*, 164.
60. Joan Martinez-Alier, *Ecological Economics* (London: Basil Blackwell, 1987).
61. Eduardo Galeano, *Open Veins of Latin America* (New York: Monthly Review Press, 1973), 72.
62. Arghiri Emmanuel, "The Socialist Project in a Disintegrated Capitalist World," *Socialist Thought and Practice* 16/9 (1976), 72–73.
63. Amin, *Imperialism and Unequal Development*, 212.
64. Amin, *Unequal Development*, 154.
65. Malthus's own work, including his well-known population theory, had nothing to do with notions of ecological carrying capacity of the earth and was in many ways anti-ecological in its thrust. See John Bellamy Foster, *Marx's Ecology*, 81–110.
66. Georg Borgstrom, *The Hungry Planet* (New York: Macmillan, 1965), 74–75.
67. Stephen Bunker, *Underdeveloping the Amazon* (Chicago: University of Chicago Press, 1985), 45.
68. Chase-Dunn, *Global Formation*, 234.
69. Andre Gunder Frank, "Entropy Generation and Displacement," in Alf Hornborg and Carole Crumley, eds., *The World System and the Earth System* (Walnut Creek, CA: Left Coast Press, 2006), 303–16.
70. William Rees, "Ecological Footprints and Appropriated Carrying Capacity," *Environment and Urbanization* 4/2 (1992): 121–30; Mathis Wackernagel and William Rees, *Our Ecological Footprint* (Gabriola Island, B.C.: New Society Publishers, 1996).
71. Richard York, Eugene Rosa, and Thomas Dietz, "Footprints on the Earth: The Environmental Consequences of Modernity," *American Sociological Review* 68 (2003): 279–300.
72. Wackernagel and Rees, *Our Ecological Footprint*, 48–55; Hornborg, *Global Ecology and Unequal Exchange*, 14–20.
73. Samir Amin, "Capitalism and the Ecological Footprint," *Monthly Review* 61/6 (2009): 19–22.
74. E. G. Jorgenson, "Unequal Ecological Exchange and Environmental Degradation"; James C. Rice, "Ecological Unequal Exchange," *Social Forces* 85 (2007): 1369–92; Lawrence, "The Thermodynamics of Unequal Exchange"; Jorgenson et al., "Ecologically Unequal Exchange and the

Resource Consumption/Environmental Degradation Paradox"; Jorgenson and Clark, "Footprints: The Division of Nations and Nature"; Eric Bonds and Liam Downey, 'Green' Technology and Ecologically Unequal Exchange. *Journal of World-Systems Research* 18/2 (2012): 167–86.
75. Jorgenson, "Unequal Ecological Exchange and Environmental Degradation," 686.
76. Rice, "Ecological Unequal Exchange," 1370.
77. Jorgenson, "Unequal Ecological Exchange and Environmental Degradation," 689; Rice, "Ecological Unequal Exchange," 1373. A step in the right direction has been the new, improved method of ecological footprint analysis developed by Wackernagel et al. where an attempt is made to include estimates of embodied energy of net-non-energy products. See Mathis Wackernagel, Lillemor Lewan, and Carina Borgström Hansson, "Evaluating the Use of Natural Capital with the Ecological Footprint Concept," *Ambio* 28 (1999): 604–12.
78. See Jorgenson, "Unequal Ecological Exchange and Environmental Degradation," 691.
79. Ibid., 691.
80. Bunker, *Underdeveloping the Amazon*, 34–37, 44–45.
81. Martinez-Alier, *The Environmentalism of the Poor*, 216–17.
82. Hornborg "Ecosystems and World-Systems," 171.
83. Joel B. Hagen, *The Entangled Bank: The Origins of Ecosystem Ecology* (New Brunswick, NJ: Rutgers University Press, 1992), 122–45.
84. Golley, *A History of the Ecosystem Concept in Ecology*, 69.
85. Hagen, *The Entangled Bank: The Origins of Ecosystem Ecology*, 126.
86. Howard Odum studied Marx's political economy very closely. Yet, although Odum played a critical role in introducing the concept of metabolism into systems ecology, he was not fully aware of Marx's own treatment of social metabolism and his theory of metabolic rift. This was because the original translation of Marx's *Capital* into English, which Odum studied, used the words "material exchange" in those instances where Marx had used *Stoffwechsel* (metabolism). Moreover, analysis of Marx's theory of metabolic rift emerged at the very end of the twentieth century with Foster's "Marx's Theory of Metabolic Rift" in sociology. Recent scholarship on Marx and thermodynamics, such as Burkett and Foster's "Metabolism, Energy, and Entropy in Marx's Critique of Political Economy," and Wendling's *Karl Marx on Technology and Alienation*, appeared only after Odum's death. Nevertheless, Odum and Scienceman were aware of some of the thermodynamic aspects of Marx's thought and even referred to the "metabolic rate of labor" in Marx's theory. See Scienceman, "EM VALUE AND LA VALUE," 27–37.

87. Odum, *Environment, Power, and Society for the Twenty-first Century: The Hierarchy of Energy*, 276.
88. Howard T. Odum and Jan E. Arding, *Emergy Analysis of Shrimp Mariculture in Ecuador*, 109.
89. Odum's approach resembled in some ways that of the physiocrats, who, according to Marx (*Theories of Surplus Value, Part 1*, 52; see also Burkett, *Marxism and Ecological Economics*, 33), "conceived value merely as use-value, merely as material substance." Nevertheless, Odum's analysis took on a much more critical form (influenced no doubt in part by his studies of Marx) whereby "use value"—for Odum "emvalue" or "emergy"—was used as a basis for a thoroughgoing critique of capitalist commodity-exchange values, which were understood as entirely based on labor or "human services," excluding nature from the calculations. Scienceman, "EM VALUE AND LA VALUE," 3.
90. Howard T. Odum, *Systems Ecology* (New York: John Wiley and Sons, 1983).
91. Odum, *Systems Ecology*, 251–68.
92. David M. Scienceman, "Emergism: A Policy for a Scientific Party," in C.A.S. Hall, ed., *Maximum Power: The Ideas and Applications of H. T. Odum* (Niwot: University Press of Colorado, 1995), 253.
93. Ibid., 253.
94. Howard T. Odum, "Self-Organization and Maximum Empower," in Hall, *Maximum Power: The Ideas and Applications of H. T. Odum*, 318.
95. Odum, *Environmental Accounting: Emergy and Environmental Decision-Making*, vii.
96. David M. Scienceman, "Energy and Emergy," in G. Pillet and T. Murota, eds., *Environmental Economics: The Analysis of a Major Interface* (Geneva: Roland Leimgruber, 1987), 262.
97. Ibid., 262.
98. Odum, "Principles of Emergy Analysis for Public Policy," in Howard T. Odum and J. E. Arding, eds., *Emergy Analysis of Shrimp Mariculture in Ecuador* (Narragansett, RI: Coastal Resources Center, University of Rhode Island, 1991), 88–111, 114; see also Odum, *Environmental Accounting: Emergy and Environmental Decision-Making*, 13, 289; Odum, "Self-Organization and Maximum Empower," 317.
99. Odum and Arding, *Emergy Analysis of Shrimp Mariculture in Ecuador*, 100.
100. However, not all goods produced in modern commodity chains are genuine use values from the standpoint of a rational production process. Odum's approach was consistent with a critical scrutiny of the "pathological waste" that was irrationally incorporated into the capitalist economy, in which the commodities sold were increasingly

"specifically capitalist use values" deriving their utility from the fact that they provided profits for the capitalist. See Howard T. Odum and Elisabeth C. Odum, *A Prosperous Way Down*; and John Bellamy Foster, "The Ecology of Marxian Political Economy," chapter 17 of this book.

101. Howard T. Odum, "Self-Organization and Maximum Empower," in Hall, *Maximum Power: The Ideas and Applications of H. T. Odum*, 317. The hierarchical logic was similar to food web analysis in ecology, in which the most efficient transformers of energy, plants carrying out photosynthesis, were at the bottom of the hierarchy and carnivorous animals at the top. The loss of efficiency from the bottom of the food web to the top was associated at the same time with the development of "dominant" species.

102. Odum and Odum, *A Prosperous Way Down*, 68; Odum, *Environmental Accounting: Emergy and Environmental Decision-Making*, 289.

103. Alfred J. Lotka, *Elements of Physical Biology* (Baltimore: Williams and Wilkins, 1925).

104. Odum, *Environmental Accounting*, 26, italics added. Odum's emphasis on *optimum* efficiency was important because it suggested that in the long run it was not maximum throughput that produced the optimum outcome, but rather the optimum could actually be a steady state. This argument is perhaps mostly clearly advanced in Odum and Odum, *A Prosperous Way Down*.

105. Odum, *Environmental Accounting*, 34.

106. Scienceman, "EM VALUE AND LA VALUE," 6.

107. Ibid.

108. Scienceman, "Emergence of Economics," 62–68.

109. Howard T. Odum, Interview conducted by Cynthia Barnett, 2001, available at http://ufdc.ufl.edu/AA00004025/00001, 40.

110. Ibid., 112.

111. Odum, "Principles of Emergy Analysis for Public Policy," 90.

112. Scienceman, "Energy and Emergy," 269.

113. Burkett and Foster, "Metabolism, Energy, and Entropy in Marx's Critique of Political Economy."

114. Marx, *Capital*, vol. 1, 215; also Scienceman, "EM VALUE AND LA VALUE," 36.

115. Rubin's discussion of abstract labor, which Odum and Scienceman studied, relates the attempts of various Marxist theorists to explain this in terms of a common energetic or physiological basis—though Rubin rightly argues against such an interpretation. See I. I. Rubin, *Essays on Marx's Theory of Value* (Montreal: Black Rose Books, 1972), 131–33.

116. Scienceman, "The Emergence of Economics," 62; Odum and

Scienceman, "An Energy Systems View of Karl Marx's Concepts of Production and Labor Value," 17; Robert L. Heilbroner, *Beyond the Veil of Economics* (New York: W. W. Norton, 1988), 132.
117. Odum and Scienceman, "An Energy Systems View," 23.
118. Ibid.
119. Karl Marx and Frederick Engels, *Collected Works* (New York: International Publishers, 1975b), vol. 46, 410–12; Paul Burkett and John Bellamy Foster, "The Podolinsky Myth: An Obituary," *Historical Materialism* 16 (2008): 131–40.
120. Martinez-Alier, *The Environmentalism of the Poor*, 218.
121. Marx and Engels, *Selected Correspondence* (Moscow: Progress Publishers, 1975), 136.
122. Scienceman, "EM VALUE AND LA VALUE," 28.
123. Odum and Scienceman, "An Energy Systems View of Karl Marx's Concepts of Production and Labor Value," 17–43.
124. David Scienceman, *Proceedings, International Society for the System Sciences* 3 (1989): 64.
125. Amin, *Unequal Development*; Becker, *Marxian Political Economy*; Guglielmo Carchedi, "The Logic of Prices as Values," *Economy and Society* 13/4 (1984): 431–55, and *Class Analysis and Social Research* (London: Basil Blackwell, 1987), and "Marxian Price Theory and Modern Capitalism," *International Journal of Political Economy* 18/3 (1988): 6–107; Harry Cleaver, *Reading Capital Politically* (Austin: University of Texas Press, 1979); Duncan K. Foley, *Understanding Capital* (Cambridge, MA: Harvard University Press, 1986); Goodwin and Punzo, *The Dynamics of a Capitalist Economy* (Boulder, CO: Westview Press, 1987); Heilbroner, *Beyond the Veil of Economics*; Howard and King, *The Economics of Marx* (London: Penguin Books, 1976), and *The Political Economy of Marx* (London: Longman, 1985); Ulrich Krause, *Money and Absract Labour* (London: Verso, 1982); Lonergan, "Theory and Measurement of Unequal Exchange"; Martinez-Alier, *Ecological Economics*; Michio Morishma, *Marx's Economics: A Dual Theory of Value and Growth* (Cambridge: Cambridge University Press, 1973); Rubin, *Essays on Marx's Theory of Value*; Paul Samuelson, "Wages and Interest," *American Economic Review* 47/6 (1957): 884–912; Francis Seton, "The Transformation Problem," *Review of Economic Studies* 24 (1957): 149–60; and Robert Paul Wolff, *Understanding Marx* (Princeton: Princeton University Press, 1984).
126. Odum and Scienceman, "An Energy Systems View," 31; Scienceman "EM VALUE AND LA VALUE," 36.
127. Scienceman, *Proceedings, International Society for the System Sciences*, 63.

128. Foster et al., *The Ecological Rift*, 61–65; Odum and Scienceman, "An Energy Systems View," 31.
129. Karl Marx, *Capital*, vol. 1, 638.
130. Foster et al., *The Ecological Rift*, 53–72.
131. Odum, "Principles of Emergy Analysis for Public Policy," 90; Scienceman "EM VALUE AND LA VALUE," 30.
132. Odum and Arding, *Emergy Analysis of Shrimp Mariculture in Ecuador*, 109; Odum and Scienceman, "An Energy Systems View."
133. Odum "Energy, Ecology and Economics," *Ambio*, 2/6 (1973): 220–27; Scienceman, "Energy and Emergy," 269–70.
134. Foster et al., *The Ecological Rift*, 61–64; Marx, *Critique of the Gotha Programme*, 3.
135. Odum and Scienceman, "An Energy Systems View," 41.
136. Ibid., 41.
137. Foster, "A Missing Chapter of *Monopoly Capital*," *Monthly Review* 64/3 (2012): 3–4; Paul A. Baran and Paul M. Sweezy, "Last Letters," *Monthly Review* 64/3 (2012): 65; Emmanuel. *Unequal Exchange*, 110–20, 127–28.
138. Frederick Engels. *The Housing Question* (Moscow: Progress Publishers, 1979), 14–15; Foster, "A Missing Chapter of *Monopoly Capital*," 13–14.
139. Levins, "Dialectics and Systems Theory," 29, 37.
140. Ibid., 48.
141. The transformity of information at the top of the energy hierarchy in Odum and Odum's 2001 analysis, depicted as 1×10^{11}, tells us nothing about the content of information. See Odum and Odum, *A Prosperous Way Down*, 69.
142. Samir Amin, *The Law of Value and Historical Materialism* (New York: Monthly Review Press, 1978), 1–18.
143. Odum and Scienceman, "An Energy Systems View," 41.
144. Lonergan, "Theory and Measurement of Unequal Exchange."
145. Ibid., 141–42.
146. Becker, *Marxian Political Economy*.
147. Ibid., 169.
148. Odum and Scienceman, "An Energy Systems View," 41.
149. William R. Catton, Jr., *Overshoot* (Urbana: University of Illinois Press, 1982); Center for Environmental Policy, National Environmental Accounting Database, 2012. Available at http://www.cep.ees.ufl.edu/nead.asp.
150. Sweeney et al., "Creation of Global Emergy Database for Standardized National Emergy Synthesis."
151. Odum and Arding, *Emergy Analysis of Shrimp Mariculture in Ecuador*, 97.

152. Odum, *Environmental Accounting: Emergy and Environmental Decision-Making*, 16.
153. Ibid., 16.
154. Odum and Odum, *A Prosperous Way Down*, 201.
155. Odum and Arding, *Emergy Analysis of Shrimp Mariculture in Ecuador*, 16.
156. Odum and Odum, *A Prosperous Way Down*, 99; Odum and Arding, *Emergy Analysis of Shrimp Mariculture in Ecuador*, 20.
157. Odum and Arding, *Emergy Analysis of Shrimp Mariculture in Ecuador*, 18.
158. Odum, *Environmental Accounting*, 201.
159. Odum and Odum, *A Prosperous Way Down*, 94.
160. Odum and Arding, *Emergy Analysis of Shrimp Mariculture in Ecuador*, 18.
161. Ibid., 22.
162. Odum, *Environment, Power, and Society for the Twenty-first Century: The Hierarchy of Energy*, 273.
163. Moore refers to the free appropriation of the non-capitalized free environment as generating an "ecological surplus" for capital. Although we do not use this terminology here, the overlap between his argument at this point and Odum's approach seems to us quite obvious. See Moore, "Transcending the Metabolic Rift," 21–22.
164. Not all developed countries have high emergy investment ratios and low emergy/money ratios. Some "rich dependencies" that heavily export raw materials, such as Canada, Australia, and New Zealand, depart from the developed/underdeveloped country norm because of their very high emergy per capita. Sweeney et al., "Creation of Global Emergy Database for Standardized National Emergy Synthesis," 13.
165. Odum, *Environmental Accounting*, 210–11, 216.
166. Odum and Odum, *A Prosperous Way Down*, 139, and Odum, *Environmental Accounting*, 210.
167. Odum and Odum, *A Prosperous Way Down*, 95.
168. Odum, *Environmental Accounting*, 276–77.
169. Odum and Odum, *A Prosperous Way Down*, 98–99.
170. Odum, *Environmental Accounting*, 274.
171. Odum and Arding, *Emergy Analysis of Shrimp Mariculture in Ecuador*, 37–39.
172. Ibid.
173. Ibid., 35.
174. Ibid., 37.
175. Ibid., 33–39.
176. Ibid., 24.
177. Ibid., 23.

178. Ibid., 89–90, 104.
179. Martinez-Alier, *The Environmentalism of the Poor*, 82.
180. Sweeney et al., "Creation of Global Emergy Database for Standardized National Emergy Synthesis."
181. David King, "Energy Accounting of the Resource Base of Nations: Human Well-Being and International Debt" (M.A. thesis, University of Florida, 2006), 77–78.
182. Ibid., 72, 86.
183. Edward D. Melillo, "The First Green Revolution: Debt Peonage and the Making of the Nitrogen Fertilizer Trade, 1840–1930" *American Historical Review* 117/3 (2012): 1028–60; Clark and Foster, "Ecological Imperialism and the Global Metabolic Rift: Unequal Exchange and the Guano/Nitrates Trade."
184. Marilyn Waring, *Counting for Nothing* (Toronto: University of Toronto Press, 1999), 65–74.
185. Odum, *Environment, Power, and Society for the Twenty-first Century: The Hierarchy of Energy*, 274, 278.
186. Odum, interview conducted by Cynthia Barnett, 87.
187. Odum and Odum, *A Prosperous Way Down*, 57.
188. Costanza, "Embodied Energy and Economic Valuation," *Science* 210 (December 12, 1980): 1219–1294; "Embodied Energy, Energy Analysis, and Economics," in *Energy, Economics, and the Environment*, ed. Herman Daly and A.F. Umaña (Boulder: Westview Press, 1981), 119–45, and "Reply: An Embodied Energy Theory of Value," in *Energy, Economics, and the Environment*, ed. Herman Daly and A.F. Umaña (Boulder: Westview Press, 1981), 187–92.
189. Odum, Interview conducted by Cynthia Barnett, 37–39. In Odum's words: "At some point, however, there was a deep struggle there [in relation to *Ecological Economics*], and one of the main issues was input-output embodied energy versus emergy. Another was market value versus other [ecological] values. So he [Costanza] took us off the board." Just prior to the July 1992 issue Odum and six others, predominantly natural scientists, were removed from the board of *Ecological Economics*.
190. Costanza, "Embodied Energy and Economic Valuation," 1223; "Embodied Energy, Energy Analysis, and Economics," 119–45, and "Reply: An Embodied Energy Theory of Value," 187–92.
191. Nicholas Georgescu-Roegen. "Man and Production," in Mauro Baranzini and Roberto Scazzieri, eds., *Foundations of Economics* (Oxford: Basil Blackwell, 1986), 270–72.
192. Nicholas Georgescu-Roegen, "Energy, Matter, and Economic Valuation," in Herman Daly and Alvaro Umaña, eds., *Energy, Economics, and the Environment* (Boulder, CO: Westview Press, 1981), 69–70.

193. Herman Daly, "Postscript: Unresolved Problems and Issues for Further Research," in Daly and Umaña, *Energy, Economics, and Environment*, 165–72.
194. Costanza, "Embodied Energy, Energy Analysis, and Economics," 140. Costanza insisted that his embodied energy theory of value in no way challenged neoclassical economics. "The results," he wrote, "indicate there is no inherent conflict between an embodied energy (or energy cost) theory of value and value theories based on utility." See also Burkett, *Marxism and Ecological Economics*, 37–41.
195. Daly, "Postscript: Unresolved Problems and Issues for Further Research," 167–68. The term "energetic [or energy] dogma" was introduced by Georgescu-Roegen to refer to views that reduced questions of economic value and all ecological/entropic issues to mere energy. Georgescu-Roegen also used the notion of "energetic dogma" in a narrower sense to refer to those who, in defiance of thermodynamics, believed that recycling could occur at a level of 100%. See "Energy, Matter, and Economic Valuation," 53. See also Mayumi, *The Origins of Ecological Economics* (London: Routledge, 2001), 60.
196. Daly, "Postscript: Unresolved Problems and Issues for Further Research," 168. Daly indicated that he thought that Odum and his associates were also gravitating toward an energy theory of economic value, though differently from Costanza. But while Daly indicated that Odum probably fell under the same stricture, and that Costanza "is representative of the Odum school," he was uncertain due to the quite different nature of Odum's argument. This was before the introduction of the emergy nomenclature two years later, which clarified the nature of Odum's theory in this respect and the differences between it and Costanza. Daly has argued for a retention of subjective value analysis in the broad neoclassical tradition—although he often draws on economic/ecological critiques from outside that tradition.
197. Burkett, *Marxism and Ecological Economics*, 37–41.
198. Odum, interview conducted by Cynthia Barnett, 37–39.
199. Odum, *Environment, Power, and Society for the Twenty-first Century: The Hierarchy of Energy*, 260–68.
200. Hornborg. "Towards an Ecological Theory of Unequal Exchange," *Ecological Economics* 25/1 (1998): 130.
201. This relates to the whole transformation process/problem in Marx's analysis, in which prices of production are transformed values. For a general discussion see Hunt and Lautzenheiser, *History of Economic Thought*, 227–31, 518–24. It is noteworthy that precisely this deviation of prices of production from values became the basis of the Marxian theory of unequal exchange.

202. Hornborg, "The Unequal Exchange of Time and Space," *Journal of Ecological Anthropology* 7 (2003): 5.
203. Hornborg, "Towards an Ecological Theory of Unequal Exchange," 130.
204. Ibid., 130–32; see also Hornborg, *The Power of the Machine* (Walnut Creek, CA: AltaMira Press, 2001), 40–43; "The Unequal Exchange of Time and Space," 5–6; *Global Ecology and Unequal Exchange* (London: Routledge, 2011), 17, 104. The same basic criticisms were repeated (though with some reservations) by Joan Martinez-Alier, who accepted Hornborg's basic definition of the problem with respect to Odum's work, and the case to be made for exergy.
205. Hornborg, *Global Ecology and Unequal Exchange*, 104.
206. Odum, *Environmental Accounting*, 60; Cutler Cleveland, "Biophysical Economics: Historical Perspective and Current Research Trends," *Economic Modeling* 38/1-2 (1987): 59; John Brolin, *The Bias of the World*, Land Studies in Human Ecology (Lund University, (2006), 262.
207. Hornborg, *Global Ecology and Unequal Exchange*, 104.
208. Hornborg, *The Power of the Machine*, 40–42. Hornborg was not the only theorist in this area to confuse Odum with Costanza. Stokes also interpreted the former in terms of the latter, and was not aware of their divergent courses. Stokes was, however, interpreting Odum only up to 1983, so his work missed the entire emergy stage of Odum's analysis. See Kenneth M. Stokes, *Man and the Biosphere* (Armonk, NY: M. E. Sharpe, 1992), 147–54.
209. Hornborg, "Towards an Ecological Theory of Unequal Exchange," 127–36.
210. Odum and Arding, *Emergy Analysis of Shrimp Mariculture in Ecuador*, 109; Brolin, "The Bias of the World," 245–46.
211. Foster et al., *The Ecological Rift*, 61–64.
212. Brolin, "The Bias of the World," 264.
213. Hornborg, "Towards an Ecological Theory of Unequal Exchange," 131–32.
214. Foster, "Marx's Theory of Metabolic Rift"; Foster, Clark, and York, *The Ecological Rift*; John Bellamy Foster and Paul Burkett, *Marx and the Earth* (Chicago: Haymarket, 2016); Kohei Saito, *Karl Marx's Ecosocialism* (New York: Monthly Review Press, 2017).
215. Sweeney et al., "Creation of Global Emergy Database for Standardized National Emergy Synthesis."
216. Sweezy, *The Theory of Capitalist Development*, 23–40.
217. Bookchin, *Toward an Ecological Society*, 88.
218. Odum and Odum, *A Prosperous Way Down*.
219. Odum, "Energy, Ecology and Economics," 222.
220. Clark and Foster, "Ecological Imperialism and the Global Metabolic Rift," 313.

221. Scienceman, "EMVALUE AND LAVALUE," 33; Odum and Scienceman, "An Energy Systems View of Karl Marx's Concepts of Production and Labor Value."
222. Foster, "Marx's Theory of Metabolic Rift"; John Bellamy Foster and Paul Burkett, *Marx and the Earth* (Chicago: Haymarket, 2016); Kohei Saito, *Karl Marx's Ecosocialism* (New York: Monthly Review Press, 2017).
223. Schneider and McMichael, "Deepening, and Repairing, the Metabolic Rift," 461.
224. Foster et al., *The Ecological Rift*, 439–40.
225. Marx, *Capital*, vol. 3, 959.
226. Marx, *Grundrisse*, 489. Italics in original.
227. Marx and Engels, *Collected Works*, vol. 30, 54–66.
228. Epicurus, *The Epicurus Reader* (Indianapolis: Hackett, 1994), 35.

11. Marxism and the Anthropocene

This chapter is adapted and revised for this book from the following article: John Bellamy Foster, "Marxism in the Anthropocene: Dialectical Rifts on the Left," *International Critical Thought* 6/3 (September 2016): 393–421.

1. Hans Joachim Schellnhuber, "'Earth System' Analysis and the Second Copernican Revolution," *Nature* 402/6761 (1999): 19–23. Schellnhuber was referring to Earth System analysis, rather than the Anthropocene. Yet the two today are inextricably related, and the phrase is therefore equally applicable to the concept of the Anthropocene.
2. Karl Marx and Frederick Engels, *Collected Works*, vol. 5 (New York: International Publishers, 1975), 40.
3. E.V. Shantser, "The Anthropogenic System (Period)," in *The Great Soviet Encyclopedia*, vol. 2 (New York: Macmillan, 1973), 140. It was in Shantser's *The Great Soviet Encyclopedia* article that Pavlov's concept of the Anthropocene first appeared in English.
4. V. I. Vernadsky, "Some Words about the Noösphere," in *The Noösphere*, vol. 2 of *150 Years of Vernadsky*, ed. J. Ross (Washington, D.C.: 21st Century Science Associates, 2014), 79–84.
5. Barry Commoner, *The Closing Circle: Nature, Man and Technology* (New York: Bantam, 1971), 39–41, 45–46; John Bellamy Foster, *The Vulnerable Planet* (New York: Monthly Review Press, 1994), 108–24.
6. G. Evelyn Hutchinson, "The Biosphere," *Scientific American* 233/3, (1970): 45–53.
7. Clive Hamilton, and Jacques Grinevald, "Was the Anthropocene Anticipated?," *The Anthropocene Review* 2/1 (2015): 67.
8. Ian Angus, "When Did the Anthropocene Begin . . . and Why Does It Matter?," *Monthly Review* 67/4 (2015): 6–7; Will Steffen., Windy

Broadgate, Lisa Deutsch, Owen Gaffney, and Cornelia Ludwig, "The Trajectory of the Anthropocene: The Great Acceleration," *Anthropocene Review* 2/1 (2015): 81–98.

9. Colin Waters, Jan Zalasiewicz, Colin Summerhayes et al., "The Anthropocene Is Functionally and Stratigraphically Distinct from the Holocene," *Science* 351/6269 (2016): 137–47. For a long-term perspective on the notion of a geological "golden spike" separating each major period of Earth geohistory, see Martin J. S. Rudwick, *Bursting the Limits of Time: The Reconstruction of Geohistory in the Age of Revolution* (Chicago: University of Chicago Press, 2005), 21–22.

10. Paul J. Crutzen and Eugene F. Stoermer, "The Anthropocene," *Global Change Newsletter* 41 (May 1, 2000): 17.

11. Rachel Carson, *Silent Spring* (Boston: Houghton Mifflin, 1962).

12. Ian Burton and Robert W. Kates, "The Great Climacteric: 1798–2048," in *Geography, Resources and Environment*, ed. Robert W. Kates and Ian Burton (Chicago: University of Chicago Press, 1986), 339–60.

13. K. William Kapp, *The Social Costs of Private Enterprise* (Cambridge, MA: Harvard University Press, 1950); Commoner, *The Closing Circle: Nature, Man and Technology*; Virginia Brodine, *Air Pollution* (New York: Harcourt Brace Jovanovich, 1971); Virginia Brodine, *Green Shoots and Red Roots* (New York: International Publishers, 2007); Herbert Marcuse, *Counter-Revolution and Revolt* (Boston: Beacon Press, 1972); Paul M. Sweezy, "Cars and Cities," *Monthly Review* 24/11 (1973): 1–18; Howard Parsons, *Marx and Engels on Ecology* (Westport, CT: Greenwood Press, 1977); Charles H. Anderson, *The Sociology of Survival* (Homewood, IL: Dorsey Press, 1976); and Allan Schnaiberg, *The Environment* (Oxford: Oxford University Press, 1980).

14. Alfred Schmidt, *The Concept of Nature in Marx* (London: New Left Books, 1971).

15. On Weber's environmental views and their relation to the Frankfurt School and the Marxian critical tradition, see John Bellamy Foster and Hannah Holleman, "Weber and the Environment," in chapter 9 of this book.

16. Schmidt, *The Concept of Nature in Marx*, 156.

17. Theodor Adorno, *Negative Dialectics* (New York: Continuum, 1973), 244.

18. John Bellamy Foster and Paul Burkett, *Marx and the Earth: An Anti-Critique* (Boston: Brill, 2016), 2–3. Also see, for example, Andre Gorz, *Capitalism, Socialism, Ecology* (London: Verso, 1994); Ted Benton, "Marxism and Natural Limits," *New Left Review* 1/178 (1989): 51–86; Robyn Eckersley, *Environmentalism and Political Theory* (Albany, NY: SUNY Press, 1992); James O'Connor, *Natural Causes* (New York:

Guilford Press, 1998); Donald Worster, *Nature's Economy: A History of Ecological Ideas* (Cambridge: Cambridge University Press, 1994); Joel Kovel, *The Enemy of Nature* (London: Zed, 2002); Daniel Bensaïd, *Marx for Our Times: Adventures and Misadventures of a Critique* (London: Verso, 2002); and Daniel Tanuro, *Green Capitalism: Why It Can't Work* (London: Merlin Press, 2013).

19. Neil Smith, *Uneven Development: Nature, Capital and the Production of Space* (Athens: University of Georgia Press, 2008); and Noel Castree, "Marxism, Capitalism, and the Production of Nature," in *Social Nature: Theory, Practice and Politics*, ed. N. Castree and B. Braun (Oxford: Blackwell, 2001), 189–207.

20. Bruno Latour, *We Have Never Been Modern* (Cambridge, MA: Harvard University Press, 1993); and Bruno Latour, *Reassembling the Social* (Oxford: Oxford University Press, 2005).

21. John Bellamy Foster, "Foreword," in P. Burkett, *Marx and Nature*, vii–xiii (Chicago: Haymarket, 2014), viii–x.

22. Paul Burkett, *Marx and Nature: A Red and Green Perspective* (Chicago: Haymarket, 2014); and John Bellamy Foster, *Marx's Ecology: Materialism and Nature* (New York: Monthly Review Press, 2000). But also Elmar Altvater, *The Future of the Market* (London: Verso, 1993); John Bellamy Foster, Brett Clark, and Richard York, *The Ecological Rift* (New York: Monthly Review Press, 2010); Peter Dickens, *Society and Nature: Changing Our Environment, Changing Ourselves* (Cambridge: Polity, 2004); Andreas Malm, "The Origins of Fossil Capital," *Historical Materialism* 21/1 (2013): 15–68, 2016); and Andreas Malm, *Fossil Capital* (London: Verso, 2016).

23. Ariel Salleh, ed., *Eco-Sufficiency and Global Justice: Women Write Political Ecology* (London: Pluto, 2009); Pamela Odih, *Watersheds in Marxist Feminism* (Newcastle upon Tyne: Cambridge Scholars Publishing, 2014).

24. Del Weston, *The Political Economy of Global Warming* (London: Routledge, 2014); Stefano Longo, Rebecca Clausen, and Brett Clark, *The Tragedy of the Commodity* (New Brunswick, NJ: Rutgers University Press, 2015); Ian Angus, *Facing the Anthropocene: Fossil Capitalism and the Crisis of the Earth System* (New York: Monthly Review Press, 2016).

25. The reference to a "modern synthesis" is meant to refer back to the synthesis of Darwinism with Mendelian genetics that occurred in the 1930s, in which geneticists like Julian Huxley and J. B. S. Haldane reached back to Darwin's original doctrines and demonstrated that the new knowledge did not displace the theory of natural selection, but gave it a new complexity and importance, bringing out more fully the

significance of Darwin's classical theory for the present. An analogous process is occurring with respect to Marx and ecology today.

26. None of the previous forms of left ecological thought have entirely gone away. The Frankfurt School's negative dialectic of the domination of nature—in which Horkheimer opined that "men cannot utilize their power over nature for the rational organization of the earth but rather must yield themselves to blind individual and national egoism under the compulsion of circumstances and of inescapable manipulation"—naturally persists in some quarters on the left, leading to a grim negativity. Quoted in William Leiss, *The Domination of Nature* (Boston: Beacon Press, 1974), 154. Also, Martin Jay characterized Horkheimer and Adorno's "fearsome anti-Enlightenment critique" as sheer "pessimism." See Jay, *Permanent Exiles: Essays on the Intellectual Migration from Germany to America* (New York: Columbia University Press, 1985), 1461. This fearsome critique had a negative effect on the interpretation of Marx that still persists in some quarters. Thus first-stage ecosocialism—which draws much of its motivation from the attempt to disgorge itself of a strong relation to Marxism—has gained a second life in recent years in its repeated attempts to demonstrate a "fundamental flaw" in Marx's ecology, e.g. in Tanuro, *Green Capitalism: Why It Can't Work*. For a critique of this tendency see Foster and Burkett, *Marx and the Earth: An Anti-Critique*, 15–50.

27. Smith, *Uneven Development: Nature, Capital and the Production of Space*; Noel Castree, "Capitalism and the Marxist Critique of Political Ecology," in *The Routledge Handbook of Political Ecology*, ed. Tom Perreault, Gavin Bridge, and James McCarthy (London: Routledge, 2015a), 279–92; Bensaïd, *Marx for Our Times: Adventures and Misadventures of a Critique*; Jason W. Moore, *Capitalism in the Web of Life* (London: Verso, 2015a); Damian F. White, Alan P. Rudy, and Brian J. Gareau, *Environments, Natures, and Social Theory* (London: Palgrave, 2015).

28. It is noteworthy that Smith, who continued to write on the production of nature up to 2008, ignored works such as Burkett's *Marxism and Ecological Economics* (Boston: Brill, 2006) and *Marx and Nature: A Red and Green Perspective*; and Foster's *Marx's Ecology: Materialism and Nature*, while Castree mentions Burkett only slightly. See Noel Castree, "Marxism and the Production of Nature," *Capital and Class* 24/3 (2000): 5–36. The implicit assumption is that Schmidt's interpretation of Marx on nature, which has been largely abandoned elsewhere, remains valid.

29. The struggle over Cartesian dualism is a long-standing one in philosophy. See Arthur Lovejoy, *The Revolt Against Dualism* (New York: W. W. Norton, 1930). It is only recently that this has been

directed against Marx and Marxism. Marx's philosophical outlook, embodying a dialectical critical realism/materialism, is hardly a likely target for those seeking to attack dogmatically dualist views. On Marx's epistemology and its relation to critical realism see Roy Bhaskar, "Knowledge," in *Dictionary of Marxist Thought,* ed. T. Bottomore (Oxford: Basil Blackwell, 1983), 254–63.
30. Latour, "Love Your Monsters."
31. Bruno Latour, *Politics of Nature: How to Bring the Sciences into Democracy* (Cambridge, MA: Harvard University Press, 2004), 246.
32. Michael Shellenberger and Ted Nordhaus, "The Monsters of Bruno Latour," Breakthrough Institute, 2012, available at http://thebreakthrough.org/index.php/journal/past-issues/online-cotent/the-monsters-of-bruno-latour. The tendency to deny natural-physical and environmental processes in social analysis, and their theoretical absorption within the social, was decried by Dunlap and Martin more than three decades ago as the rise of a "new brand of determinism—socio-cultural determinism," See Riley E. Dunlap and Kenneth E. Martin, "Bringing Environment into the Study of Agriculture," *Rural Sociology* 48/2 (1983): 201–18.
33. Bensaïd, *Marx for Our Times*; Moore, *Capitalism in the Web of Life*; White, Rudy, and Gareau, *Environments, Natures, and Social Theory*.
34. Smith, *Uneven Development: Nature, Capital and the Production of Space,* 247.
35. See Russell Jacoby, "Western Marxism," in *A Dictionary of Marxist Thought,* ed. T. Bottomore (Oxford: Blackwell, 1983), 581–84; Martin Jay, *Marxism and Totality* (Berkeley: University of California Press, 1984), 115–17; Frederic Jameson, *Valences of the Dialectic* (London: Verso, 2009), 6.
36. Bertell Ollman's influential work, interpreting Marx's dialectics in terms of "the philosophy of internal relations," accounts, in part, for this *exclusive* emphasis on internal relations. See Bertell Ollman, *Alienation* (Cambridge: Cambridge University Press, 1976), 35, and his *Dialectical Investigations* (New York: Routledge, 1993). Drawing on the metaphysical and idealist traditions of Leibniz, in particular, as well as Spinoza, Hegel, and Whitehead, together with the early Marxist Joseph Dietzgen, Ollman writes: "In the history of ideas, the view that we have been developing is known as the philosophy of internal relations. Marx's immediate philosophical influences in this regard were Leibniz, Spinoza, and Hegel. . . . What all had in common is the belief that the relations that come together to make up the whole get expressed in what are taken to be its parts. Each part is viewed as incorporating all its relations with other parts up to and

including everything that comes into the whole." This view has been questioned on essentially critical-realist grounds as inconsistent with a more open-ended materialist perspective by such Marxian theorists as Rader and Bhaskar. See Melvin Miller Rader, *Marx's Interpretation of History* (New York: Oxford University Press, 1979), 56–85; and Roy Bhaskar, *Dialectic: The Pulse of Freedom* (London: Verso, 1993), 201.

37. Georg Lukács, *Labour* (London: Merlin Press, 1980); Istvan Mészáros, *Lukács' Concept of Dialectic* (London: Merlin Press, 1972).
38. Paul Burkett, "Marx's Vision of Sustainable Human Development," *Monthly Review* 57/5 (2005): 34–62.
39. Slavoj Žižek, *Less than Nothing: Hegel and the Shadow of Dialectical Materialism* (London: Verso, 2013), 262.
40. Ibid., 261.
41. Ibid., 373.
42. Slavoj Žižek, "Censorship Today: Violence, or Ecology as a New Opium of the Masses," 2007, http://www.lacan.com/zizecologyl.htm; emphasis in original. This same phrase of ecology as "the new opium of the masses" is used in a positive way as well in Badiou and Swyngedouw. See Alain Badiou, "Live Badiou—Interview with Alain Badiou," in *Alain Badiou—Live Theory*, ed. O. Feltham (London: Continuum, 2008), 136–39; and Erik Swyngedouw, "Trouble with Nature: 'Ecology as the New Opium for the Masses,'" in *The Ashgate Research Companion to Planning Theory: Conceptual Challenges for Spatial Planning*, ed. J. Hillier and P. Healey (Burlington, VT: Ashgate, 2010), 304.
43. Žižek, *Less than Nothing: Hegel and the Shadow of Dialectical Materialism*, 373. Despite his frequent anti-ecological and even antinature statements, Žižek, who is hardly known for his consistency, is capable in certain contexts of rational discussion of the ecological crisis and its relation to capitalism. See Slavoj Žižek, *Living in the End Times* (London: Verso, 2010), 327–36.
44. Smith, *Uneven Development: Nature, Capital and the Production of Space*, 8.
45. Steven Vogel, *Against Nature: The Concept of Nature in Critical Theory* (Albany, NY: SUNY Press, 1996), 160. Vogel's strong anthropocentrism is even more clearly evident in the title to his most recent work, *Thinking like a Mall: Environmental Philosophy After the End of Nature* (Cambridge, MA: MIT Press, 2015).
46. Smith, *Uneven Development: Nature, Capital and the Production of Space*, 31.
47. Marx's concept of the "universal metabolism of nature" clearly refutes Salvatore Engel-Di Mauro's objection to Marx's use of metabolism on

the grounds that it "excludes the importance of material exchanges not involving people." See Salvatore Engel-Di Mauro, *Ecology, Soils, and the Left: An Eco-Social Approach* (London: Palgrave Macmillan, 2014), 141; and Karl Marx and Frederick Engels, *Collected Works* (New York: International Publishers, 1975), vol. 30, 54–66; and Karl Marx, *Capital*; vol. 3. (London: Penguin, 1981), 949.
48. Foster and Holleman, "Weber and the Environment"; Brett Clark and John Bellamy Foster, "Guano: The Global Metabolic Rift and the Fertilizer Trade," in *Ecology and Power*, ed. A. Hornborg, B. Clark, and K. Hermele (London: Routledge, 2012), 68–82.
49. Stephen Jay Gould, *An Urchin in the Storm*, (New York: W. W. Norton, 1987), 111; Foster, *Marx's Ecology: Materialism and Nature*, 196–207.
50. Foster and Burkett, *Marx and the Earth: An Anti-Critique*, 2016.
51. Marx and Engels, *Collected Works*, vol. 42, 558–59. In addressing Carl Fraas's discussion of desertification in pre-capitalist class society, Marx made it clear that he saw this as a problem that only worsened globally under historical capitalism—a view based on his theory of metabolic rift. See Kohei Saito, "Marx's Ecological Notebooks," *Monthly Review* 67/9 (2016): 34–39. The only real answer to such contradictions was a society of associated producers that rationally regulated the social metabolism between human beings and nature. He therefore characterized this growing ecological contradiction of civilization as an "unconscious socialist tendency." The significance of Marx's statement here was to be emphasized in late Soviet ecology. Thus E. K. Fedorov, the geophysicist and climatologist and a member of the Presidium of the Supreme Soviet, used Marx's argument here to explain why today's environmental scientists and activists "display (possibly unconsciously) certain 'socialist tendencies.'" See Evgeny Konstantinovich Fedorov, *Man and Nature* (New York: International Publishers, 1972), 146–47. Just as the dualisms and dichotomies of the post-Cartesian philosophical tradition arise from the soil of a determinate social practice, by the same token it is impossible to think of theoretically resolving them simply through the adoption of a new categorical framework, without envisaging at the same time an alternative social order from which the *practical* antinomies of capital's historically specific system can be removed.
52. See John Bellamy Foster, "William Morris' Letters on Epping Forest," *Organization and Environment* 11/ (1998): 90–92; 2000, 226–49; and Foster, Clark, and York, *The Ecological Rift*, 324–40.
53. Bensaïd, *Marx for Our Times: Adventures and Misadventures of a Critique*, 314–15.
54. Moore, *Capitalism in the Web of Life*, 85–86.

55. Burton and Kates, "The Great Climacteric: 1798–2048."
56. Jason W. Moore, "Toward a Singular Metabolism," in *Grounding Metabolism,* ed. Daniel Ibanez and Nikos Katsikis (Cambridge, MA: Harvard University Press, 2014), 10–19; Moore, *Capitalism in the Web of Life*; Jason W. Moore, "Cheap Food and Bad Climate," *Critical Historical Studies* 2/10 (2015): 1–43.
57. Istvan Mészáros, *The Social Determination of Method,* vol. 1 of *Social Structure and Forms of Consciousness* (New York: Monthly Review Press, 2010).
58. Ibid., 186.
59. Frederic Jameson, "Introduction," in John Paul Sartre, *Critique of Dialectical Reason,* vol. 1 (London: Verso, 2004), 25 and xxii. In Jameson's interpretation of Sartre's position, Sartre was arguing for a "dualism which functions as a moment in the reestablishment of monism proper."
60. Sartre, *Critique of Dialectical Reason,* 26.
61. Ibid., 180–81.
62. Ibid., 181–82.
63. Roy Bhaskar, *Dialectic: The Pulse of Freedom* (London: Verso, 1993), 394.
64. John Paul Sartre, *Literary and Philosophical Essays* (New York: Criterion, 1955), 236. In the *Critique of Dialectical Reason,* Sartre did introduce an intriguing environmental discussion at one point in his treatment of "counter-finality," or "matter as inverted practice," where he employed the example of peasant deforestation in China, leading to the counter-finality of floods, engendering an organized, collective response on the part of society. See Sartre, *Critique of Dialectical Reason,* 161–65.
65. Sartre, *Critique of Dialectical Reason,* 15.
66. Marx and Engels, *Collected Works,* vol. 1, 474; Foster, *Marx's Ecology: Materialism and Nature,* 131–32, emphasis in original.
67. Sidney Hook, *Marx and the Marxists: The Ambiguous Legacy* (Malabar, FL: Robert E. Krieger Publications, 1982), 37.
68. Monism of any variety raises serious philosophical objections. See, for example, William James, *Pragmatism* (New York: Meridian Books), 89–108; C. E. M. Joad, *Guide to Philosophy* (New York: Dover, 1936), 428–31; and Bhaskar, *Dialectic: The Pulse of Freedom,* 354–65.
69. Bensaïd, *Marx for Our Times: Adventures and Misadventures of a Critique,* 314.
70. Ibid., 320–21.
71. Ibid., 320–21.
72. Ibid., 327–41. Also see Foster and Burkett, *Marx and the Earth: An Anti-Critique,* 165–203.

73. Moore, "Toward a Singular Metabolism," 16; *Capitalism in the Web of Life,* 85.
74. Moore's "Toward a Singular Metabolism" begins with the sentence: "Metabolism is a seductive metaphor." It is in these terms that he then constructs his notion of a "singular metabolism"—itself defined in terms of the metaphor of "the web of life." This essentially idealist approach contrasts with Marx's materialist dialectic, in which metabolism was seen not as a mere metaphor but as reflecting a natural-physical process, related to material reproduction. Moore's approach, with its idealist emphasis, thus departs sharply from Marx's materialism with its deep links to physical science. See Moore, *Capitalism in the Web of Life,* 79, 86.
75. Marx and Engels, *Collected Works,* vol. 30, 63. The concepts of totality and mediation, as Lukács above all taught, are central to the Marxian dialectic, and transcend simple notions of monism and dualism. See Mészáros, *Lukács' Concept of Dialectic.*
76. Moore, "Toward a Singular Metabolism," 12, 15; Moore, *Capitalism in the Web of Life,* 15, 78–82.
77. Moore, "Toward a Singular Metabolism," 11, 13; Moore, *Capitalism in the Web of Life,* 46, 206.
78. Moore, "Toward a Singular Metabolism," 12.
79. Moore, *Capitalism in the Web of Life,* 46.
80. Moore, "Transcending the Metabolic Rift," *Journal of Peasant Studies* 38/1 (2011): 5.
81. Gulberk K. Maclean, *Bertrand Russell's Bundle Theory of Particulars* (London: Bloomsbury, 2014), 119–31; Bertrand Russell, *The Analysis of Matter* (New York: Routledge, 1992), 10, 382–93; Latour, *Reassembling the Social,* 17, 134, 139.
82. Moore, *Capitalism in the Web of Life,* 46; Moore, "Toward a Singular Metabolism,"12; Moore, "Cheap Food and Bad Climate," 28.
83. Moore, *Capitalism in the Web of Life,* 21.
84. Marx, *Capital.* vol. 3, 949; Marx and Engels, *Collected Works,* vol. 30, 63; Marx, *Capital,* vol. 1 (London: Penguin, 1976), 637. Moore's main target in this respect is Foster, Clark, and York's *The Ecological Rift.* On Marx's use of the wider conceptual framework of social metabolism, the universal metabolism of nature, and the metabolic rift, see John Bellamy Foster and Brett Clark, *The Robbery of Nature* (New York: Monthly Review Press, 2000), 204-12; and Foster, "Marxism and Ecology," *Monthly Review* 67/7 (2015): 1–13.
85. Moore, "Transcending the Metabolic Rift," 21, 80.
86. Moore, *Capitalism in the Web of Life,* 80.
87. Moore, "Toward a Singular Metabolism"; *Capitalism in the Web of Life,* 76, 79, 86.

88. Moore, "The Socio-Ecological Crises of Capitalism," in *Capitalism and Its Discontents*, ed. S. Lilley (Oakland, CA: PM Press, 2011), 139, 151.
89. Marx, *Capital*, vol. 1, 637.
90. Moore, "Transcending the Metabolic Rift," 5; *Capitalism in the Web of Life*, 76.
91. Moore, "Transcending the Metabolic Rift," 5; *Capitalism in the Web of Life*, 76.
92. Moore, "Toward a Singular Metabolism," 12; also Moore, "Cheap Food and Bad Climate," 28, emphasis added.
93. Moore, *Capitalism in the Web of Life*, 15.
94. Ibid., 118–19; Moore, "Cheap Food and Bad Climate."
95. Moore, "Transcending the Metabolic Rift," 20, 29–30.
96. Moore, "Toward a Singular Metabolism," 16–17.
97. Moore, *Capitalism in the Web of Life*, 80.
98. Ibid., 60.
99. Moore, "Toward a Singular Metabolism," 12; Moore, *Capitalism in the Web of Life*, 85; Latour, "Love Your Monsters." Bruno Latour and Noel Castree, with their philosophies of bundling, are presented by Moore as constituting the most advanced forms of "relational critiques of dualism." See Moore, "Toward a Singular Metabolism," 14, 18. The basis of Latour's analysis in "neutral monism" is noted by Morelle. See Louis Morelle, "Speculative Realism," in *Speculations III*, ed. Michael Austin, Paul Ennis, and Fabio Gironi (Brooklyn, NY: Punctum Books, 2012), 255. The concept of bundling used by Latour and other actor-network theorists, and adopted by Moore as a way of transcending dualism, has of course a long history in theories of "neutral monism," as advocated by thinkers such as Bertrand Russell. See Russell, *Philosophical Essays* (New York: Simon and Schuster, 1966), 131–46. See also Maclean, *Bertrand Russell's Bundle Theory of Particulars*, 119–31; and Leopold Stubenberg, "Neutral Monism," *Stanford Encyclopedia of Philosophy* 4 (Fall 2014), http://plato.stanford.edu/entries/neutral-monism/. In Russell's thought, the neutral monist concept of the bundling of particularities was introduced as a way of attacking dualism, while excluding dialectics (to which Russell was violently opposed). In his earlier work, Russell had himself developed a powerful critique of monism, which only led to his subsequent adoption of neutral monism as he sought to counter Marxian theory in the 1930s. See Russell, *The Analysis of Matter*, iv–5, 10, 382–83. In Russell's version of neutral monism, reality consisted of what we now call bundled entities that in large part obviated the need for the distinctions between mind and matter.
100. Moore, *Capitalism in the Web of Life*, 20–21, 85–86.

101. Moore, "Toward a Singular Metabolism," 13.
102. Marx and Engels, *Collected Works*, 461; Moore, *Capitalism in the Web of Life*, 80.
103. Mckenzie Wark, "The Capitalocene," *Public Seminar*, October 15, 2015, http://www.publicseminar.org/2015/10/the-capitalocene.
104. Joad, *Philosophical Aspects of Modern Science* (London: George Allen and Unwin, 1934), 115.
105. Wark, "The Capitalocene."
106. Bhaskar, *Dialectic: The Pulse of Freedom*, 270.
107. Ibid., 363.
108. Ibid., 205, 355, 394.
109. For Bhaskar, the "anthropic fallacy," whereby *being* is reduced to *human being*, is frequently concealed behind the "epistemic fallacy," whereby *being* is reduced to *knowledge*. Ibid., 205, 397.
110. Marx, *Grundrisse* (London: Penguin, 1973), 489, emphasis in original.
111. Neil Smith, "The Production of Nature," in *Future Natural: Nature, Science and Culture*, ed. J. Bird, B. Curtis, M. Mash, T. Putnam, G. Robertson, and L. Tuckner (London: Routledge, 1996), 49.
112. Smith, *Uneven Development: Nature, Capital and the Production of Space*, 47.
113. Neil Smith, "Nature as an Accumulation Strategy," in *Socialist Register 2007: Coming to Terms with Nature*, ed. L. Panitch and C. Leys (New York: Monthly Review Press, 2006), 27–29.
114. Smith, *Uneven Development: Nature, Capital and the Production of Space*, 81.
115. Ibid., 45–46; Smith, "The Production of Nature," 40.
116. Smith, "Nature as an Accumulation Strategy," 23.
117. Smith, *Uneven Development: Nature, Capital and the Production of Space*, 247.
118. Smith, "The Production of Nature," 40.
119. Smith, *Uneven Development: Nature, Capital and the Production of Space*, 244.
120. Ibid.
121. Smith criticized the "left romancing of nature," questioning in this regard both ecofeminism and Indigenous peoples for their conceptions of "the earth mother." See Smith, "Nature at the Millennium," in *Remaking Reality*, ed. Bruce Braun and Naul Castree (London: Routledge, 1998), 279–80.
122. Castree refers again and again in his various writings to Smith's vulnerability to the interpretation of hyper-social constructionism. He thus raises it as a fundamental issue that needs to be fixed—in Castree's case by providing a more Latourian version of Smith's production of

nature perspective. Nevertheless, Castree has recently suggested that, despite appearances to the contrary, he doesn't think that it was Smith's real intention to put forward a hyper-constructionist argument. See Castree, "Capitalism and the Marxist Critique of Political Ecology," 286. But with all the evidence that Castree has provided on Smith's tendencies on this score, coupled with his own predilections toward a strong social constructionism, this weak attempt to defend Smith in this respect has a very hollow ring to it. For example, Castree resorts to arguing that Smith was *"anthropomorphic* without being anthropocentric and Promethean." See Castree, "Marxism, Capitalism, and the Production of Nature," in *Social Nature: Theory, Practice and Politics*, ed. N. Castree and B. Braun (Oxford: Blackwell, 2001), 204–205, emphasis in original. See also Castree, "Marxism and the Production of Nature," 27–28; also "The Nature of Produced Nature," *Antipode* 27/1 (1995): 20; Erik Swyngedouw, "Modernity and Hybridity," *Annals of the American Association of Geographers* 89/3 (1999): 446–47.

123. Noel Castree and Bruce Braun, "The Construction of Nature and the Nature of Construction," in *Remaking Reality*, ed. B. Braun and N. Castree (London: Routledge, 1998), 2–41, 7.

124. Castree, "The Nature of Produced Nature," 20; Castree, "Capitalism and the Marxist Critique of Political Ecology."

125. Noel Castree, "False Antitheses? Marxism, Nature and Actor-Networks," *Antipode* 34/1 (2002): 131. Castree is somewhat ambivalent about this Latourian critique of Smith, despite putting this forward as a rational critique, and giving it more than a little credence. He suggests in the end, however, that it is overdrawn, and there is more to say in Smith's favor. Yet Castree's own disquiet in this respect, in which he signals that Smith's position has major weaknesses, has led him to seek a synthesis between the Smithian production of nature and the Latourian bundled hybrids.

126. Castree, "False Antitheses? Marxism, Nature and Actor-Networks"; Castree and Braun, "The Construction of Nature and the Nature of Construction," 6.

127. Smith, *Uneven Development: Nature, Capital and the Production of Space*, 31.

128. Castree, "Marxism and the Production of Nature," 8, emphasis in original.

129. Noel Castree and Thomas Macmillan, "Dissolving Dualisms: Actor-Networks and the Reimagination of Nature," in *Social Nature: Theory, Practice and Politics*, ed. Noel Castree and Bruce Braun (Oxford: Blackwell, 2001), 208–24; Castree, "False Antitheses? Marxism, Nature and Actor-Networks."

130. Latour, *Reassembling the Social*, 30, 49, 103, 133; Kyle McGee, *Bruno Latour: The Normativity of Networks* (London: Routledge, 2014), 48.
131. Latour, *Aramis or the Love of Technology* (Cambridge, MA: Harvard University Press, 1996), 63. On the role of Leibniz in Latour's thought, see McGee, *Bruno Latour: The Normativity of Networks*, 16–17, 42, 56, 83, 85–86.
132. Castree and Macmillan, "Dissolving Dualisms: Actor-Networks and the Reimagination of Nature," 211–18; Castree, "Environmental Issues: Relational Ontologies and Hybrid Politics," *Progress in Human Geography* 27/2 (2003): 206; White, Rudy, and Gareau, *Environments Natures and Social Theory*.
133. Smith, "Nature at the Millennium," 279.
134. Michael Shellenberger and Ted Nordhaus, *Break Through: From the Death of Environmentalism to the Politics of Possibility* (New York: Houghton Mifflin, 2007).
135. Castree, "Capitalism and the Marxist Critique of Political Ecology," 291.
136. Latour, *Politics of Nature: How to Bring the Sciences into Democracy* (Cambridge, MA: Harvard University Press, 2004), 94.
137. Ibid., 246.
138. Ibid., 58, emphasis in original.
139. Latour, "Technology Is Society Made Durable," in *Sociology of Monsters: Essays on Politics, Technology and Domination*, ed. J. Law (London: Routledge, 1991), 130.
140. Cronon, "The Trouble with Wilderness; or, Getting Back to the Wrong Nature," in *Uncommon Ground: Rethinking the Human Place in Nature*, ed. W. Cronon (New York: W. W. Norton, 1995), 89.
141. Ibid., 85.
142. Donald Worster, *Nature's Economy: A History of Ecological Ideas* (Cambridge: Cambridge University Press, 1994), 239–42, 294–304, 322.
143. Worster, *Nature's Economy: A History of Ecological Ideas*, 322, 461. See also Foster, Clark, and York, *The Ecological Rift*, 312–40. The same favorable position toward Smuts as representing ecological holism was advanced by Merchant, who also criticized Tansley as mechanistic. See Carolyn Merchant, *The Death of Nature* (New York: Harper and Row, 1980), 252, 292–93.
144. Worster, *Nature's Economy: A History of Ecological Ideas*, 426–27. For an opposing interpretation on Marx and Engels and the intrinsic value of nature, see Foster and Burkett, *Marx and the Earth: An Anti-Critique*, 34–56.
145. Marx and Engels, *Selected Correspondence* (Moscow: Progress Publishers, 1975), 38, emphasis in original.
146. See Karl Marx and Frederick Engels, *MEGA IV 26* [in German] (Berlin:

Akademie Verlag, 2011), 214–19; Joseph Beete Jukes, *The Student's Manual of Geology* (Edinburgh: Adam and Charles Black, 1872), 476–512; James Hansen, *Storms of My Grandchildren* (New York: Bloomsbury, 2009), 146–47.

147. Marx and Engels, *Collected Works*, vol. 25, 331–35; Foster and Burkett, *Marx and the Earth: An Anti-Critique*, 165–203.
148. Igor J. Polianski, "Between Hegel and Haeckel: Monistic Worldview, Marxist Philosophy, and Biomedicine in Russia and the Soviet Union," in *Monism*, ed. T. H. Weir (London: Palgrave Macmillan, 2012), 197–222.
149. Georgi Plekhanov, *Selected Philosophical Works* (Moscow: Progress Publishers, 1974), 480–697.
150. Marx and Engels, *Collected Works*, vol. 25, 489, 531, 681; V. I. Lenin, "Monism and Dualism," in *A Caricature of Marxism and Imperialist Domination*, by V. I. Lenin (1916), https://www.marxists.org/archive/lenin/works/1916/carimarx/; V. I. Lenin, *Materialism and Empirico-Criticism* (Moscow: Progress Publishers, 1964), 209–15.
151. Bhaskar, *Philosophy and the Idea of Freedom* (Oxford: Blackwell, 1991), 174; Bhaskar, "Knowledge," 259.
152. Noel Castree, "Unfree Radicals: Geoscientists, the Anthropocene, and Left Politics," *Antipode* (2015): 9. Castree has adopted a more realistic approach to climate change and other environmental changes in some of his recent work, where he is no longer simply dismissive of thinkers such as Naomi Klein. Nevertheless, he characteristically attributes society's failure to address global environmental problems to "dualism" rather than to capitalism itself, thereby excluding the reality of alienation.
153. Georg Lukács, *History and Class Consciousness* (London: Merlin Press, 1971), 24.
154. Ibid., 207.
155. G. W. F. Hegel, *Hegel's Logic* (Oxford: Oxford University Press, 1975).
156. Lukács, *History and Class Consciousness*, xvii.
157. Lukács, *Conversations with Lukács* (Cambridge, MA: MIT Press, 1974), 43; Lukács, *Labour*, 17–19.
158. Lukács, *Labour*, 56–59.
159. Ibid., 34.
160. Istvan Mészáros, *Marx's Theory of Alienation* (London: Merlin Press, 1970), 104.
161. Ibid., 104, 162–65.
162. Istvan Mészáros, *The Necessity of Social Control* (New York: Monthly Review Press, 2015), 23–51.
163. Istvan Mészáros, *Beyond Capital* (New York: Monthly Review Press, 1995), 170–77.

164. Smith, *Uneven Development: Nature, Capital and the Production of Space*, 45–46.
165. Karl Marx, *Critique of the Gotha Programme* (New York: International Publishers, 1938), 3.
166. Marx, *Capital*, vol. 3, 949.
167. David Harvey, "History versus Theory: A Comment on Marx's Method in *Capital*," *Historical Materialism* 20/2 (2012): 12–14.
168. Popper, *Conjectures and Refutations* (London: Routledge, 2002), 437; Needham, *Time: The Refreshing River* (London: George Allen and Unwin, 1943), 15.
169. Hegel, *The Phenomenology of Spirit* (Oxford: Oxford University Press, 1977), 11.
170. Richard Levins, "The Internal and External in Explanatory Theories," *Science as Culture* 7/4 (1998): 559.
171. The general views of many of these thinkers (though not their ecological analyses) are treated in Sheehan, *Marxism and the Philosophy of Science: A Critical History* (Atlantic Highlands, NJ: Humanities Press, 1985). See also George Thomson, *Aeschylus and Athens* (New York: International Publishers, 1950).
172. See Peter Ayres, *Shaping Ecology* (Oxford: Wiley-Blackwell, 2012), 41–43.
173. See Foster, Clark, and York, *The Ecological Rift*, 242–45; also A. D. Ursul, ed., *Philosophy and the Ecological Problems of Civilization* (Moscow: Progress Publishers, 1983).
174. Christopher Caudwell, *Illusion and Reality: A Study of the Sources of Poetry* (London: Lawrence and Wishart, 1946), 279, emphasis in original.
175. Paul Burkett, "Marx's Vision of Sustainable Human Development," *Monthly Review* 57/ (2005): 34–62.
176. Marx and Engels, *Collected Works*, vol. 25, 105; Lukács, *Labour*, 120–25.
177. Foster, *Ecological Revolution* (New York: Monthly Review Press, 2009).
178. Fred Magdoff, "Ecological Civilization," *Monthly Review* 62/8 (2011): 1–25.
179. Marx, *Capital*, vol. 1, 63–68; *Capital*, vol. 3, 949; Foster and Burkett, *Marx and the Earth: An Anti-Critique*, 239–40.

12. Marxism and the Dialectics of Ecology

This chapter is adapted and revised for this book from the following article: John Bellamy Foster and Brett Clark, "Marxism and the Dialectics of Ecology," *Monthly Review* 68/5 (October 2016): 1–17.

1. Epigraph: Karl Marx and Frederick Engels, *Collected Works* (New York: International Publishers, 1975), vol. 4, 150.

2. Georg Lukács, *Labour* (London: Merlin, 1980).
3. Karl Marx, *Capital*, vol. 3 (London: Penguin, 1981), 949; Marx and Engels, *Collected Works*, vol. 30, 54–66.
4. Marx and Engels, *Collected Works*, vol. 3, 732–33.
5. Jason W. Moore, *Capitalism in the Web of Life* (London: Verso, 2015), 80–81; Neil Smith, *Uneven Development* (Athens: University of Georgia Press, 2008).
6. "Capitalocentrism" refers to attempts on the left to subsume the ecological problem within the internal logic of capitalist accumulation. It can also be seen in attempts to reject scientific categories such as the Anthropocene that address the overall relations of human beings to the Earth System in favor of narrower concepts such as the Capitalocene, in which capital sets the parameters for all analysis. For an example of this tendency, see Moore, *Capitalism in the Web of Life*, 169–92.
7. Jason W. Moore, "Toward a Singular Metabolism," in Daniel Ibañez and Nikos Katsikis, eds., *Grounding Metabolism* (Cambridge, MA: Harvard University Press, 2014), 10–19.
8. István Mészáros, *Marx's Theory of Alienation* (London: Merlin, 1975), 99–114.
9. Marx, *Capital*, vol. 3, 949. Given the structure of Marx's thought, it is possible to speak, as he himself did, of a "metabolic rift" in the social metabolism, involving the specific conditions of production. Yet insofar as larger biogeochemical cycles and processes are affected by human production in ways distant from production, this involves not simply a rift in the social metabolism but more importantly between it and the universal metabolism of nature. It is this latter rift that defines the Anthropocene crisis.
10. Karl Marx, *Early Writings* (London: Penguin, 1974), 261. We owe this insight to István Mészáros, who referred to Marx's concept of "alienated mediation" in a letter to one of the authors.
11. Marx, *Capital*, vol. 3, 949; *Capital*, vol. 1 (London: Penguin, 1976), 636–39.
12. Brett Clark and John Bellamy Foster, "Guano: The Global Metabolic Rift in the Fertilizer Trade," in Alf Hornborg, Brett Clark, and Kenneth Hermele, eds., *Ecology and Power* (London: Routledge, 2012), 68–82.
13. John Bellamy Foster, Brett Clark, and Richard York, *The Ecological Rift* (New York: Monthly Review Press, 2010), 73–87.
14. Ibid., 53–72; James Maitland, Earl of Lauderdale, *An Inquiry into the Nature and Origins of Public Wealth and into the Means and Causes of Its Increase* (Edinburgh: Archibald Constable, 1819), 37–59; Marx and Engels, *Collected Works*, vol. 37, 732–33.
15. Marx, *Capital*, vol. 1, 380–81.

16. In classical value theory, only labor creates capitalist commodity value. Land and natural resources, however, are subject to rents, which constitute a form of redistribution of value, and therefore acquire prices. It should be added that if nature is not incorporated directly into value creation and is instead treated as a "free gift" in capitalist accounting, the same principle also applies to subsistence work and unpaid domestic labor.
17. Roland Daniels, *Mikrokosmos* (Frankfurt am Main: Peter Lang, 1988). We would like to thank Joseph Fracchia for translating parts of Daniels's work. We would also like to thank Kohei Saito for his comments on Daniels's work.
18. On Liebig's ecological views and their relation to Marx, see John Bellamy Foster, *Marx's Ecology* (New York: Monthly Review Press, 2000), 149–54; Kohei Saito, "Marx's Ecological Notebooks," *Monthly Review* 67/9 (February 2016): 25–33.
19. Marx and Engels, *Collected Works*, vol. 42, 558–59; Saito, "Marx's Ecological Notebooks," 34–39, fn28, fn29.
20. Marx, *Capital*, vol. 1, 638.
21. John Bellamy Foster, "Marxism and Ecology," *Monthly Review* 67/7 (December 2015): 2–3; Joel B. Hagen, *An Entangled Bank* (New Brunswick, NJ: Rutgers University Press, 1992).
22. See John Bellamy Foster, *The Return of Nature* (New York: Monthly Review Press, 2020), 11–22.
23. Georg Lukács, *History and Class Consciousness* (London: Merlin, 1968), 24.
24. Russell Jacoby, "Western Marxism," in Tom Bottomore, ed., *A Dictionary of Marxist Thought* (Oxford: Blackwell, 1983), 523–26; Maurice Merleau-Ponty, *Adventures of the Dialectic* (Evanston, IL: Northwestern University Press, 1973).
25. Smith, *Uneven Development*, 45–47, 247; Smith, "Nature as an Accumulation Strategy," *Socialist Register 2007* (New York: Monthly Review Press, 2006), 23–29.
26. Moore, *Capitalism in the Web of Life*, 152.
27. Smith, *Uneven Development*, 65–69.
28. Moore goes even further, treating the nature that precedes society as "pre-formed," because not yet produced or "co-produced" by society: "Even when environments are in some abstract sense pre-formed (the distribution of the continents, for example) historical change works through the encounters of humans with those environments, a relation that is fundamentally co-productive." See Moore, "Toward a Singular Metabolism," 15.
29. Moore, *Capitalism in the Web of Life*, 4, 19–20, 78, 152.

30. Moore, "Toward a Singular Metabolism," 16; *Capitalism in the Web of Life*, 85. What appears dualistic, when not considered dialectically, is, within dialectical discussion, often the treatment of a contradiction—the "identity of opposites"—that can only be transcended at another organizational level. Recognizing this contradiction in almost Marxian terms, Whitehead wrote: "Throughout the Universe there reigns the union of opposites which is the ground of dualism." See Alfred North Whitehead, *Adventures of Ideas* (New York: Free Press, 1933), 245, fn37, 39.
31. Moore, "Toward a Singular Metabolism," 12; "Cheap Food and Bad Climate," *Critical Historical Studies* 2/10 (2015): 28; "Putting Nature to Work," in Cecilia Wee and Olaf Arndt, eds., *Supra Markt* (Stockholm: Irene, 2015), 91.
32. Moore, "Toward a Singular Metabolism," 12; Moore, *Capitalism in the Web of Life*, 85, 179.
33. Moore, *Capitalism in the Web of Life*, 46.
34. Ibid., 37.
35. See Güberk Koç Maclean, *Bertrand Russell's Bundle Theory of Particulars* (London: Bloomsbury, 2014).
36. Erik Swyngedouw, "Modernity and Hybridity," *Annals of the Association of American Geographers* 89/3 (1999): 446.
37. See Noel Castree, "Marxism and the Production of Nature," *Capital and Class* 72 (2000): 27–28; Castree, "The Nature of Produced Nature: Materiality and Knowledge Construction in Marxism," *Antipode* 27/1 (1995): 20; "Marxism, Capitalism, and the Production of Nature," in Castree and Bruce Braun, eds., *Social Nature* (Malden, MA: Blackwell, 2001), 204–5; Castree, "Capitalism and the Marxist Critique of Political Ecology," in Tom Perreault, Gavin Bridge, and James McCarthy, eds., *The Routledge Handbook of Political Ecology* (London: Routledge, 2015), 279–92.
38. Noel Castree, "False Antitheses? Marxism, Nature and Actor-Networks," *Antipode* 34/1 (2002): 131; Bruno Latour, *Politics of Nature* (Cambridge, MA: Harvard University Press, 2004), 58.
39. Jason W. Moore, "The Capitalocene, Part II," June 2014, 34, http://jasonwmoore.com.
40. The nonexistence of nature as a referent is a basic stipulation of Bruno Latour's philosophy, a significant influence on the thinkers criticized here. See Bruno Latour, *Science in Action* (Cambridge, MA: Harvard University Press, 1987), 99, 258. See Alan Sokal's critique of Latour on this point in *Beyond the Hoax* (Oxford: Oxford University Press, 2008), 154–58, 211–16. Latour, whose work is explicitly anti-Marxist and anti-dialectical, advances what is often called a "flat ontology"

or neutral monism, in which all entities and objects are equal and intertwined and to be approached as assemblages, bundles, hybrids, or networks. Nevertheless, the extreme relationism of his views, which denies both nature and society as substantive objects, gives rise in the end to a kind of social monism, where the social is smuggled back in or "reassembled" (e.g., through technology and politics), taking the form of a capitulation to the status quo. In his recent work he has advanced a regressive political ecology that has been called "Green Schmittianism," relying on the geopolitics and political theology of the Nazi philosopher Carl Schmitt. Not surprisingly Latour has become a senior fellow of the Breakthrough Institute. See Graham Harman, *Prince of Networks* (Melbourne: re.press, 2009), 73–75, 102, 152–56, 214–15; Bruno Latour, *Reassembling the Political* (London: Pluto, 2014); Latour, *Reassembling the Social* (Oxford: Oxford University Press, 2005), 18, 116, 134–47; Latour, "Facing Gaia," Gifford Lectures, University of Edinburgh, February 18–28, 2013.

41. Jason W. Moore, "The Capitalocene, Part I," March 2014, 16, http://jasonwmoore.com; Bruce Braun, "Toward a New Earth and a New Humanity," in Noel Castree and Derek Gregory, eds., *David Harvey: A Critical Reader* (Oxford: Blackwell, 2006), 197–99; Ian Angus and Fred Murphy, "Two Views on Marxist Ecology and Jason W. Moore," *Climate and Capitalism*, June 23, 2016, http://climateandcapitalism.com.

42. Jason W. Moore, "Nature in the Limits to Capital (and Vice Versa)," *Radical Philosophy* 193 (2015): 14.

43. Erik Swyngedouw, "Trouble with Nature: 'Ecology as the New Opium for the Masses,'" in J. Hillier and P. Healey, eds., *The Ashgate Research Companion to Planning Theory: Conceptual Challenges for Spatial Planning* (Burlington, VT: Ashgate, 2010), 304.

44. Swyngedouw, "Modernity and Hybridity," 446. Ironically, in this quote Swyngedouw was purporting to present the conventional Marxist materialist view, which he then proceeded to criticize for placing *too much* emphasis on natural conditions, and indeed for seeing nature as a signifier.

45. Swyngedouw, "Trouble with Nature," 308–9.

46. Smith, *Uneven Development*, 244.

47. Moore, *Capitalism in the Web of Life*, 15; Swyngedouw, "Trouble with Nature: Ecology as the New Opium of the Masses," 309; see also Alain Badiou, "Live Badiou—Interview with Alain Badiou," in *Alain Badiou— Live Theory* (London: Continuum, 2008); Slavoj Žižek, "Censorship Today: Violence, or Ecology as a New Opium of the Masses," 2007, http://lacan.com.

48. Moore, *Capitalism in the Web of Life*, 80.

49. Jason W. Moore, "The End of Cheap Nature Or: How I Learned to Stop Worrying about 'The' Environment and Love the Crisis of Capitalism," in Christian Suter and Christopher Chase Dunn, eds., *Structures of the World Political Economy and the Future of Global Conflict and Cooperation* (Berlin: LIT, 2014), 308; "Toward a Singular Metabolism," 14. Moore flatly rejects the concept of the Anthropocene introduced by natural scientists to describe the anthropogenic rift in the Earth system. For a meaningful treatment of the Anthropocene, see Ian Angus, *Facing the Anthropocene: Fossil Capitalism and the Crisis of the Earth System* (New York: Monthly Review Press, 2016).
50. Moore, *Capitalism in the Web of Life*, 112–13. Moore's approach to ecological crisis is based on the notion that capitalism does not rely on the exploitation of labor so much as the appropriation of work or energy in a general, physical sense. This requires a post-Marxist deconstruction of Marx's value theory, and indeed of all economic theory. As Moore writes: "My argument proceeds from a certain destabilization of value as an 'economic' category." See Moore, "The Capitalocene, Part II," 29. For a critique of Moore's rejection of Marxian value theory, see Kamran Nayeri, "'Capitalism in the Web of Life'—A Critique," *Climate and Capitalism*, July 19, 2016, http://climateandcapitalism.com.fn51, fn52, fn53, fn54, fn56, fn59.
51. Moore, "Toward a Singular Metabolism," 16–17. Although Moore emphasizes capitalism's ability to transcend natural limits, he does argue, in his attack on the "apocalyptic" Green perspective, that the imminent collapse of contemporary civilization would not be "something to be feared"—using as a historical example the fall of Rome, which he says gave rise to a golden age. Quite apart from the extent of human suffering that followed the collapse of Rome, today the social destruction associated with the crossing of planetary boundaries threatens the lives and living conditions of hundreds of millions, even billions, of people, as well as innumerable other species.
52. Paul Hawken, Amory B. Lovins, L. Hunter Lovins, *Natural Capitalism* (London: Earthscan, 2010); Arthur P. J. Mol and Martin Jänicke, "The Origins and Theoretical Foundations of Ecological Modernisation Theory," in Arthur P. J. Mol, David A. Sonnenfeld, and Gert Spaargaren, eds., *The Ecological Modernisation Reader* (London: Routledge, 2009).
53. Jason W. Moore, "The Rise of Cheap Nature," in Moore, ed., *Anthropocene or Capitalocene* (Oakland, CA: PM, 2016), 111; Moore, "Putting Nature to Work," 69; Ted Nordhaus and Michael Shellenberger, *Break Through: From the Death of Environmentalism to the Politics of Possibility* (New York: Houghton Mifflin, 2007).

54. Moore, "Toward a Singular Metabolism," 11; *Capitalism in the Web of Life*, 83.
55. Moore, *Capitalism in the Web of Life*, 83–84. In substituting the "metabolic shift" for the "metabolic rift," Moore promotes one side of a dialectical process that we in our work with Richard York had earlier described as metabolic "rifts and shifts," whereby capitalism's attempt to shift the anthropogenic rifts it creates in the human relation to the environment leads to cumulatively greater rifts, universalizing ecological contradictions. See Foster, Clark, and York, *The Ecological Rift*, 73–87.
56. Smith, *Uneven Development*, 81; Marx, *Early Writings*, 261.
57. Moore, *Capitalism in the Web of Life*, 13, 37, 76, 78. Moore argues that Marx saw capitalism as capable of unifying nature. But to do so, he must distort and misread Marx's language. He writes: "Rather than ford the Cartesian divide, metabolism approaches have reinforced it. Marx's 'interdependent process of social metabolism' became 'the metabolism of nature and society.' Metabolism as 'rift' became [for ecological Marxists] a metaphor of separation, premised on material flows between nature and society" (76); and Moore, "Toward a Singular Metabolism," 13, 18. Yet Marx's actual phrase, referring to capitalism's relation to the ecology, was "*the irreparable rift in the* interdependent process of social metabolism" (emphasis added). By omitting these crucial words, Moore inverts the meaning of Marx's statement. Further, the phrase "metabolism of nature and society" as used by Foster is not a distortion of Marx, as Moore claims, but reflects Marx's own views and language, as when he famously referred in volume 1 of *Capital* to "the metabolic interaction between man and the earth." See Marx,*Capital*, vol. 3, 949; *Capital*, vol. 1, 637.
58. Lukács, *Labour*, 119–24. On integrative levels and their role in Marxian theory, see Joseph Needham, *Time: The Refreshing River* (London: George Allen and Unwin, 1943), 13–20, 233–72.
59. Karl Marx, *Grundrisse* (New York: Penguin, 1973), 489.
60. Marx, *Capital*, vol. 1, 283, 637–38, fn61, fn62, fn63, fn64, fn65, fn66, fn67, fn68, fn69.
61. Georg Lukács, *Marx* (London: Merlin, 1978), 44, 58, 107.
62. Lukács, *Labour*, 34. "As a biological being, man is a product of natural development. With his self-realization which of course even in his case means only a retreat of the natural boundary, and never its disappearance, its complete conquest, he enters into a new and self-founded being, into social being." See Lukács, *Labour*, 46.
63. Marx, *Capital*, vol. 1, 284.
64. Marx, *Capital*, vol. 3, 949–50.
65. Fredric Jameson, *Valences of the Dialectic* (London: Verso, 2009), 3–7.

66. Lukács, *Marx*, 103.
67. Moore uses "Double Internality" as a basic category of his social-monist view. He points to various "bundles," and especially the "double internality" of the capitalist world-ecology. See Moore, *Capitalism in the Web of Life*, 1.
68. Alfred North Whitehead, *Science and the Modern World* (New York: Free Press, 1925), 51. On actualism see Roy Bhaskar, *Plato Etc.* (London: Verso, 1994), 250–51.
69. Lukács, *Marx*, 107.
70. Moore, *Capitalism in the Web of Life*, 86.
71. Richard J. Evanoff, "Reconciling Realism and Constructivism in Environmental Ethics," *Environmental Values* 14 (2005): 74.
72. Naomi Klein, *This Changes Everything: Capitalism vs. the Climate* (New York: Simon and Schuster, 2014), 177, 186.
73. N. Bednaršek et al., "*Limacina Helicina* Shell Dissolution as an Indicator of Declining Habitat Suitability Owing to Ocean Acidification in the California Current Ecosystem," *Proceedings of the Royal Society B: Biological Sciences* 281/1785 (2014).
74. Evan N. Edinger et al., "Reef Degradation and Coral Biodiversity in Indonesia," *Marine Pollution Bulletin* 36/8 (1998): 617–30; Pamela Hallock, "Global Change and Modern Coral Reefs," *Sedimentary Geology* 175/1 (2005):19–33; Chris Mooney, "Scientists Say a Dramatic Worldwide Coral Bleaching Event Is Now Underway," *Washington Post*, October 8, 2015; J. P. Gattuso et al., "Contrasting Futures for Ocean and Society from Different Anthropogenic CO_2 Emissions Scenarios," *Science* 349/6243 (2015).
75. István Mészáros, "The Structural Crisis of Politics," *Monthly Review* 58/4 (2006): 34–53.
76. Karl Marx, *Theories of Surplus Value*, vol. 3 (Moscow: Progress Publishers, 1971), 309.
77. Marx, *Capital*, vol. 3, 959, fn80.
78. Mészáros, *Marx's Theory of Alienation*, 113.
79. Roy Bhaskar, *Scientific Realism and Human Emancipation* (London: Verso, 1986), 222.
80. Frederick Engels, *Anti-Dühring* (Moscow: Progress Publishers, 1969), 136–38.

13. Engels's *Dialectics of Nature* and the Anthropocene

This chapter is adapted for this book from the following article: John Bellamy Foster, "Engels's *Dialectics of Nature* in the Anthropocene," *Monthly Review* 72/6 (November 2020): 1–17.

1. Karl Marx and Frederick Engels, *Collected Works*, vol. 25 (New York: International Publishers, 1975), 459.

2. Paul Blackledge, *Friedrich Engels and Modern Social and Political Theory* (Albany: State University of New York Press, 2019), 16.
3. Walter Benjamin, *Selected Writings*, vol. 4, *1938–1940* (Cambridge, MA: Harvard University Press, 2003), 402; Michael Löwy, *Fire Alarm: Reading Walter Benjamin's "On the Concept of History"* (London: Verso, 2001), 66–67.
4. Marx and Engels, *Collected Works*, vol. 25, 145–46, 153, 270; Karl Marx and Frederick Engels, *Ireland and the Irish Question* (Moscow: Progress Publishers, 1971), 142.
5. Locomotive boiler explosions due to defective and maladjusted safety valves were common in the mid-nineteenth century. Locomotive engineers under time pressures often wedged or fastened down the safety valves, thereby jamming the safety valves on the train, which did not open, or which they were unable physically to open in time. See Christian H. Hewison, *Locomotive Boiler Explosions* (Newton Abbot, UK: David & Charles, 1983), 11, 18–19, 36, 49, 54–56, 82, 85, 110.
6. Marx and Engels, *Collected Works*, vol. 25, 459; John Bellamy Foster, "Capitalism and the Accumulation of Catastrophe," *Monthly Review* 63/7 (December 2011): 5–7; Karl Marx and Friedrich Engels, *Marx-Engels Gesamtausgabe (MEGA)* IV/31 (Amsterdam: Akadamie Verlag, 1999), 512–15.
7. Marx and Engels, *Collected Works*, vol. 25, 167; Karl Marx and Friedrich Engels, *Marx-Engels Gesamtausgabe (MEGA)* IV/18 (Berlin: Walter de Gruyter, 2019), 670–74, 731 (excerpts by Marx); Mike Davis, *Late Victorian Holocausts: El Niño Famines and the Making of the Third World* (London: Verso, 2001); Marx and Engels, *Ireland and the Irish Question*.
8. On the notion of extreme productivism and, in this sense, Prometheanism, as well as its almost complete absence in Marx and Engels's thought, see John Bellamy Foster, *The Ecological Revolution* (New York: Monthly Review Press, 2009), 226–29.
9. Marx and Engels, *Collected Works*, vol. 25, 269. For Marx and Engels, it should be noted, productive forces refer to more than simply technology. Thus, Marx insisted that the most important instrument or force of production was human beings themselves. Hence, expansion of the forces of production meant the expansion of human productive skills and powers. See Marx and Engels, *Collected Works*, vol. 6, 211; Paul A. Baran, *The Longer View* (New York: Monthly Review Press, 1969), 59.
10. Walt Rostow, *The World Economy* (Austin: University of Texas Press, 1978), 47–48, 659–62.
11. On sustainable human development as a framework governing both

Marx's and Engels's thought, see Paul Burkett, "Marx's Vision of Sustainable Human Development," *Monthly Review* 57/5 (October 2005): 34–62.
12. Eleanor Leacock, introduction to *The Origin of the Family, Private Property and the State*, by Frederick Engels (New York: International Publishers, 1972), 245.
13. Marx and Engels, *Collected Works*, vol. 4, 394, 407; Ian Angus, "Cesspools, Sewage, and Social Murder," *Monthly Review* 70/3 (July–August 2018): 38; John Bellamy Foster, *The Return of Nature* (New York: Monthly Review Press, 2020), 182–95.
14. Howard Waitzkin, *The Second Sickness* (New York: Free Press, 1983), 71–72.
15. Marx and Engels, *Collected Works*, vol. 25, 23; Foster, *The Return of Nature*, 254.
16. Marx and Engels, *Collected Works*, vol. 25, 270.
17. Marx and Engels, *Collected Works*, vol. 25, 463–64.
18. Francis Bacon, *Novum Organum* (Chicago: Open Court, 1994), 29, 43.
19. Marx and Engels, *Collected Works*, vol. 25, 461; Karl Marx, *Grundrisse* (London: Penguin, 1973), 409–10.
20. Marx and Engels, *Collected Works*, vol. 25, 461.
21. Marx and Engels, *Collected Works*, vol. 25, 460–61.
22. Marx and Engels, *Collected Works*, vol. 25, 330–31, 461.
23. Ray Lankester, *The Kingdom of Man* (New York: Henry Holt and Co., 1911), 1–4, 26, 31–33; Foster, *The Return of Nature*, 61–64.
24. Lankester, *The Kingdom of Man*, 31; Joseph Lester, *Ray Lankester and the Making of Modern British Biology* (Oxford: British Society for the History of Science, 1995), 163–64.
25. Ray Lankester, *Science from an Easy Chair* (New York: Henry Holt and Co., 1913), 365–69.
26. Marx and Engels, *Collected Works*, vol. 25, 492. The criticism of Engels on the dialectics of nature had its origins in footnote 6 of Georg Lukács's *History and Class Consciousness*, though Lukács, as he later explained, never fully abandoned the notion of a "merely objective dialectics" and was to promote such a naturalistic dialectic, based on Marx more than Engels, in his later thought. Nevertheless, the rejection of the dialectics of nature became axiomatic for Western Marxism beginning in the 1920s, taking a stronger hold in the post–Second World War period. Georg Lukács, *History and Class Consciousness* (Cambridge, MA: MIT Press, 1971), 24, 207. See also Russell Jacoby, "Western Marxism," in *A Dictionary of Marxist Thought*, ed. Tom Bottomore (Oxford: Blackwell, 1983), 523–26; Foster, *The Return of Nature*, 11–22. On the general conflict regarding Engels

within contemporary Marxism, see Blackledge, *Frederick Engels and Modern Social and Political Theory*, 1–20.
27. As Roy Bhaskar has argued, the necessity to consider the intransitive or the realm of transfactuality establishes the distinction between the epistemological and the ontological, against the tendency within much of contemporary philosophy, including the Western Marxist philosophical tradition, to promote the epistemological fallacy, characteristic of idealism, in which ontology is subsumed within epistemology. Adherence to the epistemological fallacy would make any consistent materialism or natural science impossible. Roy Bhaskar, *Dialectic: The Pulse of Freedom* (London: Verso, 1993), 397, 399–400, 405.
28. This can be seen in Alfred Schmidt's *The Concept of Nature in Marx*, published in 1962, the same year as Rachel Carson's *Silent Spring*. Schmidt's work, a product of the Frankfurt School (influenced particularly by his mentors Max Horkheimer and Theodor Adorno) for the most part denied the dialectics of nature and any reconciliation of humanity with nature on the very cusp of the emergence of the modern environmental movement. Alfred Schmidt, *The Concept of Nature in Marx* (London: Verso, 1970).
29. This and the following six paragraphs are adapted from Foster, *The Return of Nature*, 379–81.
30. Marx and Engels, *Collected Works*, vol. 25, 356.
31. Peter T. Manicas, "Engels's Philosophy of Science," in *Engels After Marx*, ed. Manfred B. Steger and Terrell Carver (University Park: Pennsylvania University Press, 1999), 77.
32. Craig Dilworth, "Principles, Laws, Theories, and the Metaphysics of Science," *Synthese* 101/2 (1994): 223–47. The principle of uniformity (or uniformitarianism), most closely associated with Charles Lyell, was challenged by Darwin's concept of evolution, though Darwin's gradualism downplayed the conflict. Stephen Jay Gould and paleontologist Niles Eldredge were to challenge uniformitarianism much more radically in their theory of punctuated equilibrium in the 1980s. See Richard York and Brett Clark, *The Science and Humanism of Stephen Jay Gould* (New York: Monthly Review Press, 2011), 28, 40–42. The traditional notion of the perpetuation of substance was challenged in Engels's day by the development of the concept of energy in physics. In relation to both of these ontological principles and the principle of causality, where he addressed the complex interchange of cause and effect, Engels's dialectical "laws" or ontological principles not only captured the revolutionary changes taking place in the science of his day, but in various ways prefigured later discoveries.

On Engels's views of causality, see Marx and Engels, *Collected Works*, vol. 25, 510.
33. J. D. Bernal, *Engels and Science* (London: Labour Monthly Pamphlets, 1936), 1–2.
34. Ibid., 5.
35. Ibid., 5–7; Marx and Engels, *Collected Works*, vol. 25, 359 (translation of this passage from Engels follows Bernal, *Engels and Science*).
36. Hyman Levy, *A Philosophy for a Modern Man* (New York: Alfred A. Knopf, 1938), 30–32, 117, 227–28.
37. This paragraph was written by Fred Magdoff. See also Fred Magdoff and Chris Williams, *Creating an Ecological Society* (New York: Monthly Review Press, 2017), 215.
38. Marx and Engels, *Collected Works*, vol. 25, 326, 507; E. Ray Lankester, "Limulus an Arachnid," *Quarterly Journal of Microscopical Science* 2 (1881): 504–48, 609–49; Foster, *The Return of Nature*, 56, 249.
39. Bernal, *Engels and Science*, 7–8; J. D. Bernal, "Dialectical Materialism," in Hyman Levy et al., *Aspects of Dialectical Materialism* (London: Watts and Co., 1934), 107–8.
40. Bernal, *Engels and Science*, 7; Foster, *The Return of Nature*, 242.
41. Bernal, *Engels and Science*, 7; Marx and Engels, *Collected Works*, vol. 25, 14.
42. All three of Engels's informal laws of dialectics can be seen as related to emergence, particularly the first and the third. Engels's third informal law, negation of the negation, as Roy Bhaskar argued in *Dialectics: Pulse of Freedom*, "raises the issue of *absenting absences* and the reassertion of lost or negated elements of reality. Bernal developed an analysis of the negation of the negation in terms of the role of residuals that reemerge and transform relations through complex evolutionary processes." Roy Bhaskar, *Dialectic: The Pulse of Freedom* (London: Verso, 1993), 150–52, 377–78; Bernal, "Dialectical Materialism," 103–4.
43. This and the following paragraph were drafted nearly in their entirety by Fred Magdoff.
44. Marx and Engels, *Collected Works*, vol. 25, 126.
45. Bernal, *Engels and Science*, 8–10; Friedrich Engels, *Ludwig Feuerbach and the Outcome of Classical German Philosophy* (New York: International Publishers, 1941), 65–69.
46. Ilya Prigogine and Isabelle Stengers, *Order Out of Chaos* (New York: Bantam, 1984), 252–53.
47. Bernal, *Engels and Science*, 4.
48. Bernal, "Dialectical Materialism," 90, 102, 107, 112–17.
49. Bernal, *Engels and Science*, 10–12. With respect to Engels on the origins of life, Richard Levins and Richard Lewontin wrote that

"dialectical materialism has focused mostly on some selected aspects of reality. At times we have emphasized the materiality of life against vitalism, as when Engels said that life was the motion of 'albuminous bodies' (i.e. proteins; now we might say macro-molecules). This seems to be in contradiction to our rejection of molecular reductionism, but simply reflects different moments in an ongoing debate where the main adversaries were first the vitalist emphasis on the discontinuity between the inorganic and living realms, and then the reductionist ersasure of the real leaps of levels." Richard Lewontin and Richard Levins, *Biology Under the Influence* (New York: Monthly Review Press, 2007), 103.
50. Bernal, *Engels and Science*, 13–14.
51. J. D. Bernal, *The Freedom of Necessity* (London: Routledge and Kegan Paul, 1949), 362.
52. Ibid., 364–65.
53. Joseph Needham, *Time, the Refreshing River* (London: George Allen, and Unwin, 1943), 214–15; Engels, *Ludwig Feuerbach*, 12.
54. Needham, *Time, the Refreshing River*, 214–15; Marx and Engels, *Collected Works*, vol. 46, 411.
55. J. B. S. Haldane, *The Marxist Philosophy and the Sciences* (New York: Random House, 1939), 199–200; Foster, *The Return of Nature*, 391.
56. Richard Levins and Richard Lewontin, *The Dialectical Biologist* (Cambridge, MA: Harvard University Press, 1985).
57. Stephen Jay Gould, *An Urchin in the Storm* (New York: W. W. Norton, 1987), 111–12.
58. Needham, *Time, the Refreshing River*, 14–15. Engels wrote: "It is precisely the *alteration of nature by* men, not solely nature as such, which is the most essential and immediate basis of human thought." Marx and Engels, *Collected Works*, vol. 25, 511.
59. See John Bellamy Foster, Brett Clark, and Richard York, *The Ecological Rift* (New York: Monthly Review Press, 2010), 13–18; Ian Angus, *Facing the Anthropocene* (New York: Monthly Review Press, 2016); Clive Hamilton, *Defiant Earth* (Cambridge: Polity, 2017).
60. Lester, *Ray Lankester*, 164.
61. John Bellamy Foster, "Capitalism and the Accumulation of Catastrophe," 1–2, 15–16; Foster, *The Return of Nature*, 64, 286–87.
62. Marx and Engels, *Collected Works*, vol. 25, 516.
63. Marx and Engels, *Collected Works*, vol. 46, 411.
64. Frederick Engels, *The Housing Question* (Moscow: Progress Publishers, 1975), 92.
65. On Engels's approach to thermodynamics, see John Bellamy Foster and Paul Burkett, *Marx and the Earth* (Chicago: Haymarket, 2016), 137–203.

66. On Marx and Engels on ecological degradation and extermination in colonial Ireland, see John Bellamy Foster and Brett Clark, *The Robbery of Nature* (New York: Monthly Review Press, 2020), 64–77.
67. Engels made it clear that the rational regulation of the human relation to nature, and thus a rational application of science, was only possible with "a complete revolution in our hitherto existing mode of production." Marx and Engels, *Collected Works*, vol. 25, 462. On the alienation of science under capitalism, see István Mészáros, *Marx's Theory of Alienation* (London: Merlin, 1975), 101–2. The role of science under capitalism is further clarified in Richard Levins's notion of the "dual nature of science." Richard Levins, "Ten Propositions on Science and Antiscience," *Social Text* 46–47 (1996): 103–4. The uncontrollability of capital is theorized in István Mészáros, *Beyond Capital* (New York: Monthly Review Press, 1995), 713.
68. Karl Marx, *On the First International*, ed. Saul Padover (New York: McGraw-Hill, 1973), 10.
69. See Foster, *The Return of Nature*, 197–204.
70. Marx and Engels, *Collected Works*, vol. 2, 95–101, 497; vol. 4, 528. Engels's admiration for Shelley led him to attempt to translate *Queen Mab*, along with *The Sensitive Plant*, into German. See John Green, *Engels: A Revolutionary Life* (London: Artery, 2008), 28–29, 59. For a fascinating treatment of Shelley's revolutionary poetry and politics, see Annette Rubinstein, *The Great Tradition in English Literature* (New York: Monthly Review Press, 1953), 516–64.
71. Percy Bysshe Shelley, *The Complete Poetical Works* (Oxford: Oxford University Press, 1914), 528.
72. Ibid., 773. Marx depicted Shelley as "essentially a revolutionist," a view that Engels shared. Edward Aveling and Eleanor Marx Aveling, *Shelley's Socialism* (London: The Journeyman Press, 1975), 4.

14. Late Soviet Ecology

This chapter is adapted for this book from the following article: John Bellamy Foster, "Late Soviet Ecology and the Planetary Crisis," *Monthly Review* 67/2 (June 2015): 1–20.

1. Murray Feshbach and Alfred Friendly Jr., *Ecocide in the USSR* (New York: Basic Books, 1992); D. J. Peterson, *Troubled Lands: The Legacy of Soviet Environmental Destruction* (Boulder, CO: Westview Press, 1993); Stephen Brain, *Song of the Forest: Russian Forestry and Stalinist Environmentalism, 1905–1953* (Pittsburgh: University of Pittsburgh Press, 2011), 2–3; Joan DeBardeleben, *The Environment and Marxism-Leninism* (Boulder: Westview Press, 1985); John Bellamy Foster, *The Vulnerable Planet* (New York: Monthly Review Press, 1994), 96–101.

The use of the term "ecocide" to describe Soviet conditions was heavily colored by the previous widespread international use of the term, beginning in the early 1970s, to criticize the U.S. employment of defoliants such as Agent Orange in its war on Vietnam.
2. Brain, *Song of the Forest*, 116; Laurent Coumel and Marc Elie, "A Belated and Tragic Ecological Revolution: Nature, Disasters, and Green Activists in the Soviet Union and Post-Soviet States, 1960s–2010s," *The Soviet and Post-Soviet Review* 40 (2013): 157–65.
3. Douglas R. Weiner, "Changing Face of Soviet Conservation," in Donald Worster, ed., *The Ends of the Earth* (Cambridge: Cambridge University Press, 1988), 258; Peterson, *Troubled Lands*, 42–44. See also Philip R. Pryde, "The 'Decade of the Environment' in the USSR," *Science* 220 (April 15, 1983): 274–79.
4. M. I. Budyko, G. S. Golitsyn, and V. A. Izrael, *Global Climatic Catastrophes* (New York: Springer-Verlag, 1988), v–vi, 39–46; Vladimir I. Vernadsky, *The Biosphere* (New York: Springer-Verlag, 1998).
5. Douglas R. Weiner, *Models of Nature* (Bloomington: Indiana University Press, 1988), 23; Weiner, "The Changing Face of Soviet Conservation," 252–56.
6. See John Bellamy Foster, *Marx's Ecology* (New York: Monthly Review Press, 2000), 121, 240–44; Roy Medvedev, *Let History Judge: The Origins and Consequences of Stalinism* (New York: Columbia University Press, 1989), 441; Peter Pringle, *The Murder of Nikolai Vavilov* (New York: Simon and Schuster, 2008), 310; Léon Rosenfeld, *Physics, Philosophy, and Politics in the Twentieth Century* (Hackensack, NJ: World Scientific Publishing, 2012), 143; Frank Benjamin Golley, *A History of the Ecosystem Concept* (New Haven: Yale University Press, 1993), 171–73; Kunai Chattopadhyay, "The Rise and Fall of Environmentalism in the Early Soviet Union," *Climate and Capitalism*, November 3, 2014, http://climateandcapitalism.com. Many of the concepts promoted by Lysenko (and by Lysenkoist thinkers in general) were perfectly rational, and even anticipated in some cases future scientific developments. Consequently, Lysenko originally had the support of Vavilov, who helped him gain a position in the scientific establishment. However, Lysenko's research methods were shoddy, if not duplicitous, and his claims regarding his research results were exaggerated. The influence of the Lysenkoism derived from the Soviet Union's pressing need to develop solutions for agriculture, given its climatic difficulties, which led to a tendency to give more credence to such ideas than their results deserved. Worse still, Lysenko and his associates took advantage of their backing from Stalin to level political accusations and promote purges in the scientific community, in violation of all scientific

ethics. On the science see Richard Levins and Richard Lewontin, *The Dialectical Biologist* (Cambridge, MA: Harvard University Press, 1985), 163–96. It should be noted that recent scientific discoveries in the field of epigenetics have shown that inheritance of a number of acquired characteristics is possible, as a result of changes in the coatings on genes. This further suggests that some of the "Lysenkoist" research in the Soviet Union was rational—even if the methods and ethics were not.

7. Weiner, *Models of Nature*, 179, 213–23, 281; Golley, *A History of the Ecosystem Concept*, 172. According to Weiner, "Sukachev had unmistakably declared to the Third All-Union Congress of Botanists that under no circumstances should the introduction of exotic plants or animals into *zapovedniki* be allowed." Loren R. Graham, *What Have We Learned About Science and Technology from the Soviet Experience?* (Stanford: Stanford University Press, 1998), 152. See also Douglas R. Weiner, *A Little Corner of Freedom: Russian Nature Protection from Stalin to Gorbachev* (Berkeley: University of California Press, 1999), 44–52; Kunal Chattopadhyay, "The Rise and Fall of Environmentalism in the Early Soviet Union," *Climate and Capitalism*, November 3, 2014, http://climateandcapitalism.com.

8. Weiner, "The Changing Face of Soviet Conservation," 255–56.

9. Sukachev first used the term geocoenosis in 1941 and then changed it to biogeocoenosis in 1944. Today biogeocoenosis is also often spelled as biogeocenosis. V. N. Sukachev, "Forest Types and their Significance for Forestry," in Institute of Forests, The Academy of Sciences of USSR, ed., *Questions of Forest Sciences* (Moscow: Academy of Sciences of the USSR, 1954), 44–54; V. Sukachev and N. Dylis, *Fundamentals of Forest Biogeocoenology* (London: Oliver and Boyd, 1964), 9. Sukachev and Dylis in recounting the history of the concept of biogeocoenosis refer briefly to the role of Stanchinskii, thereby departing from the common practice in Soviet intellectual circles of remaining silent with regard to the contributions of those who had fallen prey to the purges of the 1930s and '40s.

10. Sukachev and Dylis, *Fundamentals of Forest Biogeocoenology*, 6.

11. Ibid., 26; I. P. Gerasimov, *Geography and Ecology* (Moscow: Progress Publishers, 1983), 64–65; Golley, *A History of the Ecosystem Concept*, 173–74. Golley contends that despite the brilliance of the Vernadsky tradition and Sukachev's biogeocoenosis analysis, Soviet ecologists were so weakened by the purges and the effects of Lysenkoism that they ended up "tending local gardens" and had little effect. It is all the more ironic, then, that the Soviets with their "local gardens" revolutionized climatology and pioneered in the development of global ecology. In

fact, it was the "local gardens," and not global thinking, that ecologists in the USSR were compelled to pursue with caution. Compare Golley's view of the significance of Sukachev's work with that of the molecular geneticist Valery N. Soyfer in *Lysenko and the Tragedy of Soviet Science* (Brunswick, NJ: Rutgers University Press, 1994), who contends that Sukachev "was the first to formulate the goals and tasks of the new discipline dealing with the interrelated and interacting complex of living and inanimate nature—the discipline that has since come to be known as ecosystem ecology" (228).

12. V. N. Sukachev, "Relationship of Biogeocoenosis, Ecosystem, and Facies," *Soviet Soil Scientist* 6 (1960): 580–81.
13. Sukachev, "Relationship of Biogeocoenois, Ecosystem, and Facies," 582–83.
14. Brain, *Song of the Forest*, 139; Stephen Brain, "The Great Stalin Plan for the Transformation of Nature," *Environmental History* 15 (October 2010): 670–700.
15. Brain, *Song of the Forest*, 1–2, 116–17, 139–40, 164–67; David Moon, *The Plough that Broke the Steppes* (Oxford: Oxford University Press, 2013), 292.
16. Weiner, "Changing Face of Soviet Conservation," 257.
17. Brain, *Song of the Forest*, 157–59; David Moon, *The Plough that Broke the Steppes* (Oxford: Oxford University Press, 2013), 292–93.
18. Weiner, *A Little Corner of Freedom*, 205–7, 211–17, 250–52; Weiner, "Changing the Face of Soviet Conservation," 255–56, 260–61; Loren R. Graham, *Science and Philosophy in the Soviet Union* (New York: Alfred A. Knopf, 1993), 239–40, 244; Laurent Coumel, "A Failed Environmental Turn?: Khruschev's Thaw and Nature Protection in Soviet Russia," *Soviet and Post-Soviet Review* 40 (2013): 167–68, 170–71; Alexander Vucinich, *Empire of Reason: The Academy of Science of the USSR, 1914–1970* (Berkeley: University of California Press, 1984), 253–54, 260–62, 337–38, 359, 398; Zhores Medvedev, *Soviet Science* (New York: W.W. Norton, 1978), 89. Sukachev's criticisms of Lysenko and his associates displayed a strong grasp of both Marxian and Darwinian theory. See V. N. Sukachev and N. D. Ivanov, "Toward Problems of the Mutual Relationships of Organisms and the Theory of Natural Selection," *Current Digest of the Russian Press* 7/1 (February 16, 1955): 6–11.
19. M. I. Budyko, S. Lemeshko, and V. G. Yanuta, *The Evolution of the Biosphere* (Boston: D. Reidel Publishing, 1986), x; "Budyko, Michael I," *Encyclopedia of Global Warming and Climate Change*, vol. 1 (Thousand Oaks, CA: Sage Publishing, 2008), 143–44; Spencer Weart, "Interview with M. I. Budkyko: Oral History Transcript," March 25,

1990, http://aip.org; "Blue Planet Prize, The Laureates: Mikhail I. Budyko (1998)," http://af-info.or.jp/en; Sukachev and Dylis, *Fundamentals of Forest Biogeocoenology*, 615–16.
20. Brain, *Song of the Forest*, 170–71.fn29
21. Harry Magdoff and Paul M. Sweezy, "Perestroika and the Future of Socialism—Part One," *Monthly Review* 41/11 (March 1990): 1–13; Harry Magdoff and Paul M. Sweezy, "Perestroika and the Future of Socialism—Part Two," *Monthly Review* 41/12 (April 1990): 1–17.
22. Weiner, "The Changing Face of Soviet Conservation," 257, 264–68; Weiner, *A Little Corner of Freedom*, 368–70; Philip Micklin, "The Aral Sea Disaster," *Annual Review of Earth Planet* 35 (2007): 47–72.
23. Paul Josephson, "War on Nature as Part of the Cold War: The Strategic and Ideological Roots of Environmental Degradation in the Soviet Union," in John Robert McNeil and Corinna R. Unger, eds., *Environmental Histories of the Cold War* (New York: Cambridge University Press, 2010), 43; Joan DeBardeleben, "The New Politics in the USSR: The Case of the Environment," in John Massey Stewart, ed., *The Soviet Environment* (Cambridge: Cambridge University Press, 1992), 64–68; Weiner, "The Changing Face of Soviet Conservation," 258, 267; Coumel and Elie, "A Belated and Tragic Ecological Revolution."
24. E. K. Fedorov, *Man and Nature* (Moscow: Progress Publishers, 1972), 6, 15–19, 57–58, 74–75, 145–47, 173; Federov, "Climate Change and Human Strategy," *Environment* 21/4 (1979): 25–31; Federov, "We Have Only Begun to Mine Our Riches," *Saturday Review*, February 17, 1962, 17–19; Budyko et al., *The Evolution of the Biosphere*, 371; M. I. Budyko and Yu. A. Izrael, *Anthropogenic Climate Change* (Tucson: University of Arizona Press, 1987), xi–xii; Moon, *The Plough that Broke the Steppes*, 293; E. K. Fedorov and I. B. Novick, *Society and Environment: A Soviet View* (Moscow: Progress Publishers, 1977), 43–44; DeBardeleben, *The Environment and Marxism-Leninism*, 201; Barry Commoner, *The Closing Circle* (New York: Bantam, 1971).
25. M. I. Budyko, *Global Ecology* (Moscow: Progress Publishers, 1980), 7–14, 249; Budyko et al., *Evolution of the Biosphere*, ix–x, 163–84, 262–85, 321–30. Dismissingly, Douglas Weiner depicted Budyko as someone who had given up on conservation/ecology for "the theoretical universe of mathematical models." He thus failed to see the broad character and importance of Budyko's work. Weiner, *A Little Corner of Freedom*, 388.
26. Spencer Weart, "Interview with M. I. Budkyko"; Jonathan D. Oldfield, "Climate Modification and Climate Change Debates Among Soviet Physical Geographers, 1940s–1960s," *Advanced Physical Review* 4 (November/December 2013): 513–21; Peter E. Lydoph, "Soviet Work and Writing in Climatology," *Soviet Geography* 12/10 (1971): 637–66;

M. I. Budyko, O. A. Drozozdov, and M. I. Yudin, "The Impact of Economic Activity on Climate," *Soviet Geography* 212 (1971): 666-79; I. P. Gerasimov, *Geography and Ecology* (Moscow: Progress Publishers, 1975), 64-76; Spencer R. Weart, "The Discovery of Global Warming (Bibliography)," http://aip.org; Weart, *The Discovery of Global Warming* (Cambridge, MA: Harvard University Press, 2003), 85-88; M. I. Budyko, "The Effect of Solar Radiation on the Climate of the Earth," *Tellus* 21/5 (October 1969): 611-14; M. I. Budyko, *Climate and Life* (New York: Academic Press, 1974), 493; M. I. Budyko, *Climatic Changes* (Washington, DC: American Geophysical Union, 1977), 219-47; Budyko, *Global Ecology*, 295-304; Budyko, "Polar Ice and Climate," in J. O. Fletcher, B. Keller, and S. M. Olenicoff, *Soviet Data on the Arctic Heat Budget and Its Climatic Influence* (Santa Monica, CA: Rand Corporation, 1966), 9-23; "Budyko, Michael I," *Encyclopedia of Global Warming and Climate Change*, vol. 1, 143-44; Thayer Watkins Department of Economics, San Jose State University, "Mikhail I. Budyko's Ice-Albedo Feedback Model," http://sjsu.edu; James Lawrence Powell, *Four Revolutions in the Earth Sciences* (New York: Columbia University Press, 2015), 258-64.

27. See Ye. K. Fyodorov (E. K. Fedorov) and R. A. Novikov, *Disarmament and Environment* (Moscow: Nauka, 1981).

28. Budyko, Golitsyn, and Izrael, *Global Climatic Catastrophes*, v-vi, 39-46; Budyko, *Climatic Changes*, 241; R. P. Turco and G. S. Golitsyn, "Global Effects of Nuclear War," *Environment* 30/5 (June 1988): 8-16.

29. Budyko, *Global Ecology*, 5-15, 185, 230, 248, 258, 310; Budyko, *Climatic Catastrophes*, 26, 39, 220; Budyko et al., *The Evolution of the Biosphere*, 303-7, 323-96. See also A. I. Oparin, *Life: Its Nature, Origin, and Development* (New York: Academic Press, 1962); M. I. Budyko, A. B. Ronov, and A. L. Yanshin, *History of the Earth's Atmosphere* (New York: Springer-Verlag, 1987), 121-30; Karl Marx and Frederick Engels, *Collected Works* (New York: International Publishers, 1975), vol. 25, 452-64, vol. 42, 558-59. On Engels see Stephen Jay Gould, *An Urchin in the Storm* (New York: W.W. Norton, 1987), 111-12.

30. Budyko, *Global Ecology*, 14-15, 258, 303; Budyko et al., *Evolution of the Biosphere*, xiii, 294, 329-30; Gerasimov, *Geography and Ecology* (Moscow: Progress Publishers, 1983), 53-76; Oldfield, "Climate Modification and Climate Change Debates," 517-18, fn34

31. C. P. Snow, *The Two Cultures* (Cambridge: Cambridge University Press, 1959); Ivan Frolov, *Man, Science, Humanism: A New Synthesis* (Buffalo, NY: Prometheus Books, 1990), 9, 38.

32. I. Frolov, *Global Problems and the Future of Mankind* (Moscow: Progress Publishers, 1982); Frolov, *Man, Science, Humanism*, 19-21,

38, 103, 114–15; "I. T. Frolov," in Stuart Brown, Diane Collinson, and Robert Wilkson, eds., *Biographical Dictionary of Twentieth-Century Philosophers* (London: Routledge, 2002), 257–58; Graham, *Science and Philosophy in the Soviet Union*, 254; Karl Marx, *Early Writings* (London: Penguin, 1974), 329. Weiner deprecatingly treats Frolov and other environmental philosophers and social scientists of this period, despite their "constructive engagements," as opportunists engaged in a "double scoop: professional advancement and maintaining the appearance (not least for themselves) of engagement in 'clean' work." Much the same, though, could be said for many Western thinkers. To attack these Soviet analysts in this way is to deride what were important and theoretically engaged analyses, rooted in the traditions of Marx and Vernadsky. Frolov's role in promoting environmentalism under Gorbachev and his 1982 *Man, Science, Humanism: A New Synthesis* are not easily dismissed. See Weiner, *A Little Corner of Freedom*, 399–401.

33. A. D. Ursul, ed., *Philosophy and the Ecological Problems of Civilisation* (Moscow: Progress Publishers, 1983). Arran Gare referred to this book, but mentions only Ursul (hardly the most important thinker) saying that "some Soviet ideologists such as Ursul attempted to use environmental destruction in the West as an instrument of ideological struggle," dismissing the real ecological concerns. Oddly, the passage from a footnote that Gare cites from Ursul is not to be found in the book itself. Arran Gare, "Soviet Environmentalism: The Path Not Taken," in Ted Benton, ed., *The Greening of Marxism* (New York: Guilford Press, 1996), 111–12.

34. Ursul, ed., *Philosophy and the Ecological Problems of Civilisation*, 37–42, 212, 221, 387–88.

35. Ibid., 41.

36. Ibid., 79–97, 265–68, 369.

37. Gerasimov, *Geography and Ecology*, 26–36; Graham, *Science and Philosophy in the Soviet Union*, 253–56; DeBardeleben, *The Environment and Marxism and Leninism*, 115–16, 127–30, 135; Vucinich, *Empire of Reason*, 362.

38. DeBardeleben, *The Environment and Marxism and Leninism*, 108, 190, 214–15, 234; P. G. Oldak, "Balanced Natural Resource Utiliztion and Economic Growth," *Problems of Economic Transition* 28/ 3 (1985): 4. It is worth noting that Gerasimov's piece from the late 1970s referred to here, in which he pointed to various ecological crises and problems in the USSR, was preceded (in the same essay) by a blank statement that the state and party in the Soviet Union, especially when contrasted with capitalist societies, protected its citizens from environmental hazards. From this it is apparent that he felt he was walking a fine line.

39. Paul M. Sweezy, *Post-Revolutionary Society* (New York: Monthly Review Press, 1980).
40. Douglas R. Weiner, "A Little Reserve Raises Big Questions," *The Open Country* 4 (Summer 2002): 9; Weiner, *A Little Corner of Freedom*, 395; DeBardeleben, "The New Politics in the USSR," 67, fn45, fn46.
41. Coumel and Elie, "A Belated and Tragic Ecological Revolution."
42. DeBardeleben, "The New Politics in the USSR," 67, 73, 78, 80–81, 85; Peterson, *Troubled Lands*, 197.
43. Carbon Dioxide Emissions Analysis Center, "CO_2 Emissions from the USSR," http://cdiac.ornl.gov; U.S. Congress, Office of Technology Assessment, *Change by Degree: Steps in Reducing Greenhouse Gases* (Washington, DC: Government Printing Office, 1991), 295; Peterson, *Troubled Lands*, 49–50; UNCSTADstat, "Real GDP Growth Rates, Total and Per Capita, Annual, 1970–2013, USSR," http://unctadstat.unctad.org. The fact that economic growth continued in these years leads us to conclude that the drop in carbon emissions was primarily a result of the shift to natural gas.
44. Josephson, "War on Nature as Part of the Cold War," 43.
45. See Rachel Carson, *Lost Woods*, 230–31; John Bellamy Foster, *The Ecological Revolution* (New York: Monthly Review Press, 2009), 78–79.
46. Paul M. Sweezy, "Socialism and Ecology," *Monthly Review* 41/4 (September 1989): 7–8.

15. The Return of Nature and Marx's Ecology: Interview by Alejandro Pedregal

This chapter is adapted for this book from the following article: John Bellamy Foster, interviewed by Alejandro Pedregal, "The Return of Nature and Marx's Ecology," *Monthly Review* 72/7 (December 2020): 1–16.

16. Capitalism and Degrowth

This chapter is adapted for this book from the following article: John Bellamy Foster, 'Capitalism and Degrowth—An Impossibility Theorem," *Monthly Review* 62/8 (January 2011): 26–33.

1. James Hansen, *Storms of My Grandchildren* (New York: Bloomsbury, 2009), ix.
2. See Johan Rockström et al., "A Safe Operating Space for Humanity," *Nature* 461 (September 2009): 472–75; John Bellamy Foster, Brett Clark, and Richard York, *The Ecological Rift* (New York: Monthly Review Press, 2010), 13–19.
3. Donella Meadows, Dennis H. Meadows, Jørgen Randers, and William W. Behrens III, *The Limits to Growth: A Report for the Club of Rome's*

Project on the Predicament of Mankind (New York: Universe Books, 1972).
4. "What Is Degrowth?," http://degrowth.eu.
5. See John Bellamy Foster, Brett Clark, and Richard York, "Capitalism and the Curse of Energy Efficiency," *Monthly Review* 62/6 (November 2010): 1–12.
6. Paul M. Sweezy, "Capitalism and the Environment," *Monthly Review* 41/2 (June 1989): 6.
7. Herman E. Daly, *Beyond Growth* (Boston: Beacon Press, 1996), 3–4; John Stuart Mill, *Principles of Political Economy* (New York: Longmans, Green and Co., 1904), 452–55; Lewis Mumford, *The Condition of Man* (New York: Harcourt Brace and Jovanovich, 1973), 411–12; Nicholas Georgescu-Roegen, *The Entropy Law and the Economic Process* (Cambridge, MA: Harvard University Press, 1971).
8. Herman E. Daly, *Steady-State Economics* (Washington, D.C.: Island Press, 1991), 17.
9. Serge Latouche, "The Globe Downshifted," *Le Monde Diplomatique* (English edition), January 13, 2006.
10. Serge Latouche, "Would the West Actually Be Happier with Less?: The World Downscales," *Le Monde Diplomatique* (English edition), December 12, 2003, http://mondediplo.com; and "Can Democracy Solve All Problems?," *International Journal of Inclusive Democracy* 1/3 (May 2005): 5, http://inclusivedemocracy.org.
11. Latouche, "The Globe Downshifted."
12. See, for example, Martin O'Connor, "The Misadventures of Capitalist Nature," in Martin O'Connor, ed., *Is Capitalism Sustainable?* (New York: Guilford Press, 1994), 126–33.
13. Takis Fotopoulos, "Is Degrowth Compatible with a Market Economy?," *International Journal of Inclusive Democracy* 3/1 (January 2007), http://inclusivedemocracy.org.
14. Latouche, "Would the West Actually Be Happier with Less?"
15. "Degrowth Declaration Barcelona 2010," Second International Conference on Economic Degrowth for Ecological Sustainability and Social Equity, Barcelona, March 28–29, 2010, http://degrowth.eu.
16. Joan Martinez-Alier, "Herman Daly Festschrift: Socially Sustainable Economic Degrowth," October 9, 2009, http://eoearth.org.
17. Karl Marx, *Capital*, vol. 1 (London: Penguin, 1976), 742.
18. Serge Latouche, "Degrowth Economics," *Le Monde Diplomatique* (English edition), November 2004.
19. Daly, *Steady-State Economics*, 148–49.
20. World Wildlife Fund, *Living Planet Report*, 2006, http://panda.org.

21. See John Bellamy Foster, *Marx's Ecology* (New York: Monthly Review Press, 2000), 163–70.
22. David Harvey, *The Enigma of Capital* (New York: Oxford University Press, 2010), 228–35.
23. E. F. Schumacher, *Small Is Beautiful* (New York: Harper and Row, 1973), 254–55.
24. On the concept of plenitude, see Juliet Schor, *Plenitude* (New York: Penguin, 2010); Foster, Clark, and York, *The Ecological Rift*, 397–99.

17. The Ecology of Marxian Political Economy

This chapter is adapted for this book from the following article: John Bellamy Foster, "The Ecology of Marxian Political Economy," *Monthly Review* 63/4 (September 2011): 1–16.

1. Paul Hawken, Amory Lovins, and L. Hunter Lovins, *Natural Capitalism* (New York: Little, Brown, 1999); L. Hunter Lovins and Boyd Cohen, *Climate Capitalism* (New York: Hill and Wang, 2011).
2. Marx and Engels, *Collected Works* (New York: International Publishers, 1975), vol. 1, 224–63; Franz Mehring, *Karl Marx* (Ann Arbor: University of Michigan Press, 1979), 41–42.
3. Karl Marx, *Early Writings* (London: Penguin, 1974), 309–22; Karl Marx, *Capital*, vol. 3 (London: Penguin, 1981), 754.
4. James Maitland, Earl of Lauderdale, *An Inquiry into the Nature and Origin of Public Wealth and into the Means and Causes of Its Increase* (Edinburgh: Archibald Constable and Co., 1819), 37–59; *Lauderdale's Notes on Adam Smith*, ed. Chuhei Sugiyama (New York: Routledge, 1996), 140–41.
5. Karl Marx, *The Poverty of Philosophy* (New York: International Publishers, 1964), 35–36.
6. Marx, *Capital*, vol. 3, 180.
7. Karl Marx and Frederick Engels, *Selected Works in One Volume* (New York: International Publishers, 1968), 90; Karl Marx, *Grundrisse* (London: Penguin, 1973), 408; Allan Schnaiberg, *The Environment: From Surplus to Scarcity* (New York: Oxford University Press, 1980), 220–34.
8. Ernest Mandel, *Marxist Economic Theory* (New York: Monthly Review Press, 1968), vol. 1, 295.
9. Karl Marx, *Capital*, vol. 1 (London: Penguin, 1976), 283, 290, 348, 636–39, 860; Marx, *Capital*, vol. 3, 911, 949, 959. On Marx and thermodynamics, see Paul Burkett and John Bellamy Foster, "Metabolism, Energy, and Entropy in Marx's Critique of Political Economy," *Theory and Society* 35/1 (February 2006): 109–56.
10. On Marx's specific ecological insights in these areas see John Bellamy

Foster, *Marx's Ecology* (New York: Monthly Review Press, 2000), 165–66, 169. Engels and Marx addressed the issue of local climate change primarily in relation to changes in temperature and precipitation resulting from deforestation. See Engels's notes on Carl Fraas in Karl Marx and Frederick Engels, *MEGA* IV/31 (Amsterdam: Akadamie Verlag, 1999), 512–15; Paul Hampton, "Classical Marxism and Climate Impacts," *Workers' Liberty*, August 5, 2010, http://workersliberty.org; Clarence J. Glacken, "Changing Ideas of the Habitable World," in Carl O. Sauer, Marston Bates, and William L. Thomas, Jr., eds., *Man's Role in the Changing Face of the Earth* (Chicago: University of Chicago Press, 1956), 77–81.

11. On the relation of Marx's ecology to later scientific developments see John Bellamy Foster, *The Ecological Revolution* (New York: Monthly Review Press, 2009), 153–60. The Liebig-Marx argument on ecological metabolism was influential in Marxian political economic discussions through the end of the nineteenth century—for example, in the work of August Bebel and Karl Kautsky—but it was lost sight of during most of the twentieth century, an exception being K. William Kapp in *The Social Costs of Private Enterprise* (Cambridge, MA: Harvard University Press, 1950), 35–36.

12. Thorstein Veblen, *Absentee Ownership and Business Enterprise in Recent Times* (New York: Augustus M. Kelley, 1964), 127, 168, 171–72, 190.

13. Ibid., 285–88, 299–300.

14. Ibid., 300.

15. Ibid., 300–301.

16. Thorstein Veblen, *The Theory of the Leisure Class* (New York: New American Library, 1953), 78–80; Veblen, *Absentee Ownership*, 309.

17. Marx and Engels, *Collected Works*, vol. 25, 463.

18. Paul M. Sweezy, "Cars and Cities," *Monthly Review* 24/11 (April 1973): 1–3; Paul A. Baran and Paul M. Sweezy, *Monopoly Capital* (New York: Monthly Review Press, 1966), 131–32.

19. Baran and Sweezy, *Monopoly Capital*, 79.

20. On "The Decreasing Rate of Utilization Under Capitalism," see István Mészáros, *Beyond Capital* (New York: Monthly Review Press, 1995), 547–79.

21. Baran and Sweezy, *Monopoly Capital*, 114–15, 128. Baran and Sweezy's concept of economic waste, based on Marx's analysis of unproductive labor, was complex, taking into account (1) waste as perceived from the standpoint of capital in general (but not recognized as such by the individual capitalist); and (2) waste from the standpoint of a rational society, representing the viewpoint of society as a whole (equivalent

to Veblen's definition). For a detailed discussion see John Bellamy Foster, *The Theory of Monopoly Capitalism* (New York: Monthly Review Press, 1986), 97–101.
22. For a thorough analysis of modern marketing see Michael Dawson, *The Consumer Trap* (Urbana: University of Illinois Press, 2003).
23. Metrics 2.0 Business and Market Intelligence, "U.S. Marketing Spending Exceeded $1 Trillion in 2005," June 26, 2006, http://metrics2.com; Dawson, *The Consumer Trap*, 1. The estimate by Blackfriars Communications is clearly a vast underestimate since they are not incorporating the full effects of product management, that is, the penetration of the sales effort into the production process.
24. Baran and Sweezy, *Monopoly Capital*, 131, 137–39. It might be argued that Baran and Sweezy's argument (like Veblen's) was directed at the critique of capitalism from the standpoint of a rational socialist society, in line with what they called "the confrontation of reality with reason" (*Monopoly Capital*, 134) and was not, therefore, an ecological argument per se. Yet, it is precisely this "confrontation of reality with reason" that today unites the arguments for ecology and socialism. See, for example, Paul M. Sweezy, "Capitalism and the Environment," *Monthly Review* 41/2 (June 1989): 1–10.
25. Michael Kidron, *Capitalism and Theory* (London: Pluto Press, 1974), 35–60.
26. Henryk Szlajfer, "Waste, Marxian Theory, and Monopoly Capital," in John Bellamy Foster and Henryk Szlajfer, eds., *The Faltering Economy* (New York: Monthly Review Press, 1984), 302–4, 310–13; John Bellamy Foster, *The Theory of Monopoly Capitalism* (New York: Monthly Review Press, 1986), 39–42.
27. Paul A. Baran, *The Political Economy of Growth* (New York: Monthly Review Press, 1957), 42.
28. Paul M. Sweezy, "Comment," in Assar Lindbeck, *The Political Economy of the New Left* (New York: Harper and Row, 1977), 144–46.
29. John Kenneth Galbraith, *The Affluent Society* (New York: New American Library, 1984), 121–23; Joan Robinson, *Contributions to Modern Economics* (Oxford: Blackwell, 1978), 1–13; Barry Commoner, *The Closing Circle* (New York: Alfred A. Knopf, 1971).
30. This was the famous "IPAT formula": Impact = Population x Affluence x Technology. On the history of the IPAT formula see Marian R. Chertow, "The IPAT Equation and Its Variants: Changing Views of Technology and Environmental Impact," *Journal of Industrial Ecology* 4/4 (October 2000): 13–29.
31. Schnaiberg, *The Environment*, 245–47; John Bellamy Foster, Brett Clark, and Richard York, *The Ecological Rift* (New York: Monthly Review

Press, 2010), 193–206. Schnaiberg's analysis, while drawing heavily on Marxian political economy, never directly addressed the fundamental problem of the interpenetration of the sales effort and production raised by Veblen and Baran and Sweezy. In subsequent work, his model was de-historicized and reduced to a more reified, mechanical form, with the connection to the Marxian theory of monopoly capital, and even the critique of capitalism itself, systematically de-emphasized. Hence, in his last published book—Kenneth A. Gould, David N. Pellow, and Allan Schnaiberg, *The Treadmill of Production* (Boulder, CO: Paradigm Publishers, 2008)—capitalism makes only a cameo appearance. Nevertheless, Schnaiberg never repudiated his earliest views and continued to treat *The Environment* as his classic, fundamental contribution.

32. Paul M. Sweezy, "Socialism and Ecology," *Monthly Review* 41/4 (September 1989): 7.
33. Dawson, *The Consumer Trap*, 88–92; Douglas Dowd, *The Waste of Nations* (Boulder, CO: Westview Press, 1989), 65–66; Stephen Fox, *The Mirror Makers: A History of American Advertising and Its Creators* (New York: William Morrow, 1984), 173; "Car Mileage: 1908 Ford Model T-25 MPG 2008 EPA Average All Cars—21 MPG," http://wanttoknow.info; Research and Innovative Technology Administration, Bureau of Transportation Statistics, Table, 4-23, "Average Fuel Efficiency of U.S. Light Duty Vehicles," www.bts.gov. Baran and Sweezy referred in *Monopoly Capital* to the decline in gas mileage of U.S. automobiles from 1939 to 1961 (136–37).
34. The evolution of bread manufacture under monopoly capitalism, including the changing of wrappings for bread, was used by Baran to explain how the sales effort, waste, and unproductive expenditures are built into the production process of monopoly capital. See Baran, *The Political Economy of Growth*, xx.
35. Susan Freinkel, *Plastics: A Toxic Love Story* (Boston: Houghton Mifflin, 2011), 145–46; Annie Leonard, *The Story of Stuff* (New York: Free Press, 2010), 68–71; Heather Rogers, "Garbage Capitalism's Green Commerce," in Leo Panitch and Colin Leys, eds., *The Socialist Register, 2007* (New York: Monthly Review Press, 2007), 231.
36. Dowd, *The Waste of Nations*, 65.
37. William Morris, *News from Nowhere and Selected Writings and Designs* (London: Penguin, 1962), 121–22.
38. Commoner, *The Closing Circle*, 138–41; see also John Bellamy Foster, *The Vulnerable Planet* (New York: Monthly Review Press, 1994), 112–18.
39. Juliet Schor, *Plenitude* (New York: Penguin, 2010), 27, 40–41. See also

Raymond Williams, *Problems in Materialism and Culture* (London; Verso, 1980), 185.
40. Baran, *The Political Economy of Growth*, xv.

18. On Fire This Time

This chapter is adapted and revised for this book from John Bellamy Foster, "On Fire This Time," *Monthly Review* 71/6 (November 2019): 1–17.

1. The title to this piece is was inspired by James Baldwin, *The Fire Next Time* (New York: Dial, 1963). Revolution is viewed as a complex historical process, encompassing many actors and phases, sometimes nascent, sometimes developed, encompassing a fundamental challenging of the state, along with the property, productive, and class structure of society (which is only possible if numerous other oppressions within and across classes are also addressed). A revolution may involve actors whose intentions are not revolutionary but who are objectively part of the development of a revolutionary situation. For a historical analogue, see George Lefebvre, *The Coming of the French Revolution* (Princeton: Princeton University Press, 1947). On the concept of ecological revolution itself, see John Bellamy Foster, *The Ecological Revolution* (New York: Monthly Review Press, 2009), 11–35.
2. Naomi Klein, *On Fire: The (Burning) Case for a Green New Deal* (New York: Simon and Schuster, 2019).
3. IPCC, *Global Warming of 1.5ºC* (Geneva: IPCC, 2018).
4. John Haltiwanger, "This Is the Platform That Launched Alexandria Ocasio-Cortez, a 29-Year-Old Democratic Socialist, to Become the Youngest Woman Ever Elected to Congress," *Business Insider*, January 4, 2019.
5. Greta Thunberg, *No One Is Too Small to Make a Difference* (London: Penguin, 2019), 19–24.
6. Representative Alexandria Ocasio-Cortez, 116th Congress, 1st Session, House Resolution 109, "Recognizing the Duty of the Federal Government to Create a Green New Deal," subsequently referred to as Green New Deal Resolution, February 7, 2019, available at http://ocasio-cortez.house.gov.
7. Klein, *On Fire*, 1–7.
8. Thunberg, *No One Is Too Small to Make a Difference*, 61.
9. Bernie Sanders, "The Green New Deal," April 22, 2019, available at http://berniesanders.com.
10. Res. 109, "Recognizing the Duty of the Federal Government to Create a Green New Deal," list of co-sponsors available at http://congress.gov; S. Res. 59, "Recognizing the Duty of the Federal Government to Create a Green New Deal," list of cosponsors available at http://congress.gov.

11. Eliza Barclay and Brian Resnick, "How Big Was the Global Climate Strike? 4 Million People, Activists Estimate," *Vox*, September 20, 2019.
12. "Transcript: Greta Thunberg's Speech to UN Climate Action Summit," NPR, September 23, 2019.
13. IPCC, *Special Report on the Ocean and Cryosphere in a Changing Climate*, Summary for Policymakers (Geneva: IPCC, 2019), 22–24, 33.
14. Nicholas Stern, "We Must Reduce Greenhous Gas Emissions to Net Zero or Face More Floods," *Guardian*, October 7, 2018; "Transcript: Greta Thunberg's Speech to UN Climate Action Summit." Usually it is assumed that the world must stay below 2°C in order to avoid a point of no return with respect to the human relation to the planet. But more and more science has pointed to 1.5°C as the marker. Most IPCC-recognized climate mitigation schemes today assume a temporary overshooting of the 1.5°C boundary (or else the 2°C boundary) with negative emissions, then removing carbon from the atmosphere before the worst effects have occurred. But such a strategy, it is increasingly recognized, is worse than Russian roulette in terms of statistical odds—and more filled with illusions.
15. See http://systemchangenotclimatechange.org. See also Martin Empson, ed., *System Change Not Climate Change* (London: Bookmarks, 2019).
16. On the distinction between climate action and climate justice, see Klein, *On Fire*, 27–28.
17. The climate march was followed a few days later by the Flood Wall Street action, in which protesters engaged in civil disobedience but lacked the force of numbers.
18. Klein, *On Fire*, 27–28.
19. Thunberg, *No One Is Too Small to Make a Difference*, 16.
20. Green Party US, Green New Deal Timeline, available at http://gp.org; Green New Deal Policy Group, *A Green New Deal* (London: New Economics Foundation, 2008); Larry Elliott, "Climate Change Cannot Be Bargained With," *Guardian*, October 29, 2007.
21. Thomas Friedman, "A Warning from the Garden," *New York Times*, January 19, 2007.
22. Alexander C. Kaufman, "What's the 'Green New Deal'?," Grist, June 30, 2018, www.grist.org.
23. UNEP, *Global Green New Deal* (Geneva: UNEP, 2009).
24. Green European Foundation, *A Green New Deal for Europe* (Brussels: Green European Foundation, 2009).
25. David Milton, *The Politics of U.S. Labor* (New York: Monthly Review Press, 1982).
26. Climate Justice Alliance, "History of the Climate Justice Alliance," https://climatejusticealliance.org/cja-history/.

27. John Bellamy Foster, "Ecosocialism and a Just Transition," *MR Online*, June 22, 2019; Climate Justice Alliance, "Just Transition: A Framework for Change," https://climatejusticealliance.org.
28. Science for the People has been a leading defender of A People's Green New Deal, incorporating a just transition for workers and frontline communities as opposed to attempts to fold the Green New Deal into its previous corporatist form. See Science for the People, "People's Green New Deal," https://scienceforthepeople.org/peoples-green-new-deal/.
29. Jill Stein, "Solutions for a Country in Trouble: The Four Pillars of the Green New Deal," *Green Pages*, September 25, 2012.
30. Green Party, "We Can Build a Better Tomorrow Today, It's Time for a Green New Deal."
31. Tessa Stuart, "Sunrise Movement, the Force Behind the Green New Deal Ramps Up Plans for 2020," *Rolling Stone*, May 1, 2019. The founding Sunrise Movement activists had cut their teeth on the fossil-fuel disinvestment movement particularly in universities, which as of December 2018 claims to have succeeded in facilitating $8 trillion in disinvestments. However, activists realized that the next step was to try to tackle the state itself and to change the system through a Green New Deal. Klein, *On Fire*, 22.
32. The Green Party has moved explicitly in the direction of ecosocialism and sponsored an ecosocialism conference in Chicago on September 28, 2019. See Anita Rios, "Green Party Gears Up for Ecosocialism Conference," *Black Agenda Report*, September 10, 2019.
33. Res. 109, "Recognizing the Duty of the Federal Government to Create a Green New Deal."
34. Sanders was entirely alone among the leading Democratic candidates in the 2020 elections in promoting a genuine Green New Deal. Joe Biden's "Plan for a Clean Energy Revolution and Environmental Justice," introduced in June 2019, avoided altogether the IPCC's insistence that carbon dioxide emissions have to be reduced by nearly 50 percent by 2030 in order to stay below 1.5ºC and simply promised to promote policies that would achieve net-zero emissions by 2050, proposing to spend $1.7 trillion on combating climate change over ten years. Elizabeth Warren signed on to the Green New Deal Resolution, but her "Clean Energy Plan," introduced in September 2019, did not go beyond saying that she supported a ten-year mobilization through 2030 with the aim of reaching net-zero greenhouse gas emissions "as soon as possible." She proposed a $3 trillion investment over ten years. Her plan excluded any mention of a just transition for workers or frontline communities.

35. Sanders, "The Green New Deal."
36. While the Green New Deal Resolution introduced by Ocasio-Cortez and Markey does not address how it would be financed, the emphasis has been on the creation of public banks, green quantitative easing, and deficit financing under current low-capacity utilization—a view supported by modern monetary theory. It deliberately swerves away from funding by taxes on corporations. Ellen Brown, "The Secret to Funding a Green New Deal," *Truthdig.com*, March 19, 2019.
37. Historian David Blight, quoted in Ta-Nehisi Coates, "Slavery Made America," *Atlantic*, June 24, 2014.
38. Ben Caldecott et al., *Stranded Assets: A Climate Risk Challenge* (Washington, DC: Inter-American Development Bank, 2016), x.
39. Naomi Klein, *This Changes Everything: Capitalism vs. the Climate* (New York: Simon and Schuster, 2014), 31–63.
40. Klein, *On Fire*, 261; J. F. Mercure et al., "Macroeconomic Impact of Stranded Fossil Fuel Assets," *Nature Climate Change* 8 (2018): 588–93.
41. Klein, *This Changes Everything*, 115–16.
42. Nancy E. Rose, *Put to Work* (New York: Monthly Review Press, 2009).
43. Klein, *On Fire*, 264.
44. Kevin Anderson, "Debating the Bedrock of Climate-Change Mitigation Scenarios," *Nature*, September 16, 2019; Zeke Hausfather, "Explainer: How 'Shared Socioeconomic Pathways' Explore Future Climate Change," *Carbon Brief*, April 19, 2018.
45. These shortcomings are built directly into the SSPs and even into the IAMs. See Oliver Fricko et al., "The Marker Quantification of the Shared Socioeconomic Pathway 2: A Middle-of-the-Road Scenario for the 21st Century," *Global Environmental Change* 42 (2017): 251–67. For a general critical evaluation, see Jason Hickel and Giorgos Kallis, "Is Green Growth Possible?" *New Political Economy*, April 17, 2019.
46. Kevin Anderson and Glen Peters, "The Trouble with Negative Emissions," *Science* 354/6309 (2016): 182–83; European Academies Science Advisory Council, *Negative Emission Technologies: What Role in Meeting Paris Agreement Targets*, EASAC Policy Report 35 (Halle: German National Academy of Sciences, 2018).
47. See John Bellamy Foster, "Making War on the Planet," in chapter 6 of this book.
48. Anderson, "Debating the Bedrock of Climate-Change Mitigation Scenarios."
49. IPCC, *Global Warming of 1.5°C*, 16, 96.
50. Anderson "Debating the Bedrock of Climate-Change Mitigation Scenarios."

51. See John Bellamy Foster, Brett Clark, and Richard York, *The Ecological Rift* (New York: Monthly Review Press, 2010), 169–82.
52. IPCC, *Global Warming of 1.5°C*, 15–16, 97; Jason Hickel, "The Hope at the Heart of the Apocalyptic Climate Change Report," *Foreign Policy*, October 18, 2018. See also Arnulf Grubler, "A Low Energy Demand Scenario for Meeting the 1.5°C Target and Sustainable Development Goals Without Negative Emission Technologies," *Nature Energy* 3/6 (2018): 512–27; Joeri Rogelj et al., "Scenarios Towards Limiting Global Mean Temperature Increase Below 1.5°C," *Nature Climate Change* 8 (2018): 325–32; Christopher Bertram et al., "Targeted Policies Can Compensate Most of the Increased Sustainability Risks in 1.5°C Mitigation Scenarios," *Environmental Research Letters* 13/6 (2018).
53. Hickel and Kallis, "Is Green Growth Possible?"
54. J. D. Bernal, *The Freedom of Necessity* (London: Routledge and Kegan Paul, 1949).
55. See John Bellamy Foster, "Ecology," in *The Marx Revival*, ed. Marcelo Musto (Cambridge: Cambridge University Press, 2000), 193.
56. Klein, *On Fire*, 264.
57. See Paul A. Baran and Paul M. Sweezy, *Monopoly Capital* (New York: Monthly Review Press, 1966).
58. John Bellamy Foster, "The Ecology of Marxian Political Economy," *Monthly Review* 63/4 (September 2011): 1–16; Fred Magdoff and John Bellamy Foster, *What Every Environmentalist Needs to Know About Capitalism* (New York: Monthly Review Press, 2011), 123–44; William Morris, *News from Nowhere and Selected Writings and Designs* (London: Penguin, 1962), 121–22.
59. Kevin Anderson and Alice Bows, "Beyond 'Dangerous' Climate Change: Emission Scenarios for a New World," *Philosophical Transactions of the Royal Society* 369 (2011): 20–44.
60. For a discussion of the current ecological situation in the Global South and its relation to imperialism, see John Bellamy Foster, Hannah Holleman, and Brett Clark, "Imperialism in the Anthropocene," *Monthly Review* 71/3 (July–August 2019): 70–88. On the concept of the environmental proletariat, see Bellamy Foster, Clark, and York, *The Ecological Rift*, 440–41.
61. The issue of China and ecology is complex. See John B. Cobb, in conversation with Andre Vltchek, *China and Ecological Civilization* (Jakarta: Badak Merah, 2019); Barbara Finamore, *Will China Save the Planet?* (Cambridge: Polity, 2018); Jiahua Pan, *China's Environmental Governing and Ecological Civilization* (Berlin: Springer-Verlag, 2014); David Schwartzman, "China and the Prospects for a Global Ecological Civilization," *Climate and Capitalism*, September

17, 2019; Lau Kin Chi, "A Subaltern Perspective on China's Ecological Crisis," *Monthly Review* 70/5 (October 2018): 45–57.
62. Naomi Klein, "Only a Green New Deal Can Douse the Fires of Ecofascism," *Intercept*, September 16, 2019.
63. Roy Bhaskar, *Reclaiming Reality* (London: Routledge, 2011), 74–76.
64. Bhaskar, *Reclaiming Reality*, 76–77, 92–94.
65. Karl Marx, *Eighteenth Brumaire of Louis Bonaparte* (1852; repr. New York: International Publishers, 1963), 15.
66. Karl Marx, *Capital*, vol. 1 (London: Penguin, 1976), 283.
67. See Ian Angus, *Facing the Anthropocene* (New York: Monthly Review Press, 2016), 175–91.
68. Klein, *On Fire*, 90–91; Karl Marx, *Capital*, vol. 3 (London: Penguin, 1981), 949.
69. William James, "Proposing the Moral Equivalent of War," speech delivered at Stanford University, 1906, available at *Lapham's Quarterly* online, https://www.laphamsquarterly.org/states-war/proposing-moral-equivalent-war.

19. COVID-19 and Catastrophe Capitalism

This chapter is adapted and revised for this book from the following article: John Bellamy Foster and Intan Suwandi, "COVID-19 and Catastrophe Capitalism," *Monthly Review* 72/2 (June 2020): 1–20.

1. See John Bellamy Foster, "Late Imperialism," *Monthly Review* 71/3 (July–August 2019): 1–19; Samir Amin, *Modern Imperialism, Monopoly Finance Capital, and Marx's Law of Value* (New York: Monthly Review Press, 2018).
2. On the global labor arbitrage and commodity chains, see Intan Suwandi, *Value Chains* (New York: Monthly Review Press, 2019), 32–33, 53–54. Our statistical analysis of unit labor costs was done collaboratively with R. Jamil Jonna, also published as "Global Commodity Chains and the New Imperialism," *Monthly Review* 70/10 (March 2019): 1–24. On the global land arbitrage, see Eric Holt-Giménez, *A Foodie's Guide to Capitalism* (New York: Monthly Review Press, 2017), 102–4.
3. Evan Tarver, "Value Chain vs. Supply Chain," *Investopedia*, March 24, 2020.
4. Karl Marx, "The Value Form," *Capital and Class* 2/1 (1978): 134; Karl Marx and Frederick Engels, *Collected Works*, vol. 36 (New York: International Publishers, 1996), 63. See also Karl Marx, *Capital*, vol. 1 (London: Penguin, 1976), 156, 215; Marx, *Capital*, vol. 2 (London: Penguin, 1978), 136–37.
5. Rudolf Hilferding, *Finance Capital* (London: Routledge, 1981), 60.

6. Terence Hopkins and Immanuel Wallerstein, "Commodity Chains in the World Economy Prior to 1800," *Review* 10/1 (1986): 157–70.
7. Marx, *Capital*, vol. 1, 638.
8. Karl Marx, *Capital*, vol. 3 (London: Penguin, 1981), 949–50; Marx, *Capital*, vol. 1, 348–49.
9. Robert G. Wallace, Luke Bergmann, Richard Kock, Marius Gilbert, Lenny Hogerwerf, Rodrick Wallace, and Mollie Holmberg, "The Dawn of Structural One Health: A New Science Tracking Disease Emergence Along Circuits of Capital," *Social Science and Medicine* 129 (2015): 68–77; Rob [Robert G.] Wallace, "We Need a Structural One Health," Farming Pathogens, August 3, 2012,, https://farmingpathogens.wordpress.com/; J. Zinsstag, "Convergence of EcoHealth and One Health," *Ecohealth* 9/4 (2012): 371–73; Victor Galaz, Melissa Leach, Ian Scoones, and Christian Stein, "The Political Economy of One Health," STEPS Centre, Political Economy of Knowledge and Policy Working Paper Series (2015), http://steps-centre.org/wp-content/uploads/One-Health.pdf.
10. Rodrick Wallace, Luis Fernando Chaves, Luke R. Bergmann, Constância Ayres, Lenny Hogerwerf, Richard Kock, and Robert G. Wallace, *Clear-Cutting Disease Control: Capital-Led Deforestation, Public Health Austerity, and Vector-Borne Infection* (Cham, Switzerland: Springer, 2018), 2.
11. Wallace et al., "The Dawn of Structural One Health," 70–72; Wallace, "We Need a Structural One Health"; Rob Wallace, Alex Liebman, Luis Fernando Chaves, and Rodrick Wallace, "COVID-19 and Circuits of Capital," *Monthly Review* 72/1 (May 2020): 12; István Mészáros, *Beyond Capital* (New York: Monthly Review Press, 1995); Richard Levins and Richard Lewontin, *The Dialectical Biologist* (Cambridge, MA: Harvard University Press, 1985).
12. Rob Wallace, *Big Farms Make Big Flu* (New York: Monthly Review Press, 2016), 60–61, 118, 120–21, 217–19, 236, 332; Rob Wallace, "Notes on a Novel Coronavirus," MR Online, January 29, 2020. On the Lauderdale Paradox, see John Bellamy Foster, Brett Clark, and Richard York, *The Ecological Rift* (New York: Monthly Review Press, 2010), 53–72.
13. See John Bellamy Foster, *The Return of Nature* (New York: Monthly Review Press, 2020), 61–64, 172–204; Frederick Engels, *The Condition of the Working Class in England* (Chicago: Academy Chicago, 1984); E. Ray Lankester, *The Kingdom of Man* (New York: Henry Holt, 1911), 31–33, 159–91; Richard Levins, "Is Capitalism a Disease?," *Monthly Review* 52/4 (September 2000): 8–33. See also Howard Waitzkin, *The Second Sickness* (New York: Free Press, 1983).

14. Wallace, *Big Farms Make Big Flu*, 53.
15. Ibid., 49; Worldometer, Coronavirus Cases, accessed December 2021.
16. Ibid., 33–34.
17. Ibid., 81.
18. Mathilde Paul, Virginie Baritaux, Sirichai Wongnarkpet, Chaitep Poolkhet, Weerapong Thanapongtharm, François Roger, Pascal Bonnet, and Christian Ducrot, "Practices Associated with Highly Pathogenic Avian Influenza Spread in Traditional Poultry Marketing Chains," *Acta Tropica* 126 (2013): 43–53.
19. Wallace, *Big Farms Make Big Flu*, 306; Wallace et al., "The Dawn of Structural One Health," 69, 71, 73.
20. Wallace et al., "COVID-19 and Circuits of Capital," 11.
21. Holt-Giménez, *A Foodie's Guide to Capitalism*, 102–5.
22. Philip McMichael, "Feeding the World," in *Socialist Register 2007: Coming to Terms with Nature*, ed. Leo Panitch and Colin Leys (New York: Monthly Review Press, 2007), 180.
23. Farshad Araghi, "The Great Global Enclosure of Our Times," in *Hungry for Profit*, ed. Fred Magdoff, John Bellamy Foster, and Fredrick H. Buttel (New York: Monthly Review Press, 2000), 145–60.
24. Wallace et al., "COVID-19 and Circuits of Capital," 6; Mike Davis, *Planet of Slums* (London: Verso, 2016); Mike Davis interviewed by Mada Masr, "Mike Davis on Pandemics, Super-Capitalism, and the Struggles of Tomorrow," *Mada Masr*, March 30, 2020, madamasr.com/en.
25. Wallace, *Big Farms Make Big Flu*, 61. On the significance of the concepts of the residual and residues for dialectics, see J. D. Bernal, "Dialectical Materialism," in *Aspects of Dialectical Materialism*, ed. Hyman Levy et al. (London: Watts and Co., 1934), 103–4; Henri Lefebvre, *Metaphilosophy* (London: Verso, 2016), 299–300; Raymond Williams, *Marxism and Literature* (Oxford: Oxford University Press, 1977), 121–27.
26. Karl Marx and Frederick Engels, *Collected Works*, vol. 25 (New York: International Publishers, 1975), 460–61; Lankester, *The Kingdom of Man*, 159.
27. Matt Leonard, "What Procurement Managers Should Expect from a Bullwhip on Crack," *Supply Chain Dive*, March 26, 2020, www.supplychaindive.com.
28. On time-based competition and just-in-time production, see "What Is Time-Based Competition?," Boston Consulting Group, https://www.bcg.com/about/overview/our-history/time-based-competition.
29. Suwandi, *Value Chains*, 59–61; John Smith, *Imperialism in the Twenty-First Century* (New York: Monthly Review Press, 2016).
30. Walden Bello, "Coronavirus and the Death of 'Connectivity,'" *Foreign*

Policy in Focus, March 22, 2010, https://fpif.org/coronavirus-and-the-death-of-connectivity/; "Annual Growth in Global Air Traffic Passenger Demand from 2006 to 2020," *Statista*, https://www.statista.com/.
31. Shannon K. O'Neil, "How to Pandemic Proof Globalization," *Foreign Affairs*, April 1, 2020, https://www.foreignaffairs.com
32. Stefano Feltri, "Why Coronavirus Triggered the First Global Supply Chain Crisis," *Pro-Market*, March 5, 2020, https://promarket.org
33. Elisabeth Braw, "Blindsided on the Supply Side," *Foreign Policy*, March 4, 2020.
34. Francisco Betti and Per Kristian Hong, "Coronavirus Is Disrupting Global Value Chains. Here's How Companies Can Respond," World Economic Forum, February 27, 2020; Braw, "Blindsided on the Supply Side," https://www.weforum.org/agenda/.
35. Braw, "Blindsided on the Supply Side"; Thomas Y. Choi, Dale Rogers, and Bindiya Vakil, "Coronavirus is a Wake-Up Call for Supply Chain Management," *Harvard Business Review*, March 27, 2020, https://hbr.org/.
36. "Nearly 3 Billion People Around the Globe Under COVID-19 Lockdowns," World Economic Forum, March 26, 2020, https://www.weforum.org/.
37. Lizzie O'Leary, "The Modern Supply Chain Is Snapping," *Atlantic*, March 19, 2020, https://www.theatlantic.com/.
38. Choi et al., "Coronavirus is a Wake-Up Call for Supply Chain Management"; Willy Shih, "COVID-19 and Global Supply Chains: Watch Out for Bullwhip Effects," *Forbes*, February 21, 2020, forbes.com.
39. "Estimated March Imports Hit Five-Year Low, Declines Expected to Continue Amid Pandemic," National Retail Federation, April 7, 2020, nrf.com.
40. Emma Cosgrove, "FAA Offers Safety Guidance for Passenger Planes Ferrying Cargo," *Supply Chain Dive*, April 17, 2020.
41. Organization of Economic Cooperation and Development, "International Trade in 2020: A Look Back and a Look Ahead," *International Trade Pulse*, March 2021, https://www.oecd.org/sdd/its/OECD-International-Trade-Pulse-2020-2021.pdf./; World Economic Forum, "This Is How COVID-10 Has Affected World Trade," November 5, 2021, https://www.weforum.org/agenda/2021/11/what-happened-to-world-trade-under-covid-19/.
42. Deborah Abrams Kaplan, "Why Supply Chain Data Is King in the Coronavirus Pandemic," *Supply Chain Dive*, April 7, 2020; O'Leary, "The Modern Supply Chain Is Snapping," www.suppychaindive.com; Chad P. Bown, "COVID-19: Trump's Curbs on Exports of Medical Gear Put Americans and Others at Risk," Peterson Institute for International Economics, April 9, 2020, www.piie.com; Shefali Kapadia, "From

Section 301 to COVID-19," *Supply Chain Dive*, March 31, 2020, www.supplychaindive.com.

43. Finbarr Bermingham, Sidney Leng, and Echo Xie, "China Ramps Up COVID-19 Test Kit Exports Amid Global Shortage, as Domestic Demand Dries Up," *South China Morning Post*, March 30, 2020.

44. "Supply-Chain Finance Is New Risk in Crisis," *Wall Street Journal*, April 4, 2020; "CNE/CIS Trade Finance Survey 2017," *BNE Intellinews*, April 3, 2017, www.intellinews.com/.

45. Stephen Roach, "This Is Not the Usual Buy-on-Dips Market," *Economic Times*, March 18, 2020.

46. COVID-19 Response Team, Imperial College, *Report 12: The Global Impact of COVID-19 and Strategies for Mitigation and Suppression* (London: Imperial College, 2020), 3–4, 11.

47. Ahmed Mushfiq Mobarak and Zachary Barnett-Howell, "Poor Countries Need to Think Twice About Social Distancing," *Foreign Policy*, April 10, 2020, foreignpolicy.com; Zachary Barnett-Howell and Ahmed Mushfiq Mobarak, "The Benefits and Costs of Social Distancing in Rich and Poor Countries," ArXiv, April 10, 2020, https://arxiv.org/pdf/2004.04867.pdf.

48. Davis, "Mike Davis on Pandemics, Super-Capitalism, and the Struggles of Tomorrow."

49. "President Maduro: Venezuela Faces the COVID-19 with Voluntary Quarantine without Curfew or State of Exception," *Orinoco Tribune*, April 18, 2020; Frederico Fuentes, "Venezuela: Community Organization Key to Fighting COVID-19," *Green Left*, April 9, 2020, www.greenleft.org.

50. "Analysis: The Pandemic Is Ravaging the World's Poor Even If They Are Untouched by the Virus," *Washington Post*, April 15, 2020; Matt Leonard, "India, Bangladesh Close Factories Amid Coronavirus Lockdown," *Supply Chain Dive*, March 26, 2020, www.supplychaindive.com; Finbarr Bermingham, "Global Trade Braces for 'Tidal Wave' Ahead, as Shutdown Batters Supply Chains," *South China Morning Post*, April 3, 2020; I. P. Singh, "Punjab: 'No Orders, No Raw Material,'" *Times of India*, April 1, 2020.

51. Roach, "This Is Not the Usual Buy-On-Dips Market."

52. Kapadia, "From Section 301 to COVID-19."

53. Bown, "COVID-19: Trump's Curbs on Exports of Medical Gear."

54. David Ruccio, "The China Syndrome," *Occasional Links and Commentary*, April 14, 2020, https://anticap.wordpress.com/2020/04/14/china-syndrome-2/; Alan Rappeport, "Navarro Calls Medical Experts 'Tone Deaf' Over Coronavirus Shutdown," *New York Times*, April 13, 2020; John Bellamy Foster, *Trump in the White House* (New York: Monthly Review Press, 2017), 84–85.

NOTES TO PAGES 427–431   611

55. Cary Huang, "Is the Coronavirus Fatal for Economic Globalisation?," *South China Morning Post*, March 15, 2020; Frank Tang, "American Factory Boss Says Pandemic Will Change China's Role in Global Supply Chain," *South China Morning Post*, April 15, 2020.
56. John Reed and Song Jung-a, "Samsung Flies Phone Parts to Vietnam After Coronavirus Hits Supply Chains," *Financial Times*, February 16, 2020; Finbarr Bermingham, "Vietnam Lured Factories During Trade War, but Now Faces Big Hit as Parts from China Stop Flowing," *South China Morning Post*, February 28, 2020.
57. Fadli, "Batam Factories at Risk as Coronavirus Outbreak Stops Shipments of Raw Materials from China," *Jakarta Post*, February 18, 2020; "Covid-19: Indonesia Waives Income Tax for Manufacturing Workers for Six Months," *Star*, March 16, 2020.
58. Tang, "American Factory Boss Says Pandemic Will Change China's Role in Global Supply Chain."
59. Christopher C. Krebs, "Advisory Memorandum on Identification of Essential Critical Infrastructure Workers," U.S. Department of Homeland Security, March 28, 2020, www.cisa.gov.
60. Lauren Chambers, "Data Show that COVID-19 Is Hitting Essential Workers and People of Color Hardest," Data for Justice Project, American Civil Liberties Union, April 7, 2020, data.aclum.org.
61. Karl Marx, *Texts on Method* (Oxford: Basil Blackwell, 1975), 195.
62. Frederick Engels, "The Funeral of Karl Marx," in *Karl Marx Remembered*, ed. Philip S. Foner (San Francisco: Synthesis, 1983), 39.
63. Guy Standing, *Plunder of the Commons: A Manifesto for Sharing Public Health* (London: Pelican, 2019), 49; John Bellamy Foster and Brett Clark, *The Robbery of Nature* (New York: Monthly Review Press, 2020), 167–72.
64. John Bellamy Foster and Robert W. McChesney, *The Endless Crisis* (New York: Monthly Review Press, 2012).
65. "It's Now 100 Seconds to Midnight," *Bulletin of Atomic Scientists*, January 23, 2020, https://thebulletin.org/2020/01/press-release-it-is-now-100-seconds-to-midnight.
66. "Microbial Resistance a Global Health Emergency," *UN News*, November 12, 2018; Ian Angus, "Superbugs in the Anthropocene," *Monthly Review* 71/2 (June 2019).
67. Karl Marx and Frederick Engels, *The Communist Manifesto* (New York: Monthly Review Press, 1964), 2.
68. Karl Marx, *Capital*, vol. 3, 949.
69. Karl Marx and Frederick Engels, *Collected Works*, vol. 1 (New York: International Publishers, 1975), 173; Wallace et al., "COVID-19 and Circuits of Capital."

20. Ecological Catastrophe or Ecological Civilization

This chapter is adapted and revised for this book from the following article: John Bellamy Foster, "The Earth System Emergency and Ecological Civilization: A Marxian View," *International Critical Thought* 7/4 (2017): 439–58.

1. Civilization is often taken to mean an advanced, ordered society. The historical meaning of civilization, as understood within socialist thought, however, is much more complex and will be addressed in the following analysis.
2. See James E. Hansen, *Storms of My Grandchildren* (New York: Bloomsbury, 2009); Ian Angus, *Facing the Anthropocene* (New York: Monthly Review Press, 2016); Elizabeth Kolbert, *The Sixth Extinction* (New York: Henry Holt, 2014).
3. In referring to an Earth System crisis (or planetary emergency) the intent of course is to refer to a crisis of society (and to some extent life as it now exists) arising from the anthropogenic rift in the Earth System, rather than literally a crisis of the Earth System itself, which naturally supersedes society.
4. Johan Rockström, Will Steffen, Kevin Noone, Asa Persson, F. Stuart Chapin, Eric F. Lambin, Timothy M. Lenton et al., "A Safe Operating Space for Humanity," *Nature* 461 (24) (2009): 472–75; Clive Hamilton, and Jacques Grinevald, "Was the Anthropocene Anticipated?," *Anthropocene Review* 2/1 (2015): 67. See also Mckenzie Wark. *Molecular Red: Theory for the Anthropocene* (London: Verso, 2015). Wark has described the coming of the Anthropocene epoch as associated with a "series of metabolic rifts" (xiv).
5. Given prevailing realities in the capitalist world, an ecological revolution would need to occur in two phases: (1) an *ecodemocratic phase* based on a broad popular alliance, aimed particularly at energy transformation (though taking on other issues as well); and (2) an *ecosocialist phase* aimed at the formation of an ecological civilization, or a far-reaching transition to a socialist ecological formation. Fred Magdoff and John Bellamy Foster, *What Every Environmentalist Needs to Know About Capitalism* (New York: Monthly Review Press, 2011), 124–44; John Bellamy Foster, "Marxism and Ecology: Common Fonts of a Great Transition." *Monthly Review* 67/7 (2015): 10–12, italics in the original.
6. Ian Burton and Robert W. Kates, "The Great Climacteric, 1798–2048: The Transition to a Just and Sustainable Human Environment," in *Geography, Resources and Environment,* ed. R. W. Kates and I. Burton, vol. 2 (Chicago: University of Chicago Press, 1986), 339–60.
7. John Bellamy Foster, "The Great Capitalist Climacteric: Marxism and 'System Change Not Climate Change,'" *Monthly Review* 67/6 (2015): 1–18.

8. Burton and Kates, "The Great Climacteric, 1798–2048," 339.
9. John R. McNeill and Peter Engelke, *The Great Acceleration* (Cambridge, MA: Harvard University Press, 2014); Colin N. Waters, Jan Zalasiewicz, Colin Summerhayes, Anthony D. Barnosky, Clément Poirier, Agnieszka Galuszka, Alejandro Cearreta et al., "The Anthropocene Is Functionally and Stratigraphically Distinct from the Holocene," *Science* 351/6269 (2016): 137–47.
10. Paul Burkett, "Marx's Vision of Sustainable Human Development," *Monthly Review* 57/5 (2005): 34–62.
11. John Bellamy Foster, *Ecology Against Capitalism* (New York: Monthly Review Press, 2002), 73; McNeill and Engelke, *The Great Acceleration*, 4; Angus, *Facing the Anthropocene*, 149–51.
12. Karl Marx and Frederick Engels, *Marx/Engels Collected Works*, vol. 30 (New York: International Publishers, 1975), 54–66; Karl Marx, *Capital*, vol. 1 (London: Penguin, 1976), 949; John Bellamy Foster, "Marx and the Rift in the Universal Metabolism of Nature," *Monthly Review* 65/7 (2013): 1–19.
13. Paul M. Sweezy, "Capitalism and the Environment," *Monthly Review* 41/2 (1989): 6.
14. Paul M. Sweezy, *Post-Revolutionary Society* (New York: Monthly Review Press, 1980), 139–51.
15. Sweezy's outlook in 1989, in the context of capitalist economic crisis, growing global ecological degradation, and the destabilization of post-revolutionary societies, was fairly grim. In this situation of increasing planetary environmental peril, he wrote, "the prospect of an indefinite continuation of capitalism—a capitalism in crisis to boot—is truly terrifying. Civilization as we know it cannot survive even what a short while ago would have been considered historically a brief span of time. Socialism, if it misses out this first time, will likely never get a second chance." Paul M. Sweezy. "Socialism and Ecology," *Monthly Review* 41/4 (1989): 6.
16. William Hardy McNeill, *The Global Condition* (Princeton: Princeton University Press, 1992); John Bellamy Foster, "Capitalism and the Accumulation of Catastrophe," *Monthly Review* 63/7 (2011): 1–17.
17. Kevin Anderson, "Climate Change Going Beyond Dangerous—Brutal Numbers and Tenuous Hope," *Development Dialogue* (September2012): 16–40, http://www.whatnext.org/resources/Publications/Volume-III/Single-articles/wnv3_andersson_144.pdf, 29; italics added.
18. Raymond Williams, *Keywords* (Oxford: Oxford University Press, 1983); Fernand Braudel, *A History of Civilizations* (London: Penguin Press, 1994).
19. Strabo, *Geography*, vol. 2 (Cambridge, MA: Harvard University Press, 1923), 290.

20. Braudel, *A History of Civilizations*, 7–8; see also Gordon Childe. *What Happened in History* (London: Penguin Books, 1954), 30–31.
21. These historical distinctions with respect to the concept of civilization are not meant to set aside—especially in the historical-materialist view—the reality that some pre-capitalist societies lacking all of the above characteristics, such as the traditional Iroquois culture with its advanced form of government, were in some respects more cohesive and cultured —less "barbaric" or brutal, and less unequal—than the colonizing societies that putatively sought to "civilize" them. In classical Marxism, traditional, pre-capitalist socioeconomic formations were often seen as exhibiting more communal forms of social organization, which, if still existing in an undeveloped state, nonetheless prefigured social structures which would reemerge in more advanced modes in socialism. It was for this reason that Engels, together with Marx, displayed such high respect for the Iroquois and for other traditional societies. See Frederick Engels, *The Origin of the Family, Private Property, and the State* (New York: International Publishers, 1972), 147–61.
22. Charles Fourier, *Design for Utopia* (New York: Schocken Books, 1971), 88, italics in the original.
23. Jonathan Beecher, *Charles Fourier* (Berkeley: University of California Press, 1986), 197.
24. Sven Beckert, *The Empire of Cotton* (New York: Vintage, 2015), 244.
25. Engels, *The Origin of the Family*, 224–25.
26. Marx, *Capital*, vol. 1, 637–38; Frederick Engels, *The Housing Question* (Moscow: Progress Publishers, 1979), 92.
27. William Morris, *Signs of Change* (London: Longmans Green and Co., 1896), 116.
28. Morris, *News from Nowhere* (Oxford: Oxford University Press, 2003), 81–84, italics in the original.
29. Marx, *Capital*, vol. 2 (London: Penguin, 1978), 322.
30. Marx and Engels, *Marx/Engels Collected Works*, vol. 42, 558–59; Kohei Saito, "Marx's Ecological Notebooks." *Monthly Review* 67/9 (2016): 25–42. The above passage is often translated as "civilization . . . leaves deserts behind it." See Rudolph Bahro, *Avoiding Social and Ecological Disaster* (Bath: Gateway Books, 1994), 19.
31. E. K. Fedorov, *Man and Nature* (Moscow: Progress Publishers, 1972), 145–47.
32. Marx and Engels, *Marx/Engels Collected Works*, vol. 5, 32.
33. Marx, *Capital*, vol. 1, 637–38.
34. Marx, *Capital*, vol. 3. (London: Penguin, 1981), 949.
35. Ibid.

36. Ibid., 911.
37. Karl Marx and Frederick Engels, *Ireland and the Irish Question* (Moscow: Progress Publishers, 1971), 141. The science behind the potato blight—its proximate cause in the form of *Phytophthora infestans* was isolated in the 1860s by the German plant pathologist Anton de Bary—was poorly understood at the time Marx was writing. Yet, despite this, it was obvious to Marx that the potato blight, which had affected countries throughout Europe and engendered mass starvation in Ireland, was related, in the Irish case, to the destruction of what had earlier in the century been a more diverse agriculture, leading to the absolute dependence of the poor tenant farmers in the colonial system on potatoes (a monocrop) for their subsistence. The potato was seen as allowing the Irish peasants to eke out a bare existence on the worst land, consisting of tiny plots, and with little fertilizer, while the major commercial agricultural produce of the country controlled by the colonial plantations was being exported, primarily to England. For Marx, the deficiencies of the entire agricultural system were thus quite clearly related to the overexploitation of the land in a colonial setting. See Evan D. Fraser, "Social Vulnerability and Ecological Fragility: Building Bridges between Social and Natural Sciences Using the Irish Potato Famine as a Case Study," *Conservation Ecology* 7/2 (2003): http://www.ecologyandsociety.org/vol7/iss2/art9/; Sarah Maria Schmidt, "Anton de Bary—The Father of Plant Pathology," 2015, https://microbeseatmyfood.wordpress.com/2015/01/27/anton-de-bary-the-father-of-plant-pathology/.
38. Marx and Engels, *Ireland and the Irish Question*, 123; see also Eamonn Slater, "Marx on Nineteenth-Century Colonial Ireland: Analysing Colonialism as a Dynamic Social Process," *Irish Historical Studies* 36/142 (2008): 153–72.
39. Marx and Engels, *Ireland and the Irish Question*, 135–36.
40. Ibid., 137, italics in the original.
41. Ibid., 142, italics added.
42. Lewis Mumford, *The City in History* (New York: Harcourt Brace Jovanovich, 1961), 53.
43. Lewis Mumford, *The Condition of Man* (New York: Harcourt Brace Jovanovich, 1944), 392. Mumford was of course not referring here to the superiority of Western civilization but rather to its irrationality, and its potential catastrophic planetary effects.
44. Lewis Mumford, *Technics and Human Development*, vol. 1: *The Myth of the Machine* (New York: Harcourt Brace Jovanovich, 1967), 186.
45. E. P. Thompson, *Beyond the Cold War* (New York: Pantheon, 1982).
46. Ibid., 41–79; also see Ian Angus, "When Did the Anthropocene Begin . . . and Why Does It Matter?," *Monthly Review* 67/4 (2015): 179–80.

47. Rudolph Bahro, *Avoiding Social and Ecological Disaster* (Bath, UK: Gateway Books, 1994), 19, italics in the original.
48. Martin Rees, *Our Final Hour* (New York: Basic Books, 2003); Jared Diamond, *Collapse* (London: Penguin, 2011); James Lovelock, *The Revenge of Gaia* (New York: Basic Books, 2006); James E. Hansen, *Storms of My Grandchildren* (New York: Bloomsbury, 2009); and Naomi Oreskes and Erik M. Conway, *The Collapse of Western Civilization: A View from the Future* (New York: Columbia University Press, 2014).
49. Jared Diamond, *Collapse* (London: Penguin, 2011), 498.
50. Ibid., 7.
51. Ibid., 441–85, 555–60. Recently the emphasis on mainstream theory in the face of growing catastrophic environmental events has been on the "resilience" of individual societies, seeking to remove any responsibility from states in the center for addressing ecological devastation in the periphery. For a critique of how resilience theory has evolved in this respect, see Paul Cox and Stan Cox, *How the World Breaks* (New York: New Press, 2016).
52. Karl Marx and Frederick Engels, *The Communist Manifesto* (New York: Monthly Review Press, 1964), 2.
53. Jason W. Moore, "Toward a Singular Metabolism," in *Grounding Metabolism*, ed. D. Ibañez and N. Katsikis (Cambridge, MA: Harvard University Press, 2014), 10–19, at 17; and Jason W. Moore, *Capitalism in the Web of Life* (London: Verso, 2015), 86. Moore subsequently realized that his claim that "nearly half" the world's population was malnourished was exaggerated and in reiterating this in almost identical words a year later changed "half to "third." See Moore, "Nature in the Limits to Capital (and Vice Versa)," *Radical Philosophy* 193 (September–October 2015): 9–19, at 19.
54. Derrick Jensen, "Held Hostage to Hope: Derrick Jensen on Civilization and Its Discontents," 2015. https://bawehali.files.wordpress.com/2011/12/jensenlundberginterview.pdf; Aric McBay, Lierre Keith, and Derrick Jensen. *Deep Green Resistance* (New York: Seven Stories Press, 2011).
55. James Lovelock, *The Revenge of Gaia* (New York: Basic Books, 2006), 147, 15; James E. Hansen, *Storms of My Grandchildren* (New York: Bloomsbury, 2009), 236, 259–60.
56. Rosa Luxemburg, *The Rosa Luxemburg Reader* (New York: Monthly Review Press, 2004), 321; Ian Angus, "The Origin of Rosa Luxemburg's Slogan 'Socialism or Barbarism,'" *Climate and Capitalism*, October 24, 2014, http://climateandcapitalism.com/2014/10/22/origin-rosa-luxemburgs-slogan-socialism-barbarism/. Today the phrase "Ecosocialism or Barbarism" is frequently heard. See, for instance, Jane Kelly and Sheila

Malone, *Ecosocialism or Barbarism* (London: Socialist Resistance, 2006). This is a call for a socialist ecological civilization as opposed to an anti-ecological and anti-social barbarism.

57. See Arkadij Dmitrievič Ursul, *Philosophy and the Ecological Problems of Civilization* (Moscow: Progress Publishers, 1983). Following the 1983 publication of *Philosophy and the Ecological Problems of Civilization*, it appears that vice president of the USSR Academy of Sciences, P. N. Fedoseev (also Fedoseyev), who had written the introductory essay on ecology and the problem of civilization in the above-edited book, incorporated a treatment of "Ecological Civilization" into the second edition of his *Scientific Communism*. See also P. N. Fedoseev (Fedoseyev), *Soviet Communism* (Moscow: Progress Publishers, 1986); Jiahua Pan, *China's Environmental Governing and Ecological Civilization* (Berlin: Springer-Verlag, 2014), 35; Arran Gare, "Barbarity, Civilization and Decadence: Meeting the Challenge of Creating an Ecological Civilization," (2015), https://www.researchgate.net/publication/271169381_Barbarity_Civilization_and_Decadence; Qingzhi Huan, "Socialist Eco-Civilization and Social-Ecological Transformation," *Capitalism Nature Socialism* 27/2 (2016): 2.

58. John Cobb interviewed by Andre Vltchek, *China and Ecological Civilization* (Jakarta: Badak Merah, 2019); Jiahua Pan, *China's Environmental Governing and Ecological Civilization* (Berlin: Springer-Verlag, 2014); Xi Jinping, *The Governance of China*, vol. 3 (Beijing: Foreign Languages Press, 2020), 6, 20, 25, 417–24.

59. Ursul, "Preface," *Philosophy and the Ecological Problems of Civilization*, 15.

60. Soviet ecological thought in this period was influenced by the nuclear winter thesis, which projected the possible demise of the biosphere as a result of nuclear exchange, and which was a by-product of the research on climate change by Mikhail Budyko and others. See Michael I. Budyko, Georgi S. Golitsyn, and Yuri A. Izrael, *Global Climatic Catastrophes* (New York: Springer-Verlag, 1988), v–vi, 39–46. This was seen as linked to the whole ecological problem in a way uncommon in the West.

61. P. N. Fedoseev (Fedoseyev), "The Social Significance of the Ecological Problem," in Ursul, *Philosophy and the Ecological Problems of Civilisation*, 31.

62. Ivan T. Frolov, "The Marxist-Leninist Conception of the Ecological Problem," in Ursul, *Philosophy and the Ecological Problems of Civilisation*, 35–42.

63. Y. P. Trusov. "The Ecological Approach and Problems of Moulding the Noosphere," in Ursul, *Philosophy and the Ecological Problems of Civilization*, 70.

64. V. A. Los, "On the Road to an Ecological Culture," in Ursul, *Philosophy and the Ecological Problems of Civilization*, 339.
65. For an analysis that explains how ecological science and the critique of political economy can both be drawn upon in order to develop a conception of ecological civilization, in which socialist and ecological principles reinforce each other, see Fred Magdoff, "Ecological Civilization," *Monthly Review* 62/8 (2011): 1–25.
66. Marx and Engels, *Marx/Engels Collected Works*, vol. 46, 411, italics in the original.
67. Ryan Wishart, R. Jamil Jonna, and Jordan Besek, "The Metabolic Rift: A Selected Bibliography," *Monthly Review* (2013), http://monthlyreview.org/commentary/metabolic-rift/2013.
68. Hamilton and Grinevald, "Was the Anthropocene Anticipated?," 67.
69. William Leiss, *The Domination of Nature* (Montreal: McGill-Queen's University Press, 1972).
70. Joseph Needham, *Moulds of Understanding* (Aldershot, UK: Gower House, 1976), 300–301.
71. Ibid., 301.
72. Joseph Needham. *Science and Civilization in China*, vol. 1 (Cambridge: Cambridge University Press, 1954), 4.
73. James E. Hansen, "China and the Barbarians, Part I," 2010, http://www.columbia.edu/~jehl/mailings/2010/20101124_ChinaBarbarians1.pdf; James E. Hansen, "Wanning Workshop+Beijing Charts+Year-End Commentary," 2015, http://www.columbia.edu/~jehl/mailings/2015/20151229_Sleepless.pdf; Michael E. Mann and Lee R. Kump, *Dire Predictions* (New York: DK Publishing, 2015); Jean Chemnick, "China Takes the Climate Spotlight as U.S. Heads for Exit," *Scientific American*, 2016, https://www.scientificamerican.com/article/china-takes-the-climate-spotlight-as-u-s-heads-for-exit/. On the question of optimism versus hope see also Terry Eagleton. *Hope without Optimism* (Charlottesville: University of Virginia Press, 2015).
74. Barbara Finamore, *Will China Save the Planet?* (Cambridge: Polity, 2018); Cobb, *China and Ecological Civilization*.
75. Oreskes and Conway, *The Collapse of Western Civilization*.
76. Oreskes and Conway indicate that the reason they chose a Chinese historian from the twenty-fourth century to tell the story of "the collapse of Western civilization," and depicted China as the civilization that survived climate change, was to emphasize the importance of government regulation and government intervention, which had largely disappeared in the neoliberal West, but not in the "authoritarian societies." In the circumspect language of liberal ideology this was

meant as a reference to planning. See Oreskes and Conway, *The Collapse of Western Civilization*, 69.

77. Paul Sweezy, "Socialism and Ecology," *Monthly Review* 41/4 (1989): 8.
78. One dramatic exception to this was Cuba, which in the face of the collapse of the Soviet bloc managed to keep its own revolution going by taking a more revolutionary ecological path. This has best been explained by Richard Lewontin and Richard Levins in *Biology Under the Influence* (New York: Monthly Review Press, 2007), 343–64.
79. Tiejun Wen, Kin Chi Lau, Cunwang Cheng, Huiil He, and Jiansheng Qiu, "Ecological Civilization, Indigenous Culture, and Rural Reconstruction in China," *Monthly Review* 63/9 (2012): 29–44.
80. "Notes from the Editors," *Monthly Review* 72/ 2 (March 2021).
81. Finamore, *Will China Save the Planet?*; Deborah Seligsohn, "China's Climate Comeback," *New Scientist*, 2015, https://www.newscientist.com/article/mg22830490-400-chinas-climate-comeback-how-the-top-polluter-is-cleaning-up/; John Bellamy Foster, "Marxism, Ecological Civilization, and China," *MRzine*, June 23, 2016, http://climateandcapitalism.com/2016/06/23/two-views-on-marxist-ecology-and-jason-w-moore/.
82. Finamore, *Will China Save the Planet?* China has been playing the leading role worldwide in the development of solar power technology. The proposal of the State Grid Corporation in China to build by 2050 a $50 trillion global wind and solar power grid, called the Global Energy Interconnection, has attracted enormous attention. According to the World Economic Forum, China is proposing to construct wind farms in the North Pole and solar farms at the equator crossing international boundaries, and conceivably accounting for the majority of the world's energy generation, superseding fossil fuels. Eric Baculinao, "China Wants to Build a $50tn Global Electricity Grid," NBC News, March 31, 2016, http://www.nbcnews.com/business/energy/china-unveils-proposal-50-trillion-global-electricity-network-n548376.
83. A significant factor here is the very wide extent of environmental protest in China today, pushing the society toward more radical solutions to ecological problems. John Bellamy Foster and Robert W. McChesney, *The Endless Crisis* (New York: Monthly Review Press, 2012), 179. On the continuing struggle over the Dakota Access Pipeline at Standing Rock see Nick Estes, *Our History Is the Future: Standing Rock versus the Dakota Access Pipeline, and the Long Tradition of Indigenous Resistance* (London: Verso, 2019).
84. Noam Chomsky, *Because We Say So* (San Francisco: City Light Books, 2015), 94.
85. Noam Chomsky, "The End of History?," *In These Times*, September 4, 2014.

86. The principal basis for the notion of an environmental proletariat is Marx and Engels, and can be seen particularly in Engels's *Condition of the Working Class in England*, which concentrates on the overall environmental conditions of the working class, and sees that as constituting the basis for revolutionary action. See Frederick Engels, *The Condition of the Working Class in England* (Oxford: Oxford University Press, 1993). However, Toynbee's notion of an "internal proletariat," as well as an "external proletariat," characterized by "alienation from the dominant minority" representing much of the creative power of any given civilization, is also useful here. Arnold Toynbee and D. C. Somervell, *A Study in History: Abridgement of Volumes I–IV* (Oxford: Oxford University Press, 1946), 12; John Bellamy Foster, "Marxism and Ecology: Common Fonts of a Great Transition," *Monthly Review* 67/7 (December 2015): 11–12.

21. The Capitalinian

This chapter is adapted and revised for this book from the following article: John Bellamy Foster and Brett Clark, "The Capitalinian: The First Geological Age of the Anthropocene," *Monthly Review* 73/4 (September 2021): 1–16.

1. John R. McNeill and Peter Engelke, *The Great Acceleration: The Environmental History of the Anthropocene Since 1945* (Cambridge, MA: Harvard University Press, 2014); Ian Angus, *Facing the Anthropocene: Fossil Capitalism and the Crisis of the Earth System* (New York: Monthly Review Press, 2016), 38–47; Donald Worster, *Nature's Economy* (New York: Cambridge University Press, 1994).
2. A classic work in this regard is Paul A. Baran and Paul M. Sweezy, *Monopoly Capital: An Essay on the American Economic and Social Order* (New York: Monthly Review Press, 1966).
3. Barry Commoner, *The Closing Circle: Nature, Man, and Technology* (New York: Bantam, 1972); John Bellamy Foster, *The Vulnerable Planet: A Short Economic History of the Environment* (New York: Monthly Review Press, 1994), 112–18; Rachel Carson, *Silent Spring* (Boston: Houghton Mifflin, 1994); Murray Bookchin, *Our Synthetic Environment* (New York: Harper Colophon, 1974); Joel B. Hagen, *An Entangled Bank* (New Brunswick: Rutgers University Press, 1992), 100–21; Robert Rudd, *Pesticides and the Living Landscape* (Madison: University of Wisconsin Press, 1964).
4. Johan Rockström et al., "A Safe Operating Space for Humanity," *Nature* 461/24 (2009): 472–75; Will Steffen et al., "Planetary Boundaries," *Science* 347/6223 (2015): 736–46; John Bellamy Foster, Brett Clark, and Richard York, *The Ecological Rift* (New York: Monthly Review

Press, 2010), 13–19; Giovanni Strona and Corey J. A. Bradshaw, "Coextinctions Annihilate Planetary Life During Extreme Environmental Change," *Scientific Reports* 8/16274 (2018); James Hansen, *Storms of My Grandchildren* (New York: Bloomsbury, 2009), ix, 224–26.
5. E. V. Shantser, "Anthropogenic System (Period)," in *Great Soviet Encyclopedia*, vol. 2 (New York: Macmillan, 1973), 140; Alec Brookes and Elena Fratto, "Toward a Russian Literature of the Anthropocene," *Russian Literature* 114–115 (2020): 8. See also Anonymous (likely written by E. V. Shantser), "Anthropogenic Factors of the Environment," in *Great Soviet Encyclopedia*, vol. 2, 139.
6. John Bellamy Foster, "Late Soviet Ecology and the Planetary Crisis," *Monthly Review* 67/2 (June 2015): 1–20.
7. Vladimir I. Vernadsky, *The Biosphere* (New York: Springer-Verlag, 1998).
8. Vladimir I. Vernadsky, "Some Words About the Noösphere," in *150 Years of Vernadsky*, vol. 2: *The Noösphere*, ed. John Ross (Washington, DC: 21st Century Science Associates, 2014), 82. Vernadsky clearly meant *period* here, in geochronology, rather than *era*. See also Jan Zalasiewicz, Colin N. Waters, Mark Williams, Colin P. Summerhayes, Martin J. Head, and Reinhold Leinfelder, "A General Introduction to the Anthropocene," in *The Anthropocene as a Geological Time Unit*, ed. Jan Zalasiewicz, Colin N. Waters, Mark Williams, and Colin P. Summerhayes (Cambridge: Cambridge University Press, 2019), 6.
9. Will Steffen, "Commentary," in *The Future of Nature: Documents of Global Change*, ed. Libby Robin, Sverker Sörlin, and Paul Warde (New Haven: Yale University Press, 2013), 486; Paul J. Crutzen, "The Geology of Mankind," *Nature* 415 (2002): 23; Angus, *Facing the Anthropocene*, 27–28. Marine biologist Eugene Stoermer used the word *Anthropocene* a number of times in published articles in the 1980s to refer to the growing human impact on the earth. But unlike Pavlov in the early twentieth century (who impacted Vernadsky), as well as Crutzen in the early twenty-first century, who launched the current investigations into the Anthropocene, Stoermer's use of the term had no discernible impact on geological and Earth System discussions. See Andrew C. Revkin, "Confronting the Anthropocene," *New York Times*, May 11, 2011; Angus, *Facing the Anthropocene*, 27.
10. Will Steffen et al., "Stratigraphic and Earth System Approaches to Defining the Anthropocene," *Earth's Future* 4 (2016): 324–45.
11. Will Steffen, Paul J. Crutzen, and John R. McNeill, "Are Humans Now Overwhelming the Great Forces of Nature?" *Ambio* 36/8 (2007): 614; Angus, *Facing the Anthropocene*, 28–29.
12. Foster, *The Vulnerable Planet*, 108.

13. Clive Hamilton and Jacques Grinevald, "Was the Anthropocene Anticipated?" *Anthropocene Review* (2015): 6–7. The notion of an anthropogenic rift is closely related to the conception of a carbon rift, developed within environmental sociology, expanding on Karl Marx's early conception of a metabolic rift in the human relation to the environment through production. See Foster, Clark, and York, *The Ecological Rift*, 121–50.
14. Ian Angus, *A Redder Shade of Green: Intersections of Science and Socialism* (New York: Monthly Review Press, 2017), 70–71. As Angus explains, "*Anthropocene* names a planetary epoch that would not have begun in the absence of human activity, *not* one caused by every person on Earth."
15. C. P. Snow, *The Two Cultures* (Cambridge: Cambridge University Press, 1998).
16. Jason W. Moore, "Who Is Responsible for the Climate Crisis?," *Maize*, November 4, 2019. For a critique of such views, see Angus, *A Redder Shade of Green*, 67–85.
17. Andreas Malm, *Fossil Capital: The Rise of Steam Power and the Roots of Global Warming* (London: Verso, 2016), 391. Malm himself coined the term *Capitalocene* in 2009. See Jason W. Moore, "Anthropocene or Capitalocene?," introduction to *Anthropocene or Capitalocene?*, ed. Jason W. Moore (Oakland, CA: PM, 2016), 5.
18. Ian Burton and Robert W. Kates, "The Great Climacteric, 1798–2048: The Transition to a Just and Sustainable Human Environment," in *Geography, Resources and Environment*, vol. 2, ed. Robert W. Kates and Ian Burton (Chicago: University of Chicago Press, 1986), 393; John Bellamy Foster, "The Great Capitalist Climacteric," *Monthly Review* 67/6 (November 2015): 1–18.
19. E. Ray Lankester, *The Kingdom of Man* (New York: Henry Holt, 1911), 31–32.
20. Mike Walker et al., "Formal Ratification of the Subdivision of the Holocene Series/Epoch (Quaternary System/Period): Two New Global Boundary Stratotype Sections and Points (GSSPS) and Three New Stages/Subseries," *Episodes* 41/4 (2018): 213.
21. Walker et al., "Formal Ratification," 214.
22. International Commission on Stratigraphy, "Collapse of Civilizations Worldwide Defines Youngest Unit of the Geologic Time Scale," July 15, 2018.
23. Paul Voosen, "Massive Drought or Myth? Scientists Spar Over an Ancient Climate Event Behind Our New Geological Age," *Science*, August 8, 2018.
24. Guy D. Middleton, "Bang or Whimper?: The Evidence for Collapse of

Human Civilizations at the Start of the Recently Defined Meghalayan Age Is Equivocal," *Science* 361/6408 (2018): 1204–5.
25. Michael Walker, who chaired the geological working group that introduced the division of the Holocene into ages, insists that the designation of the Meghalayan Age in no way compromises the notion of an Anthropocene Epoch beginning in 1950. It would simply lop off seventy years from the end of the Meghalayan. "You're Living in a New Geologic Age, the Meghalayan," CBC News, July 23, 2018.
26. Jan Zalasiewicz et al., "Making the Case for a Formal Anthropocene Epoch," *Newsletters on Stratigraphy* 50/2 (2017): 210.
27. Colin N. Waters et al., "The Anthropocene Is Functionally and Stratigraphically Distinct from the Holocene," *Science* 351/6269 (2016): 137–47; Colin N. Waters, Irka Hajdas, Catherine Jeandel, and Jan Zalasiewicz, "Artificial Radionuclide Fallout Signals," in *The Anthropocene as a Geological Time Unit*, 192–99; Reinhold Leinfelder and Juliana Assunção Ivar do Sul, "The Stratigraphy of Plastics and Their Preservation in Geological Records," in *The Anthropocene as a Geological Time Unit*, 147–55. The most important thinker developing the analysis of the synthetic age was Barry Commoner. See Commoner, *The Closing Circle*; Barry Commoner, *The Poverty of Power* (New York: Alfred A. Knopf, 1976); Barry Commoner, *Making Peace with the Planet* (New York: New Press, 1972); Foster, *The Vulnerable Planet*, 108–24.
28. Zalasiewicz et al., "Making the Case for a Formal Anthropocene Epoch," 212–13.
29. On the significance of 1945 as a shift in the human relation to the earth, see Commoner, *The Closing Circle*, 49–50; Paul M. Sweezy and Harry Magdoff, "Capitalism and the Environment," *Monthly Review* 41/2 (June 1989): 3.
30. John Bellamy Foster, *The Return of Nature* (New York: Monthly Review Press, 2020), 502–3; Richard Hudson and Ben Shahn, *Kuboyama and the Saga of the Lucky Dragon* (New York: Yoseloff, 1965); Ralph E. Lapp, *The Voyage of the Lucky Dragon* (London: Penguin, 1957).
31. Zalasiewicz et al., "Making the Case for a Formal Anthropocene Epoch," 211; Waters et al., "Artificial Radionuclide Fallout," 192–99; Jan Zalasiewicz et al., "When Did the Anthropocene Begin?," *Quaternary International* 383 (2014): 196–203; "A New Geological Epoch, the Anthropocene, Has Begun, Scientists Say," CBC News, January 7, 2016.
32. Stephen Jay Gould, *Eight Little Piggies* (New York: W. W. Norton, 1993), 71; John Bellamy Foster, *Ecology Against Capitalism* (New York: Monthly Review Press, 1992), 70–72.
33. Stephen Schneider, "Whatever Happened to Nuclear Winter?," *Climatic Change* 12 (1988): 215; Richard P. Turco and Carl Sagan, *A Path Where*

No Man Thought: Nuclear Winter and the End of the Arms Race (New York: Random House, 1990), 24–27; R. P. Turco and G. S. Golitsyn, "Global Effects of Nuclear War," *Environment* 30/5 (1988): 8–16. The nuclear winter concept led to wide discussions of the actual indirect effects of a global thermonuclear exchange; the scientific consensus that emerged, as Schneider indicated, was "that the environmental and societal 'indirect' effects of a nuclear war are . . . probably more threatening for the earth as a whole than the direct blasts or radioactivity in the target zones." Schneider, "Whatever Happened to Nuclear Winter?," 217.

34. Commoner, *The Closing Circle*, 45–53.
35. Harry Braverman, *Labor and Monopoly Capital* (New York: Monthly Review Press, 1998), 107–15; Angus, *Facing the Anthropocene*, 167–69; John Bellamy Foster and Brett Clark, *The Robbery of Nature* (New York: Monthly Review Press, 2000), 247–58.
36. Roland Geyer, Jenna R. Jambeck, and Kara Lavender Law, "Production, Use, and Fate of All Plastics Ever Made," *Science Advances* 3/7 (2017).
37. Zalasiewicz et al., "Making the Case for a Formal Anthropocene Epoch," 212–13.
38. Geyer, Jambeck, and Law, "Production, Use, and Fate of All Plastics Ever Made," 1, 3.
39. Zalasiewicz et al., "The Geological Cycle of Plastics and Their Use as a Stratigraphic Indicator of the Anthropocene," *Anthropocene* 13 (2016): 4–17; Waters et al., "The Anthropocene Is Functionally and Stratigraphically Distinct from the Holocene"; Leinfelder and Ivar do Sul, "The Stratigraphy of Plastics and Their Preservation in Geological Records"; Juliana Assunção Ivar do Sul and Monica F. Costa, "The Present and Future of Microplastic Pollution in the Marine Environment," *Environmental Pollution* 185 (2014): 352–64.
40. Tamara S. Galloway, Matthew Cole, and Ceri Lewis, "Interactions of Microplastic Debris throughout the Marine Ecosystem," *Nature Ecology & Evolution* 1 (2017); Susan Casey, "Plastic Ocean," in *The Best American Science and Nature Writing 2007*, ed. Mary Roach (New York: Houghton Mifflin, 2007), 9–20.
41. Geyer, Jambeck, and Law, "Production, Use, and Fate of All Plastics Ever Made," 3.
42. Carson, *Silent Spring*; Commoner, *The Closing Circle*; Commoner, *The Poverty of Power*; John Bellamy Foster and Brett Clark, "Rachel Carson's Ecological Critique," *Monthly Review* 59/9 (2008): 1–17.
43. Baran and Sweezy, *Monopoly Capital*; Foster, Clark, and York, *The Ecological Rift*.
44. Commoner, *The Closing Circle*, 40.

45. Ulrich Beck, *The Risk Society* (London: Sage, 1992).
46. Peter Haff, "The Technosphere and Its Relation to the Anthropocene," in *The Anthropocene as a Geological Time Unit*, 143.
47. Burton and Kates, "The Great Climacteric, 1798–2048," in *Geography, Resources and Environment*, vol. 2, 393; Foster, "The Great Capitalist Climacteric"; Richard E. Leakey and Roger Lewin, *The Sixth Extinction: Patterns of Life and the Future of Humankind* (New York: Anchor Books, 1996).
48. See A. D. Ursul, ed., *Philosophy and the Ecological Problems of Civilisation* (Moscow: Progress Publishers, 1983). Following the 1983 publication of *Philosophy and the Ecological Problems of Civilisation*, the vice president of the USSR Academy of Sciences, P. N. Fedoseev (also Fedoseyev), who had written the introductory essay on ecology and the problem of civilization in the above edited book, incorporated a treatment of "Ecological Civilization" into the second edition of his *Scientific Communism*. Chinese agriculturalist Ye Qianji used the term in an article he wrote for the *Journal of Moscow University* in 1984, which was translated in Chinese in 1985. See P. N. Fedoseyev (Fedoseev), *Soviet Communism* (Moscow: Progress Publishers, 1986); Qingzhi Huan, "Socialist Eco-Civilization and Social-Ecological Transformation," *Capitalism Nature Socialism* 27/2 (2016): 52; Jiahua Pan, *China's Environmental Governing and Ecological Civilization* (Berlin: Springer-Verlag, 2014), 35; Arran Gare, "Barbarity, Civilization, and Decadence: Meeting the Challenge of Creating an Ecological Civilization," *Chromatikon* 5 (2009): 167.
49. On China and ecological civilization, see Pan, *China's Environmental Governing and Ecological Civilization*; John B. Cobb Jr. (in conversation with Andre Vitchek), *China and Ecological Civilization* (Jakarta: Badak Merah, 2019); Xi Jinping, *The Governance of China*, vol. 3 (Beijing: Foreign Languages Press, 2020), 6, 20, 25, 417–24.
50. Reuters, "Interview—Greta Thunberg Demands 'Crisis' Response to Climate Change," July 18, 2020.
51. Sweezy, "Capitalism and the Environment," 6.
52. NOAA Research News, "Carbon Dioxide Peaks Near 40 Parts Per million at Mauna Loa Observatory," July 7, 2021; James Hansen et al., "Target Atmospheric CO_2: Where Should Humanity Aim?," *Open Atmospheric Science Journal* 2 (2008): 217–31.
53. Duncan Foley, *Adam's Fallacy* (Cambridge, MA: Harvard University Press, 2006).
54. István Mészáros, *Beyond Capital* (London: Merlin, 1995); John Bellamy Foster, "The Earth-System Crisis and Ecological Civilization," *International Critical Thought* 7/4 (2017): 439–58; Foster, Clark, and

York, *The Ecological Rift*, 401–22; Foster and Clark, *The Robbery of Nature*, 269–87; Fred Magdoff, "Ecological Civilization," *Monthly Review* 62/8 (2011): 1–25.
55. Mere technological change is insufficient to effect the necessary ecological and social transformation since technology is itself constrained by the underlying social relations. In his essay "Technological Determinism Revisited," economist Robert Heilbroner indicated that modern economics ideology tends to focus on "the triadic connection of technological determinism, economic determinism, and capitalism." However, it can be argued that this triadic connection, insofar as it exists in reality, limits technological or productive rationality while often pushing it in irrational directions, since capitalism as a system promotes accumulation "by ignoring all effects of the changed environment"—and indeed all effects on the changing of the natural environment—"except those that affect our maximizing possibilities" for profit. Robert Heilbroner, "Do Machines Make History?," in *Does Technology Drive History?*, ed. Merritt Roe Smith and Leo Marx (Cambridge, MA: MIT Press, 1994), 72–73.
56. "Science for the People Statement on the People's Green New Deal," *Science-for-the-People.org*, July 23, 2021; Nick Estes, *Our History Is the Future* (London: Verso, 2019); Red Nation, *The Red Deal* (Brooklyn: Common Notions, 2021); Max Ajl, *A People's Green New Deal* (London: Pluto, 2021).
57. Karl Marx, *Capital*, vol. 3 (London: Penguin, 1981), 911.
58. Karl Marx and Frederick Engels, *Collected Works*, vol. 5 (New York: International Publishers, 1975), 141; Epicurus, *The Epicurus Reader* (Indianapolis: Hackett Publishing, 1994), 3–4.
59. Phil Ochs, "Another Age," *Rehearsals for Retirement* (A&M, 1969).
60. Frantz Fanon, *The Wretched of the Earth* (New York: Grove, 1963).

22. Conclusion: Ecology and the Future of History

1. Epigraph: Walter Benjamin, "Theses on the Concept of History," in Michael Löwy, *Fire Alarm* (London: Verso, 2016), 78.
2. Jean-Paul Sartre, "Time in Faulkner: The Sound and the Fury," in *William Faulkner: Three Decades of Criticism*, ed. Frederik J. Hoffman and Olga Villery (New York: Harcourt, Brace, and World, 1960), 230–32. Although writing about William Faulkner's metaphysic of time here, Sartre was quite consciously, as István Mészáros has explained, addressing the fundamental question of the "decapitated" time of capitalism, a problem that was to pervade his work. István Mészáros, *The Work of Sartre* (New York: Monthly Review Press, 2012), 59–61.

NOTES TO PAGES 475–476

On the metaphor of the "burning house" see Bertolt Brecht, *Tales from the Calendar* (London: Methuen, 1961), 31–32.
3. Daniel Singer, *Whose Millennium: Theirs or Ours?* (New York: Monthly Review Press, 1999), 1.
4. G. W. F. Hegel, *The Philosophy of History* (New York: Dover, 1956), 103–04. The extent to which Hegel pointed to "the end of history" in his *Phenomenology of Spirit* and his philosophy as a whole is widely debated. Certainly, the cruder versions of this need to be rejected. See Terry Pinkard, *Does History Make Sense? Hegel on the Historical Shapes of Justice* (Cambridge, MA: Harvard University Press, 2017), 2–3. Yet, in the *The Philosophy of Hstory*, in *The Philosophy of Right*, and *The Philosophy of History* in his Berlin period late in his life Hegel clearly identified bourgeois civil society and the Prussian state with the culmination of reason in history, thereby reconciling himself to his time. G. W. F. Hegel, *The Philosophy of Right* (Oxford: Oxford University Press, 1952), 155–57; István Mészáros, *The Necessity of Social Control* (New York: Monthly Review Press, 2015), 269–81.
5. Francis Fukuyama, *The End of History and the Last Man* (New York: Free Press, 1992), 338. Fukuyama relied on the conservative interpretation of Hegel, emphasizing the concept of the end of history, developed in Alexandre Kojève, *Introduction to the Reading of Hegel: Lectures on* The Phenomenology of Spirit (New York: Basic Books, 1969).
6. Frantz Fanon, *The Wretched of the Earth* (New York: Grove Press, 1963).
7. Karl Marx and Frederick Engels, *Ireland and the Irish Question* (Moscow: Progress Publishers, 1971), 142; John Bellamy Foster and Brett Clark, *The Robbery of Nature* (New York: Monthly Review Press, 2020), 64–77.
8. See L. S. Stavrianos, *Global Rift: The Third World Comes of Age* (New York: William Morrow, 1981).
9. Karl Marx, *Grundrisse* (London: Penguin, 1973), 334–35, 409–10. See also G. W. F. Hegel, *The Science of Logic* (London: George Allen and Unwin, 1969), 131–37; G. W. F. Hegel, *Hegel's Logic* (Oxford: Oxford University Press, 1975), 136–37; John Bellamy Foster, Brett Clark, and Richard York, *The Ecological Rift* (New York: Monthly Review Press, 2010), 284–86.
10. On the cumulative potential for catastrophe, or "the conservation of catastrophe" in the development of contemporary global society, see William H. McNeill, *The Global Condition* (Princeton: Princeton University Press, 1972), 143–49.
11. Karl Marx, *Capital*, vol. 3 (London: Penguin, 1981), 358.
12. John Bellamy Foster, Foreword, in Mészáros, *The Necessity of Social Control*, 16.

13. Karl Marx and Frederick Engels, *Collected Works* (New York: International Publishers, 1975), vol. 5, 59–61.
14. Sartre, "Time in Faulkner," 230; Mészáros, *The Work of Sartre*, 59.
15. Ellen Meiksins Wood, *The Origin of Capitalism* (London: Verso, 2002), 4. Also Ellen Meiksins Wood, *The Pristine Culture of Capitalism* (London: Verso, 1991), 7. To recognize the validity and importance of Wood's observation is not thereby to subscribe to the specific theory of the origins of capitalism that she advanced.
16. Wood, *The Origin of Capitalism*, 4; C. B. Macpherson, *The Political Theory of Possessive Individualism* (Oxford: Oxford University Press, 1962); Adam Smith, *The Wealth of Nations* (New York: Modern Library, 1937), 13. What is known as the Hobbesian possessive-individualist view of human nature, based on the famous quote from chapter 13 of *Leviathan* in which he wrote, with respect to "a time of Warre," that "the life of man [is] solitary, poor, brutish and short," is often taken out of context, representing a distortion of Hobbes's views. Since Hobbes did not see this as a condition inherent in all human history, and indeed sought to combat it, but rather particularly characteristic of the period of civil discord in which he lived. Thus, he wrote on the same page: "It may peradventure be thought, there was never such a time, nor condition of warre as this; and I believe it was never generally so, over all the world." Thomas Hobbes, *Leviathan* (Cambridge: Cambridge University Press, 1996), 89. Nevertheless, the "Hobbesian" view of human nature—where human government has dissolved—is commonly seen as a representation of bourgeois nature. See István Mészáros, "Preface to *Beyond Leviathan*," *Monthly Review* 69/9 (February 2018): 48.
17. Max Weber, *The Protestant Ethic and the Spirit of Capitalism* (London: Unwin Hyman, 1930), 13, 17.
18. Thorstein Veblen, *The Place of Science in Modern Civilization* (New York: Russell and Russell, 1961), 193.
19. Robert Skildelsky, "Economics and the Culture War," Project Syndicate, July 20, 2020, www.project-syndicate.org.
20. Sociological theorist Jonathan Turner, a self-styled advocate of positivism, states: "The goal of positivism is to formulate and test laws that apply to *all* societies in *all* places and at *all* times." On this basis he argues that "Marxists and others make a fundamental mistake in assuming that the laws of social organization are time bound, such that the laws governing the operation of feudalism are somehow different than those directing capitalism." In effect, thinkers such as Turner and most neoclassical economists not only dehistoricize but desocialize society, removing both human agency and social structure. Jonathan

Turner, "Explaining the Social World: Historicism vs. Positivism," *The Sociological Quarterly* 47 (2006): 453.
21. Joseph A. Schumpeter, *Capitalism, Socialism, and Democracy* (New York: Harper and Brothers, 1942), 81–86; John Bellamy Foster, "The Political Economy of Joseph Schumpeter," *Studies in Political Economy* 15 (Fall 1984): 5–42.
22. See, for example, Erik Brynjolfsson and Andrew McAffee, *The Second Machine Age* (New York: W.W. Norton, 2016); Mark Sagoff, "Schumpeter's Revolution," Breakthrough Institute, August 28, 2014, https://thebreakthrough.org.
23. Peter L. Berger, "Capitalism and the Disorder sof Modernity," First Things, January 1991, https://www.firstthings.com/article/1991/01/capitalism-and-the-disorders-of-modernity.
24. Mark Fisher, *Capitalist Realism* (Winchester, UK: Zero Books, 2009), 45.
25. "I define *postmodernism* as incredulity towards metanarratives." Jean-François Lyotard, *The Postmodern Condition* (Minneapolis: University of Minnesota Press, 1984), xxiii–xxiv.
26. Leo Marx, "The Ideology of 'Technology' and Postmodern Pessimism," in *Does Technology Drive History?* (Cambridge, MA: MIT Press, 1994), 257.
27. Terry Eagleton, "Where Do Postmodernists Come From?," in *In Defense of History*, ed. Ellen Meiksins Wood and John Bellamy Foster (New York: Monthly Review Press, 1997), 17–25.
28. Ellen Meiksins Wood, "What Is the 'Postmodern' Agenda?," in *In Defense of History*, ed. Wood and Foster, 10.
29. Keti Chukhrov, *Practicing the Good: Desire and Boredom in Soviet Socialism* (Minneapolis: e-flux/University of Minnesota Press, 2020), 20.
30. See, for example, John Asafu-Adjaye et al., *An Ecomodernist Manifesto* (April 2015), http://www.ecomodernism.org.
31. Derrick Jensen and Aric McBay, *What We Leave Behind* (New York: Seven Stories Press, 2009), 443.
32. Fredric Jameson, "The Future of the City," *New Left Review* 21 (second series) (May–June 2003): 76.
33. Fisher, *Capitalist Realism*, 2.
34. Sartre, "Time in Faulkner," 530–32; Mészáros, *The Work of Sartre*, 61.
35. See Christopher Hill, *The World Turned Upside Down* (London: Penguin, 1972); Georges Lefebvre, *The Coming of the French Revolution* (Princeton: Princeton University Press, 1947), 131–51.
36. E. P. Thompson, *The Making of the English Working Class* (New York: Vintage, 1963), 832.

37. On Shelley's critique see Amanda Jo Goldstein, *Sweet Science: Romantic Materialism and the New Logics of Life* (Chicago: University of Chicago Press, 2017), 136–208. On Ruskin and Morris see John Bellamy Foster, *The Return of Nature: Socialism and Ecology* (New York: Monthly Review Press, 2020), 75–80, 91–106, 137–63.
38. Marx and Engels, *Collected Works*, vol. 4, 390, 394; Foster, *The Return of Nature*, 184, 196.
39. Karl Marx, *Early Writings* (London: Penguin, 1970), 359–60.
40. Marx and Engels, *Collected Works*, vol. 4, 36–37. Translation modified according to Paul M. Sweezy, *Modern Capitalism and Other Essays* (New York: Monthly Review Press, 1972), 149.
41. Karl Marx, *Capital*, vol. 1 (London: Penguin, 1976), 637; John Bellamy Foster, *Marx's Ecology* (New York: Monthly Review Press, 2000), 141–77.
42. Marx and Engels, *Ireland and the Irish Question*, 142; Marx and Engels, *Collected Works*, vol. 25, 153.
43. Karl Marx, *Capital*, vol. 3 (London: Penguin, 1981), 904–5.
44. Marx, *Capital*, vol. 1, 929–30; John Bellamy Foster, Brett Clark, and Hannah Holleman, "Marx and the Commons," *Social Research* 88/1 (Spring 2021): 1–30.
45. Karl Marx, *Dispatches for the New York Tribune* (London: Penguin, 2007), 128–29; Herbert Spencer, *Social Statics* (New York: D. Appleton and Co., 1865), 13–44; Foster and Clark, *The Robbery of Nature*, 159–60.
46. Marx and Engels, *Collected Works*, vol. 40, 41.
47. Marx and Engels, *Collected Works*, vol. 24, 356; Teodor Shanin, ed., *Late Marx and the Russian Road: Marx and the Peripheries of Capitalism* (New York: Monthly Review Press, 1983), 97–126, 138–39.
48. John Bellamy Foster, Brett Clark, and Hannah Holleman, "Marx and the Indigenous," *Monthly Review* 71/9 (February 2020): 1–19; Foster and Clark, *The Robbery of Nature*, 64–77. In relation to the Americas, Marx took into account not only Indigenous struggles, but also slave revolts. See John Bellamy Foster, Hannah Holleman, and Brett Clark, "Marx and Slavery," *Monthly Review* 72/3 (July–August 2020): 96–117.
49. Marx and Engels, *Collected Works*, vol. 24, 356.
50. Marx, *Capital*, vol. 1, 917; Karl Marx and Friedrich Engels, *Marx=Engels Gesamtausgabe* (MEGA) IV/18 (Berlin: Walter de Gruyter, 2019), 670–74, 731.
51. Eric Wolf, *Peasant Wars of the Twentieth Century* (New York: Harper and Row, 1969).
52. V. I. Lenin, "For Bread and Peace," *Collected Works* (Moscow: Progress Publishers, 1972), vol. 26, 386–87.

53. Amilcar Cabral, *Return to the Source* (New York: Monthly Review Press, 1973), 41–50, 62–63.
54. Marx and Engels, *Collected Works*, vol. 5, 52–53, 73, 87.
55. Marx, *Capital*, vol. 1, 638. See also Marx and Engels, *Collected Works*, vol. 24, 357.
56. Marx, *Capital*, vol. 1, 348–49.
57. See Marx and Engels, *Collected Works*, vol. 3, 295–583; Foster, *The Return of Nature*, 172–215.
58. This can be seen, for example, in the nineteenth and early twentieth century in the work of figures such as Florence Kelley, J. B. S. Haldane, W. E. B. Du Bois, Norman Bethune, and Salvador Allende. See Foster, *The Return of Nature*, 210–15, 396–97; John Bellamy Foster, Brett Clark, and Hannah Holleman, "Capitalism and the Ecology of Disease," *Monthly Review* 73/2 (June 2021): 13–18.
59. Fred Magdoff, "Repairing the Soil Carbon Rift," *Monthly Review* 72/11 (April 2021): 1–13.
60. Marx, *Capital*, vol. 3, 754.
61. Mészáros writes: "In *History and Class Consciousness* (1923), Lukács analyzes 'possible consciousness' as the consciousness of a historically progressive class which has a future ahead of it and therefore has the possibility of objective totalization." Mészáros, *The Work of Sartre*, 59. See also Sartre, "Time in Faulkner," 231.
62. On these various movements and struggles see Michael Löwy, "The Socio-Religious Origins of Brazil's Rural Landless Workers Movement," *Monthly Review* 53/2 (June 2001): 32–40; Hannah Wittman, "Reworking the Metabolic Rift: La Via Campesina, Agrarian Citizenship, and Food Sovereignty," *Journal of Peasant Studies* 36/4 (October 2009): 805–26; John Bellamy Foster, "Chávez and the Communal State," *Monthly Review* 66/11 (April 2015): 1–17; Tricontinental.org, "Resource Sovereignty: The Agenda for Africa's Exit from the State Plunder," May 7, 2019, https://thetricontinental.org/dossier-16-resource-sovereignty-the-agenda-for-africas-exit-from-the-state-plunder/; Vijay Prashad, *The Darker Nations* (New York: New Press, 2008); Tricontinental.org, "The Farmer's Revolt in India," June 14, 2021, https://thetricontinental.org/dossier-41-india-agriculture/; John B. Cobb, Jr., in conversation with Andre Vltchek, *China and Ecological Civilization* (Jakarta: Badak Merah, 2019); Andre Vltchek, "Determined March Towards Ecological Civilization," *Investig'action*, May 12, 2018, https://www.investigaction.net/en/chinas-determined-march-towards-the-ecological-civilization/.
63. Mészáros, *The Necessity of Social* Control, 199–217; Samir Amin and Firoze Manji, "Toward the Formation of a Transnational Alliance of

Working Oppressed Peoples," *Monthly Review* 71/3 (July–August 2019): 120–26.
64. Nick Estes, *Our History Is the Future: Standing Rock versus the Dakota Access Pipeline, and the Long Tradition of Indigenous Resistance* (London: Verso, 2019), 256–57. See also *Investig'action*.
65. On "irresistible and irreversible" revolts see Vijay Prashad, *Washington Bullets* (New York: Monthly Review Press, 2020), 51.

Name Index

Adorno, Theodor, 31, 68, 192, 248, 264, 503n24, 506n45, 511n21, 527n4, 542n187, 562n17, 564n26, 585n28
Aeschylus, 342, 484
Ajl, Max, 626n56
Albrow, Martin, 157, 533n31, 534n32
Allen, Myles, 115, 516n6, 521n22
Allende, Salvador, 631n58
Almond, Gabriel A., 542n189
Altvater, Elmar, 249, 287, 511n28, 563n22
Amin, Samir, 204, 206, 208, 210, 214, 216, 225, 228, 229, 241, 354, 355, 547n14, 547n18, 548n20, 548n39, 549n40, 549n42, 549n43, 551n63, 551n64, 551n73, 555n125, 556n142, 606n1, 632n63
Anderson, Charles H., 248, 562n13
Anderson, Kevin, 63, 94, 122, 403–4, 438, 509n7, 517n7, 520n15, 523n43, 523n44, 604n44, 604n46, 605n48, 605n50, 605n59, 614n17
Anderson, Perry, 536n74, 537n78, 537n82
Angus, Ian, 83, 87, 344, 494n24, 495n26, 509n4, 514, 516n20, 562n8, 563n24, 579n41, 580n49, 584n13, 587n59, 606n67, 611n66, 612n2, 616n46, 617n56, 620n1, 621n9, 622n11, 622n14, 622n16, 624n35
Antonio, Robert J., 156, 530n3, 533n23
Araghi, Farshad, 608n23
Arding, Jan E., 230–31, 234–35, 546n6, 553n88, 553n98, 553n99, 556n132, 557n151, 557n155, 557n156, 557n157, 557n160, 557n171, 560n210
Aristotle, 352
Arndt, Olaf, 578n31
Asafu-Adjaye, John, 629n30
Ashton, T. S., 173, 538n110
Austen, Jane, 149, 529n1
Austin, Kelly, 545n4
Austin, Michael, 570n99
Aveling, Edward, 588n72

Aveling, Eleanor, 355, 588n72
; *See also* Marx, Eleanor
Ayres, Constância, 414, 607n10
Ayres, Peter, 505n37, 575n172
Ayres, R. U., 506n47

Babad, Michael, 527n25
Babones, S. J., 545n4, 549n49, 551n59
Bacon, Francis, 144, 145, 146, 301, 351, 502n12, 528n11, 528n12, 528n13, 584n18
Baculinao, Eric, 619n82
Badiou, Alain, 288, 566n42, 580n47
Baehr, Peter, 543n192
Bahro, Rudolf, 80, 444, 446, 514n64, 516n19, 614n30, 616n47
Baldwin, James, 601n1
Balfour, Arthur, 59, 508n58
Ball, John, 110
Baran, Paul A., 59, 75, 209, 354, 355, 374, 382–84, 389, 494n18, 508n59, 510n19, 513n48, 547n14, 548n35, 556n137, 583n9, 598n18, 598n19, 599n21, 599n24, 599n27, 600n31, 600n33, 600n34, 601n40, 605n57, 620n2, 625n43
Baranzini, Mauro, 559n191
Barclay, Eliza, 602n11
Baritaux, Virginie, 608n18
Barnett, Cynthia, 554n109, 558n186, 558n189, 559n198
Barnett-Howell, Zachary, 424, 610n47
Barnosky, Anthony D., 613n9
Bates, Marston, 598n10
Bauer, Otto, 207–8, 548n27
Bebel, August, 598n11
Bechmann, Roland, 539n118
Beck, Ulrich, 544n211, 625n45
Becker, James, 204, 225, 229–30, 555n125, 556n146
Beckert, Sven, 440, 500n96, 614n24
Bédarida, François, 508n1

Bednaršek, N., 582n73
Beecher, Jonathan, 439, 614n23
Behrens, William W. III, 496n36, 596n3
Beinecke, Frances, 521n19
Bell, Michael Mayerfeld, 533n25
Bello, Walden, 609n30
Bendix, Reinhard, 536n67, 536n70, 536n72
Benjamin, Walter, 297, 475, 583n3, 626n1
Bensaïd, Daniel, 249, 257–58, 264, 563n18, 564n27, 565n33, 568n53, 569n69
Benton, Ted, 68, 156, 249, 343, 511n23, 511n24, 532n18, 532n19, 563n18, 594n33
Berger, Peter L., 481, 629n23
Bergmann, Luke R., 414, 416, 607n9, 607n10
Berman, Morris, 542n188
Bermingham, Finbarr, 610n43, 610n50, 611n56
Bernal, J. D., 33, 55, 254, 274, 305, 306, 307, 308, 309, 348, 351, 496n34, 515n10, 586n33, 586n35, 586n39, 586n40, 586n41, 586n42, 586n45, 587n47, 587n48, 587n49, 587n50, 587n51, 605n54, 608n25
Bernd, Candice, 519n6
Bertram, Christopher, 605n52
Besek, Jordon, 618n67
Betancourt, Mauricio, 344
Bethune, Norman, 631n58
Betti, Francisco, 609n34
Bhaskar, Roy, 142, 262, 270, 295, 351, 408, 506n45, 527n5, 565n29, 565n36, 568n63, 568n68, 571n106, 571n109, 574n151, 582n68, 582n79, 585n27, 586n42, 606n63, 606n64
Biden, Joe, 119, 427, 603n34
Biello, David, 520n10, 520n11, 521n21, 523n35

INDEX

Bird, J., 571n111
Blackledge, Paul, 296, 583n2, 584n26
Blake, William, 13, 483, 493n3
Blaut, J. M., 532n15, 537n82, 538n93
Blight, David, 604n37
Bloch, Ernst, 31
Boisvert, Will, 107-8, 518n31, 519n32
Bolívar, Simón, 110, 519n38
Bonds, Eric, 552n74
Bonnet, Pascal, 608n18
Bookchin, Murray, 242, 528n22, 542n188, 561n217, 620n3
Bordera, Juan, 497n53
Borgstrom, Georg, 214, 551n66
Borgström Hansson, Carina, 552n77
Bottomore, Tom, 499n75, 501n4, 511n20, 565n29, 565n35, 577n24, 584n26
Bown, Chad P., 609n42, 610n53
Bows, Alice, 523n44, 605n59
Bradshaw, Corey J. A., 496n45, 621n4
Brain, Stephen, 316, 589n1, 589n2, 591n14, 591n15, 591n17, 592n20
Brandon, Pepijn, 494n15
Branson, Richard, 127
Braudel, Fernand, 439, 494n14, 494n15, 614n18, 614n20
Braun, Bruce, 287, 563n19, 571n121, 572n122, 572n123, 572n126, 573n129, 578n37, 579n41
Braverman, Harry, 624n35
Braw, Elisabeth, 420, 609n33, 609n34, 609n35
Breakthrough Institute, 565n32
Brecht, Bertolt, 31, 82, 89, 514n1, 516n21, 626n2
Brezhnev, Leonid, 324, 326, 335
Bridge, Gavin, 564n27, 578n37
Brock, William H., 540n141, 549n44
Broderick, John, 520n15
Brodine, Virginia, 248, 515n14, 562n13
Brolin, John, 548n34, 560n206, 560n210, 560n212
Brookes, Alec, 621n5
Brower, David, 324
Brown, Ellen, 604n36
Brown, Stuart, 594n32
Brynjolfsson, Erik, 629n22
Buddha, 89
Budyko, Mikhail Ivanovich, 128, 129, 274, 324, 327-32, 336, 515n4, 515n14, 525n5, 525n6, 589n4, 592n19, 592n24, 592n25, 593n26, 593n28, 593n29, 593n30, 617n60
Bukharin, Nikolai, 32, 33, 274, 318, 319, 504n33, 506n45, 542n182, 547n14
Bunker, Stephen, 215, 218, 241, 551n67, 552n80
Burger, Bruno, 522n27
Burkett, Paul, 69, 73, 134, 202, 239, 249, 260, 338, 340, 500n92, 511n22, 511n26, 511n28, 511n30, 513n42, 513n44, 514n53, 527n27, 528n22, 531n7, 545n3, 546n8, 547n13, 551n59, 552n86, 554n113, 555n119, 559n194, 559n197, 560n214, 561n222, 563n18, 563n21, 563n22, 564n26, 564n28, 566n38, 567n50, 569n72, 574n144, 574n147, 575n175, 576n179, 584n11, 588n65, 598n9, 613n10
Burton, Ian, 62, 435, 437, 471, 508n2, 562n12, 568n55, 613n6, 613n8, 622n18, 625n47
Bush, George H. W., 476
Buttel, Frederick H., 156, 530n1, 530n2, 530n4, 530n6, 531n8, 531n11, 532n12, 532n14, 532n16, 533n21, 543n202, 544n211, 608n23

Cabral, Amilcar, 488, 491, 631n53

Caldecott, Ben, 604n38
Callendar, Guy Stewart, 328
Campbell, Horace, 522n30
Canadell, J. G., 507n51
Cao Dewang, 428
Carchedi, Guglielmo, 225, 548n28, 555n125
Carey, Henry, 46
Carson, Rachel, 32, 33, 83, 85, 86, 88, 104, 246, 336, 348, 469, 499n84, 515n11, 527n8, 562n11, 585n28, 595n45, 620n3, 625n42
Carver, Terrell, 354, 585n31
Casey, Susan, 624n40
Cassirer, Ernst, 349
Castree, Noel, 249, 264, 265–67, 270, 287, 563n19, 564n27, 564n28, 570n99, 571n121, 572n122, 572n123, 572n124, 572n125, 572n126, 573n128, 573n129, 573n132, 573n135, 574n152, 578n37, 578n38, 579n41
Catton, William R., 154, 530n2, 556n149
Caudwell, Christopher, 33, 55, 254, 274, 275, 351, 352, 575n174
Cearreta, Alejandro, 613n9
Césaire, Aimé, 150
Chambers, Lauren, 611n60
Chameides, Bill, 520n15
Chandra, B., 537n82
Chapin, F. Stuart, 612n4
Chase-Dunn, Christopher, 215, 545n4, 548n38, 549n49, 551n59, 551n68, 580n49
Chattopadhyay, Kunai, 589n6, 590n7
Chaves, E. J., 549n40
Chaves, Luis Fernando, 414, 607n10, 607n11
Chemnick, Jean, 618n73
Cheng Cunwang, 619n79
Chertow, Marian R., 599n30
Childe, V. Gordon, 33, 274, 614n20
Chodorow, Marvin, 542n189

Choi, Thomas Y., 609n35, 609n38
Choldin, Harvey, 533n20
Chomsky, Noam, 100, 101, 455–56, 518n22, 620n84, 620n85
Chukhrov, Keti, 482, 629n29
Churchill, Winston, 508n1
Clark, Brett, 28, 241, 243, 249, 344, 345, 357, 498n64, 499n76, 499n77, 499n88, 499n89, 500n95, 502n12, 502n20, 502n22, 505n37, 506n44, 507n49, 512n37, 514n54, 514n56, 516, 523n36, 523n43, 523n44, 527n7, 528n22, 530n6, 531n7, 545n3, 545n4, 549n48, 550n50, 552n74, 558n183, 560n214, 561n220, 563n22, 563n24, 567n48, 568n52, 569n84, 573n143, 575n173, 576, 576n12, 577n13, 581n55, 585n32, 587n59, 588n66, 596n2, 596n5, 597n24, 600n31, 605n51, 605n60, 607n12, 611n63, 620, 621n4, 622n13, 624n35, 625n42, 626n54, 627n7, 627n9, 630n44, 630n45, 630n48, 631n58
Clausen, Rebecca, 202, 344, 514n54, 528n22, 545n3, 563n24
Cleaver, Harry, 225, 555n125
Clements, Frederick, 269
Cleveland, Cutler, 560n206
Cobb, John B., Jr., 452, 606n61, 617n58, 618n74, 625n49, 631n62
Cohen, Boyd, 597n1
Cohen, Ira J., 542n190
Cohen, Murray J., 544n211
Cole, Matthew, 624n40
Colletti, Lucio, 42–43, 501n6
Collins, Randall, 176, 177, 539n120, 539n125
Collinson, Diane, 594n32
Commoner, Barry, 19, 33, 55, 86, 135, 246, 326, 348, 385, 388, 469, 495n27, 495n28, 506n44, 515n13, 527n8, 527n28, 561n5, 562n13,

INDEX

592n24, 599n29, 601n38, 620n3, 623n27, 623n29, 624n34, 625n42, 625n44
Comte, Auguste, 189, 192
Conway, Erik M., 447, 453, 616n48, 619n75, 619n76
Cook, Deborah, 506n45
Cooper, James Fenimore, 185
Cosgrove, Emma, 609n40
Costa, Monica F., 624n39
Costanza, Robert, 237–40, 558n188, 558n189, 558n190, 559n194, 559n196, 560n208
Coumel, Laurent, 335, 589n2, 591n18, 592n23, 595n41
Cox, Oliver Cromwell, 355
Cox, Paul, 616n51
Cox, Stan, 616n51
Cronon, William, 268, 573n140
Crosby, Alfred, 149
Crowley, Joe, 396
Crumley, Carole, 551n69
Crutzen, Paul J., 21, 82, 246, 460–61, 496n40, 514n2, 562n10, 621n9, 622n11
Cullen, Heidi, 510n10, 521n21
Curtis, B., 571n111

Daly, Herman, 238, 365, 366, 370, 558n188, 559n192, 559n193, 559n195, 559n196, 596n7, 596n8, 597n19
Daniels, Roland, 27, 70, 283, 497n57, 512n35, 577n17
Darby, Andrew, 538n113
Darwin, Charles, 19, 52, 254, 303, 308, 345–46, 354, 358, 415, 463, 564n25, 585n32
Darwish, Mahmoud, 150, 529n5
Davis, Mike, 356, 418, 425, 426, 583n7, 608n24, 610n48
Dawson, Michael C., 500n96, 517n15, 599n22, 599n23, 600n33
De Bary, Anton, 615n37

De Kadt, Maarten, 501n2, 504n35
DeBardeleben, Joan, 335, 589n1, 592n23, 592n24, 594n37, 594n38, 595n40, 595n42
Defoe, Daniel, 163
Denon, M., 544n209
Descartes, René, 144, 145
Devall, Bill, 528n21
Devaney, Jacob, 520n13
Devincenzo King, Danielle Marie, 235–36
Diamond, Jared, 446, 447, 616n48, 616n49
Dick, Christopher, 545n4
Dickens, Charles, 101, 310
Dickens, Peter, 249, 530n1, 530n2, 531n7, 532n12, 532n14, 533n21, 563n22
Dietz, Thomas, 530n6, 551n71
Dietzgen, Marxist Joseph, 565n36
Dilworth, Craig, 585n32
Dobb, Maurice, 548n26
Dobrovolski, Ricardo, 202, 545n3
Dobson, Andrew, 510n18, 528n21
Dowd, Douglas, 388, 600n33, 600n36
Downey, Liam, 530n6, 552n74
Drozozdov, O. A., 593n26
Du Bois, W. E. B., 631n58
Ducrot, Christian, 608n18
Dudley, Dud, 538n113
Dühring, Eugen, 270
Dunlap, Riley E., 154, 156, 195, 530n1, 530n2, 532n12, 532n14, 533n21, 543n203, 565n32
Dunne, Daisy, 526n12
Durkheim, Emile, 154, 155, 530n4, 533n21
Dylis, N., 320, 590n9, 590n10

Eagleton, Terry, 618n73, 629n27
Echo Xie, 610n43
Eckersley, Robyn, 249, 528n21, 563n18

Edburg, Rolf, 516n20
Edinger, Evan N., 582n74
Eilperin, Juliet, 523n35
Einstein, Albert, 309
Eldredge, Niles, 585n32
Elie, Marc, 335, 589n2, 592n23, 595n41
Elliott, Larry, 394, 602n20
Ellis, Erle, 108, 519n32
Elton, Charles, 268
Elvin, Mark, 534n40, 537n88
Emmanuel, Arghiri, 208–9, 210, 213–14, 229, 547n14, 548n27, 548n28, 548n31, 548n37, 551n62, 556n137
Empson, Martin, 602n15
Engel-Di Mauro, Salvatore, 501n2, 504n35, 567n47
Engelke, Peter, 495n26, 497n58, 613n9, 613n11, 620n1
Engels, Frederick, 28, 29, 30, 31, 32, 33, 34, 35, 42, 43, 47, 54–55, 56, 74, 83, 109, 124, 133, 147, 169, 224, 225, 227, 238, 245, 248, 249, 252, 254, 258, 262, 269–70, 271, 273, 274, 276, 277, 285, 295–301, 303–15, 318, 320, 326, 329, 330, 331, 342, 349, 351, 353, 354, 355, 374, 378, 381, 415, 418, 431, 438, 440, 441, 442, 443, 448, 450, 484, 485, 486, 487, 488–89, 494n23, 498n68, 498n69, 498n73, 499n78, 499n82, 500n97, 500n100, 500n101, 501n5, 502n17, 503n26, 504n32, 504n35, 505n40, 505n41, 507n54, 512n32, 513n38, 513n45, 514n63, 515n6, 524n47, 524n49, 527n26, 528n24, 530n5, 537n75, 537n78, 538n101, 538n113, 542n180, 549n48, 551n59, 555n119, 555n121, 556n138, 561n2, 561n227, 567n47, 567n51, 568n66, 569n75, 569n84, 571n102, 574n144, 574n145, 574n146, 574n147, 574n150, 575n176, 576n1, 576n3, 576n4, 577n14, 577n19, 583n1, 583n4, 583n6, 583n7, 583n8, 583n9, 583n80, 584n11, 584n12, 584n13, 584n15, 584n16, 584n17, 584n19, 584n20, 584n21, 584n22, 584n26, 585n30, 585n32, 586n35, 586n38, 586n41, 586n42, 586n44, 586n45, 587n49, 587n54, 587n58, 588n62, 588n63, 588n64, 588n65, 588n66, 588n67, 588n70, 588n72, 593n29, 597n2, 597n7, 598n10, 598n17, 607n4, 607n13, 608n26, 611n62, 612n67, 612n69, 613n12, 614n21, 614n25, 614n30, 615n32, 615n37, 615n38, 615n39, 616n52, 618n66, 620n86, 626n58, 627n7, 628n13, 630n38, 630n40, 630n41, 630n42, 630n46, 630n47, 630n49, 630n50, 631n54, 631n55, 631n57
Engels, Jens F., 539n117
Ennis, Paul, 570n99
Epicurus, 76, 244, 342, 350, 351, 474, 502n12, 513n52, 561n228, 626n58
Essig, Michael, 420
Estes, Nick, 492, 523n39, 619n83, 626n56, 632n64
Evanoff, Richard, 294, 582n71

Fadli, 611n57
Faiz, Faiz Ahmad, 150
Fanon, Frantz, 626n60, 627n6
Farrington, Benjamin, 33, 55, 274, 348, 528n11
Faulkner, William, 626n2
Fearn, Eva, 521n21
Fedorov, Evgeni K. (Y. K. Fyodorov), 86, 324, 326, 327, 328, 331, 332, 336, 441, 515n14, 516n15, 567n51, 592n24, 615n31
Fedoseev (also Fedorseyev), Pyotr Nikolaevich, 449, 617n57, 618n61, 625n48

INDEX

Feenberg, Andrew, 498n72
Feltham, O., 566n42
Feltri, Stefano, 609n32
Ferri, Enrico, 270
Feshbach, Murray, 589n1
Feuerbach, Ludwig, 351, 502n12
Finamore, Barbara, 452, 606n61, 618n74, 619n81, 619n82
Fischer-Kowalski, Marina, 56, 506n46, 506n47
Fisher, Mark, 493n2, 629n24, 629n33
Fisher, R. A., 310
Flanders, Jonathan, 521n18
Fletcher, J. O., 515n4, 593n26
Foley, Duncan K., 225, 555n125, 626n53
Folke, Carl, 524n46
Foner, Philip S., 611n62
Fotopoulos, Takis, 368, 596n13
Fourier, Charles, 438, 439, 614n22
Fox, Stephen, 387, 600n33
Fraas, Carl, 283, 326, 441, 567n51, 598n10
Fracchia, Joseph, 541n165, 577n17
Frank, Andre Gunder, 209, 355, 551n69
Fraser, Evan D., 615n37
Fraser, Nancy, 34–35, 500n94, 500n96
Fratto, Elena, 621n5
Freedman, Andrew, 522n23
Freinkel, Susan, 600n35
Fricko, Oliver, 604n45
Friedman, Michael, 344, 526n20, 526n21
Friedman, Milton, 17, 494n19
Friedman, Thomas, 394, 602n21
Friendly, Alfred, Jr., 589n1
Fritsch, Kelly, 527
Frolov, Ivan T., 274, 330, 331, 332, 334, 336, 450, 594n31, 594n32, 618n62
Fromm, Erich, 506n45

Fuentes, Frederico, 517n19, 610n49
Fukuyama, Francis, 475–76, 510n15, 627n5
Fyodorov, Y. K., 593n27: *See* Fedorov, Evgeni K.

Galaz, Victor, 607n9
Galbraith, John Kenneth, 65–66, 97, 384–85, 510n14, 517n11, 599n29
Gale, W. K. V., 538n113
Galeano, Eduardo, 213, 356, 551n61
Galloway, Tamara S., 624n40
Galuszka, Agnieszka, 613n9
Gálvez, Jaime Paneque, 499n76
Gandhi, Mohandas, 269
Ganghofer, Ludwig, 185
Garcia Linera, Álvaro, 517n19
Gare, Arran, 594n33, 617n57, 625n48
Gareau, Brian J., 564n27, 565n33, 573n132
Gates, Bill, 127
Gattuso, J. P., 582n74
Geden, Oliver, 509n8
Georgescu-Roegen, Nicholas, 189, 238, 365, 540n150, 559n191, 559n192, 559n195, 596n7
Gerasimov, Innokenti P., 330, 331, 333, 591n11, 593n26, 593n30, 594n37, 594n38
Gerth, H. H., 542n185
Gervais, Paul, 464
Geyer, Roland, 624n36, 624n38, 625n41
Gibson, James William, 542n189
Giddens, Anthony, 343, 544n211
Gijswijt, August, 530n1, 530n2, 532n12, 532n14, 533n21
Gilbert, Marius, 607n9
Gilchrist, Glenn, 520n10
Gilly, Adolfo, 517n14
Gironi, Fabio, 570n99
Glacken, Clarence J., 598n10
Goldblatt, David, 156, 532n17, 544n205

Goldstein, Amanda Jo, 630n37
Golitsyn, Georgi S., 329, 589n4, 593n28, 617n60, 624n33
Golley, Frank Benjamin, 53, 505n38, 515n7, 527n7, 547n10, 552n84, 589n6, 590n7, 591n11
Good, John Mason, 544n209
Goodell, Jeff, 526n23
Goodwin, Richard M., 225, 555n125
Gorbachev, Mikhail, 326, 334, 594n32
Gorz, Andre, 68, 248–49, 511n23, 563n18
Gould, Kenneth A., 600n31
Gould, Stephen Jay, 33, 56, 274, 310, 348, 567n49, 585n32, 587n57, 593n29, 624n32
Gouldner, Alvin, 353
Graham, Loren R., 590n7, 591n18, 594n32, 594n37
Gramsci, Antonio, 331, 348, 501n4, 501n7
Grant, Don, 530n6
Gray, Helena, 509n6
Green, John, 588n70
Gregory, Derek, 579n41
Grigoriev, A. A., 330
Grinevald, Jacques, 87, 246, 495n32, 496n41, 514n1, 516n16, 562n7, 612n4, 618n68, 622n13
Grossman, Henryk, 208, 548n29
Grubler, Arnulf, 605n52
Guevara, Ernesto "Che," 209, 355, 548n36
Gunderson, Ryan, 202, 545n3
Guterres, António, 24

Haartsen, Adriaan, 534n40
Haberl, Helmut, 506n47
Haeckel, Ernst, 52, 84, 270, 504n35, 505n37
Haff, Peter, 471, 625n46
Hagen, Joel B., 552n83, 552n85, 577n21, 620n3

Hajdas, Irka, 623n27
Haldane, J. B. S., 33, 55, 84, 254, 274, 310, 318, 348, 351, 496n34, 515n10, 564n25, 587n55, 631n58
Hall, Charles A. S., 522n28, 553n92
Hallett, Steve, 522n28
Hallock, Pamela, 582n74
Haltiwanger, John, 601n4
Hamilton, Clive, 87, 132, 246, 496n41, 514n1, 516n16, 525n8, 526n12, 526n13, 526n14, 526n15, 526n22, 562n7, 587n59, 612n4, 618n68, 622n13
Hamilton, Peter, 542n186
Hampton, Paul, 598n10
Hansen, James E., 19, 51, 57–58, 63, 65, 79, 93, 112, 119, 125, 127, 131, 363, 447, 452, 504n32, 507n52, 507n55, 509n5, 509n9, 510n10, 510n11, 510n12, 519n2, 519n6, 520n9, 521n21, 524n51, 525n3, 526n19, 574n146, 596n1, 612n2, 616n48, 617n55, 618n73, 621n4, 625n52
Harman, Graham, 579n40
Harper Davis, W. M., 540n149
Harrington, Michael, 355
Harrison, Jane, 352
Hartshorne, Richard, 544n207
Harvey, David, 51, 78, 109, 273, 356, 371, 504n31, 514n58, 519n35, 547n14, 575n167, 597n22
Hauser, Arnold, 349
Hausfather, Zeke, 604n44
Hawken, Paul, 580n52, 597n1
He, Huiil, 619n79
Head, Martin J., 621n8
Healey, P., 566n42, 579n43
Hegel, Georg Wilhelm Friedrich, 30, 31, 42, 43, 145, 149, 252, 267, 271, 304, 305, 349, 350, 351, 475, 477, 527n4, 528n14, 565n36, 574n155, 575n169, 627n4, 627n5, 627n9
Heilbroner, Robert, 17, 22, 223,

INDEX 641

225, 494n21, 496n43, 555n116, 555n125, 626n55
Heinrich, Michael, 550n57
Helmholtz, Hermann, 308
Henderson, W. O., 504n33
Heraclitus, 311
Herman, Edward S., 518n22
Hermele, Kenneth, 502n20, 512n37, 545n4, 549n48, 567n48, 576n12
Hessen, Boris, 32, 33, 274, 319, 542n182
Hettner, Alfred, 196, 544n207
Hewison, Christian H., 583n5
Hickel, Jason, 404, 604n45, 605n52, 605n53
Hilferding, Rudolf, 379, 412, 607n5
Hill, Christopher, 630n35
Hillier, J., 566n42, 579n43
Hines, Colin, 394
Hitler, Adolf, 508n1
Hobbes, Thomas, 480, 628n16
Hobsbawm, E. J.., 539n128
Hodgskin, Thomas, 15, 493n8
Hoffman, Frederik J., 626n2
Hoffmann, Richard C., 539n117
Hogben, Lancelot, 33, 55, 254, 269, 274, 348, 351, 352
Hogerwerf, Lenny, 414, 607n9, 607n10
Holleman, Hannah, 344, 345, 357, 499n89, 505n38, 508n60, 512n37, 523n36, 529, 545, 562n15, 567n48, 605n60, 630n44, 630n48, 631n58
Holmberg, Mollie, 607n9
Holt-Giménez, Eric, 417, 606n2, 608n21
Hong, Per Kristian, 609n34
Honigsheim, Paul, 540n138
Hook, Sidney, 257, 568n67
Hooks, Gregory, 530n6
Hopkins, Terence, 412, 607n6
Horkheimer, Max, 31, 68, 192, 248, 264, 506n45, 511n21, 542n187, 564n26, 585n28

Hornborg, Alf, 219, 239–41, 502n20, 512n37, 545n4, 549n48, 549n49, 551n59, 551n69, 551n72, 552n82, 559n200, 560n202, 560n203, 560n204, 560n205, 560n207, 560n208, 560n209, 560n213, 567n48, 576n12
Howard, Michael Charles, 225, 555n125
Huan Qingzhi, 617n57, 625n48
Huang, Cary, 611n55
Hudson, Richard, 623n30
Hughes, H. Stuart, 349
Hulme, Michael, 504n33
Hunt, E. K., 547n17, 560n201
Hutchinson, G. Evelyn, 33, 496n35, 515n12, 562n6
Huxley, Julian, 564n25
Huxley, Thomas, 303, 346, 415, 463
Hyndman, H. M., 355

Ibañez, Daniel, 568n56, 576n7, 616n53
Ibsen, Hilde, 539n117
Iggers, George G., 542n189
Inman, Mason, 522n28
IPCC (Intergovernmental Panel on Climate Change), 601n3, 602n13
Ivanov, N. D., 591n18
Ivar do Sul, Juliana Assunção, 623n27, 624n39
Izrael, Yuri A., 525n5, 589n4, 592n24, 593n28, 617n60

Jaccard, Mark, 106, 107, 518n29, 518n30
Jacoby, Russell, 43, 499n75, 501n4, 510n17, 511n20, 565n35, 577n24, 584n26
Jambeck, Jenna R., 624n36, 624n38, 625n41
James, William, 409, 568n68, 606n69
Jameson, Fredric, 291, 482, 493n2,

499n75, 565n35, 568n59, 582n65, 629n32
Jänicke, Martin, 580n52
Järvikoski, Timo, 531n8
Jarvis, Brooke, 520n13
Jay, Martin, 564n26, 565n35
Jeandel, Catherine, 623n27
Jellinek, Georg, 184
Jensen, Derrick, 448, 482, 616n54, 629n31
Jevons, William Stanley, 539n128
Joad, C. E. M., 262, 568n68, 571n104
Johnston, James F. W., 46
Jones, Eric L., 537n82
Jones, Richard, 536n74
Jonna, R. Jamil, 498n65, 499n89, 606n2, 618n67
Jorgenson, Andrew, 218
Jorgenson, Andrew K., 530n6, 545n4
Jorgenson, E. G., 552n74, 552n75, 552n77, 552n78
Josephson, Paul, 336, 592n23, 595n44
Jukes, Joseph Beete, 504n32, 574n146

Kalberg, Stephen, 161, 164, 533n25, 534n34, 535n49, 536n60, 543n196
Kallis, Giorgos, 404, 604n45, 605n53
Kangal, Kaan, 500n91
Kant, Immanuel, 142, 273, 308, 349, 350
Kapadia, Shefali, 609n42, 610n52
Kaplan, Deborah Abrams, 609n42
Kapp, K. William, 70, 248, 502n16, 512n34, 562n13, 598n11
Kates, Robert W., 62, 435, 437, 471, 508n2, 562n12, 568n55, 613n6, 613n8, 622n18, 625n47
Katsikis, Nikos, 568n56, 576n7, 616n53
Kaufman, Alexander C., 602n22
Kautsky, Karl, 598n11
Keating, Kenneth M., 188, 541n169
Keith, Lierre, 616n54
Keller, B., 515n4, 593n26
Keller, David R., 527n7
Kelley, Florence, 631n58
Kelly, Jane, 617n56
Keynes, John Maynard, 385
Kharecha, Pushker A., 519n2
Khrushchev, Nikita, 325
Kidron, Michael, 384, 599n25
King, David, 558n181
King, John Edward, 225, 555n125
Klare, Michael T., 519n4, 521n19, 521n20
Klein, Naomi, 66, 78, 91–110, 148, 294, 345, 390, 400–401, 405, 409, 510n16, 514n59, 516n1, 516n2, 516n3, 516n4, 517n7, 517n8, 517n9, 517n11, 517n13, 517n16, 517n19, 518n20, 518n26, 518n31, 519n33, 519n37, 527n8, 528n22, 529n1, 574n152, 582n72, 601n2, 601n7, 602n16, 602n18, 603n31, 604n39, 604n40, 604n41, 604n43, 605n56, 606n62, 606n68
Klitgaard, Kent A., 522n28
Koch, Andrew M., 542n190
Kock, Richard, 414, 607n9, 607n10
Köhler, Gernot, 549n40, 549n42
Kojève, Alexandre, 627n5
Kolakowski, Leszek, 343
Kolbert, Elizabeth, 104, 105, 496n45, 518n26, 612n2
Korsch, Karl, 501n4
Kovel, Joel, 249, 343, 563n18
Krader, Lawrence, 169, 536n74, 537n79, 537n81
Krause, Ulrich, 225, 555n125
Krausmann, Fridolin, 506n47
Krebs, Christopher C., 611n59
Krieger, Nancy, 34
Krugman, Paul, 106–7, 518n30
Kump, Lee R., 618n73

INDEX

Lambert, Audrey M., 534n40
Lambin, Eric F., 612n4
Lange, Friedrich, 305
Lankester, E. Ray, 19, 52, 55, 254, 274, 303, 306, 311, 346, 415, 418, 463, 495n31, 504n34, 505n37, 584n23, 584n24, 586n38, 607n13, 608n26, 622n19
Laplace, Pierre-Simon, 308
Lapp, Ralph E., 623n30
Lash, Scott, 534n32, 544n211
Latouche, Serge, 364, 366–67, 368, 369–71, 596n9, 596n10, 596n11, 596n14, 597n18
Latour, Bruno, 107, 249, 250, 251, 258, 261, 266, 267, 287, 296, 518n31, 563n20, 565n30, 565n31, 569n81, 570n99, 573n130, 573n131, 573n136, 573n139, 578n38, 579n40
Lau Kin Chi, 606n61, 619n79
Lauderdale (Earl): *See* Maitland, James
Lautzenheiser, Mark, 547n17, 560n201
Lavergne, Léonce de, 513n44
Law, J., 573n139
Law, Kara Lavender, 624n36, 624n38, 625n41
Lawes, J. B., 46
Lawrence, Kirk S., 545n4, 552n74
Leach, Melissa, 607n9
Leacock, Eleanor, 299, 584n12
Leakey, Richard E., 496n39, 527n8, 625n47
Lebowitz, Michael A., 548n21, 550n55
Lefebvre, George, 601n1, 630n35
Lefebvre, Henri, 33, 499n88, 608n25
Leggett, Jeremy, 117, 394, 522n29, 523n40
Leibniz, Gottfried Wilhelm, 252, 266, 293, 565n36, 573n131
Leinfelder, Reinhold, 621n8, 623n27, 624n39

Leiss, William, 451, 527n8, 564n26, 618n69
Lemeshko, S., 592n19
Leng, Sidney, 610n43
Lenin, V. I., 16, 270, 274, 318, 319, 333, 335, 349, 351, 354, 355, 494n18, 547n14, 574n150, 631n52
Lenton, Tim, 496n38, 612n4
Leonard, Annie, 600n35
Leonard, Matt, 608n27, 610n50
Lessing, Doris, 359
Lessner, Friedrich, 504n33
Lester, Joseph, 584n23, 587n60
Levi, Michael, 118, 520n12, 522n26, 522n28, 523n32
Levins, Richard, 33, 55, 84, 273–74, 310, 348, 414, 415, 501n5, 504n29, 506n42, 506n44, 515n10, 546n9, 556n139, 575n170, 587n49, 587n56, 588n67, 589n6, 607n11, 607n13, 619n78
Levy, Hyman, 33, 53, 254, 274, 306, 348, 351, 504n29, 586n36, 586n39, 608n25
Lewan, Lillemor, 552n77
Lewin, Roger, 496n39, 527n8, 625n47
Lewis, Ceri, 624n40
Lewontin, Richard, 33, 55, 84, 274, 310, 348, 414, 501n5, 504n29, 506n42, 506n44, 515n10, 587n49, 587n56, 589n6, 607n11, 619n78
Leys, Colin, 571n113, 600n35, 608n22
Liebig, Justus von, 27, 35, 46–47, 48, 54, 55, 70–71, 177, 180, 181, 211, 212, 283, 308, 312, 377, 490, 498n60, 502n16, 549n45, 549n47, 577n18
Liebman, Alex, 414, 607n11
Lilley, S., 570n88
Lindbeck, Assar, 599n28
Lindsay, Jack, 33, 348, 351, 352–53

Lindsay, R. Bruce, 512n35
Lindwall, Courtney, 520n8, 523n38
Locke, John, 172, 538n97
Lonergan, Stephen, 225, 229, 555n125, 556n144
Longo, Stefano B., 344, 514n54, 528n22, 563n24
Lord, John, 538n113
Los, V. A., 450, 618n64
Lotka, Alfred J., 222, 228, 554n103
Lough, Joseph W. H., 540n130
Love, John, 537n84
Lovejoy, Arthur, 565n29
Lovelock, James, 20, 21, 446, 496n37, 616n48, 617n55
Lovins, Amory, 580n52, 597n1
Lovins, L. Hunter, 580n52, 597n1
Löwy, Michael, 297, 583n3, 626n1, 631n62
Lucretius, 141, 257, 342
Lukács, Georg, 30, 31, 37, 42, 43, 44, 45, 149, 151, 192, 253, 271–72, 277, 284, 291, 292, 293, 348, 349, 350–51, 352, 493n1, 498n71, 498n72, 498n73, 498n74, 500n102, 501n3, 501n4, 501n7, 502n8, 502n10, 502n12, 504n28, 542n186, 566n37, 569n75, 574n153, 574n156, 574n157, 574n158, 576n2, 577n23, 581n58, 582n61, 582n62, 582n66, 582n69, 584n26, 631n61
Lunacharskii, Anatolii Vasil'evich, 318
Luxemburg, Rosa, 29, 347, 449, 498n66, 547n14, 617n56
Lydoph, Peter E., 593n32
Lyell, Charles, 585n32
Lyotard, Jean-François, 629n25
Lysenko, Trofim Denisovich, 319, 323–24, 335, 589n6, 591n18

Machlis, Gary E., 541n169
Machlis, Rosa, 188

Mackinder, Halford, 149, 529n4
Maclean, Gulberk K., 569n81, 570n99, 578n35
Macmillan, Thomas, 573n129, 573n132
Macpherson, C. B., 528n10
Magdoff, Fred, 497n59, 505n39, 514n60, 524n50, 575n178, 586n37, 586n43, 605n58, 608n23, 612n5, 618n65, 626n54, 631n59
Magdoff, Harry, 355, 494n22, 507n57, 547n14, 592n21, 623n29
Mage, John, 347
Maine, Henry, 487
Maitland, James (Earl of Lauderdale), 226, 282, 375–76, 430, 577n14, 597n4
Malenkov, Georgy, 325
Malm, Andreas, 249, 344, 462, 494n16, 498n65, 563n22, 622n17
Malone, Sheila, 617n56
Malthus, Thomas, 62, 68, 154, 435, 511n23, 536n74, 551n65
Mamedov, N. M., 332
Mandel, Ernest, 377, 549n40, 597n8
Manicas, Peter T., 305, 585n31
Manji, Firoze, 632n63
Mann, Charles C., 519n3, 522n31
Mann, Michael E., 452, 526n11, 537n82, 618n73
Mårald, Erland, 549n45
Marchetti, Cesare, 128, 129
Marcuse, Herbert, 248, 253, 349, 452, 501n6, 506n45, 562n13
Marglin, Stephen A., 495n26
Margulis, Lynn, 20–21, 84, 496n37, 515n8
Marini, Ruy Mauro, 354
Markey, Edward, 391, 396, 397, 399, 604n56
Marsh, George Perkins, 83, 515n7
Martin, Kenneth E., 565n32
Martindale, Don, 534n38, 536n55, 542n190

INDEX

Martinez-Alier, Joan, 191, 224, 225, 368, 506n47, 533n29, 541n172, 542n174, 542n176, 542n179, 542n182, 545n4, 549n47, 551n60, 552n81, 555n120, 555n125, 558n179, 560n204, 597n16
Marx, Eleanor (Aveling), 355, 588n72
Marx, Karl, 15, 16, 19, 26–29, 31, 32, 34, 35, 41–56, 58–61, 67, 68, 69, 70, 71–74, 76–77, 78, 80, 83, 84, 86, 98, 109, 124, 145–46, 149, 150, 152, 154, 155, 168–69, 179, 180, 200, 201, 202, 203, 204, 205, 206, 207, 211–12, 213, 220, 221–29, 233, 237, 239, 240, 242, 243, 244, 245, 247, 248, 249, 250, 251, 252, 253–60, 263–66, 269–74, 277–84, 286, 289, 290–92, 294–95, 296, 297, 298, 299, 301, 303, 311, 312, 313, 314, 318, 320, 326, 330, 331, 338–47, 349–52, 353, 354, 357–58, 365, 369, 371, 373–79, 381–82, 386, 389, 408–9, 412, 415, 421, 429, 432, 436, 438, 440, 441, 442–44, 446, 448, 450, 463, 473, 476, 477–78, 481, 484, 485, 486, 487, 488–89, 490, 494n12, 494n13, 494n14, 494n17, 494n22, 494n23, 497n58, 497n59, 498n60, 498n61, 498n62, 498n67, 498n68, 498n69, 498n73, 499n78, 499n82, 499n83, 500n97, 500n100, 500n101, 500n102, 502n12, 502n13, 502n14, 502n17, 502n19, 502n21, 503n24, 503n25, 503n26, 503n27, 504n31, 504n32, 504n34, 504n35, 505n38, 505n41, 506n43, 506n45, 506n47, 507n53, 507n54, 507n55, 511n23, 511n27, 511n28, 511n29, 511n30, 512n32, 512n33, 512n36, 513n38, 513n39, 513n40, 513n41, 513n44, 513n45, 514n54, 514n55, 514n61, 514n63, 515n6, 515n7, 524n47, 524n48, 527n26, 528n15, 528n16, 528n17, 528n18, 528n19, 528n20, 528n23, 529n7, 529n10, 529n11, 530n5, 536n74, 537n75, 537n76, 537n77, 537n80, 538n113, 542n180, 547n14, 548n22, 548n23, 548n24, 548n25, 548n26, 549n46, 549n47, 549n48, 549n49, 550n51, 550n52, 550n53, 550n55, 550n56, 550n57, 550n58, 552n86, 553n89, 554n114, 555n119, 555n121, 556n129, 556n134, 560n201, 561n2, 561n225, 561n226, 561n227, 564n25, 565n29, 565n36, 567n47, 567n51, 568n66, 569n74, 569n75, 569n84, 570n89, 571n102, 571n110, 574n144, 574n145, 574n146, 574n147, 574n150, 575n165, 575n166, 575n176, 576n1, 576n3, 576n4, 576n9, 576n10, 576n11, 576n179, 577n14, 577n15, 577n18, 577n19, 577n20, 580n50, 581n57, 581n59, 581n60, 582n63, 582n64, 582n76, 582n77, 583n1, 583n4, 583n6, 583n7, 583n8, 583n9, 584n11, 584n13, 584n15, 584n16, 584n17, 584n19, 584n20, 584n21, 584n22, 584n26, 585n30, 585n32, 586n35, 586n38, 586n41, 586n44, 587n54, 587n58, 588n62, 588n63, 588n66, 588n67, 588n68, 588n70, 588n72, 593n29, 594n32, 595, 597n2, 597n3, 597n5, 597n6, 597n7, 597n17, 598n9, 598n10, 598n11, 598n17, 599n21, 606n65, 606n66, 607n4, 607n7, 607n8, 608n26, 611n61, 612n67, 612n68, 612n69, 613n12, 614n21, 614n26, 614n29, 614n30, 615n32, 615n33, 615n34, 615n37, 615n38, 615n39, 616n52, 618n66, 620n86, 622n13, 626n57, 626n58, 627n7, 627n9, 628n11,

628n13, 630n38, 630n39, 630n40,
  630n41, 630n42, 630n43, 630n44,
  630n45, 630n46, 630n47, 630n48,
  630n49, 630n50, 631n54, 631n55,
  631n56, 631n57, 631n60
Marx, Leo, 481, 626n55, 629n26
Mash, M., 571n111
Masr, Mada, 608n24
Mayer, Julius Robert, 512n35
Mayumi, Kozo, 559n195
Mazoyer, Marcel, 540n142
Mazzocchi, Tony, 395
McAffee, Andrew, 629n22
McBay, Aric, 482, 616n54, 629n31
McCall, Michael K., 499n76
McCarthy, James, 564n27, 578n37
McChesney, Robert W., 508n60,
  513n50, 518n22, 611n64, 619n83
McClellan, David, 353
McDermott, Matt, 522n31
McGee, Kyle, 573n130, 573n131
McGuckin, Gene, 524n50
McKay, Reid, 520n10
McKibben, Bill, 111–12, 120, 519n5,
  520n12, 523n41, 527n8
McMichael, Philip M., 202, 500n1,
  501n2, 503n25, 504n35, 545n2,
  545n3, 547n13, 561n223, 608n22
McNeill, John Robert, 461, 495n26,
  495n32, 539n126, 545n4, 592n23,
  613n9, 613n11, 620n1, 622n11
McNeill, William H., 613n16,
  628n10
Meadows, Dennis L., 496n36, 596n3
Meadows, Donella H., 496n36,
  596n3
Medvedev, Roy, 589n6
Medvedev, Zhores, 591n18
Mehring, Franz, 597n2
Meier, Walt, 116
Meinshausen, Malte, 521n22
Melillo, Edward D., 558n183
Melotti, Umberto, 537n80
Mendeleev, Dmitri, 306

Merchant, Carolyn, 527n8, 573n143
Mercure, J. F., 604n40
Merleau-Ponty, Maurice, 285,
  501n4, 501n6, 577n24
Mészáros, István, 44–45, 83, 255–56,
  272, 279–80, 341, 349, 352, 354,
  414, 500n101, 502n13, 504n30,
  515n5, 524n47, 525n52, 566n37,
  568n57, 569n75, 575n160,
  575n162, 575n163, 576n8,
  576n10, 582n75, 582n78, 588n67,
  598n20, 607n11, 626n2, 626n54,
  627n4, 628n16, 631n61, 632n63
Micklin, Philip, 592n22
Middleton, Guy D., 465–66, 623n24
Miliband, Ralph, 356
Mill, James, 280, 536n74
Mill, John Stuart, 168, 205, 206, 365,
  536n74, 596n7
Mills, C. Wright, 542n185
Milton, David, 603n25
Mobarak, Ahmed Mushfiq, 424,
  610n47
Möbius, Karl, 320
Mol, Arthur P. J., 544n211, 580n52
Molyneux, John, 344
Moon, David, 591n15, 591n17,
  592n24
Mooney, Chris, 582n74
Moore, Jason W., 202, 258–62,
  264, 267, 270, 285, 286, 287,
  288–90, 493n3, 500n1, 503n23,
  531n7, 545n3, 557n163, 564n27,
  568n54, 568n56, 569n73, 569n74,
  569n76, 569n77, 569n78, 569n79,
  569n80, 569n82, 569n83, 569n84,
  570n85, 570n86, 570n87, 570n88,
  570n90, 570n91, 570n92, 570n93,
  570n95, 570n96, 570n97, 570n99,
  571n100, 571n101, 576n5, 576n6,
  576n7, 577n26, 578n28, 578n29,
  578n30, 578n31, 578n32, 578n33,
  579n39, 579n41, 579n42, 580n47,
  580n48, 580n49, 580n50, 580n51,

INDEX

581n53, 581n54, 581n55, 581n57, 582n67, 582n70, 616n53, 622n16, 622n17
More, Thomas, 22, 496n44
Morelle, Louis, 570n99
Morishma, Michio, 225, 555n125
Morozov, Georgii Fedorovich, 320, 322
Morris, Adam, 516n2
Morris, May, 507n57, 508n58
Morris, William, 59, 60, 92, 101, 109–10, 254, 351, 352, 355, 388, 438, 440, 449, 483, 484, 507n57, 508n58, 516n3, 518n23, 519n36, 601n37, 605n58, 614n27, 614n28, 630n37
Morton, Oliver, 525n7
Moses (biblical character), 15, 369
Muller, H. J., 33
Mumford, Lewis, 76, 77, 80–81, 176, 365, 444–45, 513n51, 514n65, 539n124, 596n7, 615n42, 616n43, 616n44
Murota, T., 547n12, 553n96
Murphy, Fred, 579n41
Murphy, Raymond, 155, 192, 531n11, 532n12, 532n13, 542n188
Musto, Marcelo, 605n55

Naam, Ramez, 519n1, 522n26
Napoletano, Brian, 344, 499n76, 499n88
Navarro, Peter, 427
Nayeri, Kamran, 580n50
Needham, Joseph, 33, 34, 55, 254, 274, 309–10, 348, 351, 451, 452, 499n90, 575n168, 581n58, 587n53, 587n54, 587n58, 618n70, 618n72
Neruda, Pablo, 150
Newton, Isaac, 242
Nixon, Rob, 103, 518n24
Noone, Kevin, 612n4

Nordhaus, Ted, 251, 266, 267, 289, 565n32, 573n134, 581n53
Novick, I. B., 592n24
Novikov, R. A., 593n27

Obama, Barack, 119, 394, 524n51
Ocasio-Cortez, Alexandria, 391, 396, 397, 399, 601n6, 604n36
Ochs, Phil, 474, 626n59
O'Connor, Clare, 527
O'Connor, James, 249, 531n7, 563n18
O'Connor, Martin, 596n12
Odih, Pamela, 249, 563n23
Odum, Elisabeth C., 546n6, 554n100, 554n102, 554n104, 556n141, 557n154, 557n156, 557n159, 557n166, 557n167, 557n169, 558n187, 561n218
Odum, Eugene P., 53, 219, 505n38
Odum, Howard T., 53, 55, 87, 88, 201, 203–4, 219–44, 469, 505n38, 516n17, 516n18, 516n19, 546n6, 546n7, 547n11, 547n12, 550n55, 552n86, 553n87, 553n88, 553n89, 553n90, 553n91, 553n94, 553n95, 553n98, 553n99, 554n100, 554n101, 554n102, 554n104, 554n105, 554n109, 554n111, 554n115, 555n116, 555n117, 555n123, 556n126, 556n128, 556n131, 556n132, 556n133, 556n135, 556n141, 556n143, 556n148, 557n151, 557n152, 557n154, 557n155, 557n156, 557n157, 557n158, 557n159, 557n160, 557n162, 557n163, 557n165, 557n166, 557n167, 557n168, 557n169, 557n170, 557n171, 558n185, 558n186, 558n187, 558n189, 559n196, 559n198, 559n199, 560n204, 560n206, 560n208, 560n210, 561n218, 561n219, 561n221

Oldak, P. G., 333, 594n38
Oldfield, Jonathan D., 593n26, 593n30
O'Leary, Brendan, 537n76
O'Leary, Lizzie, 609n37, 609n42
Olenicoff, S. M., 515n4, 593n26
Ollman, Bertell, 350, 504n29, 546n9, 565n36
Oman, Luke, 525n8
O'Neil, Shannon K., 609n31
Oparin, Alexander I., 33, 84, 274, 318, 329, 336, 515n10, 593n29
Oreskes, Naomi, 447, 453, 495n32, 616n48, 619n75, 619n76
Ostwald, Wilhelm, 176, 182, 186–92, 199, 539n123, 541n164, 541n165

Padover, Karl, 354
Padover, Saul, 588n68
Pan, Jiahua, 606n61, 617n57, 617n58, 625n48, 625n49
Panitch, Leo, 571n113, 600n35, 608n22
Paolucci, Paul, 504n29
Park, Robert E., 544n207
Parsons, Howard, 248, 562n13
Parsons, Talcott, 533n25, 534n40, 543n192
Paul, Mathilde, 608n18
Pavlov, Aleksei Petrovich, 20, 84, 245, 274, 459, 460, 495n32, 561n3, 621n9
Pearce, Fred, 509n9
Pearce, Roy Harvey, 542n189
Pedregal, Alejandro, 339, 341–42, 344, 345–46, 348–49, 351–52, 353, 354–55, 357, 595
Pellow, David N., 600n31
Pelosi, Nancy, 391, 396
Perreault, Tom, 564n27, 578n37
Persson, Asa, 612n4
Peters, Glen, 604n46
Peterson, D. J., 589n1, 589n3, 595n43

Phillips, Matthew, 521n17
Pigou, Arthur Cecil, 367
Pillet, G., 547n12, 553n96
Pinkard, Terry, 627n4
Plekhanov, Georgi, 270, 574n149
Podolinsky, Sergei A., 224
Poirier, Clément, 613n9
Polianski, Igor J., 574n148
Poolkhet, Chaitep, 608n18
Popper, Karl, 575n168
Posdolsky, Roman, 548n21
Powell, Hugh, 526n14, 526n15
Powell, James Lawrence, 593n26
Prashad, Vijay, 529, 631n62, 632n65
Prenant, Marcel, 499n90
Preobrazhensky, Yevgeni, 548n30
Prezent, Issak Izrailovich, 319
Prigogine, Ilya, 308, 587n46
Pringle, Peter, 589n6
Proudhon, Pierre Joseph, 343
Pruett, Dave, 103–4, 518n25
Pryde, Philip R., 589n3
Punzo, Lionel F., 225, 555n125
Putin, Vladimir, 325
Putnam, T., 571n111

Qiu Jiansehng, 619n79
Quesnay, François, 225
Quetelet, Adolphe, 189

Rabinowitz, Abby, 526n16, 526n17, 526n18
Rader, Melvin Miller, 565n36
Radkau, Joachim, 156, 166, 168, 170, 533n24, 536n64, 536n73, 538n92, 538n111, 540n130, 540n146, 540n150, 541n170
Randers, Jørgen, 496n36, 596n3
Rappeport, Alan, 611n54
Reagan, Ronald, 445
Redclift, Michael R., 156, 506n43, 506n46, 532n19
Redford, Kent H., 521n21
Reed, John, 611n56

INDEX

Rees, Martin, 446, 616n48
Rees, William, 551n70, 551n72
Resnick, Brian, 602n11
Revkin, Andrew C., 621n9
Ricardo, David, 201, 205-6, 547n16, 547n19
Rice, James C., 545n4, 552n74, 552n76, 552n77
Richter, Lauren, 531n8
Riedel, Johanne, 542n190
Ringer, Fritz, 534n34, 535n46, 536n59, 541n171
Rios, Anita, 603n32
Roach, Mary, 624n40
Roach, Stephen, 423, 610n45, 610n51
Robbin, Libby, 496n40
Robertson, G., 571n111
Robin, Libby, 621n9
Robinson, Joan, 385, 599n29
Robock, Alan, 525n8, 526n9, 526n11
Rockström, Johan, 57, 123, 496n39, 507n49, 509n3, 524n46, 596n2, 612n4, 621n4
Rodney, Walter, 355
Rogelj, Joeri, 605n52
Roger, François, 608n18
Rogers, Dale, 609n35
Rogers, Heather, 600n35
Ronov, A. B., 593n29
Roosevelt, Franklin, 394
Roosevelt, Theodore, 184
Rosa, Eugene A., 530n6, 531n8, 541n169, 551n71
Rosdolsky, Roman, 548n26
Rose, Hilary, 33, 348
Rose, Nancy E., 604n42
Rose, Stephen, 33
Rose, Steven, 348
Rosen, Julia, 526n18
Rosenfeld, Léon, 589n6
Ross, Jason, 515n9, 561n4
Ross, John, 621n8

Rostow, Walt, 584n10
Roth, Guenther, 194, 196, 197, 543n197, 544n206, 544n208
Roudart, Laurence, 540n142
Rubin, Isaak Illich, 69, 225, 511n27, 554n115, 555n125
Rubinstein, Annette, 588n70
Ruccio, David, 611n54
Rudd, Robert, 620n3
Rudel, Thomas, 530n6
Rudwick, Martin J. S., 562n9
Rudy, Alan P., 564n27, 565n33, 573n132
Runciman, W. G., 535n48
Ruskin, John, 13, 101, 352, 484, 493n1, 630n37
Russell, Bertrand, 569n81, 570n99

Sagan, Carl, 624n33
Sagan, Dorian, 84, 515n8
Sagoff, Mark, 629n22
Said, Edward, 37, 148-49, 150, 151-52, 529n1, 529n5, 529n6, 529n7, 529n8, 529n9, 529n11
Saito, Kohei, 340, 497n57, 498n67, 503n24, 513n44, 560n214, 561n222, 567n51, 577n17, 577n18, 577n19, 614n30
Salleh, Ariel, 249, 528n22, 563n23
Samuelson, Paul, 225, 555n125
Sanders, Bernie, 392, 396, 397-99, 602n9, 603n34, 604n35
Sartre, Jean Paul, 256-57, 475, 479, 483, 501n6, 568n59, 568n60, 568n64, 568n65, 626n2, 628n14, 630n34, 631n61
Sassoon, David, 520n11
Sato, Makiko, 509n5, 509n9, 510n11
Sauer, Carl O., 598n10
Scaff, Lawrence A., 541n151, 541n157, 542n185
Scazzieri, Roberto, 559n191
Schellnhuber, Hans Joachim, 561n1

Schiller, Friedrich, 193, 543n194,
543n196
Schluchter, Wolfgang, 543n191,
544n206
Schmidt, Alfred, 31, 32, 44, 68,
248, 249, 252, 258, 264, 499n77,
499n78, 499n79, 499n85, 501n6,
502n11, 511n21, 562n14, 562n16,
564n28, 585n28
Schmidt, Sarah Maria, 615n37
Schmitt, Carl, 579n40
Schnaiberg, Allan, 248, 374, 377,
385–86, 513n43, 541n163,
562n13, 597n7, 600n31
Schneider, Mindi, 202, 500n1,
501n2, 503n25, 504n35, 545n2,
545n3, 547n13, 561n223
Schneider, Stephen, 624n33
Schönbein, Christian Friedrich, 47
Schopenhauer, Arthur, 349
Schor, Juliet B., 388, 495n26,
597n24, 601n39
Schroeder, Ralph, 542n185
Schumacher, E. F., 371–72, 597n23
Schumpeter, Joseph, 95, 481,
517n12, 629n21
Schwartzman, David, 606n61
Schwarzenegger, Arnold, 106
Scienceman, David, 204, 220, 221,
222–24, 225, 227, 229, 240,
505n38, 547n12, 550n55, 552n86,
553n92, 553n96, 554n106,
554n108, 554n112, 554n115,
555n116, 555n117, 555n122,
555n123, 555n124, 556n126,
556n127, 556n128, 556n131,
556n135, 556n143, 556n148,
561n221
Scoones, Ian, 607n9
Seligsohn, Deborah, 619n81
Sellars, William D., 515n4
Sellers, William, 328
Sessions, George, 528n21
Seton, Francis, 225, 555n125

Shahn, Ben, 623n30
Shanin, Teodor, 630n47
Shantser, E. V., 20, 459, 460, 495n32,
515n9, 561n3, 621n5
Shaw, George Bernard, 355
Sheasby, Walt, 343
Sheehan, Helena, 347, 499n87,
575n171
Shellenberger, Michael, 251, 266,
267, 289, 565n32, 573n134,
581n53
Shelley, Mary, 107, 251, 343
Shelley, Percy Bysshe, 142, 314–15,
484, 588n70, 588n71, 588n72,
630n37
Shigeto Tsuru, 511n27
Shih, Willy, 609n38
Signer, Michael, 105–6, 518n28
Simonis, U. E., 506n47
Simson, Amanda, 526n16, 526n17,
526n18
Singer, Daniel, 494n20, 627n3
Singh, I. P., 610n50
Skildelsky, Robert, 628n19
Skinner, Quentin, 348
Skuce, Andy, 526n24
Slater, Eamonn, 344, 615n38
Smelser, Neil J., 160, 535n43, 535n44
Smil, Vaclav, 132, 174, 526n23,
539n114
Smith, Adam, 15, 168, 205, 472, 480,
493n6, 536n74, 628n16
Smith, Chad L., 530n6
Smith, John, 549n41, 549n43,
609n29
Smith, Mark J., 510n18, 528n21
Smith, Merritt Roe, 626n55
Smith, Neil, 149, 249, 251, 253, 263–
67, 270, 285, 286, 287–88, 289,
499n76, 529n3, 563n19, 564n27,
564n28, 565n34, 566n44, 567n46,
571n111, 571n112, 571n113,
571n114, 571n115, 571n116,
571n117, 571n118, 571n119,

INDEX

571n121, 572n122, 572n125, 572n127, 573n133, 575n164, 576n5, 577n25, 578n27, 580n46, 581n56
Smith, Richard, 519n34
Smith, Tony, 350, 546n9
Smuts, Jan Christiaan, 269, 270, 573n143
Snow, C. P., 330, 594n31, 622n15
Sokal, Alan, 579n40
Solomon, Susan, 521n21
Solvay, Ernest, 189, 192
Sombart, Werner, 182, 183, 191, 194, 543n200
Somervell, D. C., 620n86
Song Jung-a, 611n56
Sonnenfeld, David A., 544n211, 580n52
Sörlin, Sverker, 496n40, 621n9
Soyfer, Valery N., 591n11
Spence, Theresa, 113
Spencer, Herbert, 188, 305, 486–87, 630n45
Spinoza, Baruch, 252, 565n36
Stalin, Joseph, 33, 317, 323, 325, 335, 346, 353, 589n6
Stammler, Rudolf, 163
Stanchinskii, Vladimir Vladimirovich, 32, 33, 319, 590n9
Standing, Guy, 430, 611n63
Stanley, John L., 501n5
Stavrianos, L. S., 627n8
Steffen, Will, 461, 495n26, 495n32, 496n39, 496n40, 496n42, 510n13, 612n4, 621n4, 621n9, 622n10, 622n11
Steger, Manfred B., 585n31
Stein, Christian, 607n9
Stein, Jill, 395, 603n29
Stenchikov, Georgiy L., 525n8
Stengers, Isabelle, 587n46
Stern, Nicholas H., 122, 523n43, 602n14
Stewart, John Massey, 592n23

Stoermer, Eugene F., 514n2, 562n10, 621n9
Stokes, Kenneth M., 513n45, 539n123, 541n172, 560n208
Stoner, Alexander M., 500n1, 503n24
Stover, Dawn, 509n7
Strabo (ca. 64 BCE–AD 24), 439, 614n19
Strona, Giovanni, 496n45, 621n4
Stuart, Tessa, 603n31
Stubenberg, Leopold, 570n99
Suess, Edward, 515n8
Sugiyama, Chuhei, 597n4
Sukachev, Vladimir Nikolaevich, 274, 318–27, 330–31, 335–36, 590n9, 590n10, 591n11, 591n12, 591n13, 591n18, 592n19
Summerhayes, Colin P., 495n26, 495n32, 562n9, 613n9, 621n8
Suter, Christian, 580n49
Suwandi, Intan, 606, 606n2, 609n29
Sweeney, Sharlynn, 557n150, 557n164, 558n180, 560n215
Sweezy, Paul M., 59, 69, 95, 209, 248, 337, 354, 355, 365, 374, 382–84, 386, 436–37, 453–54, 494n18, 494n22, 508n59, 511n27, 517n10, 548n35, 550n54, 556n137, 561n216, 562n13, 592n21, 595n39, 595n46, 596n6, 598n18, 598n19, 599n21, 599n24, 599n28, 600n31, 600n32, 600n33, 605n57, 613n13, 613n14, 613n15, 619n77, 620n2, 623n29, 625n43, 625n51, 630n40
Swyngedouw, Erik, 265, 286, 288, 566n42, 572n122, 578n36, 579n43, 579n44, 580n45, 580n47
Szlajfer, Henryk, 599n26

Tang, Frank, 611n55, 611n58
Tansley, Arthur, 52–53, 55, 254, 268, 269, 274, 320, 505n37, 573n143

Tanuro, Daniel, 249, 563n18
Tarver, Evan, 606n3
Taylor, Charles, 543n194
Tester, Keith, 143, 527n6
Thackeray, William Makepeace, 15, 494n11
Thanapongtharm, Weerapong, 608n18
Thatcher, Margaret, 17, 475
Thiebaud, Lisa, 530n6
Thomas, William L., Jr., 598n10
Thompson, A. K., 527
Thompson, E. P., 80, 88, 101, 348, 350, 352, 356, 444, 445–46, 483, 514n64, 516n3, 516n19, 518n23, 616n45, 630n36
Thompson, Lindsay, 348
Thomson, George, 33, 274, 348, 352, 575n171
Thunberg, Greta, 390, 391, 392, 393, 410, 601n5, 602n8, 602n19
Tillerson, Rex, 118–19, 133, 527n25
Timpanaro, Sebastiano, 498n70, 501n7
Toles, Tom, 526n11
Tollefson, Jeff, 521n16
Toynbee, Arnold, 620n86
Trautner, Mary Nell, 530n6
Troeltsch, Ernst, 182
Trump, Donald, 119, 127, 407, 427, 454
Trusov, Yu. P., 450, 618n63
Tuckner, L., 571n111
Turco, Richard P., 593n28, 624n33
Turner, Bryan S., 538n106
Turner, Frederick Jackson, 183, 196, 541n151, 541n153
Turner, Jonathan, 628n20
Tyndall, John, 51–52, 504n33

Ulin, David, 105, 518n27
Umaña, A. F., 558n188, 559n192, 559n193
Unger, Corinna R., 592n23

Uranovsky, Y. M., 318, 504n33
Urquijo, Pedro S., 499n76, 499n88
Ursul, Arkadij Dmitrievič, 331, 332, 449–50, 575n173, 594n33, 594n34, 617n57, 617n59, 618n62, 618n63, 618n64, 625n48

Vakil, Bindiya, 609n35
Van Es, Harold, 505n39
van Marrewijk, Dre, 534n40
Vaughan, Adam, 522n23
Vavilov, Nikolai Ivanovich, 32, 33, 274, 318, 319, 496n34, 589n6
Veblen, Thorstein, 59, 374, 379–83, 480, 508n59, 523n44, 598n12, 598n16, 599n21, 599n24, 600n31, 628n18
Vernadsky, Vladimir I., 20, 33, 84, 86, 245, 246, 274, 318, 320, 324, 327, 330, 331, 332, 336, 460, 495n33, 496n34, 505n36, 515n8, 515n9, 561n4, 589n4, 591n11, 594n32, 621n7, 621n8
Verstegen, Wybren, 539n117
Vidal, John, 522n23
Villery, Olga, 626n2
Virchow, Rudolf, 300
Vltchek, Andre, 606n61, 617n58, 631n62
Vogel, Steven, 253, 501n6, 566n45
Voltaire, 252, 342
Voosen, Paul, 623n23
Vucinich, Alexander, 591n18, 594n37

Wackernagel, Mathis, 551n70, 551n72, 552n77
Waitzkin, Howard, 34, 584n14, 607n13
Walker, Mike, 622n20, 623n21, 623n25
Wallace, Alfred Russell, 346
Wallace, Robert "Rob" G., 34, 414, 415–18, 496n42, 499n89, 500n93, 607n9, 607n10, 607n11, 607n12,

INDEX

608n14, 608n19, 608n20, 608n24, 608n25, 612n69
Wallace, Rodrick, 414, 607n9, 607n10, 607n11
Wallerstein, Immanuel, 209–10, 355, 412, 494n14, 548n37, 607n6
Warde, Paul, 496n40, 621n9
Waring, George, 46
Waring, Marilyn, 512n31, 558n184
Wark, McKenzie, 262, 571n103, 571n105, 612n4
Warner, R. Stephen, 160, 535n43, 535n44
Warren, Bill, 356
Warren, Elizabeth, 603n34
Washington, Booker T., 182
Waters, Colin N., 495n26, 514n2, 562n9, 613n9, 621n8, 623n27, 623n31, 624n39
Weart, Spencer R., 495n30, 504n33, 515n4, 525n5, 592n19, 593n26
Weber, Marianne, 184, 540n150
Weber, Max, 35, 141, 153, 154, 155–73, 175–200, 349, 480, 527n2, 529, 530n4, 531n11, 532n13, 533n21, 533n25, 533n26, 533n27, 533n28, 533n30, 534n34, 534n35, 534n36, 534n38, 534n40, 535n41, 535n42, 535n45, 535n46, 535n47, 535n48, 536n50, 536n51, 536n54, 536n55, 536n56, 536n57, 536n58, 536n59, 536n61, 536n62, 536n63, 536n65, 536n67, 536n68, 536n72, 537n80, 537n83, 537n85, 537n86, 537n88, 538n89, 538n90, 538n91, 538n94, 538n95, 538n96, 538n98, 538n99, 538n100, 538n101, 538n102, 538n106, 538n107, 538n109, 538n113, 539n115, 539n119, 539n121, 539n122, 539n127, 539n129, 540n130, 540n131, 540n132, 540n137, 540n139, 540n140, 540n142, 540n144, 540n145, 540n147, 540n148, 540n150, 541n151, 541n152, 541n154, 541n156, 541n158, 541n160, 541n173, 542n175, 542n177, 542n180, 542n181, 542n182, 542n183, 542n184, 542n190, 543n191, 543n192, 543n193, 543n195, 543n196, 543n198, 543n199, 543n201, 544n204, 544n207, 544n208, 544n209, 544n210, 544n212, 544n213, 544n214, 545n216, 549n47, 562n15, 628n17
Wee, Cecilia, 578n31
Weiner, Douglas, 318, 323, 589n3, 589n5, 590n7, 590n8, 591n16, 591n18, 592n22, 592n23, 592n25, 594n32, 595n40
Weir, T. H., 574n148
Wells, Gordon C., 543n192
Wells, H. G., 254
Wen Tiejun, 619n79
Wendling, Amy E., 551n59, 552n86
West, Patrick, 155, 156, 159, 160, 531n9, 531n10, 531n11, 532n13, 533n22, 535n42, 535n43, 536n67, 536n74
Weston, Del, 77, 344, 514n55, 514n57, 563n24
Whewell, William, 305
Whimster, Sam, 534n32
White, Curtis, 119, 523n37
White, Damian F., 564n27, 565n33, 573n132
Whited, Tamara L., 539n117
Whitehead, Alfred North, 142, 293, 527n3, 565n36, 578n30, 582n68
Wideband, Wilhelm, 544n207
Wilkinson, John, 174
Wilkson, Robert, 594n32
Williams, Chris, 497n59, 586n37
Williams, Mark, 495n26, 621n8
Williams, Raymond, 15, 68, 141, 356, 493n7, 494n9, 511n25, 527n1, 601n39, 608n25, 614n18

653

Wilson, E. O., 144, 528n9
Winch, Donald, 536n74, 547n15
Winiwarter, Verena, 506n47
Wishart, Ryan, 498n65, 499n89, 502n22, 507n48, 618n67
Wittfogel, Karl, 168, 170, 536n74
Wittman, Hannah, 202, 545n3, 631n62
Wolf, Eric, 631n51
Wolff, Robert Paul, 225, 555n125
Wongnarkpet, Sirichai, 608n18
Wood, Ellen Meiksins, 479–80, 482, 510n15, 510n17, 514n62, 628n15, 628n16, 629n27, 629n28
Wood, Neal, 514n62
Woodgate, Graham, 506n43, 506n46
Wordsworth, William, 142
World Wildlife Fund, 597n20
Worster, Donald, 249, 268–69, 563n18, 573n142, 573n143, 574n144, 589n3, 620n1
Wundt, Wilhelm, 159, 200, 534n38, 545n215

Xi Jinping, 617n58, 625n49

Yablokov, Alexei, 516n20
Yanshin, A. L., 593n29

Yanuta, V. G., 592n19
Yates, Michael D., 500n95
Ye Qianji, 625n48
Yeats, William Butler, 149, 150
Yergin, Daniel, 504n33
York, Richard, 28, 118, 249, 344, 498n64, 500n95, 502n12, 502n22, 505n37, 506n44, 507n49, 514n56, 523n33, 523n43, 524n50, 527n7, 530n6, 531n7, 545n3, 551n71, 560n214, 563n22, 568n52, 569n84, 573n143, 575n173, 577n13, 581n55, 585n32, 587n59, 596n2, 596n5, 597n24, 600n31, 605n51, 605n60, 607n12, 621n4, 622n13, 626n54, 627n9
Young, Arthur, 15
Yudin, M. I., 593n26

Zalasiewicz, Jan, 495n26, 495n32, 562n9, 613n9, 621n8, 623n26, 623n27, 623n28, 623n31, 624n37, 624n39
Zavadovsky, B. M., 32–33
Zencey, Eric, 121, 522n28, 523n42
Zinsstag, J., 607n9
Žižek, Slavoj, 252–53, 288, 566n39, 566n42, 566n43, 580n47

Subject Index

Absentee Ownership and Business Enterprise in Recent Times (Veblen), 379
abstract labor, 223, 273, 278, 554n115
accumulation: bioaccumulation, 65, 469, 510n13; with treadmill of production, 58; wealth, 430; *see also* capital accumulation
activism: anti-shrimp-farming, 235; Black Lives Matter, 491; on bridges over Thames River, 391; chronology of events, 2018 to 2019, 390–92; Climate Emergency declaration, 392; Climate Justice Alliance, 393, 395; environmental protests in China, 619n83; Extinction Rebellion, 24, 391, 392–93, 406; Flood Wall Street action, 602n17; fossil fuels disinvestment movement, 603n31; Keystone XL Pipeline, 113, 118–19; People's Climate Movement, 393; Sanders and, 392, 397–98, 399, 603n34; *Special Report on Global Warming of 1.5°C*, 390–91; *Special Report on the Ocean and Cryosphere*, 392; Standing Rock Sioux, 64, 119–20, 396, 455, 492; strikes, 391, 392, 489; Sunrise Movement, 391, 392–93, 396, 603n31; System Change Not Climate Change, 81, 92, 109, 250, 345, 392; Thunberg with, 390, 391, 392, 393, 410; *see also* Green New Deal; revolutions
actor-network-theory (ANT), 251, 266, 267
"Adam's Fallacy," 472
Adventures of Ideas (Whitehead), 578n30
Adventures of the Dialectic (Merleau-Ponty), 285, 501n4
advertising: automobile industry, 383, 387; corporate, 381; marketing and, 382–83
aerosol loading, 57, 328, 329
affluence, IPAT formula, 599n30
The Affluent Society (Galbraith), 384–85

Africa, 25, 117; North, 424; paleoclimatic change, 330; partition of, 16; South, 78, 269, 340, 487; Sub-Saharan, 235–36, 424
Against Nature (Vogel), 253
Age of Ecological Enlightenment, 247
The Age of Imperialism and Imperialism (Magdoff), 355
The Agrarian Sociology of Ancient Civilizations (Weber), 164, 165
agribusiness: antibiotics and, 431; with cheap labor and land, 416, 430; commodity chains and, 416, 418; crop plantations, 131; disease and, 413, 415, 418; epidemics and, 414; global land arbitrage and, 411–12; "The Metabolic Rift of Livestock Agribusiness," 202; monocultures, 132, 136, 423, 431; multinational corporations., 417–18
agricultural chemistry, 27, 46–47, 70
Agricultural Chemistry (Liebig), 46–47
agriculture: Asian hydraulic civilization and, 168–71, 537n88; *Capital* on soil and, 341; capitalist, 46–48, 54, 70–71, 145–46, 165, 415, 442; crops, 25, 131; famine, 298, 313, 356, 443–44, 615n37; industrial, 27, 46–47, 70, 179–81, 211, 281, 442; Lenin All-Union Academy of Agricultural Sciences, 323; potato blight, 443, 615n37; rain-fed, 165, 166, 168–71; soil, humans and, 281; *see also* fertilizer; soil
agroecology, 128, 132, 136, 243, 344, 418
air: pollution, 333, 384, 453; quality index, 338; travel, 420, 422
albedo effect, 64, 128, 274, 327, 328
alienation: capitalist class and, 279–80; from dominant minority, 620n86; of humans, 280; of labor, 341, 351, 357; *Marx's Theory of Alienation*, 272, 341, 502n13; of nature, 144, 244, 251, 259, 261, 262, 263–64, 268, 280, 284–85, 290–91, 310, 315, 341, 449; from nature and humans, 28, 31–32, 34–35, 150–51; theory of, 44–45, 272
All-Russian Conservation Society (VOOP), 318, 320, 324, 326, 335, 336
All-Union Conference on the Problem of Climate Modification by Man, Leningrad (1961), 327
Alternative Assessment principle, 134
Amazon: drought in, 130; oil, 235; *Underdeveloping the Amazon*, 215
Amu Darya River, 325
ancient civilizations, fall of, 298, 302, 444, 448, 449, 465–66
Ancient Judaism (Weber), 161, 164, 165, 166–68
animals: disease and, 416–17, 418; ecology, 268, 308; endangered species, 136; feeding operations for confined, 398; habitat destruction, 423; hogs and poultry, 415, 416; Livestock Revolution, 417; monocultures, 415, 417; wet markets, 415
"Another Age," 474
Antarctica, 57, 116
anthropic fallacy, 262, 571n109
Anthropocene, Marxism in: ANT, 251, 266, 267; Bensaïd and Moore, 257–63; biosphere and, 245, 246; Cronon and Worster, 267–71; dialectical ecology versus radical ecological monism, 251–57; ecological thought in, 247–51; first-stage ecosocialism and, 248–49; "left apocalypticism"

and, 251, 264; return of dialectics of nature/ecology, 271–76; Sartre and, 256–57; Smith and Castree, 249, 263–67; System Change Not Climate Change, 250; Western Marxism and, 251, 271, 275
Anthropocene, Stoermer and, 621n9
Anthropocene crisis, 82–90, 246–47, 434–35, 451, 576n9
Anthropocene Epoch, 63, 143, 495n32; Angus on, 622n14; Capitalinian Age, 14, 18, 22, 35, 83, 464, 466–72; capitalism before, 14–18; capitalism in, 18–22; Capitalocene controversy versus, 459–64; Crutzen and, 460–61; Earth System change and, 461–62; ecological civilization and, 434; extinction event, 14, 19, 22, 459, 463, 473; *Facing the Anthropocene*, 83, 344; Great Acceleration and, 495n26; Great Climacteric and, 247, 255, 433, 434, 435, 451, 463; nuclear weapons testing and, 467–68; Planetary Precautionary principle, 133–35; stratigraphic record and, 458; Walker on Meghalayan Age and, 622n25; "Was the Anthropocene Anticipated?," 622n13
Anthropocene Working Group, 18, 288, 466, 467
anthropogenic desertification, 254
"The Anthropogenic System (Period)" (Shantser), 459–60
antibiotics, agribusiness and, 431
Anti-Dühring (Engels), 29, 169, 297–98, 300, 354
anti-nuclear movement, 329
apartheid, 88, 269, 270
après moi le déluge philosophy, 136, 283
Aral Sea, drying up of, 316, 325, 333

Arctic: amplification, 116; with ice loss, 116, 128; oil, 115
art: nature and, 142; social character of, 352
Asia: Central, 117, 424; deaths in East and South, 424; production mode, 536n74
Asia Minor, 302
Asian hydraulic civilization, 168–71, 537n88
"Atmospheric Homeostasis By and For the Biosphere" (Lovelock and Margulis), 20
atomic bombs, 17
austerity, 97, 104, 105, 414–15, 419
Australia, 114, 116, 557n164
Australopithecines, 330
automobile industry: advertising, 383, 387; commodity chains, 421; economic waste in, 384; gas mileage decline in U.S., 600n33; growth of, 470; in World War II, 96
Avoiding Social and Ecological Disaster (Bahro), 80

Baconian principle, nature and, 301
Baikalsk Pulp and Paper Plant, 336
barbarism, 445; bourgeois civilization and, 440; capitalist, 36; ecological, 407; "Ecosocialism or Barbarism," 617n56; socialism or, 449; Strabo and, 439
Barcelona Degrowth Declaration (2010), 364, 368
basic communism, 76, 77, 365
Bedouins, 161, 166–67
Berlin Conference (1884), 16
Beyond Capital (Mészáros), 45, 272, 414
Big Farms Make Big Flu (Wallace, R. G.), 415
Bikini Atoll, nuclear weapons test, 467

billionaires, 127, 135, 428, 430
bioaccumulation, 65, 469, 510n13
biodiversity, 363; crisis, 23; loss, 123, 470; marine, 294
Bio-Energy with Carbon Capture and Storage (BECCS), 127, 130–31, 402
biofilms, 306
biofuels, 118, 120
biogeochemical cycles, 52, 86, 261, 320, 463, 576n9; capital accumulation and, 65, 73; chemical fertilizer and, 282; Earth System and, 20–21, 75, 87, 143, 263, 311, 364, 436, 461, 505n36
biogeocoenosis, 274, 320–22, 324, 330, 332, 335, 590n9, 591n11
biological diversity, loss of, 21, 65
Biology Under the Influence (Lewontin and Levins), 274
biomagnification, 65, 469, 510n13
biomass, 118, 120, 131, 177
bionomics, 52
biosphere: "Atmospheric Homeostasis By and For the Biosphere," 20; dialectical materialism and, 33; Earth System science and, 20–21; *The Evolution of the Biosphere*, 336; extinction of, 449; International Geosphere-Biosphere Program, 460; with limits of capitalism, 261; Marxism in Anthropocene and, 245, 246; Vernadsky and, 20, 84, 86, 245, 246, 274, 318, 336, 460
The Biosphere (Vernadsky), 20, 84, 86, 245, 246, 274, 318, 336, 460
Black Death (bubonic plague), 159–60, 199
Black Lives Matter, 491
Black movement, in U.S., 355, 491
Blackfriars Communications, 383, 599n23
"blank sailings," shipping, 422

"Blockadia," 78, 98
Blue Planet Prize (1998), 324
boiler explosions, locomotives, 583n5
Bolivarian communes, Venezuela, 473
Bolivia, 99, 371, 406
boni patres familias (good heads of the household), 72, 146, 378, 443, 474
The Book of Nature (Good), 544n209
Botanical Journal, 323
Brazil, 115, 233, 310, 345, 473, 491
bread production, 600n34
Break Through (Shellenberger and Nordhaus), 266–67
Breakthrough Institute, 95, 107–8, 289; with geoengineering, 127; Klein on, 518n31; Latour and, 251, 266, 267, 565n32, 579n40
British East India Company, 536n74
British Isles, 33, 175
British Medical Journal, 25
broadband, 420
"The Buddha's Parable of the Burning House" (Brecht), 89
budgets: global carbon, 63, 119, 126, 399, 472, 525n1; solar, 77, 79; *see also* costs
"Budyko energy balance models," 327
Bulletin of Atomic Scientists, doomsday clock, 430–31
Bulletin of the Academy of Sciences, 327
bullwhip effect, commodity chain disruption and, 418–23
bureaucracy, non-industrial society and hydraulic, 168–71, 537n88

Caesium-137, nuclear fallout, 467
California, climate bill, 106
Canada, 119, 120; emergy/money ratio, 557n164; oil, 114, 115

INDEX

Canadian Medical Association Journal, 25
Candide (Voltaire), 252
Capital (Marx), 15, 16, 52, 59, 84, 146, 169, 280, 339, 374; on agriculture and soil, 341; deforestation, 441; ecological Marxism, 71–72; exchange and use values, 412; international trade and, 206–7; social metabolism, 27; soil nutrient cycle, 54
capital accumulation, 64, 65; Earth System threatened by, 74; through environmental crisis, 73; expropriation and, 244; insidious nature of, 95
Capitalinian Age: Anthropocene Epoch, 14, 18, 22, 35, 83, 464, 466–72; Capitalocene controversy versus Anthropocene, 459–64; Communian Age and, 472–74; Meghalayan to, 464–66; nuclear weapons testing and, 467–68; stratigraphic record and, 458
capitalism: before Anthropocene, 14–18; in Anthropocene, 18–22; as barrier to future of history, 477–82; biosphere with limits of, 261; catastrophe, 22–26, 79, 411–13; climate, 373; climate change as threat to, 66, 100, 103, 104, 108, 128; creative destruction of, 28, 80, 108, 459, 477–79; with criticism as taboo, 65–66, 67, 99–101; decapitated time of, 626n2; defined, 17–18, 498n63; degrowth and, 363–72; desertification and, 567n51; dialectics of nature in Anthropocene and, 312; dissolving effects of, 180, 184, 200; with Earth System change, 13–14, 87; eco-compatible, 366–67; ecological ruin or revolution, 35–37; formula, 376; fossil, 35, 87–88, 495n26; freedom and, 17; golden age of, 18; imperial, 53, 88, 203, 220, 233, 243; *Imperialism, the Highest Stage of Capitalism*, 355; industrial, 28, 158, 164–65, 173, 175–86, 192, 211, 458, 482; "Is Capitalism a Disease?," 415; Marxism, universal metabolism of nature and, 26–35; modernity and, 481; natural, 373; nature, public wealth and, 374; nature and, 287, 374; *The Protestant Ethic and the Spirit of Capitalism*, 157, 165, 176, 193, 194, 195, 197; science and, 588n67; Sweezy on, 436–37; waste and, 406, 523n44; Weber and, 173, 176, 180–81, 480; Wood on circular nature of, 479–80; *see also* Great Capitalist Climacteric; monopoly capitalism; *Raubbau*
Capitalism, Nature, Socialism (journal), 343, 347
"Capitalism and the Disorders of Modernity" (Berger), 481
Capitalism and Theory (Kidron), 384
Capitalism in the Web of Life (Moore), 580n50, 616n53
capitalist agriculture, 46–48, 54, 70–71, 145–46, 165, 415, 442
capitalist class, 29, 407; alienation and, 279–80; civilization, 440; consumption and, 382; Engels on, 297; inequality and, 431; law of value and, 376; material conditions of, 300
capitalist income accounting, 512n31
capitalist law of value, 282–84
capitalist production, 58, 133, 242, 298, 375, 400, 418; destructive character of, 488–89; labor power, nature and, 512n33; Marx and, 47–48, 213, 225, 229, 376–77,

478; real wealth and, 241; universal metabolism of nature and, 27, 47–48
capitalocentrism, 279, 576n6
carbon: BECCS, 127, 130–31, 402; fee and dividend system, 79; metabolism, 28, 49, 57–58, 450; war and fossil fuels, 117–21
carbon capture and sequestration (CCS), 132
carbon dioxide removal (CDR), 129, 130, 131, 404
carbon emissions: coal and, 520n15; drop in, 336, 595n43; emergency reduction plan, 79; from fossil capital, 471; fracking and lower, 114; global budget, 63, 119, 126, 399, 472, 525n1; increase, 404; IPCC on, 23, 603n34; mitigation, 401; net-zero to avoid catastrophe, 23, 94, 95, 106, 398, 603n34; reduction opposition, 523n43; rising, 63–64
"Carbon Metabolism" (Clark and York), 28
Carbon Tracker, 523n40
Carbon-14, nuclear fallout, 467
The Carbon War (Leggett), 117
Caribbean, deaths, 424
"Cartesian binary," 49, 259, 260, 261
Cartesian dualism, 41, 50, 250, 284–85, 565n29; Kant and, 273; Marx and, 258; Western Marxism and, 252, 262
Castle Bravo nuclear test, Bikini Atoll, 467
catastrophe capitalism, 22–26, 79, 411–13; *see also* COVID-19, catastrophe capitalism and
Cellular Pathology (Virchow), 300
Cenozoic Era, 457
Center for Environmental Policy, 230
Center for Health, Environment and Justice, 387–88

Centers for Disease Control and Prevention, 413
Central Asia, 117, 424
Ceylon, 170, 171
charcoal: with iron and coal, 538n113; smelting, 173, 174, 175, 197
chemicals: fertilizer, 35, 282; lithium battery, 421; Oil, Chemical and Atomic Workers Union, 395; synthetic, 19, 65, 143, 177, 246, 388, 490; toxic, 75, 136, 388; *see also* biogeochemical cycles; petrochemicals
chemistry: agricultural, 27, 46–47, 70; Nobel Prize, 187, 189, 308, 461; organic, 19, 468
Chernobyl nuclear accident, 316, 336, 468
Chevron, 447
Chile, 119, 355
China, 119, 127, 171, 444, 619n76; with alternative-energy technologies, 407; coal consumption, 178; *The Collapse of Western Civilization*, 447, 453, 619n76; commodity chains, 421; ecological civilization and, 345, 407, 434, 451, 455, 472, 491; environmental protests, 619n83; with historical materialism, 434; hope and, 452–55; labor costs, 419; manufacturing in, 427–28; New Cold War on, 476; peasants, 180; population growth, 179–80; *The Religion of China*, 164, 165, 170; rise of, 477; *Science and Civilization in China*, 452; South China Sea, 115; U.S. coal exports to, 520n15; *Will China Save the Planet?*, 452, 619n82; with wind and solar energy, 619n82; Yangtze River Valley, 465
China and Ecological Civilization (Cobb), 452

INDEX

Chinese Marxism, 449
Chinese Science Bulletin (medical journal), 25
circular economies, promotion of, 136
The City in History (Mumford), 444
Civil War, U.S., 72, 182
Civilian Conservation Corps (CCC), 398
civilization: barbarism and bourgeois, 440; Beecher on, 439; capitalist class and, 440; Chomsky on end of, 456; *The Collapse of Western Civilization*, 447, 453, 619n76; critique of, 438–44; Engels on, 440; fall of ancient, 298, 302, 444, 448, 449, 465–66; Fourier on, 439; historical meaning, 612n1; *A History of Civilizations*, 439; in Holocene, 433–34; Morris on revolution and, 440; Mumford on, 444–45; "Notes on Exterminism, the Last Stage of Civilization," 80, 445–46; *Philosophy and the Ecological Problems of Civilisation*, 331, 332, 449–50, 594n33, 625n48; *Philosophy and the Ecological Problems of Civilization*, 449–50, 617n57; *Science and Civilization in China*, 452; socialism with narrow window for, 613n15; *see also* ecological civilization
class: capitalist, 29, 279–80, 297, 300, 376, 382, 407, 431, 440; *The Condition of the Working Class in England*, 124, 299–300, 354, 415, 489, 620n86; *History and Class Consciousness*, 30, 42, 43, 44, 271, 284–85, 349, 631n61; imperialism, pandemic and, 423–29; *The Making of the English Working Class*, 483; *The Theory of the Leisure Class*, 381

"Clean Energy Plan" (2019), 603n34
Clear-Cutting Disease Control (Wallace, R., Chaves, Bergmann, Ayres, Hogerwerf, Kock and Wallace, R. G.), 414
climacteric: climate change and global, 92–99; Great Capitalist, 68–81
Climate and Life (Budyko), 328, 336
climate capitalism, 373
climate change: best case scenarios, 23–24; capitalism threatened by, 66, 100, 103, 104, 108, 128; Castree and, 574n152; deforestation and, 598n10; eco-revolution for, 135–37; global climacteric and, 92–99; global weirding and, 23, 116; Intergovernmental Panel on Climate Change, 93; IPCC, 19, 23–24, 93, 127, 130–31, 390–91, 392, 401–8, 602n14, 603n34; irreversible, 63, 75, 92, 93, 115, 126, 510n11; with liberal critics in denial, 99–108; mitigation, 24, 401–8, 602n14; *Nature Climate Change*, 118; Peoples' Agreement on Climate Change, 79; with planetary boundaries crossed, 57; rich and poor countries with, 98; solutions through people power, 96–97; strikes, 391, 392; System Change Not Climate Change, 64, 81, 92, 109, 250, 345, 392; tipping points, 22, 26, 64, 115, 301–2; Tyndall Center for Climate Change, 63, 94, 122, 403, 438, 520n15; ultimate line of defense, 108–10; *see also* activism
Climate Change 2021 (IPCC), 23–24
"Climate Change and the Means of Its Transformation" (Budyko), 327
"Climate Change Going Beyond Dangerous—Brutal Numbers and Tenuous Hope" (Anderson), 438

Climate Code Red, 127
Climate Emergency declaration, 392
Climate Justice Alliance, 393, 395
climate justice movement, 78, 400
climate refugees, 357
climatic conditions, *Ancient Judaism* and, 166–68
The Closing Circle (Commoner), 86, 143, 246, 385, 470
cloud computing, 420
Club of Rome, 20, 45, 272, 326, 336, 364
coal: carbon emissions and, 520n15; clean, 118; coked, 158, 174–75; consumption, 177–78; fracking and, 114; iron, industrial capitalism and, 175–86; iron and, 538n113; mountaintop removal mining, 399; plant shut downs, 79; pollution, 184; support for, 119
Cochabamba (Bolivia), 79
coffee, 300, 310
Cold War, 17, 18, 325, 329, 445; China and New, 476
Collapse (Diamond), 446, 447
"Collapse of Civilizations Worldwide Defines Youngest Unit of the Geologic Time Scale," 465
The Collapse of Western Civilization (Oreskes and Conway), 447, 453, 619n76
Collected Works (Marx and Engels), 342
Colombia, 119
colonialism, 205, 298, 300, 356, 370, 444, 486–87, 492
commodity chains (labor-value chains): agribusiness and, 416, 418; disruption and global bullwhip effect, 418–23; imperialism and, 411; metabolic rift and, 415; of metamorphoses, 421; pandemics and, 414; poor countries,

death and, 425; reorganization of, 136; supply and, 421; use and exchange values with, 412–13
Communian Age, 14, 19, 22, 36, 459, 472–74
communication technologies, 420
communism: basic, 76, 77, 365; *Scientific Communism*, 617n57, 625n48
Communist Manifesto (Marx), 343, 448
community-cultural life, rich, 97
The Concept of Nature (Whitehead), 142
The Concept of Nature in Marx (Schmidt), 31, 44, 68, 248, 585n28
The Condition of Man (Mumford), 76, 80–81, 365, 444, 616n43
The Condition of the Working Class in England (Engels), 124, 299–300, 354, 415, 489, 620n86
"confrontation of reality with reason," 599n24
conservative critics, in denial, 102
consumer marketing, limits on destructive corporate, 136
consumer sovereignty theory, 385
consumption: capitalist class and, 382; environmental costs of, 218; fossil fuels, 177–78; of natural resources, 157, 180–86; outrage and over-, 370
contrapuntal reading, 148, 529n1
Conversations with Lukács (Lukács), 43, 271, 349, 502n10
coral bleaching, 294
co-revolutionary movement, 371
corporations: with advertising, 381; with commodity chain disruption, 422; Credit Suisse, debt and, 423; defense of, 447; energy, 400; Fortune 1000 multinational, 421; with limits on destructive marketing, 136; with manufacturing

INDEX

in China, 427–28; multinational agribusiness., 417–18; taxes on polluters, 398; with "territorial restructuring," 418; *see also* agribusiness; automobile industry costs: carbon dividend system and fees, 79; consumption-based environmental, 218; efficiencies, 423–24, 430; fossil fuels industry losses, 400; health care, 398; labor, 227, 419, 428; military spending, 79, 136, 382, 384, 395, 399; shipping, 420
cotton crisis, British, 72
Council on Foreign Relations, 118
"counter-finality," 568n64
country, town and, 49, 54–55, 281, 440, 442, 450
COVID-19, 26, 34, 199, 314, 390, 476; catastrophe capitalism and: capital and ecological-epidemiological crises, 413–18; commodity chain disruption and bullwhip effect, 418–23; imperialism, class and, 423–29; social production, planetary metabolism and, 429–32; supply chain and, 421; World Economic Forum and, 421; Davis on deaths, 425–26; racism and, 500n93; Response Team of Imperial College, 424; test kits, 423
"COVID-19 and Circuits of Capital" (Wallace, R., Liebman, Chaves and Wallace, R. G.), 414, 416–17
creative destruction: of capitalism, 28, 80, 108, 459, 477–79; Schumpeter and, 481
Credit Suisse, 423
credit-debt system, 75
critical epochs theory, 327
Critique of Stammler (Weber), 141, 162, 163

crops: global yield decrease, 25; plantations, 131
Cuba, 300, 344, 371, 406, 491, 619n78
culture: community-cultural life, 97; *Energetic Foundations of a Science of Culture*, 187; "'Energetic' Theories of Culture," 189; environment and, 161–62, 166–71; gene-culture coevolution, 254, 310, 329; Iroquois, 614n21; nature and, 450; of resistance, 483; slavery and soil, 181–82; socio-cultural determinism, 565n32; *Transition Culture*, 122; *see also* monocultures
Culture and Imperialism (Said), 148, 151–52

Daily Beast (newspaper), 105
Dakota Access Pipeline, 119–20, 396, 455
De rerum natura (*On the Nature of Things*) (Lucretius), 141
deaths (mortality rates): Davis on COVID-19, 425–26; global estimates for pandemics, 424–25; H1N1, 415–16; heat-related, 25; poor countries, racism and, 424–25; SARS-CoV-2, 416; social murder, 299, 313, 431, 484
The Death of Nature (Merchant), 143
"Debates on the Law on Theft of Wood" (Marx), 374
debt: climate, 98; credit-debt system, 75; Sub-Saharan Africa with ecological, 235–36; supply-chain finance and, 423
decapitated: future, 481; time of capitalism, 626n2
"Deepening, and Repairing, the Metabolic Rift" (Schneider and McMichael), 202, 503n25, 504n35
Deepwater Horizon oil disaster, 114

Defense Production Act, 427
deforestation, 302, 378; climate change and, 598n10; desertification and, 73, 198; Marx on, 441; from wood epoch to age of iron, 173–75
degrowth (*décroissance*): Barcelona Degrowth Declaration, 364, 368; capitalism and, 363–72; co-revolutionary movement and, 371; defined, 364, 365–66; eco-compatible capitalism with, 366–67; "Economic De-Growth for Ecological Sustainability and Social Equity," 364; economics, 364; Green New Deal and, 368; impossibility theorem and, 370; Jevons Paradox, 364; Latouche and, 364, 366–67, 368, 369–70, 371; poor countries and, 369–71; reformist measures and, 367; rich countries and, 366, 369–70, 372; rise of, 364; steady-state economy and, 76, 122, 365, 366, 370; Sweezy on reversal of disequilibrium, 365; unemployment and, 369; "voluntary simplicity" movement, 364
demagification, 178, 193, 543n196
Democratic Socialists of America (DSA), 396
denial, climate change, 99–108
depeasantization, 418
dependence effect, Galbraith and, 95, 384–85
desertification, 283, 330, 333, 378, 490; anthropogenic, 254; capitalism and, 567n51; deforestation and, 73, 198; with fall of ancient civilizations, 298; with inroads of sand, 544n209; Marx on, 441, 446; production and, 312
Deutscher Prize Lecture: Harvey and, 51; Mészáros, 45, 272

development: Inter-American Development Bank, 400; OECD, 422; socialism with sustainable human, 76–77, 278; sustainable, 79; *The Theory of Capitalist Development*, 95; *Uneven Development*, 253; World Summit for Sustainable Development, 517n8
Dialectic (Bhaskar), 262–63, 351, 586n42
dialectic of continuity and change, 36, 83, 345, 346
dialectic of sensuous certitude, 351
dialectical ecology, 251–57, 505n36
dialectical integral unity, 331
dialectical materialism: biosphere and, 33; dialectics of nature and, 32, 42, 274; Engels and, 252, 353, 587n49; fundamental importance of, 34; Hook and, 257; Levins and Lewontin on, 587n49; monism and, 270, 285; Soviet-style, 68, 330, 353; transformation and, 308; Western Marxism and, 43, 68
dialectical realism, Marxism and, 290–95
dialectical-ecological synthesis, Odum, Marx and, 222–28
The Dialectical Biologist (Levins and Lewontin), 310, 414–15, 501n5
dialectics of ecology, Marxism and: capitalist law of value and destruction of nature, 282–84; conceptual structure of metabolism theory, 279–82; with labor as metabolism of society and nature, 277; nature as free gift and, 278; realism and reunification of, 290–95; subsumption of nature and, 284–90
dialectics of nature, 41, 45; dialectical materialism and, 32, 42, 274;

INDEX

history and, 304–11; humans and, 44; Lukács and, 349; rejection of, 501n7; return of, 271–76; Western Marxism and, 42–43, 55, 501n4, 584n26
Dialectics of Nature (Engels), 29, 42, 147, 353–54
dialectics of nature in Anthropocene, Engels and: "all things are sold," 314–15; capitalism and, 312; epidemics and, 313; history and, 304–11; influence of, 296; racing to ruin, 297–99; with revenge of nature, 299–303
Dialektik der Aufklärung (*Dialectic of Enlightenment*) (Horkheimer and Adorno), 248
discursive bundling, 259
disease: agribusiness and, 413, 415, 418; animals and, 416–17, 418; *Big Farms Make Big Flu*, 415; Black Death, 159–60, 199; *Clear-Cutting Disease Control*, 414; Ebola, 415; epidemics, 313, 414; H1N1, 413, 415, 416; H5N1, 415, 416; imperialism, class and pandemic, 423–29; "Is Capitalism a Disease?," 415; MERS, 413, 415; monopoly-finance capital with, 429; One Health-One World and, 413–14; radiation sickness, 467; SARS-CoV-2, 415–16, 421, 423, 424, 426; zoonotic, 19, 21, 413, 416, 423; *see also* COVID-19
disenchantment of the world (*Entzauberung der Welt*), 166, 192–95, 543n192, 543n196
disinvestment movement, fossil fuels, 603n31
displacement: environmental load, 216–18; humans, 26, 65, 131, 200, 357
dividend system, carbon fee and, 79
Doctrine of Being, Hegel, 271

Doctrine of Essence, Hegel, 271, 351
Dollard basin, flooding, 160, 199, 534n40
The Domination of Nature (Leiss), 143, 451
doomsday clock, *Bulletin of Atomic Scientists*, 430–31
"double Internality," 259, 582n67
drama, origins of, 352
A Dream of John Ball (Morris, W.), 109–10
drought, 23, 57, 129, 167; in Amazon, 130; megadrought, 465–66; in Palestine, 148
Dun & Bradstreet, 421
dust bowls, 325, 344
Dust Bowls of Empire (Holleman), 344
Dutch colonists, private wealth, 375

Earth Day, 57, 299
Earth Summit, Rio (1993), 517n8
Earth System: albedo effect, 64, 128, 274, 327, 328; Alternative Assessment principle, 134; analysis, 87, 496n34, 561n1; biogeochemical cycles and, 20–21, 75, 87, 143, 263, 311, 364, 436, 461, 505n36; capital accumulation as threat to, 74; change, 13–14, 36, 87, 490; change and Anthropocene Epoch, 461–62; crisis, 148, 612n3; defense in lieu of military spending, 79, 136; eco-revolution for, 135–37; emergency, 23, 434, 435, 456; expropriation of, 28, 36, 151; geoengineering and, 127–35; global carbon budget, 63, 119, 126, 399, 472, 525n1; Great Rift in, 42, 57–58; heat balance of, 324, 327; humans endangering, 82, 430, 433, 436; metabolic rift, 52–58, 436; ownership of, 357–58, 378,

473; Planetary Precautionary principle, 133–35; *Raubbau* and, 549n47; Reverse Onus principle and, 134; science, 20–21; social production and planetary metabolism, 429–32

The Earthly Paradise (Morris, W.), 101

Earthmasters (Hamilton), 132

earthquakes, 114, 421

East African Medical Journal, 25

East Asia, deaths, 424

East India College, 536n74

"An Eastern Perspective on Western Anti-Science" (Needham), 451–52

Ebola, 415

ecocide, 316, 334, 588n1, 589n1

eco-compatible capitalism, 366–67

ecodemocratic phase, revolution, 78–79, 405, 612n5

ecofascism, 407

ecofeminism, 144, 146, 147, 249, 512n31, 571n121

ecological barbarism, 407

ecological civilization, 14, 22, 36, 276, 433; Anthropocene Epoch and, 434; China and, 345, 407, 434, 451, 455, 472, 491; *China and Ecological Civilization*, 452; east and west, north and south, 451–56; exterminism or, 444–51; with transformational change, 408–10; Ye Qianji and, 625n48

Ecological Economics (journal), 237–38, 239, 558n189

ecological footprint analysis, 214, 216–17, 218, 236, 242, 552n77

ecological imperialism: criticism of, 242; Earth System crisis and, 148; *Ecological Imperialism*, 149; emergy exchange and, 234; metabolic rift and, 202, 450; poor countries and, 456; theory of, 201, 356; unequal ecological exchange and, 56, 202, 236, 250, 345, 356; Western, 487–88

Ecological Imperialism (Crosby), 149

ecological Marxism, 286, 317, 331, 340; ecosocialism and, 249, 284; freedom and, 72, 359; left ecological monism and, 253; radical social monism and, 251; three critical breakthroughs, 67, 68–74

ecological materialism, 42, 228, 532n13

ecological racism, 269

ecological revolution, 65; action for, 136; for Earth System, 135–37; ecological ruin or, 35–37; Marx and socio-, 58–61; new wave, 483–90

ecological ruin, revolution or, 35–37

ecological surplus, 557n163

ecological thought, in Anthropocene, 247–51

Ecological Threat Register 2020 report, 357

ecological value-form analysis, 69, 74, 250, 278–79, 282–83, 511n27

ecological-epidemiological crises, capital and, 413–18

The Ecological Rift (Foster, Clark and York), 344, 523n43, 524n50, 569n84

ecologism, 66, 146

ecologization, of modern science, 333

ecology: Age of Ecological Enlightenment, 247; agroecology, 128, 132, 136, 243, 344, 418; animal, 268, 308; bionomics and, 52; dialectical, 251–57, 505n36; *Geography and Ecology*, 330, 333; global, 327, 329; Haeckel and, 84; *Marx's Ecology*, 338, 339, 344, 345–46, 347, 348–51; as new opium of masses, 253,

288, 566n42; *oikeios*, 259; political, 251, 267, 313, 579n40; social monism and, 284–90; socialism and, 599n24; "Socialism and Ecology," 337, 453–54; *see also* dialectics of ecology, Marxism and; political economy, ecology of Marxian; Soviet ecology
Economic and Philosophic Manuscripts of 1844 (Marx), 257–58, 330, 331, 374
Economic Bill of Rights, Green Party New Deal, 395
"Economic De-Growth for Ecological Sustainability and Social Equity," Paris (2008), 364
economic growth, moratorium, 79, 135
Economic Manuscripts of 1861–1862 (Marx), 71
economic waste, 75, 79, 379, 381–84, 386, 599n21
The Economics of Innocent Fraud (Galbraith), 66
economy: costs, 79, 136, 218, 227, 382, 384, 395, 398, 399, 400, 419, 420, 423–24, 428, 430; debt, 75, 98, 235–36, 423; *Economy and Society*, 157, 160, 164, 170, 194–98, 534n40; Great Depression, 17, 368, 394; Great Financial Crisis, 418, 422; Great Recession, 368; Greenpeace International Economics unit, 394; International Society of Ecological Economics, 237; *Marxian Political Economy*, 229–30; *Marxist Economic Theory*, 377; *Nature's Economy*, 268; OECD, 422; *The Political Economy of Global Warming*, 77; *Principles of Political Economy*, 365; "Second Crisis of Economic Theory," 385; steady-state, 76, 122, 365, 366, 370; World Economic Forum, 391, 421, 619n82; *see also* political economy, ecology of Marxian
Economy and Society (Weber), 157, 160, 164, 170, 194–98, 534n40
ecosocialism: action steps for, 136–37; conference in Chicago, 603n32; ecological Marxism and, 249, 284; *Karl Marx's Ecosocialism*, 340; second-stage, 69, 72, 247, 249, 253
"Ecosocialism or Barbarism," 617n56
ecosocialist phase, revolution, 78, 79–80, 88, 405, 434, 612n5
ecosystem: Carson and, 85; destruction, 23; restoration and Indigenous peoples, 136; Tansley and, 53
"ecotechnic" phase of civilization, 176
Ecuador, 233–35
"The Effacement of Nature by Man" (Lankester), 303, 504n34
"The Effect of Cuban Agroecology in Mitigating the Metabolic Rift" (Betancourt), 344
"The Effect of Solar Radiation Variation on the Climate of the Earth" (Budyko), 328
Egypt, 165, 166, 167, 168–69, 170, 171, 465, 544n209
electricity, 77, 105, 114, 116, 130–31, 222, 398, 619n82
embodied energy (emergy), 201, 203–4, 220–22, 229, 238–40, 552n77, 558n189, 559n194; calculations for natural resources, 230; emergy received/emergy exported ratio, 234–35; emergy/money ratio, 231, 233, 557n164; investment ratio, 231
"Embodied Energy, Energy Analysis, and Economics" (Costanza), 559n194

embodied energy (emergy) exchange, 232, 233, 234
emergency: carbon emissions reduction plan, 79; Climate Emergency declaration, 392; Earth System, 23, 434, 435, 456
emergent properties, 307
emergy-transformity framework, 222; *see also* embodied energy
The Empire of Cotton (Beckert), 440
emvalue, 220, 222–23, 226, 229, 240, 553n89
Enbridge Northern Gateway Pipelines Project, 113
endangered species, protection of, 136
end-Anthropocene extinction event, 14, 19, 22, 459, 463, 473
end-Quaternary extinction event, 459, 463
The End of History and the Last Man (Fukuyama), 476, 627n5
energetic dogma, 238, 559n195
Energetic Foundations of a Science of Culture (Ostwald), 187
"'Energetic' Theories of Culture" (Weber), 189
energy: alternative infrastructure, 136; BECCS, 127, 130–31, 402; "Budyko energy balance models," 327; China with alternative-energy technology, 407; "Clean Energy Plan," 603n34; corporations, 400; fire and, 342; Global Energy Interconnection, 619n82; International Energy Agency report, 111; "Plan for a Clean Energy Revolution and Environmental Justice," 603n34; poverty, 133; renewable, 116–17, 397–98; scarcity, 190; sociology of, 157, 186–95; solar, 116, 118, 187–88, 189, 619n82; TC Energy, 119; theory of economic value, 220, 238, 559n196; theory of value, 191, 203, 238–39, 559n194; wind, 116, 118, 619n82; *see also* fossil fuels
energy return on energy investment (EROEI), 117, 121
Enewetak Atoll, nuclear weapons test, 467
Engels as a Scientist (Bernal), 305–9
"Engels's Philosophy of Science" (Manicas), 305
English Revolution, 483
Enlightenment: Age of Ecological Enlightenment, 247; de-, 481; dualist worldview of, 257, 268; Horkheimer, Adorno and, 248, 564n26; nature dominated by dialectic of, 32, 56, 68, 192, 248; Prometheus and, 342, 343
entropy law, 189–90, 287, 365
The Entropy Law and the Economic Process (Georgescu-Roegen), 365
environment, 326, 600n31; Center for Environmental Policy, 230; Center for Health, Environment and Justice, 387–88; consumption-based environmental costs, 218; crisis and capital accumulation, 73; culture and, 161–62, 166–71; defined, 530n3; *Global Environmental Change*, 344; invisibility of, 195–97; monopoly capitalism and, 378–89; monopoly capitalism and destruction of, 374; *Organization and Environment*, 343; "Our Polluted Environment," 85; UN Environment Programme, 394; *see also* non-industrial society, environment and
Environment, Power and Society (Odum), 87
environment, Weber and: coal, iron and industrial capitalism,

INDEX

175–86; culture and, 161–62, 166–71; environmental sociology and, 153–58; interpretive sociology and, 158–64; non-industrial society and, 164–73; post-exemptionalist sociology and, 195–200; sociology of energy, 186–95; wood epoch to age of iron, 173–75
Environmental Defense Fund, 127
environmental history, nature of, 268–71
environmental justice, 79, 201, 395, 398, 450, 473, 490–91; "Plan for a Clean Energy Revolution and Environmental Justice," 603n34
environmental load displacement, 216–18
environmental proletariat, 124–25, 243, 313, 407, 620n86; capitalist barbarism and, 36; defined, 77; *The Ecological Rift* and, 524n50; emerging, 477, 483, 490–93; mandate, 37; movements, 473; new, 456
environmental sociology: defined, 530n3; environment, Weber and, 153–58
environmentalism, Soviet Union and, 594n32, 594n38
The Environment (Schnaiberg), 600n31
epidemics, 313, 414
epigenetics, 589n6
epistemic fallacy, 571n109
epistemic rift, 49, 251, 503n25
epochal crisis, 75, 76, 79, 476
equator, solar farms at, 619n82
An Essay on the Principle of Population (Malthus), 62, 435
Ethiopia, 318
ethnocentrism, racism and, 183
Ethnological Notebooks (Marx), 169
Europe, deaths, 424

European Green New Deal, 394
European Heart Journal, 25
evolution, theory of, 308, 346
The Evolution of the Biosphere (Budyko), 336
exchange values: with ecology of Marxian political economy, 375; private riches, 226, 282, 375–76, 379, 389, 415, 430; use and, 212, 226, 227, 412–13; value-analytic and, 511n30
exergy, 220, 231, 241, 560n204
exploitation, expropriation and, 409, 420, 431–32, 488
expropriation: capital accumulation and, 244; of Earth System, 28, 36, 151; exploitation and, 409, 420, 431–32, 488; fossil fuels, 397; of guano, 356; of land, 35, 150, 345, 406, 487, 488; Marx and, 345, 486, 489; metabolic rift and, 35; of natural resources, 36, 341, 356; of people, 28, 35, 298, 374, 406, 487, 488; racism and, 136, 500n95
exterminism, 88, 444–51
external proletariat, 620n86
extinction: of biosphere, 449; end-Anthropocene event, 14, 19, 22, 459, 463, 473; end-Quaternary event, 459, 463; humans, 425; mass, 270, 294, 392, 463; "Notes on Exterminism, the Last Stage of Civilization," 80, 445–46; sixth, 21, 143, 262, 471, 490; species, 57, 270, 466
Extinction Rebellion, 24, 391, 392–93, 406
ExxonMobil, 105, 118, 127, 133

Facing the Anthropocene (Angus), 83, 344
factory farming, 54, 55
famine: India, 298, 313, 356; Ireland, 313, 443–44, 615n37

Farmer's Revolt, India, 491
farms: *Big Farms Make Big Flu*, 415; factory, 54, 55; National Farmers Union, 524n50; offshore, 415; poultry, 415, 416; shrimp, 235; solar, 619n82; "territorial restructuring," 418; *see also* agriculture
feedback mechanisms, global warming, 93, 327, 328
feeding operations, confined animals, 398
fees, carbon dividend system, 79
"Fellowship on Earth," 110
feminism (and ecofeminism), 11, 35, 144, 146–47, 249, 345, 511–12n31, 565, 571n121; *see also* gender, social reproduction, and women
fertilizer, 49, 73, 300, 443, 450, 615n37; artificial, 180–81; chemical, 35, 282; commercial, 54, 214, 281–82; guano, 46, 71, 211, 212, 236, 281, 356, 412
feudalism, 171
fiber-optic cables, 420
Finance Capital (Hilferding), 412
fires, 338; emergencies, 23, 79, 392, 434, 435; *On Fire*, 390, 405; management, 450; Prometheus myth and, 342
The Fire Next Time (Baldwin), 601n1
First Nations people: with Idle No More movement, 98, 113, 120, 524n50; Red New Deal, 473, 491
First New Left, 356
first-stage ecosocialism, 68, 248–49, 253, 564n26
Five Civilized Tribes, 185
flat ontology, 579n40
Flood Wall Street action, 602n17
floods, 160, 199, 534n40, 602n14
food web, 53, 294, 554n101
A Foodie's Guide to Capitalism (Holt-Giménez), 417

Foreign Policy (journal), 420, 424
foreign trade, 208
forests: clearances with rainfed agriculture, and land, 168–71; deforestation, 73, 173–75, 198, 302, 378, 441, 598n10; *Fundamentals of Forest Biogeocoenology*, 320, 336; Germany, 182; Main Administration of Forest Protection and Afforestation, 322; Ministry of Forest Management, 323, 325; planting new, 322–23; *The Silence of the Forest*, 185; taiga, 333; U.S., 182; wood epoch to age of iron, 173–75
Fortune 1000 multinational corporations, 421
fossil capital: carbon budget and, 399; carbon emissions from, 471; climate justice movement and, 400; geoengineering planet, 127, 128–32; labor and, 16; Malm on, 462; petrochemical industry and, 468–69; poverty of power and, 470; system, 133, 135
fossil capitalism, 35, 87–88, 495n26
fossil fuels, 16, 79, 158, 233–34; burning all, 58; combustion cessation, 93–94; consumption, 177–78; disinvestment movement, 603n31; expropriation, 397; global warming and, 363; industry and financial losses, 400; No Fossil Fuel Money pledge, 396; people power against, 98; phase-out of, 136; unconventional, 79, 111, 115, 116, 119, 519n2; *see also* oil
fossil fuels war: carbon and, 117–21; fracking, 112–14, 118–20; revolution against system, 121–25; tar sands oil, 112–13, 117–19; Unconventionals Era, 111–12, 116

INDEX

4°C future, dangers of, 438
"Fourier's critique of civilization" (Beecher), 439
fracking, 107, 112–14, 118–20, 398–99
France, 33, 174
Frankenstein, or, the Modern Prometheus (Shelley, M.), 343
Frankfurt School, 56, 68, 264, 331, 501n4, 506n45, 564n26, 585n28
freedom: capitalism and, 17; ecological Marxism and, 72, 359; Engels on, 276, 295; Marx on, 48; of necessity, 276, 405; struggle for, 35; Third World liberation movements, 488
French Revolution, 483
Fukushima nuclear disaster, 421, 468
Functioning Democracy, Green Party New Deal, 395
Fundamentals of Forest Biogeocoenology (Sukachev and Dylis), 320, 336
Fuyao Glass Industry, 428

Gaia hypothesis: Earth System science and, 21; European Green New Deal, 394; Lovelock and, 20; UK Green New Deal Group, 394
gas mileage decline, in U.S., 600n33
gender, 11, 35, 78, 136, 396, 477; *see also* feminism, social reproduction, and women
gene-culture coevolution, 254, 310, 329
General Mills, 423
General Strike (1842), 489
The General Economic History (Weber), 164, 165, 176, 178–79, 195–96, 531n11, 538n113
genocide, 16, 447
geocoenosis, 274, 320–22, 324, 330, 332, 335, 590n9, 591n11
geoengineering, 88, 107, 127–35, 289, 328

"The Geographic Conditions of the Human Economy" (Hettner), 196
Geography and Ecology (Gerasimov), 330, 333
Germanic villages, 172, 173
The German Ideology (Marx and Engels), 35, 488–89
Germany, 116, 180, 181, 248; emergy exchange ratio, 233; forests, 182
ghost acreage, 214
glasnost, 334
global catastrophes, 327, 329
global ecology, 327, 329
Global Ecology (Budyko), 329–30, 336
Global Ecosocialist Network, 344–45
Global Energy Interconnection, 619n82
Global Environmental Change (journal), 344
Global Green New Deal (UN Environment Programme), 394
global heating, 24, 25
global labor arbitrage, 411, 417, 420, 423, 427, 606n2
global land arbitrage, 411–12, 417, 606n2
Global Problems and the Future of Mankind (Frolov), 331
global warming, 77, 86, 264; Commoner and, 495n27; feedback mechanisms, 93, 327, 328; fossil fuels and, 363; irreversible, 83; isotherms and, 51
global weirding, 23, 116
"The Globe Downshifted" (Latouche), 366–67
Goddard Institute for Space Studies, NASA, 19, 112
golden ages, 18, 103, 448, 580n51
golden spike, 246, 465, 467, 562n9
The Golden Notebook (Lessing), 359
GoogleScholar, 531n11

Great Acceleration, 18, 83, 255, 311, 409, 466, 495n26; Anthropocene crisis and, 246; Great Capitalist Climacteric and, 435–36; indicators, 469–70; radionuclides and, 82
Great Britain: cotton crisis, 72; Industrial Revolution, 16
Great Capitalist Climacteric: defined, 62–63; ecological Marxism and, 67, 68–74; ecological revolution and, 65; Great Acceleration and, 435–36; Green theory and, 66–67, 68; historical conditions, 434–38; Marxism and, 74–81; metabolic rift and, 70–71, 73–74; 2°C guardrail and, 63–64, 65
Great Climacteric: Age of Ecological Enlightenment and, 247; Anthropocene and, 247, 255, 433, 434, 435, 451, 463; Bédarida and, 508n1; Burton and Kates on, 62, 435, 471; *An Essay on the Principle of Population*, 435; with reversal, 472; social-historical transition and, 436
Great Depression, 17, 368, 394
Great Financial Crisis (2007–2009), 418
Great Financial Crisis (2008–2010), 422
Great Irish Famine (1845–46), 313, 443–44, 615n37
Great Recession (2008–2009), 368
Great Rift, 42, 57–58
Great Stalin Plan for the Transformation of Nature (1948), 322–23, 325
"the Great Climacteric," 508n1
The Great Soviet Encyclopedia, 20, 459, 495n32
Greece, 173, 302, 441, 444, 465
Green Building Council, U.S., 388

Green Climate Fund, 398
Green European Foundation, 394
Green Keynesianism, 394
Green New Deal: climate Keynesianism and, 401; degrowth and, 368; demand for Select Committee for, 391; European Green New Deal, 394; fossil fuel disinvestment movement and, 603n31; funds for, 398; *Global Green New Deal* report, 394; *A Green New Deal for Europe*, 394; Indigenous peoples and, 397; Obama proposal and, 394; People's Green New Deal, 395, 399, 473, 603n28; reform or revolution, 393–401; Sanders and, 392, 397–98, 399, 603n34; support for, 393; UK Green New Deal Group, 394
Green New Deal Resolution: based on "10-year national mobilization," 397; cosponsors, 392, 602n10; monetary theory supporting, 604n36; Ocasio-Cortez and Markey with, 391, 396, 399, 601n6, 604n36; supporters, 392, 603n34
Green Party: ecosocialism conference, 603n32; New Deal, 395–96, 397; Stein and, 395
Green theory, 66–67, 68, 358
Green Transition, Green Party New Deal, 395, 396
A Green New Deal for Europe (Green European Foundation), 394
greenhouse effect, 23, 52
greenhouse gas emissions, 106, 294; mitigation strategies, 401; net-zero, 397, 602n14, 603n34; reduction, 128, 136, 399; tipping point, 64, 126
Greenland, 116, 326
the Greenlandian, 464, 465

INDEX 673

Greenpeace, 235, 391; International Economics unit, 394
gross social wealth, 333
Grundriss für Sozialökonomik (Outline of Social Economics) (Weber), 196
Grundrisse (Marx), 47, 70, 145, 169, 254, 263, 290, 330, 477–78
guano, 46, 71, 211, 212, 236, 281, 356, 412
The Guardian (newspaper), 394
Gulf of Guinea, 115

H1N1, 413, 415, 416
H5N1, 415, 416
habitat destruction, animals, 423
Harvard Business Review, 421
"Healing the Rift" (Clausen), 202
health: care costs, 398; Center for Health, Environment and Justice, 387–88; commodity chain disruptions, 422; One Health-One World, 413–14; Structural One Health, 414, 415, 431; superbugs, 431; World Health Organization, 413, 431; *see also* COVID-19; disease
heat balance, of Earth System, 324, 327
Heat Balance of the Earth Surface (Budyko), 324
"The Heat and Water Balance Theory of the Earth's Surface" (Budyko), 327
Himalayas, 57
Hiroshima (Japan), 17, 468
Historical Materialism (Bukharin), 506n45
history: capitalism as barrier to future of, 477–82; *The City in History*, 444; decapitation of, 479; dialectics of nature and, 304–11; *The General Economic History*, 164, 165, 176, 178–79, 195–96,

531n11, 538n113; nature of environmental, 268–71; "On the Concept of History," 297; with revolution as complex process, 601n1; revolutions as locomotive of world, 297; Sartre on future and, 475, 479, 483
History and Class Consciousness (Lukács), 30, 42, 43, 44, 271, 284–85, 349, 631n61
History of Creation (Haeckel), 52
A History of Civilizations (Braudel), 439
A History of the Ecosystem Concept in Ecology (Golley), 53
hogs, 415, 416
holism, 253, 255, 256, 269, 573n143
Holocene Epoch, 14, 21, 57, 63, 82, 123, 311, 457; ages of, 464–65; civilization in, 433–34; Gervais and, 464; Late Holocene Meghalayan Age, 466
The Holy Family (Marx and Engels), 485
hominins, early, 330
Homo sapiens, 329
hope, China and, 452–55
horseshoe crab, 306
hospitals, commodity chain disruptions, 422
The Housing Question (Engels), 54–55, 227
The Huffington Post (newspaper), 103–4
human-exemptionalist paradigm, 154, 179, 198
humans: agency, 77, 481, 483, 628n20; agriculture and soil, 281; alienation from nature and, 28, 31–32, 34–35, 150–51; alienation of, 280; All-Union Conference on the Problem of Climate Modification by Man, 327; capitalist barbarism and, 36;

community-based management of oceans, 136; *The Condition of Man*, 76, 80–81, 365, 444, 616n43; dialectics of nature and, 44; displacement, 26, 65, 131, 200, 357; Earth System endangered by, 82, 433, 436; *The End of History and the Last Man*, 476, 627n5; as expendable, 18; extinction, 425; with "Fellowship on Earth," 110; impact of, 14, 19, 82, 143, 621n9; *The Kingdom of Man*, 19, 303, 415, 463; *Man, Science, Humanism*, 331; *Man and Nature*, 336; Marx with definition of, 341; metabolism and, 47; naturalization of, 248; nature and, 303, 319, 357–58, 378, 409, 587n58, 588n67; nature degraded by, 143, 144–45; origin of society, 308–9; people power with climate change solutions, 96–98; planetary boundaries and safe operating space for, 123, 363; with precarious lives, 78; as productive force, 583n9; solar energy and, 187–88; with sustainable development, 76–77, 278; with universal metabolism of nature and social metabolism, 280; *see also* people

The Hungry Planet (Borgstrom), 214
hydraulic bureaucracy, non-industrial society and, 168–71, 537n88
hydroelectric power, 188
hyper-social constructionism, 249, 265, 572n122

ice: albedo feedback, 327, 328; cover destruction, 327; Greenland sheet, 326; loss in Arctic, 116, 128; polar, 317, 327, 328; "Polar Ice and Climate," 328
Ice Age, 464, 465

ice-free world, 57, 58, 115–16, 125
Idle No More movement, 98, 113, 120, 524n50
Illusion and Reality (Caudwell), 275
"The Impact of Economic Activity on the Climate" (Budyko), 328
imperial capitalism, 53, 88, 203, 220, 233, 243
Imperial College, London, 424, 425
imperialism: *The Age of Imperialism and Imperialism*, 355; class, pandemic and, 423–29; commodity chains and, 411; de-, 152; *Imperialism, the Highest Stage of Capitalism*, 355; land and, 150–51; metabolic rift and, 545n1; planetary ecological crisis and, 357; reversed, 356; unequal ecological exchange and, 547n14; *see also* ecological imperialism
Imperialism, the Highest Stage of Capitalism (Lenin), 355
"Imperialism in the Anthropocene" (Foster, Holleman and Clark), 357
impossibility theorem, degrowth and, 370
India, 444; British East India Company, 536n74; emergy exchange ratio, 233; famine in, 298, 313, 356; Farmer's Revolt, 491; Kerala, 237; labor costs, 419; Mawmluh cave, 465; *The Religion of India*, 162–63, 164, 165, 170, 535n42, 537n80
Indian Territory, 184–85, 186, 197, 200
Indians, privatization of land, 185
Indigenous peoples: colonialism and, 486; ecosystem restoration and, 136; First Nations, 98, 113, 120, 473, 491, 524n50; Green New Deal and, 397; Marx and, 630n48; Native Americans,

119–20, 184–85, 186, 197, 200, 396, 455, 492, 614n21; nature and, 571n121; populations, 16, 78, 79; revolts, 487, 492; Third World liberation movements, 488; tribal sovereignty, 398; Yanomamo Indians, 237
Indonesia, 419, 428
Indus Valley, 465
industrial agriculture, 27, 46–47, 70, 179–81, 211, 281, 442
industrial capitalism, 28, 158, 164–65, 173, 192, 211, 458, 482; coal, iron and, 175–86
industrial metabolism, 56, 506n47
Industrial Revolution, 16, 82, 98, 131, 175, 177, 198, 298–99; fossil capitalism and, 495n26; Prometheanism and, 343
information: energy hierarchy and transformy of, 556n141; real wealth, 235
An Inquiry into the Nature of Public Wealth and into the Means and Causes of Its Increase (Maitland), 375
inroads of sand, desertification and, 544n209
Institute of Applied Geophysics, 326
Institute of Economics and Peace, 357
integrated assessment models (IAMs), 401, 403–4, 604n45
Inter-American Development Bank, 400
Intergovernmental Panel on Climate Change, 93
internal proletariat, 620n86
International Commission on Stratigraphy, 457, 464, 465, 466, 467
International Energy Agency report (2010), 111
International Geographic Congress, 464

International Geosphere-Biosphere Program, Mexico, 460
International Society of Ecological Economics, 237
international trade, 205–7, 208, 229, 235, 470
International Union of Geological Sciences, 18, 457
Internet, 420
interpenetration effect, 60
interpretive sociology, 158–64
invisibility, of environment, 195–97
Iodine-131, nuclear fallout, 467
IPAT formula (Population x Affluence x Technology), 599n30
Iraq, 117, 119
Ireland: expropriation of people, 298, 487; famine in, 313, 443–44, 615n37
Ireland and the Irish Question (Marx and Engels), 443, 444, 615n37
iron: coal, industrial capitalism and, 175–86; coal and, 538n113; from wood epoch to age of, 173–75
Iroquois culture, 614n21
"Is Capitalism a Disease?" (Levins), 415
isotherms, 51, 270
Ivy Mike test, Enewetak Atoll, 467

Jacobin (magazine), 127
Japan, 233, 470; Fukushima nuclear disaster, 421, 468; Hiroshima and Nagasaki, 17, 468
jet stream, redirection of, 116
Jevons Paradox, 364, 404
Jordan, 119

Karl Marx's Ecosocialism (Saito), 340
Kellogg, 423
Kerala (India), 237
Keystone XL Pipeline, 112–13, 118–19, 519n6
Keywords (Williams), 141

The Kingdom of Man (Lankester), 19, 303, 415, 463
Kommunist, 334
Kyoto Protocol, 93

labor: abstract, 223, 273, 278, 554n115; agribusiness with cheap, 416, 430; alienation of, 341, 351, 357; costs, 227, 419, 428; fossil capital and, 16; global labor arbitrage, 411, 417, 420, 423, 427, 606n2; international immobility of, 206, 208; international trade and, 229; Marx on, 291, 331, 507n54; metabolic rift and, 357; as metabolism of society and nature, 277; power, 18, 58, 69, 226, 227, 278, 282, 284, 376, 512n33; production and, 378; servile, 172; theory of value, 218, 222, 225, 238, 239–40; unproductive, 599n21; *see also* commodity chains
Labour (Lukács), 582n62
Labour Defended Against the Claims of Capital (Hodgskin), 15
Lake Baikal, 316, 325, 326, 333
Lancet (medical journal), 25
land: agribusiness with cheap, 416, 430; deforestation, 73, 173–75, 198, 302; ecological footprint analysis and, 214, 216–17, 218, 236, 242, 552n77; expropriation of, 35, 150, 345, 406, 487, 488; ghost acreage, 214; global land arbitrage, 411–12, 417, 606n2; grabs, 49, 77, 418; imperialism and, 150–51; Indian Territory, 184–85; Louisiana Purchase, 182; "Marx's Ecology and the Understanding of Land Cover Change," 202; peasants and, 172, 179; with population growth, 183; privatization of Indian, 185; rain-fed agriculture, forest clearances and, 168–71; rent, 417; robbery of, 165; "territorial restructuring," 418; "Virgin Lands" program, 325
Landless Workers' Movement (MST), 345, 473, 491
landowners, property rights, 374
Late Victorian Holocausts (Davis), 356
Latin America, deaths, 424
Lauderdale Paradox, 226, 282, 375–76, 379, 415, 430
"left apocalypticism," 251, 264, 285
left ecological monism, 253
"The Left vs. the Climate" (Boisvert), 107–8
Lenin All-Union Academy of Agricultural Sciences, 323
Lenin Prize, 324
"Let Them Drown" (Klein), 148
Letters of Karl Marx (Padover), 354
Leviathan (Hobbes), 628n16
Levi's, 427
liberal critics, with *This Changes Everything*, 99–108
Liberia, emergy exchange ratio, 233
The Limits to Growth (Meadows, D., Meadows, D. H., Randers and Behrens), 20, 45, 272, 326, 336, 364
Literary Review of Canada, 106
lithium battery chemicals, 421
Livestock Revolution, 417
Living Planet Report, 371
locomotives: boiler explosions, 583n5; fireboxes and coffee, 310; safety-valves, 297, 583n5; of world history and revolutions, 297
London Stock Exchange, 391
"long sixteenth century," 16, 494n14
Los Angeles Times (newspaper), 105, 106
Louisiana Purchase, 182

INDEX

"love your monsters," 107, 251
"A Lover from Palestine" (Darwish), 150
Lucky Dragon (fishing boat), 467

Main Administration of Forest Protection and Afforestation, 322
Main Geophysical Observatory, Leningrad, 324
"Makeshift" (Morris, W.), 59
Making Peace with the Planet (Commoner), 135
The Making of the English Working Class (Thompson), 483
Man, Science, Humanism (Frolov), 331, 336
Man and Nature (Fedorov), 326, 336
Man and Nature (Marsh), 84
manors, 172
Mansfield Park (Austen), 529n1
manufacturing, in China, 427–28
marine biodiversity, 294
marine calcifiers, 294
marine cloud brightening, 129, 130
marketing: advertising and, 382–83; corporations with limits on destructive, 136; Power Marketing Authorities, 398
Marshall Islands, nuclear fallout, 467
Marx and Nature (Burkett), 69, 338, 340, 564n28
Marx for Our Times (Bensaïd), 257
Marx-Engels Gesamtausgabe (Engels and Marx), 340
Marxian Political Economy (Becker), 229–30
Marxism: Chinese, 449; dialectical realism and reunification of, 290–95; ecological, 67, 68–74, 249, 251, 253, 284, 286, 317, 331, 340, 359; Great Capitalist Climacteric and, 74–81; metabolism theory, 251, 259, 279–82; universal metabolism of nature and, 26–35; *see also* Anthropocene, Marxism in; dialectics of ecology, Marxism and; political economy, ecology of Marxian; Western Marxism
Marxism and Ecological Economics (Burkett), 202, 564n28
Marxism and Hegel (Colletti), 42–43
Marxism and the Philosophy of Science (Sheehan), 347
Marxist Economic Theory (Mandel), 377
Marx-Odum dialectic, 204, 222–28, 243–44
Marx's Ecology (Foster), 202, 338, 344, 345–46, 350–51, 564n28; criticism of, 347; validation for, 339–41; Western Marxism and, 348–49
"Marx's Ecology and the Understanding of Land Cover Change" (Dobrovolski), 202
Marx's Theory of Alienation (Mészáros), 272, 341, 502n13
"Marx's Theory of Metabolic Rift" (Foster), 202
The Masculine Birth of Time (Bacon), 145
mass extinction, 270, 294, 392, 463
mass migration, 418
materialism: biosphere and dialectical, 33; China with historical, 434; ecological, 42, 228, 532n13; historical, 486; *Historical Materialism*, 506n45; Marx and, 341, 485; *see also* dialectical materialism
"Materialism and Revolution" (Sartre), 256–57
materiality paradox, waste and, 388–89
Mawmluh cave, India, 465
maximum empower, 228, 554n101
media, with Chomsky as *persona non grata*, 101

Medical Journal of Australia, 25
Medicare for All, 398
The Mediterranean (Braudel), 494n14
megadrought, 465–66
Meghalayan Age, 457, 458, 464–66, 622n25
mercantilism, 16
MERS, 413, 415
Mesopotamia, 165, 166, 168, 170, 171, 302, 441, 446; megadrought and, 465
metabolic restoration, 77, 295
metabolic rift (social metabolism), 41, 345, 598n11; Ayres and Simonis with, 506n46; commodity chains and, 415; Earth System and, 52–58, 436; ecological imperialism and, 202, 450; "The Effect of Cuban Agroecology in Mitigating the Metabolic Rift," 344; expropriation and, 35; imperialism and, 545n1; labor and, 357; Marx and, 45, 48–49, 52, 70–71, 73–74, 77, 503n24, 506n46, 576n9; NASA monitoring, 57; nature and, 145–46; soil nutrient cycle and, 54, 211; treadmill of production and, 73; universal metabolism of nature and, 26–35; with universal metabolism of nature and humans, 280
"The Metabolic Rift of Livestock Agribusiness" (Gunderson), 202
metabolism (*Stoffwechsel*), 47, 53, 70, 212, 341, 552n86; Budyko on, 330; carbon, 28, 49, 57–58, 450; industrial, 56, 506n47; social production and planetary, 429–32; of society and nature with labor, 277; "Toward a Singular Metabolism," 569n74, 570n99; translation issues, 341
metabolism theory, Marx and, 251; conceptual structure, 259, 279–82; triadic relationship, 279–80
metaphysical belief, 534n38
methane, 64, 114, 118, 126, 519n2
Mexico, 119, 318, 322, 419, 427, 460, 488
microorganisms, populations of, 306
Middle East, 117, 424
Mikrokosmos (Daniels), 70, 283
military incursions, for oil, 117
military spending: cuts, 395, 399; Earth System defense in lieu of, 79, 136; economic waste and, 382, 384
mimesis, Aristotle, 352
minerals, real wealth, 235
Ministry of Forest Management, Soviet Union, 323, 325
minority, dominant, 445, 620n86
Mirror Makers (Fox), 387
mitigation, climate change: carbon emissions, 401; *Climate Change 2021*, 23–24; IAMs, 401, 403–4, 604n45; IPCC, 23–24, 401–8, 602n14; scenarios, 402–3; strategies, 401–8
mobile phones, 420
"modern synthesis," 250, 564n25
modernity, capitalism and, 481
"Modernity and Hybridity" (Swyngedouw), 579n44
Molecular Red (Wark), 262
monetary theory, Green New Deal Resolution, 604n36
monism: dialectical materialism and, 270, 285; ecology and social, 284–90; left ecological, 253; neutral, 259, 266–67, 286, 570n99, 579n40; radical ecological, 251–57; radical social, 251, 255; Social Darwinism and, 270; social-constructionist, 250, 263; world ecology and social, 257–63

INDEX

monocultures: agribusiness, 132, 136, 423, 431; livestock, 415, 417
Monopoly Capital (Baran and Sweezy): "confrontation of reality with reason," 599n24; economic waste, 382, 599n21
monopoly capitalism: environment and, 378–89; environmental destruction and, 374; Hilferding and Veblen with, 379; Morris on, 388; rise of, 16–17, 18; technology and, 470–71
monopoly-finance capital: disease and pandemic with, 429; global austerity with, 419; negative use values of, 58–59; supply chains, 421; as threat, 74, 405, 406, 411
monsoon system, 129
monsters, 107, 250–51, 565n32
"The Monsters of Bruno Latour" (Shellenberger and Nordhaus), 565n32
Mont Blanc (Shelley, P. B.), 314
Monthly Review (magazine), 338, 339, 347–48, 354, 355, 357, 414
moratorium, on economic growth, 79, 135
mortality rates: *see* deaths
Moscow Society of Naturalists (MOIP), 324, 325–26, 335
mountaintop removal coal mining, 399
Movement Toward Socialism, 99
The Myth of the Machine (Mumford), 444–45

N95 respirators, 422
Nagasaki (Japan), 17, 468
NASA, 19, 57, 112
National Farmers Union, 524n50
National Medical Journal of India, 25
National Retail Federation, U.S., 422
National Snow and Ice Data Center, U.S., 116

Native Americans: Indian Territory and, 184–85, 186, 197, 200; Iroquois culture, 614n21; Standing Rock Sioux, 119–20, 396, 455, 492
NATO, 117
natural capitalism, 373
natural resources: consumption of, 157, 180–86; depletion, 183; emergy calculations for, 230; expropriation of, 36, 341, 356; extraction, 99, 173; Natural Resources Defense Council, 127, 452; scarcity, 188, 189; Weber and, 182–83
Natural Resources Defense Council, 127, 452
nature: *Against Nature*, 253; alienation from humans and, 28, 31–32, 34–35, 150–51; alienation of, 144, 244, 251, 259, 261, 262, 263–64, 268, 280, 284–85, 290–91, 310, 315, 341, 449; Baconian principle and, 301; *The Book of Nature*, 544n209; *Capitalism, Nature, Socialism*, 343, 347; capitalism, public wealth and, 374; capitalism and, 287, 374; capitalist barbarism and, 36; capitalist law of value and destruction of, 282–84; capitalist production and, 512n33; *Concept of Nature in Marx*, 44; *The Concept of Nature*, 142; culture and, 450; *The Death of Nature*, 143; defined, 141–42, 530n3; dialectic of Enlightenment and, 32, 56, 68, 192, 248; dialectics of, 41, 42–45, 271–76, 501n7; *Dialectics of Nature*, 29, 42, 147, 353–54; *The Domination of Nature*, 143, 451; ecofeminism and, 146, 147; Epicurus on, 244; as free gift, 18, 70, 213, 226, 278, 282, 341, 431, 512n32, 550n57, 577n16; Great

Stalin Plan for the Transformation of Nature, 322–23, 325; human degradation of, 143, 144–45; humanization of, 248; humans and, 303, 319, 357–58, 378, 409, 587n58, 588n67; Indigenous peoples and, 571n121; *An Inquiry into the Nature of Public Wealth and into the Means and Causes of Its Increase*, 375; labor as metabolism of society and, 277; *Man and Nature*, 84, 326, 336; *Marx and Nature*, 338, 340; metabolic rift and, 145–46; *On Human Nature*, 144; philosophy and, 142–43; *Philosophy of Nature*, 145; as pre-formed, 578n28; *The Return of Nature*, 339, 345–49, 351–52, 358–59; revenge of, 299–303; *The Robbery of Nature*, 345, 357; science, art and, 142; Smith, Castree and production of, 249, 253, 263–67; society and, 146–47, 264–65, 267, 268, 270, 441, 502n10, 502n12, 503n25; socionature, 286; subsumption of, 284–90; third, 37, 148–52; with unforeseen consequences, 133; use values and, 511n30; value excluding, 512n31; views on, 141–47; as wealth, 77; Weber on, 162–64; *see also* universal metabolism of nature

Nature (magazine), 142
Nature Climate Change (York), 118
Nature's Economy (Worster), 268
necessity: freedom of, 276, 405; struggle of, 35
needs: to each according to, 76, 77, 80, 365; hierarchy of, 429
negation of the negation, 66, 305, 307–8, 317, 335, 486, 586n42
neoliberalism, 17, 97, 102, 103, 104, 429, 453
Netherlands, 214, 233

net-zero: carbon emissions, 23, 94, 95, 106, 398, 603n34; greenhouse gas emissions, 397, 602n14, 603n34
neutral monism, 259, 266–67, 286, 570n99, 579n40
New Deal: first, 394, 398, 400–401; of Green Party, 395–96, 397; second, 394, 395, 399, 401; *see also* Green New Deal
New England Journal of Medicine, 25
New Left hybrid theories, 266
New Left Review, 356
New York Review of Books, 104
New York Times (newspaper), 103, 106, 394
New York Tribune (newspaper), 168–69
New Zealand, emergy/money ratio, 557n164
The Newcomes (Thackeray), 15
News from Nowhere (Morris, W.), 352, 507n57
The New Atlantis (Bacon), 145
"The New Politics in the USSR" (DeBardeleben), 335
nitrogen cycle, phosphorus and, 21, 57, 65, 123, 466, 490
No Fossil Fuel Money pledge, Sunrise Movement, 396
No Logo (Klein), 92, 97, 517n11
Nobel Prize, 187, 189, 308, 461
non-industrial society, environment and: *Ancient Judaism*, climatic conditions and, 166–68; hydraulic bureaucracy and, 168–71, 537n88; with rain-fed agriculture, forest clearances and landholdings, 168–71; Weber and, 164–73
noosphere, 330, 496n34
North Africa, deaths, 424
North America, deaths, 424
North Dakota, 114, 396, 455
North Pole, 619n82

INDEX 681

the Northgrippian, 464, 465
Notes on Adolph Wagner (Marx), 70
"Notes on Exterminism, the Last Stage of Civilization" (Thompson, E. P.), 80, 445–46
Novum Organum (Bacon), 145
nuclear power: anti-nuclear movement and, 329; Chernobyl accident, 316, 336, 468; Fukushima disaster, 421, 468; support for, 119; Three Mile Island disaster, 468
nuclear weapons, 83, 143; aerosol loading and, 329; testing and radionuclides, 18, 19, 82, 246, 436, 467–69, 471
nuclear winter theory, 329, 445, 446, 468, 617n60, 624n33
nutrient cycles, soil, 53–54, 55, 190, 442, 497n59, 505n39

Occupy Wall Street movement (2011), 98
oceans: acidification, 21, 24, 57, 65, 123, 130, 294, 363, 470, 490; biodiversity, 294; coral bleaching, 294; human community-based management of, 136; marine calcifiers, 294; marine cloud brightening and, 129, 130; *Special Report on the Ocean and Cryosphere*, 392
offshore farming, 415
offshore oil drilling, 99, 112, 114, 115
Ogallala aquifer, 113
oikeios, 259
oil: Amazon, 235; Dakota Access Pipeline, 119–20, 396, 455; Deepwater Horizon disaster, 114; fracking, 107, 112–14, 118–20, 398–99; Keystone XL Pipeline, 112–13, 118–19, 519n6; offshore drilling, 99, 112, 114, 115; peak, 111, 121, 519n2; petrochemicals and, 18, 21, 385, 458, 467–71; safeguarding supplies, 399; spills, 114; tar sands, 79, 112–13, 117–19, 519n6, 524n50; West Texas Intermediate, 113
Oil, Chemical and Atomic Workers Union, 395
On Fire (Klein), 390, 405
On Human Nature (Wilson), 144
"On the Concept of History" (Benjamin), 297
One Health-One World approach, disease and, 413–14
1.5°C: IPCC carbon dioxide emissions and, 603n34; *Special Report on Global Warming of 1.5ºC*, 390–91; target with mitigation strategies, 404; 2°C guardrail and true marker of, 602n14
Ontology of Social Being (Lukács), 292–93, 351
Oparin-Haldane theory, 85
open-system thermodynamics, 74
The Open Veins of Latin America (Galeano), 356
opium: new opium of masses, 253, 288, 566n42; trade, 15
optimum efficiency, 222, 228, 554n104
Oregon, fires in, 338
organic chemistry, 19, 468
Organization and Environment (journal), 343
Organization of Economic Cooperation and Development (OECD), 422
origin of human society, 308–9
origin of the family, 309
origin of the universe, 308
origins of life, 308, 318, 329, 336, 496n34, 587n49
Our Final Hour (Rees), 446
Our History Is the Future (Estes), 492

"Our Polluted Environment" (Carson), 85
overconsumption outrage, 370
ownership: absentee, 379; of earth, 357–58, 378, 473
ozone depletion, 57

packaging: bread manufacture and, 600n34; for saleable goods, 380–81, 387; wasteful, 383
paleoclimatic change, 330
"paleotechnic" phase of civilization, 177
Palestine, 148, 165, 166–67, 465
pandemics: commodity chains and, 414; global death estimates, 424; imperialism, class and, 423–29; monopoly-finance capital with, 429; new, 428–29; *see also* COVID-19
Paris Climate Agreement, 403, 404, 407
"The Part Played by Labour in the Transformation from Ape to Man" (Engels), 329
patriarchy, racism and, 371
Patriot Act, 395
peak oil, 111, 121, 519n2
peasants: China, 180; depeasantization, 418; with ecological ruin, 444; expropriation of, 374; international mobilisation, 202; Irish, 615n37; land and, 172, 179; movements, 473, 491; simple existence of, 165; wars, 487–88; with wood, 374
people: expropriation of, 28, 35, 298, 374, 406, 487, 488; mass migration, 418; power with climate change solutions, 96–98; Science for the People, 395, 603n28; self-mobilization of, 60, 127, 135, 405, 409; slaves, 16, 181–82, 473, 630n48; *see*

also activism; Indigenous peoples
Peoples' Agreement on Climate Change (2010), 79
People's Climate Movement, New York (2014), 393
People's Commissariat of Education, 318
People's Green New Deal, 395, 399, 473, 603n28
perestroika, 334
permafrost, methane from, 64
"perpetuum mobile", 190
Persia, 168–69, 441
Perspectives on the Nature of Geography (Hartshorne), 544n207
Peru, 236, 318
petrochemical industry, 467, 468–69
petrochemicals, 18, 21, 385, 458, 470–71
Phanerozoic Eon, 457
Phenomenology of Spirit (Hegel), 627n4
philosophy: *après moi le déluge*, 136, 283; *Economic and Philosophic Manuscripts of 1844*, 257–58, 330, 331, 374; "Engels's Philosophy of Science," 305; of internal relations, 565n36; *Marxism and the Philosophy of Science*, 347; nature and, 142–43; *The Poverty of Philosophy*, 145, 343; *Problems of Philosophy*, 330
Philosophy and the Ecological Problems of Civilisation (Ursul), 331, 332, 449–50, 594n33, 617n57, 625n48
Philosophy of Nature (Hegel), 145
The Philosophy of Poverty (Proudhon), 343
phosphorus cycle, nitrogen and, 21, 57, 65, 123, 466, 490
Phytophthora infestans, 615n37

INDEX

phytoplankton, 130
"Plan for a Clean Energy Revolution and Environmental Justice," 603n34
planetary boundaries: approach, 123; crossing of, 21, 57, 363–64, 580n51; safe operating space within, 123, 363
planetary metabolism, social production and, 429–32
plantation system, slaves, 181–82
plastics: petrochemical industry and, 467, 468–69; threat of, 470; waste, 387–88, 469
Pleistocene Epoch, 464
Plug Plot Riots, 489
Plunder of the Commons (Standing), 430
poetry, origins of, 352
Poland, 119
polar ice, 317, 327, 328
"Polar Ice and Climate" (Budyko), 328
political ecology, 251, 267, 313, 579n40
political economy, ecology of Marxian: with advertising and marketing, 382–83; Baran, Sweezy and, 382–84; *boni patres familias* and, 378; capitalist *Raubbau* and, 374–78, 389; degrowth, poor countries and, 369–71; dependence effect and, 384–85; Lauderdale Paradox and, 375–76, 379; materiality paradox and, 388–89; monopoly capital, environment and, 378–89; packaging for saleable goods and, 380–81, 387; revolution and, 389; Schnaiberg and, 377, 385–86; with use value and exchange value, 375; Veblen and, 379–82; waste, 379, 381–84, 386

The Political Economy of Global Warming (Weston), 77
The Political Economy of Growth (Baran), 384
pollution: air, 333, 384, 453; fracking and toxic, 114; industrial, 298; nuclear fallout, 467–68; Soviet ecology and air, 317; taxes and corporate, 398; in U.S. and Weber, 184, 186
poor countries: climate change and, 98; commodity chains, death and, 425; consumption-based environmental costs, 218; as "counting for nothing," 236–37; degrowth and, 369–71; ecological imperialism and, 456; ghost acreage in, 214; Green Climate Fund and, 398; racism and deaths in, 424–25; Third World, 25–26, 78–79, 205, 236–37, 354, 447, 488, 537n80
"Poor Countries Need to Think Twice About Social Distancing" (Mobarak and Barnett-Howell), 424–25
populations: density, 214; *An Essay on the Principle of Population*, 62, 435; growth, 179–80, 183; IPAT formula, 599n30; of microorganisms, 306; shift to cities, 181; theory and Malthus, 68, 551n65
positivism, 157, 189, 191, 304, 350, 628n20
post-exemptionalist sociology, 153, 155, 197–200
potash mines, 180, 181
potato blight, 443, 615n37
poultry farms, 415, 416
poverty of power, 470
The Poverty of Philosophy (Marx), 145, 343
power: climate change solutions through people, 96–98; fire and,

342; fossil capital and poverty of, 470; hydroelectric, 188; labor, 18, 58, 69, 226, 227, 278, 282, 284, 376, 512n33; maximum empower, 228, 554n101; poverty of, 470; redistribution, 96; *see also* nuclear power
Power Marketing Authorities, 398
The Power of the Machine (Hornborg), 560n208
Pravda (newspaper), 334
Precautionary Principle, planetary, 133–35
Premier, 422
Presidium of the Supreme Soviet of the USSR, 326
primordial soup theory, 84
Principles of Political Economy (Mill), 365
private riches, 226, 282, 375–76, 379, 389, 415, 430
privatization, 77, 136, 185, 415, 430
Problems of Philosophy (*Voprosy filosofi*) (journal), 330
production: Asian mode of, 536n74; bread manufacture, 600n34; Defense Production Act, 427; desertification and, 312; increase in global, 298–99; labor and, 378; less wasteful, more collective forms of, 136; with manufacturing in China, 427–28; of nature, 249, 253, 263–67; planetary metabolism and social, 429–32; treadmill of, 58, 73, 76, 96, 99, 186, 200, 377, 385–86, 513n43, 600n31; *The Treadmill of Production*, 600n31; *see also* commodity chains
productive forces, 35, 297, 298, 339, 488, 489, 583n9
Prometheanism, 34, 341–43, 583n8
Prometheus Bound (Aeschylus), 342
"Promises Are Not Enough," 24–25

property rights, landowners, 374
The Protestant Ethic and the Spirit of Capitalism (Weber), 157, 164, 165, 176, 192, 193, 194, 195, 533n25
public transportation, 96, 97, 136
public wealth, 226, 282, 389, 415, 430; capitalism, nature and, 374; private riches destroying, 375–76, 379, 430
Pugwash conferences, 328
punctuated equilibrium theory, 585n32

qualitative value, 69, 213, 242, 511n27, 550n55
quantitative value problem, 213, 550n55
quantity into quality, 304
Quaternary Period, 20, 245, 457, 459, 460, 463, 464
Queen Mab (Shelley, P. B.), 315

racial capitalism, 34-35, 345, 500n95; *see also* racism, slavery, genocide, Indigenous
racism: Black Lives Matter and, 491; COVID-19 and, 500n93; ecological, 269; ethnocentrism and, 183; expropriation and, 136, 500n95; patriarchy and, 371; social distancing and, 424–26
radiation sickness, 467
radical ecological monism, dialectical ecology versus, 251–57
radical social monism, 251, 255
radionuclides, 18, 19, 82, 246, 436, 467–69, 471
rain-fed agriculture, 165, 166, 168–71
rational-inorganic era, 164, 165, 175–86
rationalization, 193–94
rationing, 79, 96, 123, 400

INDEX

Raubbau (robbery system of capitalism), 35, 70, 181, 200; capitalist, 374–78, 389; Earth System and, 549n47; emergy exchange ratio, 233; with free gifts of nature, 550n57; industrial agriculture, 211
Real Financial Reform, Green Party New Deal, 395
real labour, Marx on, 507n54
real wealth: analysis and Odum, 219–28; capitalist production and, 241; dialectical-ecological synthesis and, 222–28; emergy analysis and, 222; minerals and information, 235; rich countries with, 237; use values and, 213
reciprocal action (universal interaction), 31
"Record of a Speech on the Irish Question Delivered by Karl Marx to the German Workers' Educational Association in London on December 16, 1867," 443
Red New Deal, First Nations people, 473, 491
A Redder Shade of Green (Angus), 622n14
reformist measures, degrowth and, 367
refraction, 160–61, 164, 535n42
refugees, climate, 357
The Religion of China (Weber), 164, 165, 170
The Religion of India (Weber), 162–63, 164, 165, 170, 535n42, 537n80
rent, land, 417
resilience: of individual societies with catastrophe, 616n51; Stockholm Resilience Center, 57, 123
The Return of Nature (Foster): after death of Marx, 339; after deaths of Marx and Darwin, 345, 346, 358; dialectics of nature and, 351; Lessing and, 359; Lukács and, 349; Skinner in, 348; writing process, 347–48, 352
Reuleauxian concept of technology, 191–92
The Revenge of Gaia (Lovelock), 446
Reverse Onus principle, 134
revolutionary transcendence (*Aufhebung*), 448
revolutions: as complex historical process, 601n1; co-revolutionary movement, 371; Cuba, 619n78; ecodemocratic phase, 78–79, 405, 612n5; ecological, 35–37, 58–61, 65, 135–37, 483–90; ecosocialist phase, 78, 79–80, 88, 405, 434, 612n5; English, 483; Farmer's Revolt, 491; French, 483; Green New Deal as reform or, 393–401; Indigenous revolts, 487, 492; Industrial Revolution, 16, 82, 98, 131, 175, 177, 198, 298–99, 343, 495n26; Livestock Revolution, 417; as locomotive of world history, 297; Marx and socio-ecological, 58–61; Marxian political economy and, 389; Morris on civilization and, 440; "Plan for a Clean Energy Revolution and Environmental Justice," 603n34; Plug Plot Riots, 489; ruin or, 35–37, 444, 476, 490; slaves revolts, 630n48; socialism with strategy for co-, 109; against system and fossil fuels war, 121–25
"Reworking the Metabolic Rift" (Wittman), 202
Rheinische Zeitung (newspaper), 342, 374
Rhineland Diet, 374
rich countries: climate change and,

98; consumption-based environmental costs, 218; degrowth and, 366, 369–70, 372; emissions cuts for, 403; with real wealth, 237; social distancing for, 424–26
rights: Green Party Voter's Bill of Rights, 395; for workers, 395
"the river of fire," crossing, 92, 101, 109
The Robbery of Nature (Clark and Foster), 345, 357
Robinson Crusoe (fictional character), 163–64
Rome, ancient, 444, 448, 449
Roscher and Knies (Weber), 159, 160
Royal Society of London, 123
ruin: racing to, 297–99; revolution or, 35–37, 444, 476, 490
Russia, 127

safe operating space, planetary boundaries, 123, 363
Samsung, 428
SARS-CoV-2, 415–16, 421, 423, 424, 426
science: *Bulletin of Atomic Scientists* with doomsday clock, 430–31; *Bulletin of the Academy of Sciences*, 327; capitalism and, 588n67; Earth System, 20–21; "An Eastern Perspective on Western Anti-Science," 451–52; ecologization of modern, 333; *Engels as a Scientist*, 305–9; "Engels's Philosophy of Science," 305; epigenetics, 589n6; International Union of Geological Sciences, 18, 457; with labor and nature, 341; Lenin All-Union Academy of Agricultural Sciences, 323; *Man, Science, Humanism*, 331, 336; *Marxism and the Philosophy of Science*, 347; nature and, 142; for profit, 301; Soviet Union, 459–60

Science (magazine), 465–66
Science and Civilization in China (Needham), 452
Science and Survival (Commoner), 19
Science for the People, 395, 603n28
Science of Logic (Hegel), Doctrine of Essence in, 271, 351
Scientific American (magazine), 86, 246
Scientific Communism (Fedoseev), 617n57, 625n48
Scientist Rebellion, 24
Scotland, with Climate Emergency declaration, 392
sea level rise, 24, 64–65, 116, 148, 326, 453
Seattle (1991), 98
"Second Crisis of Economic Theory" (Robinson), 385
Second New Left, 356
second-stage ecosocialism, 69, 72, 247, 249, 253
Select Committee, for Green New Deal, 391
self-mobilization, of people, 60, 127, 135, 405, 409
servile labor, 172
shared socioeconomic pathways (SSPs), 23, 24, 401, 604n45
Shell, 127
shipping: "blank sailings," 422; costs, 420
The Shock Doctrine (Klein), 92, 103
shrimp farming, 235
The Silence of the Forest (Ganghofer), 185
Silent Spring (Carson), 32, 83, 103, 104, 143, 246, 336, 585n28
sixth extinction, 21, 143, 262, 471, 490
The Sixth Extinction (Kolbert), 143
slavery, 16, 181–82, 473, 630n48
Small Is Beautiful (Schumacher), 371–72

social activity, transformational model of, 408–9
Social Darwinism, monism and, 270
social distancing, racism and, 424–26
social metabolism: *see* metabolic rift
social monism: ecology and, 284–90; radical, 251, 255; as world ecology, 257–63
social murder, 299, 313, 431, 484
social production, planetary metabolism and, 429–32
social reproduction, 34-35, 79, 272, 278, 283, 345, 408, 473, 489, 500n95; *see also* feminism, gender, and women
social science, Marxism, universal metabolism of nature and, 26–35
social-constructionist monism, 250, 263
social-historical transition, 436
socialism: barbarism or, 449; *Capitalism, Nature, Socialism*, 343, 347; civilization with narrow window for, 613n15; with co-revolutionary strategy, 109; defined, 22, 358, 378; DSA, 396; ecology and, 599n24; Movement Toward Socialism, 99; Schumacher on, 371–72; with sustainable human development, 76–77, 278; as ultimate line of defense, 108–10; *see also* ecosocialism
"Socialism and Ecology" (Sweezy), 337, 453–54
Socialist League, 355
The Social Determination of Method (Mészáros), 255–56
society: *Economy and Society*, 157, 160, 164, 170, 194–98, 534n40; labor as metabolism of nature and, 277; nature and, 146–47, 264–65, 267, 268, 270, 441, 502n10, 502n12, 503n25; nonindustrial, 164–73

"Sociobiophysicality and the Necessity of Critical Theory" (Stoner), 503n24
socio-cultural determinism, 565n32
socio-ecological revolution, Marx and, 58–61
sociology: of energy, 157, 186–95; environmental, 153–58, 530n3; interpretive, 158–64; post-exemptionalist, 153, 155, 197–200; "Some Categories of Interpretive Sociology," 159, 193
Sociology of Religion (Weber), 175–76
socionature, 286
soil: *Capital* on agriculture and, 341; crisis, 27; culture and slavery, 181–82; depletion, 25, 46, 47, 48, 49, 54, 71, 145, 181, 211, 298, 300, 377, 443, 490; humans, agriculture and, 281; nutrient cycles, 53–54, 55, 190, 211, 442, 497n59, 505n39; *see also* desertification
solar budget, 77, 79
solar energy, 118, 189; China with, 619n82; growth of, 116; humans and, 187–88
solar radiation management (SRM), 129, 130
"Some Categories of Interpretive Sociology" (Weber), 159, 193
Some Unsettled Questions (Mill), 206
South Africa, 78, 269, 340, 487
South Asia, deaths, 424
South China Sea, 115
Soviet ecology: air pollution, 317; Baikalsk Pulp and Paper Plant and, 336; biogeocoenosis and, 320–22, 324, 330, 332, 335; Lake Baikal and, 316, 325, 326, 333; late, 325–34; under Lenin and Stalin, 318–24; MOIP, 324, 325–26, 335; sea-level rise and, 326; three periods of, 317; in

twenty-first century, 334–37; "Virgin Lands" program, 325; VOOP, 318, 320, 324, 326, 335, 336; *zapovedniki*, 316, 318–20, 323–25, 590n7
Soviet Union, 17, 128; environmentalism and, 594n32, 594n38; *The Great Soviet Encyclopedia*, 20, 459, 495n32; Marxism and, 32–33; Ministry of Forest Management, 323, 325; science, 459–60
The Soviet and Post-Soviet Review (Coumel and Elie), 335
Spanish Civil War, 275
Special Report on Global Warming of 1.5°C, IPCC, 390–91
Special Report on the Ocean and Cryosphere (IPCC), 392
species extinction, 57, 270, 466
standard of living: natural wealth and high, 235; reversion to 1970s, 97
Standing Rock Sioux, 119–20, 396, 455, 492
State Committee of the USSR on Hydrometeorology and Control of the Natural Environment, 326
State Grid Corporation, China, 619n82
steady-state economy, 76, 122, 365, 366, 370
steam engines, 177, 191
Stern Report, 122
Stern Review, 523n43
Stockholm Resilience Center, 57, 123
Storms of My Grandchildren (Hansen), 363, 447
stratospheric aerosol injection, 128–30
strikes: climate: global, 392; youth-led, 391, 392; general (1842), 489
Strontium-9, nuclear fallout, 467

Structural One Health, 414, 415, 431
Sub-Saharan Africa: deaths, 424; ecological debt, 235–36, 424
"Summary for Policymakers" (IPCC), 24
Sunrise Movement, 391, 392–93, 396, 603n31
superbugs, 431
Superstorm Sandy (2012), 116
supply chains: finance, 423; monopoly-finance capital with global, 421; value and, 412, 419; *see also* commodity chains
Swamps (Sukachev), 318
Swedish Parliament, Thunberg at, 390
synthetic age, 18, 19, 458, 467, 469, 623n27
synthetic chemicals, 19, 65, 143, 177, 246, 388, 490
Syr Darya River, 325
Syria, 465
SYRIZA Party (Greece), 99
System Change Not Climate Change, 64, 81, 92, 109, 250, 345, 392
systems science, 203

taiga forests, 333
Tailism and the Dialectic (Lukács), 30–31, 44, 502n12
tar sands oil, 79, 112–13, 117–19, 519n6, 524n50
taxes, 399; corporate polluters, 398; war, 427
TC Energy, 119
Technocracy movement, 239–40
"Technological Determinism Revisited" (Heilbroner), 626n55
technology: BECCS, 127, 130–31, 402; China with alternative-energy, 407; communication, 420; IPAT formula, 599n30; monopoly capitalism and, 470–71; nuclear

INDEX

weapons, 468; productive forces, 583n9; Reuleauxian concept of, 191–92; with treadmill of production, 58
Tellus (journal), 21
termination problem, 130
"territorial restructuring," 418
terrorism, 117, 367
test kits, COVID-19, 423
Thames River, activism over, 391
Theory and Society (journal), 28
The Theory of Capitalist Development (Sweezy), 95
The Theory of the Leisure Class (Veblen), 381
The Theory of the Novel (Lukács), 151
thermodynamics, 308; first law of, 512n35; Marx and, 551n59, 552n86
Theses on Feuerbach (Marx), 145
Thinking like a Mall (Vogel), 566n45
Third Congress of the Geographical Society of the USSR, 327
third nature, 37, 148–52
Third World countries, 354, 537n80; with brunt of catastrophe, 25–26, 79; collapse, 447; liberation movements, 488; with unequal exchange, 205, 236–37; workers, 78; *see also* poor countries
This Changes Everything (Klein), 66, 91, 92, 97, 143, 517n8; with liberal critics in denial, 99–108; socialism and, 109
Three Mile Island nuclear disaster, 468
350.org, 115
"Thus Far and No Further" principle, 100, 103
Tibet, 318
tipping points, 22, 26, 64, 115, 301–2
"Toward a Singular Metabolism" (Moore), 569n74, 570n99, 580n51, 581n57

town, country and, 49, 54–55, 281, 440, 442, 450
toxic waste, 333
trade: foreign, 208; international, 205–7, 208, 229, 235, 470; labor and international, 229; opium, 15; slave, 16; wars, 427; world merchandise, 422; WTO, 94, 517n8
traditional-organic phase, 164, 165, 176, 177
The Tragedy of the Commodity (Longo, Clausen and Clark), 344
"Transcending the Metabolic Rift" (Moore), 202
transfactuality, 585n27
transformational change, age of, 408–10
transformational model of social activity, 408–9
Transition Culture, 122
transportation: air travel, 420, 422; public, 96, 97, 136; shipping and, 420, 422
Travels in France (Young), 15
treadmill of production, 76, 186, 200; accumulation and technology with, 58; advancement with expansion of, 99; metabolic rift and, 73; Schnaiberg and, 377, 385, 386, 513n43, 600n31; slowdown of, 96
The Treadmill of Production (Gould, Pellow and Schnaiberg), 600n31
tribal sovereignty, Indigenous peoples, 398
Trinity detonation, nuclear weapons test, 467
tsunami, 199, 421
Turkey, 318
Turn Down the Heat (World Bank), 509n8
Tuskegee Institute, 182
twenty-first century, Soviet ecology in, 334–37

2°C guardrail, 63–65, 93–94, 115, 122, 125, 126; dangers with 4°C future, 438; with 1.5°C as true marker, 602n14
Tyndall Center for Climate Change, 63, 94, 122, 403, 438, 520n15

UK Green New Deal Group, 394
Ukraine, 119
"unconscious socialist tendency," 254, 263, 283, 441–42, 567n51
Unconventionals Era, fossil fuels war, 111–12, 116
Underdeveloping the Amazon (Bunker), 215
unemployment, 78, 368–69, 420, 425
unequal ecological exchange, 71, 254, 450, 545n1; dialectical-ecological synthesis and, 222–28; ecological footprint analysis and, 214, 216–17, 218, 236, 242, 552n77; ecological imperialism and, 56, 202, 236, 250, 345, 356; embodied energy and, 201, 203–4, 220–21, 229, 238–40; imperial capitalism and, 203, 220, 233, 243; imperialism and, 547n14; Marx, Odum and theoretical challenges, 237–44, 550n55; Marx-Odum dialectic and, 204, 222–28, 243–44; Odum and, 228–37; with Odum and real wealth analysis, 219–28; systems science and, 203; theory of, 210–19, 356; with unequal economic exchange, 205–10; use values and, 215; wage-based, 208
unequal economic exchange, with unequal ecological exchange, 205–10
Uneven Development (Smith), 253
uniformity principle, 305, 585n32
United Nations (UN): Environment Programme, 394; Thunberg at, 392, 393, 410
United Nations Intergovernmental Panel on Climate Change (IPCC), 19, 93; with BECCS, 130–31; on carbon dioxide emissions, 603n34; *Climate Change 2021*, 23–24; geoengineering and, 127; mitigation strategies, 23–24, 401–8, 602n14; *Special Report on Global Warming of 1.5°C*, 390–91; *Special Report on the Ocean and Cryosphere*, 392
United States (U.S.), 166; Black movement in, 355, 491; China with coal imports from, 520n15; Civil War, 72, 182; decline of, 476; emergy exchange ratio, 233; emergy received/emergy exported ratio, 234; forests, 182; gas mileage decline in, 600n33; Green Building Council, 388; labor costs, 419; Marxism and, 33–34; National Retail Federation, 422; National Snow and Ice Data Center, 116; Weber on, 183–86
United Steel Workers, 395
unity of opposites, 304–8, 312, 317, 335, 486, 586n42
Universal Exposition, Chicago, 541n151
Universal Exposition of the Congress of Arts and Science, St. Louis, 182–83, 540n150
universal interaction (reciprocal action), 31
universal metabolism of nature: dialectics of nature and, 41, 42–45; with humans and social metabolism, 280; Marx and, 27, 45–52, 567n47; Marxism and, 26–35; rift in, 52–58; socio-ecological revolution and, 58–61

INDEX

universities, fossil fuel disinvestment, 603n31
University of Oregon, 338
use values: defined, 223; with ecology of Marxian political economy, 375; exchange and, 212, 226, 227, 412–13; monopoly-finance capital with negative, 58–59; nature and, 511n30; public wealth, 226, 282, 374–76, 379, 389, 415, 430; real wealth and, 213; unequal ecological exchange and, 215

value chains: analysis, 416; supply and, 412, 419; *see also* commodity chains
value-analytic, exchange value and, 511n30
values: destruction of nature and capitalist law of, 282–84; emvalue, 220, 222–23, 226, 229, 240, 553n89; energy theory of, 191, 203, 238–39, 559n194; energy theory of economic, 220, 238, 559n196; exchange, 212, 226, 227, 282, 375–76, 389, 412–13, 415, 430, 511n30; labor theory of, 218, 222, 225, 238, 239–40; nature excluded from, 512n31; qualitative, 69, 213, 242, 511n27, 550n55; quantitative problem, 213, 550n55; theory and Marx, 49, 69, 220, 224–25, 228, 254, 503n27, 580n50; use, 58–59, 212–13, 215, 223, 226, 227, 282, 374–76, 379, 389, 412–13, 415, 430, 511n30; wealth and, 550n55
Venezuela, 99, 237, 371, 406, 426, 473
Venus syndrome, 510n11
Verstehen, 534n34
verstehende Soziologie, 158
La Vía Campesina, 473, 491

"Victorian Holocausts," 298
video conferencing, 420
Vietnam, 428
Vietnam War, 355
villages, Germanic, 172, 173
violence, 78, 256, 448, 462, 529n1
"Virgin Lands" program, 325
Volga-Chograi canal, 336
Volkswagen, 421
"voluntary simplicity" movement, 364
Voter's Bill of Rights, Green Party New Deal, 395

Wales, with Climate Emergency declaration, 392
Wall Street: Flood Wall Street action, 602n17; Occupy movement, 98
Wall Street Journal (newspaper), 106, 423
Walmart, 105
"Was the Anthropocene Anticipated?" (Hamilton and Grinevald), 622n13
waste: bread manufacture, 600n34; capitalism and inherent, 406, 523n44; economic, 75, 79, 379, 381–84, 386, 599n21; materiality paradox and, 388–89; Odum on pathological, 554n100; plastic, 387–88, 469; toxic, 333; zero-waste systems, 136
The Waste of Nations (Dowd), 388
water: contaminated, 114; drought, 23, 57, 129, 130, 148, 167, 465–66; floods, 160, 199, 534n40, 602n14; loss of fresh, 21, 57, 65; Ogallala aquifer, 113; privatization, 77, 136; Volga-Chograi canal, 336
We Have Never Been Modern (Latour), 258
"We Must Reduce Greenhouse Gas Emissions to Net Zero or Face More Floods" (Stern), 602n14

wealth: accumulation, 430; billionaires, 127, 135, 428, 430; funds, 418; gross social, 333; high standard of living and natural, 235; inequality, 75, 428, 429–30, 513n49; IPAT formula and, 599n30; nature as, 77; private riches, 226, 282, 375–76, 379, 389, 415, 430; public, 226, 282, 374–76, 379, 389, 415, 430; real, 213, 219–28, 235, 237, 241; redistribution, 79, 96, 97, 135; rich countries, 98, 218, 237, 366, 369–70, 372, 403; value and, 550n55; *see also* poor countries

The Wealth of Nations (Smith, A.), 15

weather, extreme, 23, 25, 64, 116

Web of Science, 531n11

West Germany, 233

West Texas Intermediate oil, 113

Western Marxism, 32, 284–85, 291–92, 304, 354, 506n45; Cartesian dualism and, 252, 262; defined, 43; dialectical materialism and, 43, 68; dialectics of nature and, 42–43, 55, 501n4, 584n26; Engels and, 353; Marxism in Anthropocene and, 251, 271, 275; *Marx's Ecology* and, 348–49; Merleau-Ponty and, 501n4

wet markets, animals, 415

What Is Life? (Margulis and Sagan), 84

What We Leave Behind (Jensen and McBay), 482

Will China Save the Planet? (Finamore), 452, 619n82

William Morris (Morris, M.), 507n57, 508n58

wind energy, 118; China with, 619n82; growth of, 116

women, 78–79, 146, 249, 397, 512n3, 563n23; *see also* feminism, gender, and social reproduction

wood: "Debates on the Law on Theft of Wood," 374; epoch to age of iron, 173–75; peasants jailed for collecting, 374; *see also* deforestation; forests

workers: capitalist class and, 29; *The Condition of the Working Class in England*, 124, 299–300, 354, 415, 489, 620n86; just transition for, 398; Landless Workers' Movement, 345; *The Making of the English Working Class*, 483; Marks on environmental conditions of, 484; MST, 345, 473, 491; Oil, Chemical and Atomic Workers Union, 395; "Record of a Speech on the Irish Question Delivered by Karl Marx to the German Workers' Educational Association in London on December 16, 1867," 443; rights for, 395; social murder of, 313; Third World countries, 78; United Steel Workers, 395; *see also* environmental proletariat; labor

World Bank, 64, 413, 509n8

World Conference on Climate, Geneva (1979), 326–27

World Council of Peace, 328–29

world ecology, social monism as, 257–63

World Economic Forum: on China with wind and solar energy, 619n82; on COVID-19 and supply chain, 421; Thunberg, activism and, 391

World Health Organization, 413, 431

world merchandise trade, 422

World Summit for Sustainable Development, Johannesburg (2002), 517n8

World Trade Organization (WTO), 94, 517n8

INDEX

World War I, 17
World War II, 18, 246, 409; automobile industry in, 96; Hiroshima and Nagasaki, 17, 468

Yangtze River Valley, 465
Yanomamo Indians, 237
The Young Hegel (Lukács), 351

zapovedniki (ecological reserves), 316, 318–20, 323–25, 590n7
zero-waste systems, 136
zoonotic diseases, 19, 21, 413, 416, 423
zooplankton, 130